Advances in Computer Vision and Pattern Recognition

More information about this series at http://www.springer.com/series/4205

Kenichi Kanatani · Yasuyuki Sugaya
Yasushi Kanazawa

Guide to 3D Vision Computation

Geometric Analysis and Implementation

Springer

Kenichi Kanatani
Okayama University
Okayama
Japan

Yasushi Kanazawa
Toyohashi University of Technology
Toyohashi, Aichi
Japan

Yasuyuki Sugaya
Toyohashi University of Technology
Toyohashi, Aichi
Japan

ISSN 2191-6586 ISSN 2191-6594 (electronic)
Advances in Computer Vision and Pattern Recognition
ISBN 978-3-319-48492-1 ISBN 978-3-319-48493-8 (eBook)
DOI 10.1007/978-3-319-48493-8

Library of Congress Control Number: 2016955063

Printed on acid-free paper

This Springer imprint is published by Springer Nature
The registered company is Springer International Publishing AG
The registered company address is: Gewerbestrasse 11, 6330 Cham, Switzerland

Preface

Today, computer vision techniques are used for various purposes, and there exist many textbooks and references that describe principles of programming and system organization. At the same time, ever new research is going on all over the world, and the achievements are offered in the form of open source code on the Web. It appears, therefore, that sufficient environments already exist for students and researchers for embarking on computer vision research.

However, although executing a public source code may be easy, improving or modifying it for other applications is rather difficult, because the intent of the code author is difficult to discern simply by reading the code. On the other hand, many computer vision textbooks focus on theoretical principles coupled with application demos. As a result, one is often at a loss as to how to write a program oneself. Actual implementation of algorithms requires many small details that must be carefully taken into consideration, which a ready-to-use code does not provide. This book intends to fill that gap, describing in detail the computational procedures for programming 3D geometric tasks. The algorithms presented in this book are based on today's state of the art, yet arranged in a form simple and easy enough to understand. The authors also believe that they are the most appropriate form for practical use in real situations.

In this book, the mathematical background of the presented algorithms is mostly omitted for the ease of reading, but for theoretically minded readers detailed derivations and justifications are given in the form of Problems in each chapter; their Solutions are given at the end of the volume. Also, historical notes and related references are discussed in the Supplemental Note at the end of each chapter. In this sense, this book can also serve as a theoretical reference of computer vision research. To help readers implement the algorithms in this book, sample codes of typical procedures are placed on the publisher's Web page.[1]

This book is based on the teaching materials that the authors used for student projects at Okayama University and Toyohashi University of Technology, Japan. Every year, new students with little background knowledge come to our labs to do computer vision work. According to our experience, the most effective way for them to learn is to let them implement basic algorithms such as those given here.

[1]http://www.springer.com/book/9783319484921

Through this process, they learn the basic know-how of programming and at the same time gradually understand the theoretical background as their interest deepens. Thus we are hoping that this book can serve not only as a reference of the latest computer vision techniques but also as useful material for introductory courses of computer vision.

The authors thank Takayuki Okatani of Tohoku University, Japan, Mike Brooks and Wojciech Chojnacki of the University of Adelaide, Australia, Peter Meer of Rutgers University, the United States, Wolfgang Förstner, of the University of Bonn, Germany, Michael Felsberg of Linköping University, Sweden, Rudolf Mester of the University of Frankfurt, Germany, Prasanna Rangarajan of Southern Methodist University, the United States, Ali Al-Sharadqah of California State University, Northridge, the United States, Alexander Kukush of the University of Kiev, Ukraine, and Chikara Matsunaga of For-A, Co. Ltd., Japan. Special thanks go to (the late) Prof. Nikolai Chernov of the University of Alabama at Birmingham, the United States, without whose inspiration and assistance this work would not have been possible. Parts of this work are used with permission from the authors' work *Ellipse Fitting for Computer Vision: Implementation and Applications*, ©Morgan & Claypool,[2] 2016.

Okayama, Japan — Kenichi Kanatani
Toyohashi, Japan — Yasuyuki Sugaya
Toyohashi, Japan — Yasushi Kanazawa
October 2016

[2]http://www.morganclaypool.com.

Contents

Introduction 1

Abstract

This chapter states the background and organization of this book and describes distinctive features of the volume.

1.1 Background

The study of computer vision, also known as image understanding, which aims to extract information contained in 3D scenes by analyzing their images, began in the United States in the 1960s. Computer vision is a natural extension of pattern recognition, whose study had started earlier for analyzing 2D images, typically handwritten or printed letters and symbols. For understanding a 3D scene using pattern recognition techniques, many image processing operations, including edge detection, thresholding, and thinning, were extensively studied in the 1970s and used for extracting various features that help classify the images and understand their content. At first, computer vision researchers thought that it would suffice merely to combine and apply such image processing operations to 3D scene images. Various types of research were done in this line in the 1980s, mainly in the United States. At the time, computer vision was also thought of as a typical artificial intelligence problem, solving tasks by combining the if-then-else type propositional logic with the knowledge and common sense of humans about the outside world.

It was soon found, however, that understanding 3D scenes is impossible by merely combining 2D pattern recognition techniques with a knowledge database and that mathematical analysis of imaging geometry of the cameras that project a 3D scene onto a 2D image is indispensable. Around the 1980s, a new mathematical approach occurred. The resulting mathematical framework was called *epipolar geometry*, which allowed one to reconstruct the 3D shape of an object from its images. This

K. Kanatani et al., *Guide to 3D Vision Computation*, Advances in Computer Vision and Pattern Recognition, DOI 10.1007/978-3-319-48493-8_1

marked a turning point from artificial intelligence research based on knowledge-based logic to geometric analysis using mathematical means.

In the 1980s, computer vision researchers using mathematical methods were relatively few worldwide, including one of the authors, Kanatani, who was staying in the United States at the time. In the 1990s, however, the use of mathematical disciplines, projective geometry in particular, quickly spread to European countries, mainly the United Kingdom, France, and Sweden. Today, it is no exaggeration to say that the mathematical approach is the mainstream of computer vision research, playing a central role in such applications as intelligent robots, autonomous vehicles, and virtual reality coupled with 3D computer graphics.

One of the authors, Kanatani, since the 1980s, has energetically endorsed the mathematical approach of computer vision and published such textbooks as *Group-Theoretical Methods in Image Understanding* (Springer 1990), *Geometric Computation for Machine Vision* (Oxford 1993), and *Statistical Optimization for Geometric Computation* (Elsevier 1996). As compared with these, this book describes the research development thereafter and presents the state-of-the-art techniques for 3D reconstruction from multiple images with full consideration of programming and implementation aspects.

1.2 Organization

As mentioned above, this book can be viewed as a supplement to Kanatani's earlier books. However, two decades have passed, and the advance since then is remarkable. This is, of course, mostly due to the dramatic performance increase of today's computers, but also significant theoretical progress has been made. This book consists of three parts. In Part I (Chaps. 2–9), fundamental algorithms underlying computer vision and their implementation are described. Most of these topics were already treated in earlier books, but since then there have been significant performance improvements due to newly introduced careful statistical error analysis. We present their up-to-date versions. Part II (Chaps. 10–13) then describes techniques for 3D reconstruction from multiple images based on the developments of the last two decades. Finally, we summarize in Part III (Chaps. 14–16) mathematical theories of statistical error analysis for general geometric estimation problems.

In the following, we briefly give an outline of each chapter. More detailed historical notes including the references are given in the Supplemental Note at the end of each chapter.

Chapter 2. **Ellipse Fitting**

Because circular objects are projected to ellipses in images, ellipse fitting is necessary for 3D analysis of circular objects. For this reason, the study of ellipse fitting began as soon as computers came into use for image analysis in the 1970s. The basic principle was to compute the parameters such that the sum of squares of expressions that should ideally be zero is minimized, which today is called *least squares* or *algebraic distance minimization*. In the 1990s, the notion of optimal computation based on the statistical properties of noise was introduced by the authors and other researchers. The first notable technique was the authors' *renormalization*, which was improved in the 2000s as *FNS* and *HEIVE* by researchers in Australia and the United States. Later, further improvements, called *hyperaccurate correction*, *HyperLS*, and *Hyper-renormalization*, were presented in the 2000s and 2010s by the authors. Chapter 2 describes the implementation of these techniques.

Chapter 3. Fundamental Matrix Computation

The *fundamental matrix* is a matrix determined from point correspondences between two images; from it one can compute the relative positions of the two cameras that took the images. In the 1980s, camera focal lengths were assumed to be known (they can be calibrated beforehand), and this matrix was called the *essential matrix*. In the 1990s, it was found that the matrix can be computed without the knowledge of the focal lengths and that the focal lengths can be computed from that matrix. Since then, it has been called the fundamental matrix.

Computing the fundamental or essential matrix from point correspondences has the same mathematical structure as ellipse fitting. Thus ellipse fitting methods such as renormalization, FNS, and hyper-renormalization can also be applied. However, the fundamental matrix has an additional constraint: it has determinant 0 with rank 2. This chapter mainly focuses on this *rank constraint*. A simple method popular since the 1980s is the *a posteriori correction*: the fundamental matrix is computed without regard to the rank constraint, and then the smallest singular value of its singular value decomposition is replaced by 0. In the 1990s, the *optimal correction* was found by the authors: the reliability of the fundamental matrix computation was evaluated from the statistical properties of the data noise, and the rank was corrected in a statistically optimal manner. In the 2000s, alternative methods were proposed by the authors and others, including iterating the fundamental matrix computation such that it automatically converges to an optimal solution of determinant 0 (*extended FNS*); an alternative is to parameterize the fundamental matrix such that its determinant is identically 0 and doing a search in that parameter space. Chapter 3 describes the implementation of these methods.

Chapter 4. Triangulation

The *triangulation* for computing the 3D position of a point using two images is necessary in the final stage of 3D shape reconstruction. In the 1980s, a simple least squares was used, which is practically sufficient in real situations. In the 1990s and the 2000s, a new light was shed on this problem from a theoretical point of view. To be specific, the authors presented an iterative procedure for computing a theoretically optimal solution by considering the statistical properties of image noise, and Hartley and Sturm presented an alternative algebraic method for it. The resulting solutions are identical, but the latter emphasizes global optimality at the cost of a heavier computational burden. This chapter describes the practically more efficient method of the authors.

Chapter 5. 3D Reconstruction from Two Images

The mathematical framework for this problem was already established in the 1980s. However, the camera focal lengths were assumed to be known. The fact that they can be computed from the fundamental matrix of the two images was found in the 1990s by many researchers including Bougnoux and the authors of this book. Another advance was that in the 1980s the essential matrix was computed by least squares, but today the fundamental matrix is computed by a statistically optimal method, as described in Chap. 3.

Chapter 6. Homography Computation

The fact that the 3D position of a planar surface can be reconstructed from its two projection images by computing the homography between them has been known since the early days before the advent of computers. In this chapter, we discuss optimal homography computation from point correspondences by extending the optimal computation of ellipses and fundamental matrices described in Chaps. 3 and 4. Specifically, we describe the implementation of renormalization, hyper-renormalization, FNS, geometric distance minimization, and hyperaccurate correction.

Chapter 7. Planar Triangulation

This is the triangulation task described in Chap. 4 with an additional constraint that the 3D position to be computed should be on a specified plane. An optimal computation scheme for this problem was first presented by the authors in 1995. This is a first approximation that omits higher-order terms in image noise but is sufficient in practical situations. Later, in 2011, the authors showed an iterative scheme that computes the exact solution. This chapter describes its implementation. Chum et al., on the other hand, in 1997 presented from a theoretical motivation an algebraic procedure that solve an eight-degree polynomial. This corresponds to the algebraic method of Hartley and Sturm in 1997 for the standard triangulation. As in that case, the resulting solutions are identical, but the procedure described here is much more efficient than the method of Chum et al.

Chapter 8. 3D Reconstruction of a Planar Surface

This is the task of computing the 3D position from point correspondences between two images with the knowledge that the object is a planar surface; the camera positions need not be known. Analytical procedures for this problem were already obtained in the 1960s, before computer vision research began, in relation to perceptual psychology. In the 1980s, elegant mathematical formulations were presented by many researchers including Longuet-Higgins. The computation consisted of two stages: computing the homography between the two images and then decomposing it to the 3D positions of the cameras and the plane. This chapter describes an up-to-date procedure for this. For the first stage, we optimally compute the homography as described in Chap. 4; for the second stage, we adopt the procedure of Longuet-Higgins.

Chapter 9. Analysis of Ellipses and 3D Computation of Circles

Circular objects are projected onto images as ellipses, and from the observed ellipses one can compute their 3D positions and orientations. The analytical procedure was found in the 1990s: the 3D computation of circular objects was first shown by Forsyth et al. in 1991 and extended to elliptic objects by Kanatani and Liu in 1993. This chapter shows various examples of 3D analysis based on ellipse images, combining the optimal ellipse fitting techniques of Chap. 2 and mathematical facts of projective geometry.

Chapter 10. **Multiview Triangulation**

This is an extension of the two-view triangulation of Chap. 4 to multiple views, computing the 3D position of a point from its multiple images taken by cameras in known positions. For this task, many papers were published from the late 1990s to the early 2000s, mainly from a theoretical point of view in relation to global optimization techniques. This chapter describes an iterative procedure that the authors presented in 2010, extending the two-view triangulation procedure of Chap. 4 to multiple images. Being computationally very efficient, this is considered to be the best method in practical applications.

Chapter 11. **Bundle Adjustment**

This is a technique for computing, from multiple images of a scene, not only its 3D shape but also, simultaneously, the positions and orientations of all the cameras that take the images as well as their intrinsic parameters. This is done by iteratively updating all the unknowns, starting from given initial values, such that the computed 3D shape and the observed images better satisfy the assumed perspective projection relationship. The principle was well known in photogrammetry before the advent of computers, but the computation requires a vast amount of time and memory for a huge number of unknowns. Only in the 2000s were various computational tools made available due to the dramatic progress in hardware. The main focus is on how to store the vast number of unknowns efficiently and speed up the computation of large-scale matrices, the majority of whose elements are 0. This chapter presents a specific computational procedure based on the work of the authors.

Chapter 12. **Self-calibration of Affine Cameras**

Bundle adjustment requires initial values. One way for computing an approximate 3D is to ignore the foreshortening effects of the camera imaging. This is called *affine camera* modeling. In the 1990s, Kanade and his group introduced the orthographic, weak perspective, and paraperspective camera models. Because the computation involves factorizing a matrix into the product of two, the method came to be generally known as *factorization*. However, the orthographic, weak perspective, and paraperspective models are mutually unrelated; for example, none is a special case of another. In 2007, the authors pointed out that all these are special cases of the *symmetric affine camera* and showed that 3D reconstruction is possible only using symmetric affine camera modeling coupled with factorization. This chapter describes the algorithm in detail and shows how it reduces to the paraperspective, weak perspective, and orthographic models in that order by restricting the parameters.

Chapter 13. **Self-calibration of Perspective Cameras**

Self-calibration is a process of computing 3D from images without knowledge of the camera parameters, the cameras calibrating themselves. In the 1990s, efforts were made to extend the self-calibration process from affine cameras to perspective cameras. The resulting techniques are built on highly mathematical theories of projective geometry and are regarded as one of the most significant achievements of computer vision study. The basic principle is to assume the parameters of foreshortening, called the *projective depths*, and regard the cameras as affine. Then the factorization technique is applied to update projective depths iteratively such that the resulting solution better satisfies the perspective projection relationship. The obtained 3D shape is deformed from the true shape by an unknown projective transformation. The process of rectifying this *projective reconstruction* into the correct shape is called *Euclidean upgrading*. These computations require a large amount of computation for many unknowns, therefore the computational efficiency is the main focus. Many different formulations have been proposed for both projective reconstruction and Euclidean upgrading. This chapter describes their combination in the way that the authors think is the best.

Chapter 14. **Accuracy of Geometric Estimation**

This chapter generalizes the algebraic methods for ellipse fitting, fundamental matrix computation, and homography computation described in Chaps. 2, 3, and 6 in a unified mathematical framework. Then we give a detailed error analysis in general terms and derive explicit expressions for the covariance and bias of the solution. The *hyper-renormalization* procedure of the authors is obtained in this framework.

Chapter 15. **Maximum Likelihood of Geometric Estimation**

This chapter discusses maximum likelihood estimation and Sampson error minimization in the general mathematical framework of Chap. 14. We present here higher-order error analysis for deriving explicite expressions of the covariance and bias of the solution. The *hyperaccurate correction* procedure of the authors is derived in this framework.

Chapter 16. **Theoretical Accuracy Limit**

This chapter presents the derivation of a theoretical accuracy limit of the geometric estimation problems of Chaps. 2, 3, and 6 in the general mathematical framework of Chaps. 14 and 15. It is given in the form of a bound, called the *KCR* (*Kanatani-Cramer-Rao*) *lower bound*, on the covariance matrix of the solution. The resulting form indicates that all iterative algebraic and geometric methods achieve this bound up to higher-order noise terms, meaning that these are all optimal with respect to covariance. The mathematical relationship of the KCR lower bound with the *Cramer-Rao lower bound*, well known in statistics, is also explained.

1.3 Features

The uniqueness of this book is the order of description. Most textbooks on computer vision begin with mathematical fundamentals followed by resulting computational procedures. This is naturally a logical order, but this would give the impression that one is reading a sophisticated book on mathematics. This book, in contrast, *immediately* describes actual computational procedures after a brief statement of the purpose and the basic principle. This is to emphasize that one can obtain the desired result by simply computing as instructed without knowing its derivation. All the computational procedures described in this book are based on the state of the art of the domain and constructed in a manner that is faithful to the principle and yet is easy to understand. The authors believe that they are in the most appropriate form for practical applications. *After* the procedure is described, its theoretical background is briefly explained in the *Comments*. Thus the reader need not worry about mathematical details, which are almost always annoying in most computer vision textbooks for actually building a computer vision system. This *omission* of mathematical details is the biggest feature of this book.

Yet, there may certainly be some who want to know the details of the derivation and justification of the procedure. For them, all the derivations and justifications are given in the section, *Problems*, at the end of each chapter and in *Solutions* at the end of the volume. This means that theorems and propositions and their proofs in other computer vision textbooks are replaced by the problems and their solutions in this book. Also, the background mathematical theories of statistical optimization that underlie this book are deferred to the final Part III (Chaps. 14–16). Thus, this book can serve as both a practical programming guidebook and a mathematics reference for computer vision, satisfying practitioners who want to implement computer vision algorithms without regard to mathematical details and also theoreticians who want to know the underlying mathematical details.

In the *Examples* section of each chapter, a few experimental results are shown to give one a *feeling* of what can be done using the procedures in that chapter; they are not intended as comparative evidence for justification of the computation. At the end of each section is the *Supplemental Note* section, explaining the historical background behind the subject of that chapter along with the reference literature. Also, related topics and mathematical knowledge not directly discussed in that chapter are briefly introduced. In this respect, the description order of this book is opposite to most computer vision textbooks, where general backgrounds are given first followed by particular topics. This is to give priority to the desire of those who want to implement the algorithms as quickly as possible.

The description of the computational procedures in this book consists of explicit lists of mathematical expressions to be evaluated. They can be immediately implemented in any computer language such as C, C++, and MATLAB®. Today, various packages are provided on the Web for basic mathematical operations including vector and matrix manipulation and eigenvalue computation; any of them can be used.

For the convenience of the readers, however, sample codes that implement typical procedures of each chapter have been placed on the publisher's website.[1]

[1] http://www.springer.com/book/9783319484921.

Part I
Fundamental Algorithms for Computer Vision

Ellipse Fitting

2

Abstract

Extracting elliptic edges from images and fitting ellipse equations to them is one of the most fundamental tasks of computer vision. This is because circular objects, which are very common in daily scenes, are projected as ellipses in images. We can even compute their 3D positions from the fitted ellipse equations, which along with other applications are discussed in Chap. 9. In this chapter, we present computational procedures for accurately fitting an ellipse to extracted edge points by considering the statistical properties of image noise. The approach is classified into algebraic and geometric. The algebraic approach includes least squares, iterative reweight, the Taubin method, renormalization, HyperLS, and hyper-renormalization; the geometric approach includes FNS, geometric distance minimization, and hyper-accurate correction. We then describe the ellipse-specific method of Fitzgibbon et al. and the random sampling technique for avoiding hyperbolas, which may occur when the input information is insufficient. The RANSAC procedure is also discussed for removing nonelliptic arcs from the extracted edge point sequence.

2.1 Representation of Ellipses

An ellipse observed in an image is described in terms of a quadratic polynomial equation in the form

$$Ax^2 + 2Bxy + Cy^2 + 2f_0(Dx + Ey) + f_0^2 F = 0, \tag{2.1}$$

where f_0 is a constant for adjusting the scale. Theoretically, we can let it be 1, but for finite-length numerical computation it should be chosen that x/f_0 and y/f_0 have approximately the order of 1; this increases the numerical accuracy, avoiding the loss of significant digits. In view of this, we take the origin of the image xy coordinate

K. Kanatani et al., *Guide to 3D Vision Computation*, Advances in Computer Vision and Pattern Recognition, DOI 10.1007/978-3-319-48493-8_2

system at the center of the image, rather than the upper-left corner as is customarily done, and take f_0 to be the length of the side of a square that we assume to contain the ellipse to be extracted. For example, if we assume there is an ellipse in a 600×600 pixel region, we let $f_0 = 600$. Because Eq. (2.1) has scale indeterminacy (i.e., the same ellipse is represented if A, B, C, D, E, and F were multiplied by a common nonzero constant), we normalize them to

$$A^2 + B^2 + C^2 + D^2 + E^2 + F^2 = 1. \tag{2.2}$$

If we define the 6D vectors

$$\boldsymbol{\xi} = \begin{pmatrix} x^2 \\ 2xy \\ y^2 \\ 2f_0x \\ 2f_0y \\ f_0^2 \end{pmatrix}, \qquad \boldsymbol{\theta} = \begin{pmatrix} A \\ B \\ C \\ D \\ E \\ F \end{pmatrix}, \tag{2.3}$$

Equation (2.2) can be written as

$$(\boldsymbol{\xi}, \boldsymbol{\theta}) = 0, \tag{2.4}$$

where and hereafter we denote the inner product of vectors $\mathbf{a}$ and $\mathbf{b}$ by $(\mathbf{a}, \mathbf{b})$. The vector $\boldsymbol{\theta}$ in Eq. (2.4) has scale indeterminacy, and the normalization of Eq. (2.2) is equivalent to vector normalization $\|\boldsymbol{\theta}\| = 1$ to the unit norm.

2.2 Least-Squares Approach

Fitting an ellipse in the form of Eq. (2.1) to a sequence of points $(x_1, y_1), \ldots, (x_N, y_N)$ in the presence of noise (Fig. 2.1) is to find A, B, C, D, E, and F such that

$$Ax_\alpha^2 + 2Bx_\alpha y_\alpha + Cy_\alpha^2 + 2f_0(Dx_\alpha + Ey_\alpha) + f_0^2F \approx 0, \qquad \alpha = 1, \ldots, N. \tag{2.5}$$

If we write $\boldsymbol{\xi}_\alpha$ for the value of $\boldsymbol{\xi}$ of Eq. (2.3) for $x = x_\alpha$ and $y = y_0$, Eq. (2.5) can be equivalently written as

$$(\boldsymbol{\xi}_\alpha, \boldsymbol{\theta}) \approx 0, \qquad \alpha = 1, \ldots, N. \tag{2.6}$$

Our task is to compute such a unit vector $\boldsymbol{\theta}$. The simplest and most naive method is the following *least squares (LS)*.

Procedure 2.1 (Least squares)

1. Compute the 6×6 matrix

$$\mathsf{M} = \frac{1}{N} \sum_{\alpha=1}^{N} \boldsymbol{\xi}_\alpha \boldsymbol{\xi}_\alpha^\top. \tag{2.7}$$

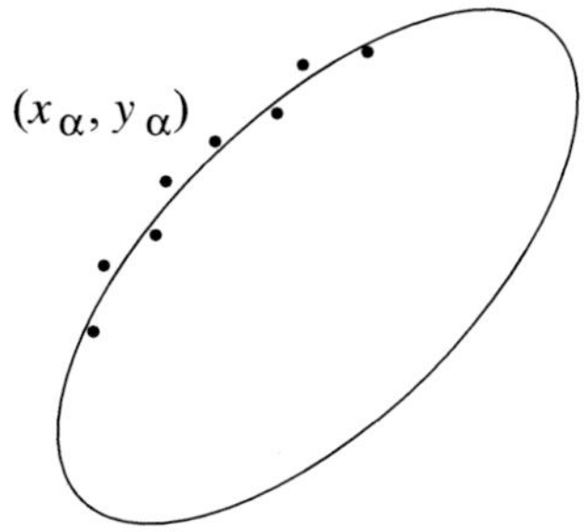

Fig. 2.1 Fitting an ellipse to a noisy point sequence

2. Solve the eigenvalue problem

$$\mathsf{M}\boldsymbol{\theta} = \lambda\boldsymbol{\theta}, \tag{2.8}$$

and return the unit eigenvector $\boldsymbol{\theta}$ for the smallest eigenvalue λ.

Comments This is a straightforward generalization of line fitting to a point sequence ($\hookrightarrow$ Problem 2.1); we minimize the sum of squares

$$J = \frac{1}{N}\sum_{\alpha=1}^{N}(\boldsymbol{\xi}_\alpha, \boldsymbol{\theta})^2 = \frac{1}{N}\sum_{\alpha=1}^{N}\boldsymbol{\theta}^\top\boldsymbol{\xi}_\alpha\boldsymbol{\xi}_\alpha^\top\boldsymbol{\theta} = (\boldsymbol{\theta}, \mathsf{M}\boldsymbol{\theta}) \tag{2.9}$$

subject to $\|\boldsymbol{\theta}\| = 1$. As is well known in linear algebra, the minimum of this quadratic form in $\boldsymbol{\theta}$ is given by the unit eigenvector $\boldsymbol{\theta}$ of M for the smallest eigenvalue. Equation (2.9) is often called the *algebraic distance* (although it does not have the dimension of square length), and Procedure 2.1 is also known as *algebraic distance minimization*. It is sometimes called *DLT* (*direct linear transformation*). Inasmuch as the computation is very easy and the solution is immediately obtained, this method has been widely used. However, when the input point sequence covers only a small part of the ellipse circumference, it often produces a small and flat ellipse very different from the true shape (we show such examples in Sect. 2.8). Still, this is a prototype of all existing ellipse-fitting algorithms. How this can be improved has been a major motivation of many researchers.

2.3 Noise and Covariance Matrices

The reason for the poor accuracy of Procedure 2.1 is that the properties of image noise are not considered; for accurate fitting, we need to take the statistical properties of noise into consideration. Suppose the data x_α and y_α are disturbed from their true values $\bar{x}_\alpha$ and $\bar{y}_\alpha$ by Δx_α and Δy_α, and write

$$x_\alpha = \bar{x}_\alpha + \Delta x_\alpha, \qquad y_\alpha = \bar{y}_\alpha + \Delta y_\alpha. \tag{2.10}$$

Substituting this into $\boldsymbol{\xi}_\alpha$, we obtain

$$\boldsymbol{\xi}_\alpha = \bar{\boldsymbol{\xi}}_\alpha + \Delta_1\boldsymbol{\xi}_\alpha + \Delta_2\boldsymbol{\xi}_\alpha, \tag{2.11}$$

where $\bar{\boldsymbol{\xi}}_\alpha$ is the value of $\boldsymbol{\xi}_\alpha$ for $x_\alpha = \bar{x}_\alpha$ and $y_\alpha = \bar{y}_\alpha$ and $\Delta_1\boldsymbol{\xi}_\alpha$, and $\Delta_2\boldsymbol{\xi}_\alpha$ are, respectively, the first-order noise terms (i.e., linear expressions in Δx_α and Δy_α) and the second-order noise terms (i.e., quadratic expressions in Δx_α and Δy_α). Specifically, they are

$$\Delta_1\boldsymbol{\xi}_\alpha = \begin{pmatrix} 2\bar{x}_\alpha \Delta x_\alpha \\ 2\Delta x_\alpha \bar{y}_\alpha + 2\bar{x}_\alpha \Delta y_\alpha \\ 2\bar{y}_\alpha \Delta y_\alpha \\ 2 f_0 \Delta x_\alpha \\ 2 f_0 \Delta y_\alpha \\ 0 \end{pmatrix}, \qquad \Delta_2\boldsymbol{\xi}_\alpha = \begin{pmatrix} \Delta x_\alpha^2 \\ 2\Delta x_\alpha \Delta y_\alpha \\ \Delta y_\alpha^2 \\ 0 \\ 0 \\ 0 \end{pmatrix}. \tag{2.12}$$

Regarding the noise terms Δx_α and Δy_α as random variables, we define the *covariance matrix* of $\boldsymbol{\xi}_\alpha$ by

$$V[\boldsymbol{\xi}_\alpha] = E[\Delta_1\boldsymbol{\xi}_\alpha \Delta_1\boldsymbol{\xi}_\alpha^\top], \tag{2.13}$$

where $E[\,\cdot\,]$ denotes the expectation over the noise distribution, and $\top$ denotes the vector transpose. We assume that Δx_α and Δy_α are subject to an independent Gaussian distribution of mean 0 and standard deviation σ. Thus

$$E[\Delta x_\alpha] = E[\Delta y_\alpha] = 0, \quad E[\Delta x_\alpha^2] = E[\Delta y_\alpha^2] = \sigma^2, \quad E[\Delta x_\alpha \Delta y_\alpha] = 0. \tag{2.14}$$

Substituting Eq. (2.12) and using this relationship, we obtain the covariance matrix of Eq. (2.13) in the form

$$V[\boldsymbol{\xi}_\alpha] = \sigma^2 V_0[\boldsymbol{\xi}_\alpha], \quad V_0[\boldsymbol{\xi}_\alpha] = 4\begin{pmatrix} \bar{x}_\alpha^2 & \bar{x}_\alpha\bar{y}_\alpha & 0 & f_0\bar{x}_\alpha & 0 & 0 \\ \bar{x}_\alpha\bar{y}_\alpha & \bar{x}_\alpha^2 + \bar{y}_\alpha^2 & \bar{x}_\alpha\bar{y}_\alpha & f_0\bar{y}_\alpha & f_0\bar{x}_\alpha & 0 \\ 0 & \bar{x}_\alpha\bar{y}_\alpha & \bar{y}_\alpha^2 & 0 & f_0\bar{y}_\alpha & 0 \\ f_0\bar{x}_\alpha & f_0\bar{y}_\alpha & 0 & f_0^2 & 0 & 0 \\ 0 & f_0\bar{x}_\alpha & f_0\bar{y}_\alpha & 0 & f_0^2 & 0 \\ 0 & 0 & 0 & 0 & 0 & 0 \end{pmatrix}. \tag{2.15}$$

All the elements of $V[\boldsymbol{\xi}_\alpha]$ have the multiple σ^2, therefore we factor it out and call $V_0[\boldsymbol{\xi}_\alpha]$ the *normalized covariance matrix*. We also call the standard deviation σ the *noise level*. The diagonal elements of the covariance matrix $V[\boldsymbol{\xi}_\alpha]$ indicate the noise susceptibility of each component of the vector $\boldsymbol{\xi}_\alpha$ and the off-diagonal elements measure the correlation between its components.

The covariance matrix of Eq. (2.13) is defined in terms of the first-order noise term $\Delta_1\boldsymbol{\xi}_\alpha$ alone. It is known that incorporation of the second-order term $\Delta_2\boldsymbol{\xi}_\alpha$ will have little influence over the final results. This is because $\Delta_2\boldsymbol{\xi}_\alpha$ is very small as compared with $\Delta_1\boldsymbol{\xi}_\alpha$. Note that the elements of $V_0[\boldsymbol{\xi}_\alpha]$ in Eq. (2.15) contain true values $\bar{x}_\alpha$ and $\bar{y}_\alpha$. They are replaced by their observed values x_α and y_α in actual computation. It is known that this replacement has practically no effect on the final results.

2.4 Algebraic Methods

2.4.1 Iterative Reweight

The following *iterative reweight* is an old and well-known method.

Procedure 2.2 (Iterative reweight)

1. Let $\boldsymbol{\theta}_0 = \mathbf{0}$ and $W_\alpha = 1, \alpha = 1, \dots, N$.
2. Compute the 6×6 matrix

$$\mathsf{M} = \frac{1}{N}\sum_{\alpha=1}^{N} W_\alpha \boldsymbol{\xi}_\alpha \boldsymbol{\xi}_\alpha^\top. \tag{2.16}$$

3. Solve the eigenvalue problem

$$\mathsf{M}\boldsymbol{\theta} = \lambda\boldsymbol{\theta}, \tag{2.17}$$

and compute the unit eigenvector $\boldsymbol{\theta}$ for the smallest eigenvalue λ.

4. If $\boldsymbol{\theta} \approx \boldsymbol{\theta}_0$ up to sign, return $\boldsymbol{\theta}$ and stop. Else, update

$$W_\alpha \leftarrow \frac{1}{(\boldsymbol{\theta}, V_0[\boldsymbol{\xi}_\alpha]\boldsymbol{\theta})}, \qquad \boldsymbol{\theta}_0 \leftarrow \boldsymbol{\theta}, \tag{2.18}$$

and go back to Step 2.

Comments As is well known in linear algebra, computing the unit eigenvector for the smallest eigenvalue of a symmetric matrix M is equivalent to computing the unit vector $\boldsymbol{\theta}$ that minimizes the quadratic form $(\boldsymbol{\theta}, \mathsf{M}\boldsymbol{\theta})$. Because

$$(\boldsymbol{\theta}, \mathsf{M}\boldsymbol{\theta}) = \Big(\boldsymbol{\theta}, \Big(\frac{1}{N}\sum_{\alpha=1}^{N} W_\alpha \boldsymbol{\xi}_\alpha \boldsymbol{\xi}_\alpha^\top\Big)\boldsymbol{\theta}\Big) = \frac{1}{N}\sum_{\alpha=1}^{N} W_\alpha (\boldsymbol{\theta}, \boldsymbol{\xi}_\alpha \boldsymbol{\xi}_\alpha^\top \boldsymbol{\theta}) = \frac{1}{N}\sum_{\alpha=1}^{N} W_\alpha (\boldsymbol{\xi}_\alpha, \boldsymbol{\theta})^2, \tag{2.19}$$

we minimize the sum of squares $(\boldsymbol{\xi}_\alpha, \boldsymbol{\theta})^2$ weighted by W_α. This is commonly known as *weighted least squares*. According to statistics, the weights W_α are optimal if they are inversely proportional to the variance of each term, being small for uncertain terms and large for certain terms. Inasmuch as $(\boldsymbol{\xi}_\alpha, \boldsymbol{\theta}) = (\bar{\boldsymbol{\xi}}_\alpha, \boldsymbol{\theta}) + (\Delta_1\boldsymbol{\xi}_\alpha, \boldsymbol{\theta}) + (\Delta_2\boldsymbol{\xi}_\alpha, \boldsymbol{\theta})$ and $(\bar{\boldsymbol{\xi}}_\alpha, \boldsymbol{\theta}) = 0$, the variance is given from Eqs. (2.13) and (2.15) by

$$E[(\boldsymbol{\xi}_\alpha, \boldsymbol{\theta})^2] = E[(\boldsymbol{\theta}, \Delta_1\boldsymbol{\xi}_\alpha \Delta_1\boldsymbol{\xi}_\alpha^\top \boldsymbol{\theta})] = (\boldsymbol{\theta}, E[\Delta_1\boldsymbol{\xi}_\alpha \Delta_1\boldsymbol{\xi}_\alpha^\top]\boldsymbol{\theta}) = \sigma^2(\boldsymbol{\theta}, V_0[\boldsymbol{\xi}_\alpha]\boldsymbol{\theta}), \tag{2.20}$$

omitting higher-order noise terms. Thus we should let $W_\alpha = 1/(\boldsymbol{\theta}, V_0[\boldsymbol{\xi}_\alpha]\boldsymbol{\theta})$, but $\boldsymbol{\theta}$ is not known yet. Therefore we instead use the weights W_α determined in the preceding iteration to compute $\boldsymbol{\theta}$ and update the weights as in Eq. (2.18). Let us call the solution $\boldsymbol{\theta}$ computed in the initial iteration the *initial solution*. Initially, we set $W_\alpha = 1$, therefore Eq. (2.19) implies that we are starting from the least-squares solution. The phrase "up to sign" in Step 4 reflects the fact that eigenvectors have sign indeterminacy; we align $\boldsymbol{\theta}$ and $\boldsymbol{\theta}_0$ by reversing the sign $\boldsymbol{\theta} \leftarrow -\boldsymbol{\theta}$ when $(\boldsymbol{\theta}, \boldsymbol{\theta}_0) < 0$.

2.4.2 Renormalization and the Taubin Method

It has been well known that the accuracy of least squares and iterative reweight is rather low with large bias when the input elliptic arc is short; they tend to fit a smaller ellipse than expected. The following *renormalization* was introduced to reduce the bias.

Procedure 2.3 (Renormalization)

1. Let $\boldsymbol{\theta}_0 = \mathbf{0}$ and $W_\alpha = 1$, $\alpha = 1, \ldots, N$.
2. Compute the 6×6 matrices

$$\mathsf{M} = \frac{1}{N}\sum_{\alpha=1}^{N} W_\alpha \boldsymbol{\xi}_\alpha \boldsymbol{\xi}_\alpha^\top, \qquad \mathsf{N} = \frac{1}{N}\sum_{\alpha=1}^{N} W_\alpha V_0[\boldsymbol{\xi}_\alpha]. \tag{2.21}$$

3. Solve the generalized eigenvalue problem

$$\mathsf{M}\boldsymbol{\theta} = \lambda \mathsf{N}\boldsymbol{\theta}, \tag{2.22}$$

   and compute the unit generalized eigenvector $\boldsymbol{\theta}$ for the generalized eigenvalue λ of the smallest absolute value.
4. If $\boldsymbol{\theta} \approx \boldsymbol{\theta}_0$ up to sign, return $\boldsymbol{\theta}$ and stop. Else, update

$$W_\alpha \leftarrow \frac{1}{(\boldsymbol{\theta}, V_0[\boldsymbol{\xi}_\alpha]\boldsymbol{\theta})}, \qquad \boldsymbol{\theta}_0 \leftarrow \boldsymbol{\theta}, \tag{2.23}$$

   and go back to Step 2.

Comments As is well known in linear algebra, solving the generalized eigenvalue problem of Eq. (2.22) for symmetric matrices M and N is equivalent to computing the unit vector $\boldsymbol{\theta}$ that minimizes the quadratic form $(\boldsymbol{\theta}, \mathsf{M}\boldsymbol{\theta})$ subject to the constraint $(\boldsymbol{\theta}, \mathsf{N}\boldsymbol{\theta}) = $ constant. Initially $W_\alpha = 1$, therefore the first iteration minimizes the sum of squares $\sum_{\alpha=1}^{N}(\boldsymbol{\xi}_\alpha, \boldsymbol{\theta})^2$ subject to $(\boldsymbol{\theta}, \left(\sum_{\alpha=1}^{N} V_0[\boldsymbol{\xi}_\alpha]\right)\boldsymbol{\theta}) = $ constant. This is known as the *Taubin method* for ellipse fitting (↪ Problem 2.2). Standard numerical tools for solving the generalized eigenvalue problem in the form of Eq. (2.22) assume that N is positive definite, but Eq. (2.15) implies that the sixth column and the sixth row of the matrix $V_0[\boldsymbol{\xi}_\alpha]$ all consist of zero, therefore N is not positive definite. However, Eq. (2.22) is equivalently written as

$$\mathsf{N}\boldsymbol{\theta} = \frac{1}{\lambda}\mathsf{M}\boldsymbol{\theta}. \tag{2.24}$$

If the data contain noise, the matrix M is positive definite, therefore we can apply a standard numerical tool to compute the unit generalized eigenvector $\boldsymbol{\theta}$ for the generalized eigenvalue $1/\lambda$ of the largest absolute value. The matrix M is not positive definite only when there is no noise. We need not consider that case in practice, but if M happens to have eigenvalue 0, which implies that the data are exact, the corresponding unit eigenvector $\boldsymbol{\theta}$ gives the true solution.

2.4.3 Hyper-Renormalization and HyperLS

According to experiments, the accuracy of the Taubin method is higher than least squares and iterative reweight, and renormalization has even higher accuracy. The accuracy can be further improved by the following *hyper-renormalization*.

Procedure 2.4 (Hyper-renormalization)

1. Let $\boldsymbol{\theta}_0 = \mathbf{0}$ and $W_\alpha = 1$, $\alpha = 1, \ldots, N$.
2. Compute the 6×6 matrices

$$\mathsf{M} = \frac{1}{N}\sum_{\alpha=1}^{N} W_\alpha \boldsymbol{\xi}_\alpha \boldsymbol{\xi}_\alpha^\top, \tag{2.25}$$

$$\mathsf{N} = \frac{1}{N}\sum_{\alpha=1}^{N} W_\alpha \Big(V_0[\boldsymbol{\xi}_\alpha] + 2\mathscr{S}[\boldsymbol{\xi}_\alpha \mathbf{e}^\top] \Big) - \frac{1}{N^2}\sum_{\alpha=1}^{N} W_\alpha^2 \Big((\boldsymbol{\xi}_\alpha, \mathsf{M}_5^- \boldsymbol{\xi}_\alpha) V_0[\boldsymbol{\xi}_\alpha] + 2\mathscr{S}[V_0[\boldsymbol{\xi}_\alpha]\mathsf{M}_5^- \boldsymbol{\xi}_\alpha \boldsymbol{\xi}_\alpha^\top] \Big), \tag{2.26}$$

   where $\mathscr{S}[\,\cdot\,]$ is the symmetrization ($\mathscr{S}[\mathsf{A}] = (\mathsf{A} + \mathsf{A}^\top)/2$), and $\mathbf{e}$ is the vector

$$\mathbf{e} = (1, 0, 1, 0, 0, 0)^\top. \tag{2.27}$$

   The matrix M_5^- is the pseudoinverse of M of truncated rank 5.
3. Solve the generalized eigenvalue problem

$$\mathsf{M}\boldsymbol{\theta} = \lambda \mathsf{N}\boldsymbol{\theta}, \tag{2.28}$$

   and compute the unit generalized eigenvector $\boldsymbol{\theta}$ for the generalized eigenvalue λ of the smallest absolute value.
4. If $\boldsymbol{\theta} \approx \boldsymbol{\theta}_0$ up to sign, return $\boldsymbol{\theta}$ and stop. Else, update

$$W_\alpha \leftarrow \frac{1}{(\boldsymbol{\theta}, V_0[\boldsymbol{\xi}_\alpha]\boldsymbol{\theta})}, \qquad \boldsymbol{\theta}_0 \leftarrow \boldsymbol{\theta}, \tag{2.29}$$

   and go back to Step 2.

Comments The vector $\mathbf{e}$ is defined in such a way that $E[\Delta_2\boldsymbol{\xi}_\alpha] = \sigma^2\mathbf{e}$ for the second-order noise term $\Delta_2\boldsymbol{\xi}_\alpha$ in Eq. (2.12). The pseudoinverse M_5^- of truncated rank is computed by

$$\mathsf{M}_5^- = \frac{1}{\mu_1}\boldsymbol{\theta}_1\boldsymbol{\theta}_1^\top + \cdots + \frac{1}{\mu_5}\boldsymbol{\theta}_5\boldsymbol{\theta}_5^\top, \tag{2.30}$$

where $\mu_1 \geq \cdots \geq \mu_6$ are the eigenvalues of M, and $\boldsymbol{\theta}_1, \ldots, \boldsymbol{\theta}_6$ are the corresponding eigenvectors (note that $\mathbf{m}_6$ and $\boldsymbol{\theta}_6$ are not used). The derivation of Eq. (2.26) is given in Chap. 14.

The matrix N in Eq. (2.26) is not positive definite, but we can use a standard numerical tool by rewriting Eq. (2.28) in the form of Eq. (2.24) and computing the unit generalized eigenvector for the generalized eigenvalue $1/\lambda$ of the largest absolute value. The initial solution minimizes the sum of squares $\sum_{\alpha=1}^{N}(\boldsymbol{\xi}_\alpha, \boldsymbol{\theta})^2$ subject to the constraint $(\boldsymbol{\theta}, \mathsf{N}\boldsymbol{\theta}) =$ constant for the matrix N obtained by letting $W_\alpha = 1$ in Eq. (2.26). This corresponds to the method called *HyperLS* (↪ Problem 2.3).

2.4.4 Summary of Algebraic Methods

We have seen that all the above methods compute the $\boldsymbol{\theta}$ that satisfies

$$\mathsf{M}\boldsymbol{\theta} = \lambda\mathsf{N}\boldsymbol{\theta}, \tag{2.31}$$

where the matrices M and N are defined from the data and contain the unknown $\boldsymbol{\theta}$. Different choices of them lead to different methods:

$$\mathsf{M} = \begin{cases} \dfrac{1}{N}\displaystyle\sum_{\alpha=1}^{N}\boldsymbol{\xi}_\alpha\boldsymbol{\xi}_\alpha^\top, & \text{(least squares, Taubin, HyperLS)} \\ \dfrac{1}{N}\displaystyle\sum_{\alpha=1}^{N}\dfrac{\boldsymbol{\xi}_\alpha\boldsymbol{\xi}_\alpha^\top}{(\boldsymbol{\theta}, V_0[\boldsymbol{\xi}_\alpha]\boldsymbol{\theta})}, & \text{(iterative reweight, renormalization, hyper-renormalization)} \end{cases} \tag{2.32}$$

$$\mathsf{N} = \begin{cases} \mathsf{I} \text{ (identity)}, & \text{(least squares, iterative reweight)} \\ \dfrac{1}{N}\displaystyle\sum_{\alpha=1}^{N}V_0[\boldsymbol{\xi}_\alpha], & \text{(Taubin)} \\ \dfrac{1}{N}\displaystyle\sum_{\alpha=1}^{N}\dfrac{V_0[\boldsymbol{\xi}_\alpha]}{(\boldsymbol{\theta}, V_0[\boldsymbol{\xi}_\alpha]\boldsymbol{\theta})}, & \text{(renormalization)} \\ \dfrac{1}{N}\displaystyle\sum_{\alpha=1}^{N}\Big(V_0[\boldsymbol{\xi}_\alpha] + 2\mathcal{S}[\boldsymbol{\xi}_\alpha\mathbf{e}^\top]\Big) & \\ \quad -\dfrac{1}{N^2}\displaystyle\sum_{\alpha=1}^{N}\Big((\boldsymbol{\xi}_\alpha, \mathsf{M}_5^-\boldsymbol{\xi}_\alpha)V_0[\boldsymbol{\xi}_\alpha] + 2\mathcal{S}[V_0[\boldsymbol{\xi}_\alpha]\mathsf{M}_5^-\boldsymbol{\xi}_\alpha\boldsymbol{\xi}_\alpha^\top]\Big), & \\ & \text{(HyperLS)} \\ \dfrac{1}{N}\displaystyle\sum_{\alpha=1}^{N}\dfrac{1}{(\boldsymbol{\theta}, V_0[\boldsymbol{\xi}_\alpha]\boldsymbol{\theta})}\Big(V_0[\boldsymbol{\xi}_\alpha] + 2\mathcal{S}[\boldsymbol{\xi}_\alpha\mathbf{e}^\top]\Big) & \\ \quad -\dfrac{1}{N^2}\displaystyle\sum_{\alpha=1}^{N}\dfrac{1}{(\boldsymbol{\theta}, V_0[\boldsymbol{\xi}_\alpha]\boldsymbol{\theta})^2}\Big((\boldsymbol{\xi}_\alpha, \mathsf{M}_5^-\boldsymbol{\xi}_\alpha)V_0[\boldsymbol{\xi}_\alpha] + 2\mathcal{S}[V_0[\boldsymbol{\xi}_\alpha]\mathsf{M}_5^-\boldsymbol{\xi}_\alpha\boldsymbol{\xi}_\alpha^\top]\Big). & \\ & \text{(hyper-renormalization)} \end{cases} \tag{2.33}$$

For least squares, Taubin, and HyperLS, the matrices M and N do not contain the unknown $\boldsymbol{\theta}$, thus Eq. (2.31) is a generalized eigenvalue problem that can be directly solved without iterations. For other methods (iterative reweight, renormalization, and hyper-renormalization), the unknown $\boldsymbol{\theta}$ is contained in the denominators in the expressions of M and N. We let the part that contains $\boldsymbol{\theta}$ be $1/W_\alpha$, compute W_α using the value of $\boldsymbol{\theta}$ obtained in the preceding iteration, and solve the generalized eigenvalue problem in the form of Eq. (2.31). We then use the resulting $\boldsymbol{\theta}$ to update W_α and repeat this process.

According to experiments, HyperLS has accuracy comparable to renormalization, and hyper-renormalization has even higher accuracy. Because the iteration starts from the HyperLS solution, the convergence is very fast; usually, three to four iterations are sufficient.

2.5 Geometric Methods

2.5.1 Geometric Distance and Sampson Error

Geometric fitting refers to computing an ellipse such that the sum of squares of the distance of observed points (x_α, y_α) to the ellipse is minimized (Fig. 2.2). Let $(\bar{x}_\alpha, \bar{y}_\alpha)$ be the point on the ellipse closest to (x_α, y_α). Let d_α be the distance between them. Geometric fitting computes the ellipse that minimizes

$$S = \frac{1}{N}\sum_{\alpha=1}^{N}\Big((x_\alpha - \bar{x}_\alpha)^2 + (y_\alpha - \bar{y}_\alpha)^2\Big) = \frac{1}{N}\sum_{\alpha=1}^{N} d_\alpha^2. \tag{2.34}$$

This is called the *geometric distance* (although this has the dimension of square length, this term is widely used). It is also sometimes called the *reprojection error*. If we identify $(\bar{x}_\alpha, \bar{y}_\alpha)$ with the true position of the datapoint (x_α, y_α), we see from Eq. (2.10) that Eq. (2.34) can also be written as $S = (1/N)\sum_{\alpha=1}^{N}(\Delta x_\alpha^2 + \Delta y_\alpha^2)$, where we regard Δx_α and Δy_α as unknown variables rather than stochastic quantities. When point (x_α, y_α) is close to the ellipse, the square distance d_α^2 can be written (except for high-order terms in Δx_α and Δy_α) as follows ($\hookrightarrow$ Problem 2.4).

$$d_\alpha^2 = (x_\alpha - \bar{x}_\alpha)^2 + (y_\alpha - \bar{y}_\alpha)^2 \approx \frac{(\boldsymbol{\xi}_\alpha, \boldsymbol{\theta})^2}{(\boldsymbol{\theta}, V_0[\boldsymbol{\xi}_\alpha]\boldsymbol{\theta})}. \tag{2.35}$$

Hence, the geometric distance in Eq. (2.34) can be approximated by

$$J = \frac{1}{N}\sum_{\alpha=1}^{N}\frac{(\boldsymbol{\xi}_\alpha, \boldsymbol{\theta})^2}{(\boldsymbol{\theta}, V_0[\boldsymbol{\xi}_\alpha]\boldsymbol{\theta})}. \tag{2.36}$$

This is known as the *Sampson error*.

2.5.2 FNS

A well-known method to minimize the Sampson error of Eq. (2.36) is the *FNS* (*Fundamental Numerical Scheme*):

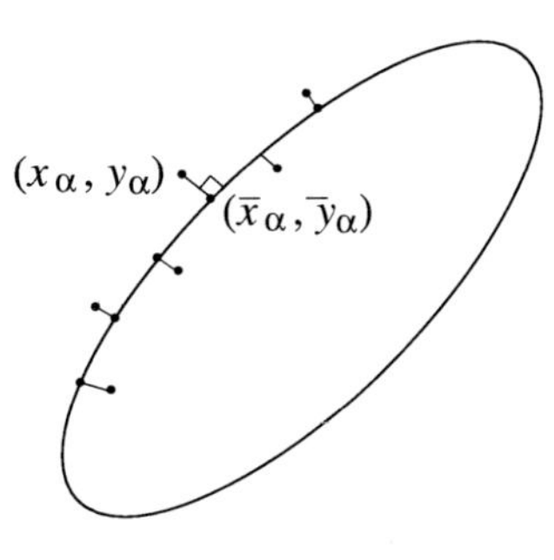

Fig. 2.2 We fit an ellipse such that the geometric distance (i.e., the average of the square distance of each point to the ellipse) is minimized

Procedure 2.5 (FNS)

1. Let $\boldsymbol{\theta} = \boldsymbol{\theta}_0 = \mathbf{0}$ and $W_\alpha = 1, \alpha = 1, \ldots, N$.
2. Compute the 6×6 matrices

$$\mathsf{M} = \frac{1}{N}\sum_{\alpha=1}^{N} W_\alpha \boldsymbol{\xi}_\alpha \boldsymbol{\xi}_\alpha^\top, \qquad \mathsf{L} = \frac{1}{N}\sum_{\alpha=1}^{N} W_\alpha^2 (\boldsymbol{\xi}_\alpha, \boldsymbol{\theta})^2 V_0[\boldsymbol{\xi}_\alpha]. \tag{2.37}$$

3. Compute the 6×6 matrix

$$\mathsf{X} = \mathsf{M} - \mathsf{L}. \tag{2.38}$$

4. Solve the eigenvalue problem

$$\mathsf{X}\boldsymbol{\theta} = \lambda\boldsymbol{\theta}, \tag{2.39}$$

   and compute the unit eigenvector $\boldsymbol{\theta}$ for the smallest eigenvalue λ.
5. If $\boldsymbol{\theta} \approx \boldsymbol{\theta}_0$ up to sign, return $\boldsymbol{\theta}$ and stop. Else, update

$$W_\alpha \leftarrow \frac{1}{(\boldsymbol{\theta}, V_0[\boldsymbol{\xi}_\alpha]\boldsymbol{\theta})}, \qquad \boldsymbol{\theta}_0 \leftarrow \boldsymbol{\theta}, \tag{2.40}$$

   and go back to Step 2.

Comments This scheme solves $\nabla_{\boldsymbol{\theta}} J = \mathbf{0}$ for the Sampson error J, where $\nabla_{\boldsymbol{\theta}} J$ is the gradient of J, that is, the vector whose ith component is $\partial J/\partial \theta_i$. Differentiating Eq. (2.36), we obtain the following expression (↪ Problem 2.5(1)).

$$\nabla_{\boldsymbol{\theta}} J = 2(\mathsf{M} - \mathsf{L})\boldsymbol{\theta} = 2\mathsf{X}\boldsymbol{\theta}. \tag{2.41}$$

Here, M, L, and X are the matrices given by Eqs. (2.37) and (2.38). It can be shown that $\lambda = 0$ after the iterations have converged (↪ Problem 2.5(2)). Thus from Eq. (2.41) we obtain the value $\boldsymbol{\theta}$ that satisfies $\nabla_{\boldsymbol{\theta}} J = \mathbf{0}$. Because we start with $\boldsymbol{\theta} = \mathbf{0}$, the matrix L in Eq. (2.37) is initially O, therefore Eq. (2.39) reduces to $\mathsf{M}\boldsymbol{\theta} = \lambda\boldsymbol{\theta}$. This means that the FNS iterations start from the least squares solution.

2.5.3 Geometric Distance Minimization

Once we have computed $\boldsymbol{\theta}$ by the above FNS, we can iteratively modify it such that it strictly minimizes the geometric distance of Eq. (2.34). To be specific, we modify the data $\boldsymbol{\xi}_\alpha$, using the current solution $\boldsymbol{\theta}$, and minimize a modified Sampson error to obtain a new solution $\boldsymbol{\theta}$. After a few iterations, the Sampson error coincides with the geometric distance. The procedure is as follows.

Procedure 2.6 (Geometric distance minimization)

1. Let $J_0 = \infty$ (a sufficiently large number), $\hat{x}_\alpha = x_\alpha$, $\hat{y}_\alpha = y_\alpha$, and $\tilde{x}_\alpha = \tilde{y}_\alpha = 0$, $\alpha = 1, \ldots, N$.
2. Compute the normalized covariance matrix $V_0[\hat{\boldsymbol{\xi}}_\alpha]$ obtained by replacing $\bar{x}_\alpha$ and $\bar{y}_\alpha$ in the definition of $V_0[\boldsymbol{\xi}_\alpha]$ in Eq. (2.15) by $\hat{x}_\alpha$ and $\hat{y}_\alpha$, respectively.
3. Compute the following modified data $\boldsymbol{\xi}_\alpha^*$.

$$\boldsymbol{\xi}^*_\alpha = \begin{pmatrix} \hat{x}^2_\alpha + 2\hat{x}_\alpha \tilde{x}_\alpha \\ 2(\hat{x}_\alpha \hat{y}_\alpha + \hat{y}_\alpha \tilde{x}_\alpha + \hat{x}_\alpha \tilde{y}_\alpha) \\ \hat{y}^2_\alpha + 2\hat{y}_\alpha \tilde{y}_\alpha \\ 2f_0(\hat{x}_\alpha + \tilde{x}_\alpha) \\ 2f_0(\hat{y}_\alpha + \tilde{y}_\alpha) \\ f_0^2 \end{pmatrix}. \tag{2.42}$$

4. Compute the value $\boldsymbol{\theta}$ that minimizes the following *modified Sampson error*.

$$J^* = \frac{1}{N} \sum_{\alpha=1}^{N} \frac{(\boldsymbol{\xi}^*_\alpha, \boldsymbol{\theta})^2}{(\boldsymbol{\theta}, V_0[\hat{\boldsymbol{\xi}}_\alpha]\boldsymbol{\theta})}. \tag{2.43}$$

5. Update $\tilde{x}_\alpha$ and $\tilde{y}_\alpha$ as follows.

$$\begin{pmatrix} \tilde{x}_\alpha \\ \tilde{y}_\alpha \end{pmatrix} \leftarrow \frac{2(\boldsymbol{\xi}^*_\alpha, \boldsymbol{\theta})}{(\boldsymbol{\theta}, V_0[\hat{\boldsymbol{\xi}}_\alpha]\boldsymbol{\theta})} \begin{pmatrix} \theta_1 & \theta_2 & \theta_4 \\ \theta_2 & \theta_3 & \theta_5 \end{pmatrix} \begin{pmatrix} \hat{x}_\alpha \\ \hat{y}_\alpha \\ f_0 \end{pmatrix}. \tag{2.44}$$

6. Update $\hat{x}_\alpha$ and $\hat{y}_\alpha$ as follows.

$$\hat{x}_\alpha \leftarrow x_\alpha - \tilde{x}_\alpha, \qquad \hat{y}_\alpha \leftarrow y_\alpha - \tilde{y}_\alpha. \tag{2.45}$$

7. Compute

$$J^* = \frac{1}{N} \sum_{\alpha=1}^{N} (\tilde{x}^2_\alpha + \tilde{y}^2_\alpha). \tag{2.46}$$

If $J^* \approx J_0$, return $\boldsymbol{\theta}$ and stop. Else, let $J_0 \leftarrow J^*$, and go back to Step 2.

Comments Because $\hat{x}_\alpha = x_\alpha$, $\hat{y}_\alpha = y_\alpha$, and $\tilde{x}_\alpha = \tilde{y}_\alpha = 0$ in the initial iteration, the value $\boldsymbol{\xi}^*_\alpha$ of Eq. (2.42) is the same as $\boldsymbol{\xi}_\alpha$. Therefore the modified Simpson error J^* in Eq. (2.43) is the same as the function J in Eq. (2.36), and the value $\boldsymbol{\theta}$ that minimizes J is computed initially. Now, we can rewrite Eq. (2.34) in the form

$$\begin{aligned} S &= \frac{1}{N} \sum_{\alpha=1}^{N} \Big((\hat{x}_\alpha + (x_\alpha - \hat{x}_\alpha) - \bar{x}_\alpha)^2 + (\hat{y}_\alpha + (y_\alpha - \hat{y}_\alpha) - \bar{y}_\alpha)^2 \Big) \\ &= \frac{1}{N} \sum_{\alpha=1}^{N} \Big((\hat{x}_\alpha + \tilde{x}_\alpha - \bar{x}_\alpha)^2 + (\hat{y}_\alpha + \tilde{y}_\alpha - \bar{y}_\alpha)^2 \Big), \end{aligned} \tag{2.47}$$

where

$$\tilde{x}_\alpha = x_\alpha - \hat{x}_\alpha, \qquad \tilde{y}_\alpha = y_\alpha - \hat{y}_\alpha. \tag{2.48}$$

In the next iteration step, we regard the corrected values $(\hat{x}_\alpha, \hat{y}_\alpha)$ as the input data and minimize Eq. (2.47). Let $(\hat{\hat{x}}_\alpha, \hat{\hat{y}}_\alpha)$ be the solution. If we ignore high-order small terms in $\hat{x}_\alpha - \bar{x}_\alpha$ and $\hat{y}_\alpha - \bar{y}_\alpha$ and rewrite Eq. (2.47), we obtain the modified Sampson error J^* in Eq. (2.43) ($\hookrightarrow$ Problem 2.6). We minimize this, regard $(\hat{\hat{x}}_\alpha, \hat{\hat{y}}_\alpha)$ as $(\hat{x}_\alpha, \hat{y}_\alpha)$, and repeat the same process. Because the current $(\hat{x}_\alpha, \hat{y}_\alpha)$ are the best approximation to $(\bar{x}_\alpha, \bar{y}_\alpha)$, Eq. (2.48) implies that $\tilde{x}^2_\alpha + \tilde{y}^2_\alpha$ is the corresponding approximation of $(\bar{x}_\alpha - x_\alpha)^2 + (\bar{y}_\alpha - y_\alpha)^2$. Therefore we use Eq. (2.46) to evaluate the geometric distance S. The ignored higher-order terms decrease after each iteration, therefore we obtain in the end the value $\boldsymbol{\theta}$ that minimizes the geometric distance S, and Eq. (2.46) coincides with S. According to experiments, however, the correction of $\boldsymbol{\theta}$ by this procedure is very small: the three or four significant digits are unchanged in typical problems. Therefore the corrected ellipse is indistinguishable from the original one when displayed or plotted. Thus, we can practically identify FNS with a method to minimize the geometric distance.

2.5.4 Hyper-Accurate Correction

The geometric method, whether it is geometric distance minimization or Sampson error minimization, is known to have small statistical bias, which means that the expectation $E[\hat{\boldsymbol{\theta}}]$ of the computed solution $\hat{\boldsymbol{\theta}}$ does not completely agree with the true value $\bar{\boldsymbol{\theta}}$. We express the computed solution $\hat{\boldsymbol{\theta}}$ in the form

$$\hat{\boldsymbol{\theta}} = \bar{\boldsymbol{\theta}} + \Delta_1\boldsymbol{\theta} + \Delta_2\boldsymbol{\theta} + \ldots, \tag{2.49}$$

where $\Delta_k\boldsymbol{\theta}$ is the kth-order term in the noise components Δx_α and Δy_α. Inasmuch as $\Delta_1\boldsymbol{\theta}$ is a linear expression in Δx_α and Δy_α, we have $E[\Delta_1\boldsymbol{\theta}] = \mathbf{0}$. However, $E[\Delta_2\boldsymbol{\theta}] \neq \mathbf{0}$ in general. If we are able to evaluate $E[\Delta_2\boldsymbol{\theta}]$ in an explicit form by doing error analysis, it is expected that the subtraction

$$\tilde{\boldsymbol{\theta}} = \hat{\boldsymbol{\theta}} - E[\Delta_2\boldsymbol{\theta}] \tag{2.50}$$

will yield higher accuracy than $\hat{\boldsymbol{\theta}}$, and its expectation is $E[\tilde{\boldsymbol{\theta}}] = \bar{\boldsymbol{\theta}} + O(\sigma^4)$, where σ is the noise level. Note that $\Delta_3\boldsymbol{\theta}$ is a third-order expression in Δx_α and Δy_α, and hence $E[\Delta_3\boldsymbol{\theta}] = \mathbf{0}$. The operation of increasing the accuracy by subtracting $E[\Delta_2\boldsymbol{\theta}]$ is called *hyper-accurate correction*. The actual procedure is as follows.

Procedure 2.7 (Hyper-accurate correction)

1. Compute $\boldsymbol{\theta}$ by FNS (Procedure 2.5).
2. Estimate σ^2 in the form
$$\hat{\sigma}^2 = \frac{(\boldsymbol{\theta}, \mathsf{M}\boldsymbol{\theta})}{1 - 5/N}, \tag{2.51}$$
where M is the value of the matrix M in Eq. (2.37) after the FNS iterations have converged.
3. Compute the correction term
$$\Delta_c\boldsymbol{\theta} = -\frac{\hat{\sigma}^2}{N}\mathsf{M}_5^-\sum_{\alpha=1}^{N} W_\alpha(\mathbf{e}, \boldsymbol{\theta})\boldsymbol{\xi}_\alpha + \frac{\hat{\sigma}^2}{N^2}\mathsf{M}_5^-\sum_{\alpha=1}^{N} W_\alpha^2(\boldsymbol{\xi}_\alpha, \mathsf{M}_5^- V_0[\boldsymbol{\xi}_\alpha]\boldsymbol{\theta})\boldsymbol{\xi}_\alpha, \tag{2.52}$$
where W_α is the value of W_α in Eq. (2.40) after the FNS iterations have converged, and $\mathbf{e}$ is the vector in Eq. (2.27). The matrix M_5^- is the pseudoinverse of truncated rank 5 given by Eq. (2.30).
4. Correct $\boldsymbol{\theta}$ into
$$\boldsymbol{\theta} \leftarrow \mathcal{N}[\boldsymbol{\theta} - \Delta_c\boldsymbol{\theta}], \tag{2.53}$$
where $\mathcal{N}[\,\cdot\,]$ denotes normalization to unit norm ($\mathcal{N}[\mathbf{a}] = \mathbf{a}/\|\mathbf{a}\|$).

Comments The derivation of Eqs. (2.51) and (2.52) is given in Chap. 15. According to experiments, the accuracy of FNS and geometric distance minimization is higher than renormalization but lower than hyper-renormalization. It is known, however, that after this hyper-accurate correction the accuracy improves and is comparable to hyper-renormalization.

2.6 Ellipse-Specific Methods

2.6.1 Ellipse Condition

Equation (2.1) does not necessarily represent an ellipse; depending on the coefficients, it can represent a hyperbola or a parabola. In special cases, it may degenerate into two lines, or (x, y) may not satisfy the equation. It is easy to show that Eq. (2.1) represents an ellipse when

$$AC - B^2 > 0. \tag{2.54}$$

We discuss this in Chap. 9. Usually, an ellipse is obtained when Eq. (2.1) is fit to a point sequence obtained from an ellipse, but we may obtain a hyperbola when the point sequence is very short in the presence of large noise; a parabola results only when the solution exactly satisfies $AC - B^2 = 0$, but this possibility can be excluded in real numerical computation using noisy data. If a hyperbola results, we can ignore it in practice, inasmuch as it indicates that the data do not have sufficient information to define an ellipse. From a theoretical interest, however, schemes to force the fit to be an ellipse have also been studied.

2.6.2 Method of Fitzgibbon et al.

A well-known method for fitting only ellipses is the following method of Fitzgibbon et al.

Procedure 2.8 (Method of Fitzgibbon et al.)

1. Compute the 6×6 matrices

$$\mathsf{M} = \frac{1}{N}\sum_{\alpha=1}^{N} \boldsymbol{\xi}_\alpha \boldsymbol{\xi}_\alpha^\top, \qquad \mathsf{N} = \begin{pmatrix} 0 & 0 & 1 & 0 & 0 & 0 \\ 0 & -2 & 0 & 0 & 0 & 0 \\ 1 & 0 & 0 & 0 & 0 & 0 \\ 0 & 0 & 0 & 0 & 0 & 0 \\ 0 & 0 & 0 & 0 & 0 & 0 \\ 0 & 0 & 0 & 0 & 0 & 0 \end{pmatrix}. \tag{2.55}$$

2. Solve the generalized eigenvalue problem

$$\mathsf{M}\boldsymbol{\theta} = \lambda \mathsf{N}\boldsymbol{\theta}, \tag{2.56}$$

and return the unit generalized eigenvector $\boldsymbol{\theta}$ for the generalized eigenvalue λ of the smallest absolute value.

Comments This is an algebraic method, minimizing the algebraic distance $\sum_{\alpha=1}^{N}(\xi_\alpha, \theta)^2$ subject to $AC - B^2 = 1$ such that Eq. (2.54) is satisfied. The matrix N in Eq. (2.56) is not positive definite, as in the case of renormalization and hyper-renormalization, therefore we rewrite Eq. (2.56) in the form of Eq. (2.24) and use a standard numerical tool to compute the unit eigenvector θ for the generalized eigenvalue $1/\lambda$ of the largest absolute value.

2.6.3 Method of Random Sampling

One way always to end up with an ellipse is first to do ellipse fitting without considering the ellipse condition (e.g., by hyper-renormalization) and modify the result to an ellipse if it is not an ellipse. A simple and effective method is the use of random sampling: we randomly sample five points from the input sequence, fit an ellipse to them, and repeat this many times to find the best ellipse. The actual procedure is as follows.

Procedure 2.9 (Random sampling)

1. Fit an ellipse without considering the ellipse condition. If the solution θ satisfies
$$\theta_1\theta_3 - \theta_2^2 > 0, \tag{2.57}$$
return it as the final result.
2. Else, randomly select five points from the input sequence. Let $\xi_1, \ldots, \xi_5$ be the corresponding vectors of the form of Eq. (2.3).
3. Compute the unit eigenvector θ of the following matrix for the smallest eigenvalue.
$$\mathsf{M}_5 = \sum_{\alpha=1}^{5} \xi_\alpha \xi_\alpha^\top. \tag{2.58}$$
4. If that θ does not satisfy Eq. (2.57), discard it and select a new set of five points randomly for fitting an ellipse.
5. If θ satisfies Eq. (2.57), store it as a candidate. Also, evaluate the corresponding Sampson error J of Eq. (2.36), and store its value.
6. Repeat this process many times, and return the value θ whose Sampson error J is the smallest.

Comments In Step 2, we select five integers over $[1, N]$ by sampling from a uniform distribution; overlapped values are discarded. Step 3 computes the ellipse passing through the given five points. We may solve the simultaneous linear equations in $A, B, \ldots, F$ obtained by substituting the point coordinates to Eq. (2.1); the solution is determined up to scale, which is then normalized. However, using least squares as shown above is easier to program; we compute the unit eigenvector of the matrix M in the first row of Eq. (2.32). The coefficient $1/N$ is omitted in the above matrix M_5, but the computed eigenvectors are the same. In Step 4, we judge whether the solution is an ellipse. Theoretically, any selected five points may define a hyperbola, for example, when all the points are on a hyperbola. We ignore such an extreme case. In Step 5, if the evaluated J is larger than the stored J, we go directly on to the next sampling without storing $\boldsymbol{\theta}$ and J. We stop the sampling if the current value of J is not updated consecutively after a fixed number of iterations.

In practice, however, if a standard method returns a hyperbola, it forces the solution to be an ellipse by an ellipse-specific method such as the method of Fitzgibbon et al. and random sampling does not have much meaning, as shown in the example below.

2.7 Outlier Removal

For fitting an ellipse to real image data, we must first extract a sequence of edge points from an elliptic arc. However, edge segments obtained by an edge detection filter may not belong to the same ellipse; some may be parts of other object boundaries. We call points of such segments *outliers*; those coming from the ellipse under consideration are called *inliers*. Note that we are not talking about random dispersions of edge pixels due to image processing inaccuracy, although such pixels, if they exist, are also called outliers. Our main concern is the parts of the extracted edge sequence that do not belong to the ellipse in which we are interested.

Methods that are not sensitive to the existence of outliers are said to be *robust*. The general idea of robust fitting is to find an ellipse such that *the number of points close to it is as large as possible*. Then those points that are not close to it are removed as outliers. A typical such method is *RANSAC*. The procedure is as follows.

Procedure 2.10 (RANSAC)

1. Randomly select five points from the input sequence. Let $\boldsymbol{\xi}_1, \ldots, \boldsymbol{\xi}_5$ be their vector representations in the form of Eq. (2.3).
2. Compute the unit eigenvector $\boldsymbol{\theta}$ of the matrix
$$\mathsf{M}_5 = \sum_{\alpha=1}^{5} \boldsymbol{\xi}_\alpha \boldsymbol{\xi}_\alpha^\top, \tag{2.59}$$
for the smallest eigenvalue, and store it as a candidate.
3. Let n be the number of points in the input sequence that satisfy
$$\frac{(\boldsymbol{\xi}, \boldsymbol{\theta})^2}{(\boldsymbol{\theta}, V_0[\boldsymbol{\xi}]\boldsymbol{\theta})} < d^2, \tag{2.60}$$
where $\boldsymbol{\xi}$ is the vector representation of the point in the sequence in the form of Eq. (2.3), $V_0[\boldsymbol{\xi}]$ is the normalized covariance matrix defined by Eq. (2.15), and d is a threshold for admissible deviation from the fitted ellipse; for example, $d = 2$ (pixels). Store that n.
4. Select a new set of five points from the input sequence, and do the same. Repeat this many times, and return from among the stored candidate ellipses the one for which n is the largest.

Comments The random sampling in Step 1 is done in the same way as in Sect. 2.6.3. Step 2 computes the ellipse that passes through the five points. In Step 3, we are measuring the distance of the points from the ellipse by Eq. (2.35). As in Sect. 2.6.3, we can go directly on to the next sampling if the count n is smaller than the stored count, and we can stop if the current value of n is not updated over a fixed number of samplings. In the end, those pixels that do not satisfy Eq. (2.60) for the chosen ellipse are removed as outliers.

2.8 Examples

Figure 2.3a shows edges extracted from an image that contains a circular object. An ellipse is fitted to the 160 edge points indicated there, and Fig. 2.3b shows ellipses obtained using various methods superimposed on the original image, where the occluded part is artificially composed for visual ease. We can see that least squares and iterative reweight produce much smaller ellipses than the true shape. All other fits are very close to the true shape, and the geometric distance minimization gives the best fit in this example.

Figure 2.4a is another edge image of a scene with a circular object. An ellipse is fitted to the 140 edge points indicated there, using different methods. Figure 2.4b shows the resulting fits. In this case, hyper-renormalization returns a hyperbola, and the method of Fitzgibbon et al. fits a small and flat ellipse. As an alternative ellipse-specific method, Szpak et al. minimized the Sampson error of Eq. (2.36) with a penalty term such that it diverges to infinity as the solution approaches a hyperbola. In this case, their method fits a large ellipse close to the hyperbola. The random sampling fit is somewhat in between, but it is still not very close to the true shape. In general, if a standard method returns a hyperbola, it indicates data insufficiency

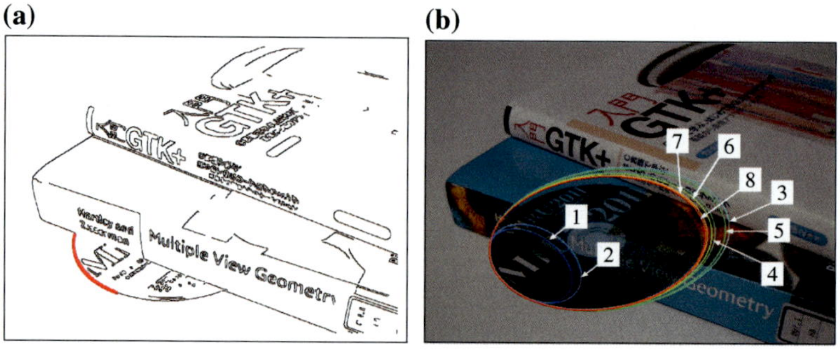

Fig. 2.3 **a** An edge image of a scene with a circular object. An ellipse is fitted to the 160 edge points indicated. **b** Fitted ellipses superimposed on the original image. The occluded part is artificially composed for visual ease: (1) least squares, (2) iterative reweight, (3) Taubin method, (4) renormalization, (5) HyperLS, (6) hyper-renormalization, (7) FN, and (8) FNS with hyper-accurate correction

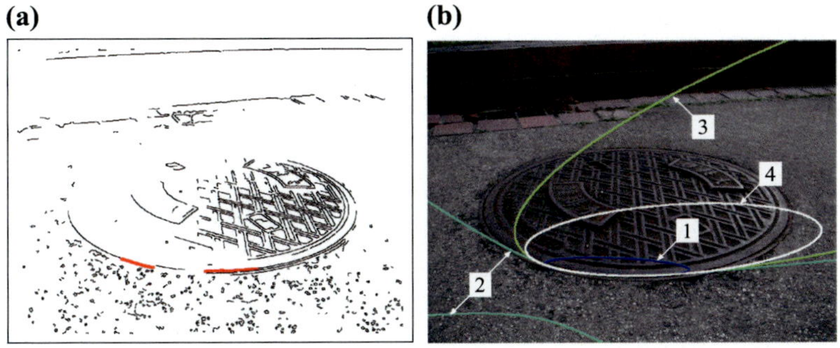

Fig. 2.4 **a** An edge image of a scene with a circular object. An ellipse is fitted to the 140 edge points indicated. **b** Fitted ellipses superimposed on the original image. (1) Fitzgibbon et al., (2) hyper-renormalization, (3) Szpak et al., (4) random sampling

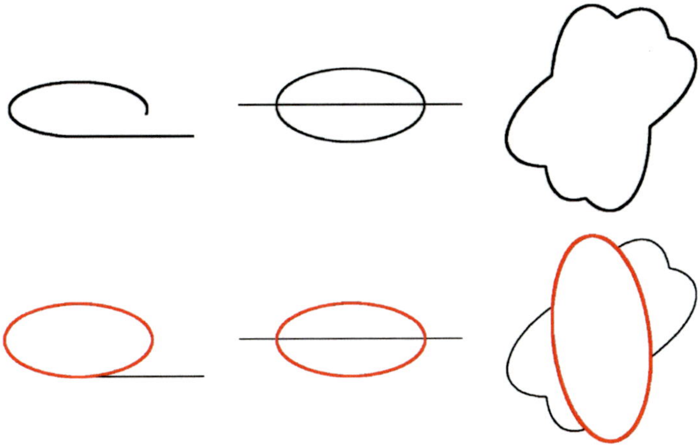

Fig. 2.5 *Upper row:* input edge segments. *Lower row:* ellipses fitted by RANSAC

for ellipse fitting. Hence, although ellipse-specific methods have theoretical interest, they are not suitable for practical applications.

The upper row of Fig. 2.5 shows examples of edge segments that do not entirely consist of elliptic arcs. Regarding those nonelliptic parts as outliers, we fit an ellipse to inlier arcs, using RANSAC. The lower row of Fig. 2.5 depicts fitted ellipses. We see that the segments that constitute an ellipse are automatically selected.

2.9 Supplemental Note

The description in this chapter is rather sketchy. For a more comprehensive description, see the authors' book [17], where detailed numerical experiments for accuracy comparison of different methods are also shown. The renormalization method was proposed by Kanatani in [5], and details are discussed in [6]. The Taubin method was

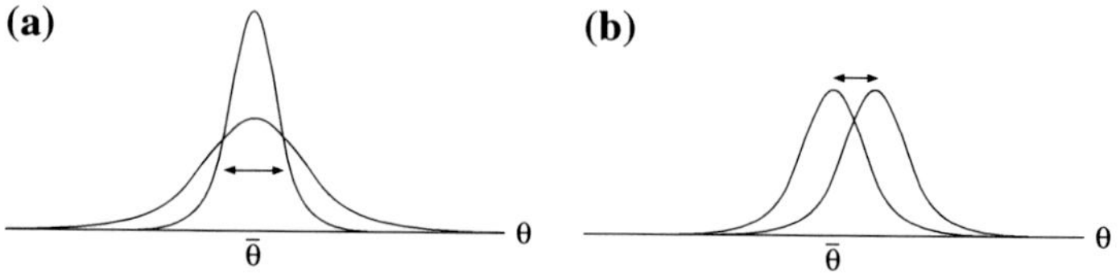

Fig. 2.6 **a** The matrix M controls the covariance of $\boldsymbol{\theta}$. **b** The matrix N controls the bias of $\boldsymbol{\theta}$

proposed in [28], but his derivation was heuristic without considering the statistical properties of image noise in terms of the covariance matrix. However, the procedure can be interpreted as described in this chapter. The HyperLS method was proposed by the author's group in [11, 12, 24], which was later extended to hyper-renormalization in [9, 10].

The fact that all known algebraic methods can be generalized in the form of Eq. (2.31) was pointed out in [10]. Because noise is assumed to occur randomly, the noisy data are random variables having probability distributions. Therefore the value $\boldsymbol{\theta}$ computed from them is also a random variable and has its probability distribution, and the choice of the matrices M and N affects the probability distribution of the computed $\boldsymbol{\theta}$. According to the detailed error analysis in [10], the choice of M controls the covariance of $\boldsymbol{\theta}$ (Fig. 2.6a), whereas the choice of N controls its bias (Fig. 2.6b). If we choose M to be the second row of Eq. (2.32), we can show that the covariance matrix of $\boldsymbol{\theta}$ attains the theoretical accuracy bound called the *KCR* (*Kanatani-Cramer-Rao*) *lower bound* (this is discussed in Chap. 16) except for $O(\sigma^4)$ terms. If we choose N to be the fifth row of Eq. (2.33), we can show that the bias of $\boldsymbol{\theta}$ is 0 except for $O(\sigma^4)$ terms (this is discussed in Chap. 14). Hence, the accuracy of hyper-renormalization cannot be improved any further except for higher-order terms of σ. In this sense, hyper-renormalization is theoretically optimal. It has also been shown in [16] that the Sampson error minimization with hyper-accurate correction has theoretically the same order of accuracy (this is discussed in Chap. 15). For detailed performance comparison of various ellipse fitting methods, see [17].

The FNS scheme for minimizing the Sampson error was introduced by Chojnacki et al. [1]. They called Eq. (2.36) the *ALM* (*approximated maximum likelihood*), but later the term "Sampson error" came to be used widely. This name originates from Sampson [26] who studied ellipse fitting in early days. The minimization scheme that Sampson used is essentially the iterative reweight (Procedure 2.2), but it does not strictly minimize Eq. (2.36), because the numerator part and the denominator part are updated separately. In Step 4 of Procedure 2.5, we can compute the unit eigenvector for the eigenvalue of the smallest absolute value, but it was experimentally found by Chojnacki et al. [1] and the authors' group [13] that the iterations converge faster using the smallest eigenvalue. There exists an alternative iterative scheme closely resembling FNS called *HEIV* (*heteroscedastic errors-in-variables*) [18, 21]; the computed solution is the same as FNS. The general strategy of minimizing the geometric distance by repeating Sampson error minimization was presented by the authors in [15], and its use for ellipse fitting was presented in [14].

When the noise distribution is Gaussian, the geometric distance minimization is equivalent to what is called in statistics *maximum likelihood* (ML; we discuss this in Chap. 15). The fact that the solution has statistical bias has been pointed out by many people. The bias originates from the fact that an ellipse is a convex curve: if a point on the curve is randomly displaced, the probability of falling outside is larger than the probability of falling inside. Given a datapoint, its closest point on the ellipse is the foot of the perpendicular line from it to the ellipse, and the probability of on which side its true position is more likely depends on the shape and the curvature of the ellipse in the neighborhood. Okatani and Deguchi [22] analyzed this for removing the bias. They also used a technique called the *projected score* to remove bias [23]. The hyper-accurate correction described in this chapter was introduced by the authors in [7,8,16].

The ellipse-specific method of Fitzgibbon et al. was published in [3]. The method of Szpak et al. was presented in [27], and the method of random sampling was given in [19]. For their detailed performance comparison, see [17].

The RANSAC technique for outlier removal was proposed by Fischler and Bolles [2]. Its principle is to select randomly a minimum set of datapoints to fit a curve (or any shape) many times, each time counting the number of datapoints nearby, and to output the curve to which the largest number of datapoints is close. Hence, a small number of datapoints are allowed to be far apart, and they are regarded as outliers. Alternatively, instead of counting the number of points nearby, we may sort the square distances of all the datapoints from the curve, evaluate the median, and output the curve for which the median is the smallest. This is called *LMedS* (*least median of squares*) [25]. This also ignores those datapoints far apart from the curve. Instead of ignoring far apart points, we may use a distance function that does not penalize far apart points very much. Such a method is generally called *M-estimation*, for which various types of distance functions have been studied [4]. An alternative approach for the same purpose is the use of the L_p-norm for $p < 2$, typically $p = 1$. These methods make no distinction regarding whether the datapoints are continuously connected, but techniques for classifying edge segments into elliptic and nonelliptic arcs have also been studied [20].

Problems

2.1 Fitting a line in the form of $n_1 x + n_2 y + n_3 f_0 = 0$ to points (x_α, y_α), $\alpha = 1, \ldots, N$, can be viewed as the problem of computing $\mathbf{n}$, which can be normalized to a unit vector, such that $(\mathbf{n}, \boldsymbol{\xi}_\alpha) \approx 0$, $\alpha = 1, \ldots, N$, where

$$\boldsymbol{\xi}_\alpha = \begin{pmatrix} x_\alpha \\ y_\alpha \\ f_0 \end{pmatrix}, \qquad \mathbf{n} = \begin{pmatrix} n_1 \\ n_2 \\ n_3 \end{pmatrix}. \tag{2.61}$$

(1) Write down the least-squares procedure for computing this.
(2) If the noise terms Δx_α and Δy_α are subject to a mutually independent Gaussian distribution of mean 0 and variance σ^2, how is their covariance matrix $V[\boldsymbol{\xi}_\alpha]$ defined?

2.2 Specifically write down the procedure for the Taubin method as the computation of the initial solution of Procedure 2.3.

2.3 Specifically write down the HyperLS procedure as the computation of the initial solution of Procedure 2.4.

2.4 (1) Define the vector $\mathbf{x}$ and the symmetric matrix Q by

$$\mathbf{x} = \begin{pmatrix} x/f_0 \\ y/f_0 \\ 1 \end{pmatrix}, \qquad \mathsf{Q} = \begin{pmatrix} A & B & D \\ B & C & E \\ D & E & F \end{pmatrix}. \tag{2.62}$$

Show that the ellipse equation of Eq. (2.1) is written in the form

$$(\mathbf{x}, \mathsf{Q}\mathbf{x}) = 0. \tag{2.63}$$

(2) Show that the square distance d_α^2 of point (x_α, y_α) from the ellipse of Eq. (2.63) is written, up to high-order small noise terms, in the form

$$d_\alpha^2 \approx \frac{f_0^2}{4} \frac{(\mathbf{x}_\alpha, \mathsf{Q}\mathbf{x}_\alpha)^2}{(\mathsf{Q}\mathbf{x}_\alpha, \mathsf{P}_\mathbf{k}\mathsf{Q}\mathbf{x}_\alpha)}, \tag{2.64}$$

where $\mathsf{P}_\mathbf{k}$ is the matrix defined by

$$\mathsf{P}_\mathbf{k} = \begin{pmatrix} 1 & 0 & 0 \\ 0 & 1 & 0 \\ 0 & 0 & 0 \end{pmatrix}. \tag{2.65}$$

(3) Show that in terms of the vectors $\boldsymbol{\xi}$ and $\boldsymbol{\theta}$ Eq. (2.64) is written as

$$d_\alpha^2 \approx \frac{(\boldsymbol{\xi}_\alpha, \boldsymbol{\theta})^2}{(\boldsymbol{\theta}, V_0[\boldsymbol{\xi}_\alpha]\boldsymbol{\theta})}. \tag{2.66}$$

2.5 (1) Show that the derivative of Eq. (2.36) is given by Eq. (2.41).
(2) Show that the eigenvalue λ in Eq. (2.39) is 0 after the FNS iterations have converged.

2.6 (1) Show that for the ellipse given by Eq. (2.63), the point $(\bar{x}_\alpha, \bar{y}_\alpha)$ that minimizes Eq. (2.47) is written up to high-order terms in $\hat{x}_\alpha - \bar{x}_\alpha$ and $\hat{y}_\alpha - \bar{y}_\alpha$ as

$$\begin{pmatrix} \hat{\hat{x}}_\alpha \\ \hat{\hat{y}}_\alpha \end{pmatrix} = \begin{pmatrix} x_\alpha \\ y_\alpha \end{pmatrix} - \frac{(\hat{\mathbf{x}}_\alpha, \mathsf{Q}\hat{\mathbf{x}}'_\alpha) + 2(\mathsf{Q}\hat{\mathbf{x}}'_\alpha, \tilde{\mathbf{x}}_\alpha)}{2(\mathsf{Q}\hat{\mathbf{x}}_\alpha, \mathsf{P}_\mathbf{k}\mathsf{Q}\hat{\mathbf{x}}_\alpha)} \begin{pmatrix} A & B & D \\ B & C & E \end{pmatrix} \begin{pmatrix} \hat{x}_\alpha \\ \hat{y}_\alpha \\ f_0 \end{pmatrix}, \tag{2.67}$$

where A, B, C, D, and E are the elements of the matrix Q in Eq. (2.62), and $\hat{\mathbf{x}}_\alpha$ and $\tilde{\mathbf{x}}_\alpha$ are defined as follows (the point $(\tilde{x}_\alpha, \tilde{y}_\alpha)$ is defined by Eq. (2.48)).

$$\hat{\mathbf{x}}_\alpha = \begin{pmatrix} \hat{x}_\alpha/f_0 \\ \hat{y}_\alpha/f_0 \\ 1 \end{pmatrix}, \qquad \tilde{\mathbf{x}}_\alpha = \begin{pmatrix} \tilde{x}_\alpha/f_0 \\ \tilde{y}_\alpha/f_0 \\ 0 \end{pmatrix}. \tag{2.68}$$

(2) Show that if we define the 9D vector $\boldsymbol{\xi}^*_\alpha$ by Eq. (2.42), Eq. (2.67) can be written in the form

$$\begin{pmatrix} \hat{\hat{x}}_\alpha \\ \hat{\hat{y}}_\alpha \end{pmatrix} = \begin{pmatrix} x_\alpha \\ y_\alpha \end{pmatrix} - \frac{2(\boldsymbol{\xi}^*_\alpha, \boldsymbol{\theta})}{(\boldsymbol{\theta}, V_0[\hat{\boldsymbol{\xi}}_\alpha]\boldsymbol{\theta})} \begin{pmatrix} \theta_1 & \theta_2 & \theta_4 \\ \theta_2 & \theta_3 & \theta_5 \end{pmatrix} \begin{pmatrix} \hat{x}_\alpha \\ \hat{y}_\alpha \\ f_0 \end{pmatrix}, \tag{2.69}$$

where $V_0[\hat{\boldsymbol{\xi}}_\alpha]$ is the normalized covariance matrix obtained by replacing $(\bar{x}, \bar{y})$ in Eq. (2.15) by $(\hat{x}_\alpha, \hat{y}_\alpha)$.

(3) Show that by replacing $(\bar{x}_\alpha, \bar{y}_\alpha)$ in Eq. (2.34) by $(\hat{\hat{x}}_\alpha, \hat{\hat{y}}_\alpha)$, we can rewrite the geometric distance S in the form of Eq. (2.43).

References

1. W. Chojnacki, M.J. Brooks, A. van den Hengel, D. Gawley, On the fitting of surfaces to data with covariances. IEEE Trans. Pattern Anal. Mach. Intell. **22**(11), 1294–1303 (2000)
2. M.A. Fischler, R.C. Bolles, Random sample consensus: a paradigm for model fitting with applications to image analysis and automated cartography. Commun. ACM **24**(6), 381–395 (1981)
3. A. Fitzgibbon, M. Pilu, R.B. Fisher, Direct least squares fitting of ellipses. IEEE Trans. Pattern Anal. Mach. Intell. **21**(5), 476–480 (1999)
4. P.J. Huber, *Robust Statistics*, 2nd edn. (Wiley, Hoboken, 2009)
5. K. Kanatani, Renormalization for unbiased estimation, in *Proceedings of 4th International Conference on Computer Vision*, Berlin, Germany (1993), pp. 599–606
6. K. Kanatani, *Statistical Optimization for Geometric Computation: Theory and Practice* (Elsevier, Amsterdam, The Netherlands, 1996) (Reprinted by Dover, New York, U.S., 2005)
7. K. Kanatani, Ellipse fitting with hyperaccuracy. IEICE Trans. Inf. Syst. **E89-D**(10), 2653–2660 (2006)
8. K. Kanatani, Statistical optimization for geometric fitting: theoretical accuracy bound and high order error analysis. Int. J. Comput. Vis. **80**(2), 167–188 (2008)
9. K. Kanatani, A. Al-Sharadqah, N. Chernov, Y. Sugaya, Renormalization returns: hyper-renormalization and its applications, in *Proceedings of 12th European Conference on Computer Vision*, Firenze, Italy, October 2012
10. K. Kanatani, A. Al-Sharadqah, N. Chernov, Y. Sugaya, Hyper-renormalization: non-minimization approach for geometric estimation. IPSJ Trans. Comput. Vis. Appl. **6**, 143–159 (2014)
11. K. Kanatani, P. Rangarajan, Hyper least squares fitting of circles and ellipses. Comput. Stat. Data Anal. **55**(6), 2197–2208 (2011)
12. K. Kanatani, P. Rangarajan, Y. Sugaya, H. Niitsuma, HyperLS and its applications. IPSJ Trans. Comput. Vis. Appl. **3**, 80–94 (2011)
13. K. Kanatani, Y. Sugaya, Performance evaluation of iterative geometric fitting algorithms. Comput. Stat. Data Anal. **52**(2), 1208–1222 (2007)
14. K. Kanatani, Y. Sugaya, Compact algorithm for strictly ML ellipse fitting, in *Proceedings of 19th International Conference on Pattern Recognition*, Tampa, FL, U.S. (2008)
15. K. Kanatani, Y. Sugaya, Unified computation of strict maximum likelihood for geometric fitting. J. Math. Imaging Vis. **38**(1), 1–13 (2010)

16. K. Kanatani, Y. Sugaya, Hyperaccurate correction of maximum likelihood for geometric estimation. IPSJ Trans. Comput. Vis. Appl. **5**, 19–29 (2013)
17. K. Kanatani, Y. Sugaya, K. Kanazawa, *Ellipse Fitting for Computer Vision: Implementation and Applications* (Morgan and Claypool, San Rafael, 2016)
18. Y. Leedan, P. Meer, Heteroscedastic regression in computer vision: problems with bilinear constraint. Int. J. Comput. Vis. **37**(2), 127–150 (2000)
19. T. Masuzaki, Y. Sugaya, K. Kanatani, High accuracy ellipse-specific fitting, in *Proceedings of 6th Pacific-Rim Symposium on Image and Video Technology*, Guanajuato, Mexico (2013), pp. 314–324
20. T. Masuzaki, Y. Sugaya, K. Kanatani, Floor-wall boundary estimation by ellipse fitting, in *Proceedings IEEE 7th International Conference on Robotics, Automation and Mechatronics*, Angkor Wat, Cambodia (2015), pp. 30–35
21. J. Matei, P. Meer, Estimation of nonlinear errors-in-variables models for computer vision applications. IEEE Trans. Pattern Anal. Mach. Intell. **28**(10), 1537–1552 (2006)
22. T. Okatani, K. Deguchi, On bias correction for geometric parameter estimation in computer vision, in *Proceedings of IEEE Conference on Computer Vision and Pattern Recognition*, Miami Beach, FL, U.S. (2009), pp. 959–966
23. T. Okatani, K. Deguchi, Improving accuracy of geometric parameter estimation using projected score method, in *Proceedings of 12th International Conference on Computer Vision*, Kyoto, Japan (2009), pp. 1733–1740
24. P. Rangarajan, K. Kanatani, Improved algebraic methods for circle fitting. Electron. J. Stat. **3**(2009), 1075–1082 (2009)
25. P.J. Rousseeuw, A.M. Leroy, *Robust Regression and Outlier Detection* (Wiley, New York, 1987)
26. P.D. Sampson, Fitting conic sections to "very scattered" data: an iterative refinement of the Bookstein Algorithm. Comput. Vis. Image Process. **18**(1), 97–108 (1982)
27. Z.L. Szpak, W. Chojnacki, A. van den Hengel, Guaranteed ellipse fitting with a confidence region and an uncertainty measure for centre, axes, and orientation. J. Math. Imaging Vis. **52**(2), 173–199 (2015)
28. G. Taubin, Estimation of planar curves, surfaces, and non-planar space curves defined by implicit equations with applications to edge and range image segmentation. IEEE Trans. Pattern Anal. Mach. Intell. **13**(11), 1115–1138 (1991)

Fundamental Matrix Computation

3

Abstract

Two images of the same scene are related by what is called the epipolar equation. It is specified by a matrix called the fundamental matrix. By computing the fundamental matrix between two images, one can analyze the 3D structure of the scene, which we discuss in Chaps. 4 and 5. This chapter describes the principle and typical computational procedures for accurately computing the fundamental matrix by considering the statistical properties of the noise involved in correspondence detection. As in ellipse fitting, the methods are classified into algebraic and geometric approaches. However, the fundamental matrix has an additional property called the rank constraint: it is required to have determinant 0. Three approaches for enforcing it are introduced here: a posteriori rank correction, hidden variables, and extended FNS. We then describe the procedure of repeatedly using them to compute the geometric distance minimization solution. The RANSAC procedure for removing wrong correspondences is also described.

3.1 Fundamental Matrices

Suppose we take images of the same scene using two cameras, and suppose a particular point in the scene is imaged at point (x, y) by one camera and at (x', y') by the other. It can be shown (the details are given in Chaps. 4 and 5) that the two points satisfy

$$\left(\begin{pmatrix} x/f_0 \\ y/f_0 \\ 1 \end{pmatrix}, \mathsf{F}\begin{pmatrix} x'/f_0 \\ y'/f_0 \\ 1 \end{pmatrix}\right) = 0, \tag{3.1}$$

where f_0 is a constant for adjusting the scale as in the ellipse fitting case. It can be theoretically set to 1, but it is better to choose it such that x/f_0 and y/f_0 have order

K. Kanatani et al., *Guide to 3D Vision Computation*, Advances in Computer Vision and Pattern Recognition, DOI 10.1007/978-3-319-48493-8_3

1 considering finite-length numerical computation. For example, we let $f_0 = 600$ if the scene of interest is in an image region of 600×600 pixels. We also take the origin of the image xy coordinate system at the center of the image, as in the ellipse fitting case.

The 3×3 matrix in Eq. (3.1), called the *fundamental matrix*, is determined only by the relative position of the two cameras and their internal parameters, such as the focal length, independent of the scene content, and Eq. (3.1) is called the *epipolar constraint* or the *epipolar equation*. Because any nonzero scalar multiple of the fundamental matrix F defines the same epipolar equation, we normalize it to the unit matrix norm:

$$\|\mathsf{F}\| \left(\equiv \sqrt{\sum_{i,j=1}^{3} F_{ij}^2} \right) = 1. \tag{3.2}$$

If we define 9D vectors

$$\boldsymbol{\xi} = \begin{pmatrix} xx' \\ xy' \\ f_0 x \\ yx' \\ yy' \\ f_0 y \\ f_0 x' \\ f_0 y' \\ f_0^2 \end{pmatrix}, \qquad \boldsymbol{\theta} = \begin{pmatrix} F_{11} \\ F_{12} \\ F_{13} \\ F_{21} \\ F_{22} \\ F_{23} \\ F_{31} \\ F_{32} \\ F_{33} \end{pmatrix}, \tag{3.3}$$

we can see that the left-hand side of Eq. (3.1) is written as $(\boldsymbol{\xi}, \boldsymbol{\theta})/f_0^2$. Hence, the epipolar equation of Eq. (3.1) is written in the form

$$(\boldsymbol{\xi}, \boldsymbol{\theta}) = 0. \tag{3.4}$$

The vector $\boldsymbol{\theta}$ in this equation has scale indeterminacy, and the normalization in the form of Eq. (3.2) is equivalent to normalization to a unit vector: $\|\boldsymbol{\theta}\| = 1$. Thus, the epipolar equation has the same form as the ellipse equation of Eq. (2.4).

3.2 Covariance Matrices and Algebraic Methods

Computing the fundamental matrix from point correspondences over two images is the first step of many computer vision applications (Fig. 3.1). Finding a matrix F that satisfies Eq. (3.1) for noisy pairs of corresponding points (x_α, y_α) and (x'_α, y'_α), $\alpha = 1, ..., N$, can be mathematically formulated as finding a unit vector $\boldsymbol{\theta}$ such that

$$(\boldsymbol{\xi}_\alpha, \boldsymbol{\theta}) \approx 0, \qquad \alpha = 1, ..., N, \tag{3.5}$$

where $\boldsymbol{\xi}_\alpha$ is the value of $\boldsymbol{\xi}$ for $x = x_\alpha$, $y = y_\alpha$, $x' = x'_\alpha$, and $y' = y'_\alpha$. We regard the data x_α, y_α, x'_α, and y'_α as disturbed from their true values $\bar{x}_\alpha$, $\bar{y}_\alpha$, $\bar{x}'_\alpha$, and $\bar{y}'_\alpha$ by

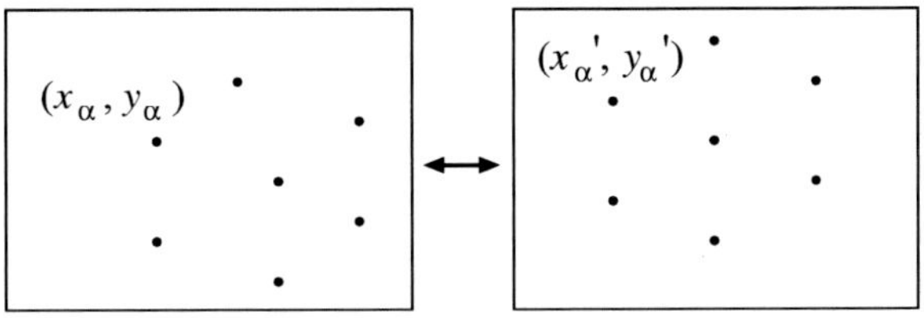

Fig. 3.1 Computing the fundamental matrix from noisy point correspondences

noise terms Δx_α, Δy_α, $\Delta x'_\alpha$, and $\Delta y'_\alpha$, respectively:

$$x_\alpha = \bar{x}_\alpha + \Delta x_\alpha, \qquad y_\alpha = \bar{y}_\alpha + \Delta y_\alpha, \qquad x'_\alpha = \bar{x}'_\alpha + \Delta x'_\alpha, \qquad y'_\alpha = \bar{y}'_\alpha + \Delta y'_\alpha. \tag{3.6}$$

Substituting these into $\boldsymbol{\xi}_\alpha$, we obtain

$$\boldsymbol{\xi}_\alpha = \bar{\boldsymbol{\xi}}_\alpha + \Delta_1 \boldsymbol{\xi}_\alpha + \Delta_2 \boldsymbol{\xi}_\alpha, \tag{3.7}$$

where $\bar{\boldsymbol{\xi}}_\alpha$ is the value of $\boldsymbol{\xi}_\alpha$ for $x_\alpha = \bar{x}_\alpha$, $y_\alpha = \bar{y}_\alpha$, $x'_\alpha = \bar{x}'_\alpha$, and $y'_\alpha = \bar{y}'_\alpha$, and $\Delta_1 \boldsymbol{\xi}_\alpha$ and $\Delta_2 \boldsymbol{\xi}_\alpha$ are, respectively, the first- and the second-order noise terms. After expansion, we obtain

$$\Delta_1 \boldsymbol{\xi}_\alpha = \begin{pmatrix} \bar{x}'_\alpha \Delta x_\alpha + \bar{x}_\alpha \Delta x'_\alpha \\ \bar{y}'_\alpha \Delta x_\alpha + \bar{x}_\alpha \Delta y'_\alpha \\ f_0 \Delta x_\alpha \\ \bar{x}'_\alpha \Delta y_\alpha + \bar{y}_\alpha \Delta x'_\alpha \\ f_0 \Delta y_\alpha \\ f_0 \Delta x'_\alpha \\ f_0 \Delta y'_\alpha \\ 0 \end{pmatrix}, \qquad \Delta_2 \boldsymbol{\xi}_\alpha = \begin{pmatrix} \Delta x_\alpha \Delta x'_\alpha \\ \Delta x_\alpha \Delta y'_\alpha \\ 0 \\ \Delta y_\alpha \Delta x'_\alpha \\ \Delta y_\alpha \Delta y'_\alpha \\ 0 \\ 0 \\ 0 \\ 0 \end{pmatrix}. \tag{3.8}$$

Regarding Δx_α and Δy_α as random variables, we define the *covariance matrix* of $\boldsymbol{\xi}_\alpha$ by

$$V[\boldsymbol{\xi}_\alpha] = E[\Delta_1 \boldsymbol{\xi}_\alpha \Delta_1 \boldsymbol{\xi}_\alpha^\top], \tag{3.9}$$

where $E[\,\cdot\,]$ denotes the expectation over their distribution. If Δx_α, Δy_α, $\Delta x'_\alpha$, and $\Delta y'_\alpha$ are independent Gaussian variables of mean 0 and variance σ^2, we have

$$E[\Delta x_\alpha] = E[\Delta y_\alpha] = E[\Delta x'_\alpha] = E[\Delta y'_\alpha] = 0,$$

$$E[\Delta x_\alpha^2] = E[\Delta y_\alpha^2] = E[\Delta x'^2_\alpha] = E[\Delta y'^2_\alpha] = \sigma^2,$$

$$E[\Delta x_\alpha \Delta y_\alpha] = E[\Delta x'_\alpha \Delta y'_\alpha] = E[\Delta x_\alpha \Delta y'_\alpha] = E[\Delta x'_\alpha \Delta y_\alpha] = 0. \tag{3.10}$$

Thus we can write from Eq. (3.8) the covariance matrix of Eq. (3.9) in the form

$$V[\boldsymbol{\xi}_\alpha] = \sigma^2 V_0[\boldsymbol{\xi}_\alpha], \tag{3.11}$$

where we have removed the common multiplier σ^2 and written the remaining part as $V_0[\boldsymbol{\xi}_\alpha]$. We call this the *normalized covariance matrix*; it has the form

$$V_0[\boldsymbol{\xi}_\alpha] = \begin{pmatrix} \bar{x}_\alpha^2+\bar{x}_\alpha'^2 & \bar{x}_\alpha'\bar{y}_\alpha' & f_0\bar{x}_\alpha' & \bar{x}_\alpha\bar{y}_\alpha & 0 & 0 & f_0\bar{x}_\alpha & 0 & 0 \\ \bar{x}_\alpha'\bar{y}_\alpha' & \bar{x}_\alpha^2+\bar{y}_\alpha'^2 & f_0\bar{y}_\alpha' & 0 & \bar{x}_\alpha\bar{y}_\alpha & 0 & 0 & f_0\bar{x}_\alpha & 0 \\ f_0\bar{x}_\alpha' & f_0\bar{y}_\alpha' & f_0^2 & 0 & 0 & 0 & 0 & 0 & 0 \\ \bar{x}_\alpha\bar{y}_\alpha & 0 & 0 & \bar{y}_\alpha^2+\bar{x}_\alpha'^2 & \bar{x}_\alpha'\bar{y}_\alpha' & f_0\bar{x}_\alpha' & f_0\bar{y}_\alpha & 0 & 0 \\ 0 & \bar{x}_\alpha\bar{y}_\alpha & 0 & \bar{x}_\alpha'\bar{y}_\alpha' & \bar{y}_\alpha^2+\bar{y}_\alpha'^2 & f_0\bar{y}_\alpha' & 0 & f_0\bar{y}_\alpha & 0 \\ 0 & 0 & 0 & f_0\bar{x}_\alpha' & f_0\bar{y}_\alpha' & f_0^2 & 0 & 0 & 0 \\ f_0\bar{x}_\alpha & 0 & 0 & f_0\bar{y}_\alpha & 0 & 0 & f_0^2 & 0 & 0 \\ 0 & f_0\bar{x}_\alpha & 0 & 0 & f_0\bar{y}_\alpha & 0 & 0 & f_0^2 & 0 \\ 0 & 0 & 0 & 0 & 0 & 0 & 0 & 0 & 0 \end{pmatrix}. \tag{3.12}$$

As in the case of ellipse fitting, we need not consider the second-order term $\Delta_2\boldsymbol{\xi}_\alpha$ in the covariance matrix of Eq. (3.9). Also, the true values $\bar{x}_\alpha$, $\bar{y}_\alpha$, $\bar{x}_\alpha'$, and $\bar{y}_\alpha'$ in Eq. (3.12) are, in actual computation, replaced by their observed values x_α, y_α, x_α', and y_α', respectively.

Thus, computing the fundamental matrix F from noisy correspondences can be formalized as computing the unit vector $\boldsymbol{\theta}$ that satisfies Eq. (3.5) by considering the noise properties described by the covariance matrices $V[\boldsymbol{\xi}_\alpha]$. Mathematically, this is the same problem as ellipse fitting. Therefore the algebraic methods described there (least squares, the Taubin method, HyperLS, iterative reweight, renormalization, and hyper-renormalization) can be directly applied. All of them are modifications of least squares, and the simplest is the least squares of Procedure 2.1.

Procedure 3.1 (Least squares)

1. Compute the 9×9 matrix

$$\mathsf{M} = \frac{1}{N}\sum_{\alpha=1}^{N} \boldsymbol{\xi}_\alpha \boldsymbol{\xi}_\alpha^\top. \tag{3.13}$$

2. Solve the eigenvalue problem $\mathsf{M}\boldsymbol{\theta} = \lambda\boldsymbol{\theta}$, and return the unit eigenvector $\boldsymbol{\theta}$ for the smallest eigenvalue λ.

The next simplest is the Taubin method (↪ Problem 2.2).

Procedure 3.2 (Taubin)

1. Compute the 9×9 matrices

$$\mathsf{M} = \frac{1}{N}\sum_{\alpha=1}^{N} \boldsymbol{\xi}_\alpha \boldsymbol{\xi}_\alpha^\top, \qquad \mathsf{N} = \frac{1}{N}\sum_{\alpha=1}^{N} V_0[\boldsymbol{\xi}_\alpha]. \tag{3.14}$$

2. Solve the generalized eigenvalue problem $\mathsf{M}\boldsymbol{\theta} = \lambda\mathsf{N}\boldsymbol{\theta}$, and return the unit generalized eigenvector $\boldsymbol{\theta}$ for the smallest generalized eigenvalue λ.

The matrix N is not positive definite, therefore we rewrite the problem in the form of Eq. (2.24) and use a standard numerical tool. It is known that the Taubin method has higher accuracy than least squares. The procedures of iterative reweight and renormalization are identical to those of ellipse fitting. The procedures of HyperLS and hyper-renormalization are also identical except that M_5^- in Eq. (2.26) is replaced by M_8^-, and $\mathbf{e} = \mathbf{0}$ (see Chap. 14 for the general formulation of hyper-renormalization).

3.3 Geometric Distance and Sampson Error

The geometric approach for fundamental matrix computation is to optimally displace the observed corresponding points (x_α, y_α) to (x'_α, y'_α) to new positions $(\bar{x}_\alpha, \bar{y}_\alpha)$ and $(\bar{x}'_\alpha, \bar{y}'_\alpha)$ that satisfy the epipolar equation of Eq. (3.1) for some fundamental matrix F. This displacement is done in such a way that the average square distance

$$S = \frac{1}{N}\sum_{\alpha=1}^{N}\Big((x_\alpha - \bar{x}_\alpha)^2 + (y_\alpha - \bar{y}_\alpha)^2 + (x'_\alpha - \bar{x}'_\alpha)^2 + (y'_\alpha - \bar{y}'_\alpha)^2\Big) \tag{3.15}$$

is minimized subject to the epipolar equation

$$(\begin{pmatrix} \bar{x}_\alpha/f_0 \\ \bar{y}_\alpha/f_0 \\ 1 \end{pmatrix}, \mathsf{F}\begin{pmatrix} \bar{x}'_\alpha/f_0 \\ \bar{y}'_\alpha/f_0 \\ 1 \end{pmatrix}) = 0. \tag{3.16}$$

The desired fundamental matrix F is the one that minimizes Eq. (3.15). If we identify $(\bar{x}_\alpha, \bar{y}_\alpha)$ and $(\bar{x}'_\alpha, \bar{y}'_\alpha)$ with the true positions of the observation (x_α, y_α) and (x'_α, y'_α), we can write Eq. (3.15), using Eq. (3.6) in the form $S = (1/N)\sum_{\alpha=1}^{N}(\Delta x_\alpha^2 + \Delta y_\alpha^2 + \Delta x'^2_\alpha + \Delta y'^2_\alpha)$, where we regard Δx_α, Δy_α, $\Delta x'_\alpha$, and $\Delta y'_\alpha$ as unknown variables rather than stochastic quantities. Equation (3.15) is widely known as the *geometric distance* (although it has the dimension of square length) or the *reprojection error*.

The values of $(\bar{x}_\alpha, \bar{y}_\alpha)$ and $(\bar{x}'_\alpha, \bar{y}'_\alpha)$ that minimize

$$S_\alpha = \Delta x_\alpha^2 + \Delta y_\alpha^2 + \Delta x'^2_\alpha + \Delta y'^2_\alpha \tag{3.17}$$

subject to Eq. (3.16) are, omitting high-order terms in Δx_α, Δy_α, $\Delta x'_\alpha$, and $\Delta y'_\alpha$, approximated in the form ($\hookrightarrow$ Problem 3.1)

$$S_\alpha \approx \frac{f_0^2(\mathbf{x}_\alpha, \mathsf{F}\mathbf{x}'_\alpha)^2}{\|\mathsf{P}_\mathbf{k}\mathsf{F}\mathbf{x}'_\alpha\|^2 + \|\mathsf{P}_\mathbf{k}\mathsf{F}^\top\mathbf{x}_\alpha\|^2}, \tag{3.18}$$

where vectors $\mathbf{x}_\alpha$ and $\mathbf{x}'_\alpha$ are defined by

$$\mathbf{x}_\alpha = \begin{pmatrix} x_\alpha/f_0 \\ y_\alpha/f_0 \\ 1 \end{pmatrix}, \qquad \mathbf{x}'_\alpha = \begin{pmatrix} x'_\alpha/f_0 \\ y'_\alpha/f_0 \\ 1 \end{pmatrix}, \tag{3.19}$$

and the matrix $\mathsf{P}_\mathbf{k}$ is defined by

$$\mathsf{P}_\mathbf{k} \equiv \begin{pmatrix} 1 & 0 & 0 \\ 0 & 1 & 0 \\ 0 & 0 & 0 \end{pmatrix}. \tag{3.20}$$

Multiplying the matrix $\mathsf{P}_{\mathbf{k}}$ from the left means replacing the third component of the vector by 0. The following identity can be confirmed by substituting Eq. (3.3) to the left-hand side.

$$(\boldsymbol{\theta}, V_0[\boldsymbol{\xi}_\alpha]\boldsymbol{\theta}) = f_0^2\Big(\|\mathsf{P}_{\mathbf{k}}\mathsf{F}\mathbf{x}'_\alpha\|^2 + \|\mathsf{P}_{\mathbf{k}}\mathsf{F}^\top\mathbf{x}_\alpha\|^2\Big). \tag{3.21}$$

Here, $V_0[\boldsymbol{\xi}_\alpha]$ is the normalized covariance matrix in Eq. (3.12), where $\bar{x}_\alpha$, $\bar{y}_\alpha$, $\bar{x}'_\alpha$, and $\bar{y}'_\alpha$ are replaced by x_α, y_α, x'_α, and y'_α, respectively. Because $(\boldsymbol{\xi}_\alpha, \boldsymbol{\theta}) = f_0^2(\mathbf{x}_\alpha, \mathsf{F}\mathbf{x}'_\alpha)$ holds from the definition of $\boldsymbol{\xi}$ and $\boldsymbol{\theta}$ in Eq. (3.3), we can rewrite Eq. (3.18) in the form

$$S_\alpha \approx \frac{(\boldsymbol{\xi}_\alpha, \boldsymbol{\theta})^2}{(\boldsymbol{\theta}, V_0[\boldsymbol{\xi}_\alpha]\boldsymbol{\theta})}. \tag{3.22}$$

Using this, we can approximate the geometric distance of Eq. (3.15) in the form:

$$J = \frac{1}{N}\sum_{\alpha=1}^{N} \frac{(\boldsymbol{\xi}_\alpha, \boldsymbol{\theta})^2}{(\boldsymbol{\theta}, V_0[\boldsymbol{\xi}_\alpha]\boldsymbol{\theta})}. \tag{3.23}$$

This is nothing but the Sampson error of Eq. (2.36). It is known that this is a very accurate approximation of the geometric distance of Eq. (3.15).

Thus, the formulation is completely parallel to the ellipse fitting problem. Hence, we can minimize the Sampson error by FNS described in Sect. 2. As described in Sect. 2.5.3, we can also minimize the geometric distance by repeating Sampson error minimization, and we can apply hyper-accurate correction as described in Sect. 2.5.4. The procedure is the same except that for fundamental matrix computation the numerator on the right-hand side of Eq. (2.51) is $1 - 8/N$, the matrix M_5^- in Eq. (2.52) is replaced by M_8^-, and $\mathbf{e} = \mathbf{0}$ (see Chap. 15 for the general formulation of hyper-accurate correction).

3.4 Rank Constraint

As pointed out above, fundamental matrix computation from noisy point correspondences is mathematically the same as ellipse fitting, and hence the algebraic methods and the geometric methods described in Chap. 2 can be applied. However, there is one additional aspect that does not exist for ellipse fitting. It is the requirement that the fundamental matrix F that satisfies Eq. (3.1) must have rank 2 and hence

$$\det \mathsf{F} = 0. \tag{3.24}$$

This is easily seen from a geometric consideration, as shown later in Sect. 3.11. We call Eq. (3.24) the *rank constraint*. Generally, there exist three approaches for enforcing this rank constraint:

- We compute $\boldsymbol{\theta}$ without considering the rank constraint, using a standard (algebraic or geometric) method. If the data are exact, Eq. (3.24) is automatically satisfied whatever method is used, and if the point correspondences are sufficiently accurate,

we obtain a solution such that $\det \mathsf{F} \approx 0$. Therefore we do *a posteriori correction*, slightly modifying the solution F so that Eq. (3.24) is satisfied.
- Introducing a *hidden variable* $\mathbf{u}$, we express each component of $\boldsymbol{\theta}$ in terms of $\boldsymbol{\theta}$ in such a way that $\boldsymbol{\theta}(\mathbf{u})$ satisfies the rank condition identically in $\mathbf{u}$. Then, we search for the value of $\mathbf{u}$ such that it will satisfy Eq. (3.5), for example, minimizing the Sampson error of Eq. (3.23).
- We do iterations without using hidden variables in such a way that $\boldsymbol{\theta}$ approaches the true value at each iteration and at the same time the rank constraint is automatically satisfied when the iterations have converged.

We now describe the details of these approaches.

3.5 A Posteriori Correction

We first compute the vector $\boldsymbol{\theta}$, that is, the fundamental matrix F, using some method without considering the rank constraint. Then, we correct it such that its determinant is 0. The best-known procedure for this is the use of the *singular value decomposition (SVD)*, which is as follows.

Procedure 3.3 (Rank correction by SVD)

1. Compute the fundamental matrix F using a standard method.
2. Compute the SVD of the computed F in the form

$$\mathsf{F} = \mathsf{U} \begin{pmatrix} \sigma_1 & 0 & 0 \\ 0 & \sigma_2 & 0 \\ 0 & 0 & \sigma_3 \end{pmatrix} \mathsf{V}^\top, \tag{3.25}$$

   where $\sigma_1 \geq \sigma_2 \geq \sigma_3$ (> 0) are the singular values and U and V are orthogonal matrices.
3. Correct the F as follows.

$$\mathsf{F} \leftarrow \mathsf{U} \begin{pmatrix} \sigma_1/\sqrt{\sigma_1^2+\sigma_2^2} & 0 & 0 \\ 0 & \sigma_2/\sqrt{\sigma_1^2+\sigma_2^2} & 0 \\ 0 & 0 & 0 \end{pmatrix} \mathsf{V}^\top. \tag{3.26}$$

Comments The division of the two diagonal elements by $\sqrt{\sigma_1^2+\sigma_2^2}$ is to enforce the normalization $\|\mathsf{F}\| = 1$ ($\hookrightarrow$ Problem 3.2).

The above procedure does not take the noise property described by Eq. (3.12) into consideration. A more reasonable method is first to evaluate the reliability of F, or $\boldsymbol{\theta}$, using Eq. (3.12) and correct it *optimally* so that its reliability is least affected. The procedure is as follows.

Procedure 3.4 (Optimal rank correction)

1. Compute the vector $\boldsymbol{\theta}$ that represents the fundamental matrix using a standard method.
2. Compute the 9×9 matrix

$$\hat{\mathsf{M}} = \frac{1}{N}\sum_{\alpha=1}^{N} \frac{(\mathsf{P}_{\boldsymbol{\theta}}\boldsymbol{\xi}_\alpha)(\mathsf{P}_{\boldsymbol{\theta}}\boldsymbol{\xi}_\alpha)^\top}{(\boldsymbol{\theta}, V_0[\boldsymbol{\xi}_\alpha]\boldsymbol{\theta})}, \tag{3.27}$$

where we define

$$\mathsf{P}_{\boldsymbol{\theta}} \equiv \mathsf{I} - \boldsymbol{\theta}\boldsymbol{\theta}^\top. \tag{3.28}$$

3. Compute the eigenvalues $\lambda_1 \geq \cdots \geq \lambda_9 \ (= 0)$ of $\hat{\mathsf{M}}$ and the corresponding unit eigenvectors $\mathbf{u}_1$, ..., $\mathbf{u}_9 (= \boldsymbol{\theta})$. Define the following matrix $V_0[\boldsymbol{\theta}]$.

$$V_0[\boldsymbol{\theta}] = \frac{1}{N}\Big(\frac{\mathbf{u}_1\mathbf{u}_1^\top}{\lambda_1} + \cdots + \frac{\mathbf{u}_8\mathbf{u}_8^\top}{\lambda_8}\Big). \tag{3.29}$$

4. Compute the 9D vector

$$\boldsymbol{\theta}^\dagger = \begin{pmatrix} \theta_5\theta_9 - \theta_8\theta_6 \\ \theta_6\theta_7 - \theta_9\theta_4 \\ \theta_4\theta_8 - \theta_7\theta_5 \\ \theta_8\theta_3 - \theta_2\theta_9 \\ \theta_9\theta_1 - \theta_3\theta_7 \\ \theta_7\theta_2 - \theta_1\theta_8 \\ \theta_2\theta_6 - \theta_5\theta_3 \\ \theta_3\theta_4 - \theta_6\theta_1 \\ \theta_1\theta_5 - \theta_4\theta_2 \end{pmatrix}. \tag{3.30}$$

5. Update $\boldsymbol{\theta}$ and $V_0[\boldsymbol{\theta}]$ in the form

$$\boldsymbol{\theta} \leftarrow \mathscr{N}[\boldsymbol{\theta} - \frac{(\boldsymbol{\theta}^\dagger, \boldsymbol{\theta})V_0[\boldsymbol{\theta}]\boldsymbol{\theta}^\dagger}{3(\boldsymbol{\theta}^\dagger, V_0[\boldsymbol{\theta}]\boldsymbol{\theta}^\dagger)}], \qquad V_0[\boldsymbol{\theta}] \leftarrow \mathsf{P}_{\boldsymbol{\theta}} V_0[\boldsymbol{\theta}]\mathsf{P}_{\boldsymbol{\theta}}, \tag{3.31}$$

where $\mathscr{N}[\,\cdot\,]$ denotes normalization to the unit norm ($\mathscr{N}[\mathbf{a}] = \mathbf{a}/\|\mathbf{a}\|$), and $\mathsf{P}_{\boldsymbol{\theta}}$ in the second equation is the matrix defined by Eq. (3.28) using the value $\boldsymbol{\theta}$ updated in the first equation.
6. If $(\boldsymbol{\theta}^\dagger, \boldsymbol{\theta}) \approx 0$, return $\boldsymbol{\theta}$ and stop. Else, go back to Step 4.

Comments The matrix $\mathsf{P}_{\boldsymbol{\theta}}$ in Eq. (3.28) projects an arbitrary vector $\mathbf{v}$ onto a plane perpendicular to the unit vector $\boldsymbol{\theta}$ (↪ Problem 3.3). Thus we have $\hat{\mathsf{M}}\boldsymbol{\theta} = \mathbf{0}$ for the matrix $\hat{\mathsf{M}}$ in Eq. (3.27), and $\boldsymbol{\theta}$ is the unit eigenvector of $\hat{\mathsf{M}}$ for eigenvalue 0. Evidently, $\hat{\mathsf{M}}$ is a symmetric positive semi-definite matrix of rank 8, and $V_0[\boldsymbol{\theta}]$ in Eq. (3.29) equals its pseudoinverse $\hat{\mathsf{M}}^-$ (↪ Eq. (2.30)) divided by N. We can see that $V_0[\boldsymbol{\theta}]\boldsymbol{\theta} = \mathbf{0}$ identically holds. It is known that $\sigma^2 V_0[\boldsymbol{\theta}]$ coincides with the covariance matrix $V[\boldsymbol{\theta}]$ of an optimally computed $\boldsymbol{\theta}$ except for $O(\sigma^4)$ terms (↪ Problem 3.4).

The vector $\boldsymbol{\theta}^{\dagger}$ in Eq. (3.30) is obtained by vertically arranging the elements of the transpose $\mathsf{F}^{\dagger\top}$ of the *cofactor matrix* $\mathsf{F}^{\dagger}$ in the same order as obtaining $\boldsymbol{\theta}$ in Eq. (3.3) from F. From Eq. (3.30), we can see that the following identity holds.

$$(\boldsymbol{\theta}^{\dagger}, \boldsymbol{\theta}) = 3 \det \mathsf{F}. \tag{3.32}$$

The details are omitted here, but the increment $\Delta\boldsymbol{\theta}$ of $\boldsymbol{\theta}$, that is, of F, to satisfy $\det \mathsf{F} = 0$, that is, $(\boldsymbol{\theta}^{\dagger}, \boldsymbol{\theta}) = 0$, in an optimal manner is given by the value that minimizes $(\Delta\boldsymbol{\theta}, V_0[\boldsymbol{\theta}]^{-}\Delta\boldsymbol{\theta})$ when $\Delta\boldsymbol{\theta}$ is small, where $V_0[\boldsymbol{\theta}]^{-}$ is the pseudoinverse of $V_0[\boldsymbol{\theta}]$ (of rank 8) in Eq. (3.29). If we ignore high-order terms in $\Delta\boldsymbol{\theta}$ and normalize the corrected $\boldsymbol{\theta}$ to unit norm, we obtain the first of Eq. (3.31) ($\hookrightarrow$ Problem 3.5). Because Eq. (3.31) is a first approximation, we repeat the same process by regarding the corrected $\boldsymbol{\theta}$ as the original $\boldsymbol{\theta}$. The update of $V_0[\boldsymbol{\theta}]$ in Eq. (3.31) is to ensure that $V_0[\boldsymbol{\theta}]\boldsymbol{\theta} = \mathbf{0}$ always holds. Repeating this once or twice is usually sufficient; no significant change occurs if this is repeated more often.

3.6 Hidden Variables Approach

There are many different ways of parameterizing F to make it rank 2. The most direct one is the use of the SVD

$$\mathsf{F} = \mathsf{U} \begin{pmatrix} \sigma_1 & 0 & 0 \\ 0 & \sigma_2 & 0 \\ 0 & 0 & 0 \end{pmatrix} \mathsf{V}^{\top}, \tag{3.33}$$

where we regard the singular values σ_1 and σ_2 and the orthogonal matrices U and V as independent variables. Inasmuch as the normalization $\|\mathsf{F}\| = 1$ is equivalent to $\sigma_1^2 + \sigma_2^2 = 1$($\hookrightarrow$ Problem 3.2), we parameterize σ_1 and σ_2 in the form

$$\sigma_1 = \cos\phi, \qquad \sigma_2 = \sin\phi. \tag{3.34}$$

Then, we search the space of the orthogonal matrices U and V and the angle ϕ so that the Sampson error J decreases. Many algorithms exist for this search, but the most widely used scheme for computer vision applications is the *Levenberg-Marquardt method* (*LM*). Using this, we obtain the following procedure.

Procedure 3.5 (Hidden variable LM)

1. Find an initial value of F such that $\det \mathsf{F} = 0$ and $\|\mathsf{F}\| = 1$, using a simple method, for example,, computing F by least squares of Procedure 3.1, or better by the Taubin method of Procedure 3.2, followed by the SVD rank correction of Procedure 3.3. Let its SVD be

$$\mathsf{F} = \mathsf{U} \begin{pmatrix} \cos\phi & 0 & 0 \\ 0 & \sin\phi & 0 \\ 0 & 0 & 0 \end{pmatrix} \mathsf{V}^{\top}. \tag{3.35}$$

2. Compute the Sampson error J of Eq. (3.23), and let $c = 0.0001$.

3. Compute the 9×3 matrices

$$\mathsf{F}_U = \begin{pmatrix} 0 & F_{31} & -F_{21} \\ 0 & F_{32} & -F_{22} \\ 0 & F_{33} & -F_{23} \\ -F_{31} & 0 & F_{11} \\ -F_{32} & 0 & F_{12} \\ -F_{33} & 0 & F_{13} \\ F_{21} & -F_{11} & 0 \\ F_{22} & -F_{12} & 0 \\ F_{23} & -F_{13} & 0 \end{pmatrix}, \qquad \mathsf{F}_V = \begin{pmatrix} 0 & F_{13} & -F_{12} \\ -F_{13} & 0 & F_{11} \\ F_{12} & -F_{11} & 0 \\ 0 & F_{23} & -F_{22} \\ -F_{23} & 0 & F_{21} \\ F_{22} & -F_{21} & 0 \\ 0 & F_{33} & -F_{32} \\ -F_{33} & 0 & F_{31} \\ F_{32} & -F_{31} & 0 \end{pmatrix}, \tag{3.36}$$

where F_{ij} is the (ij) element of F.

4. Compute the 9D vector

$$\boldsymbol{\theta}_\phi = \begin{pmatrix} \sigma_1 U_{12} V_{12} - \sigma_2 U_{11} V_{11} \\ \sigma_1 U_{12} V_{22} - \sigma_2 U_{11} V_{21} \\ \sigma_1 U_{12} V_{32} - \sigma_2 U_{11} V_{31} \\ \sigma_1 U_{22} V_{12} - \sigma_2 U_{21} V_{11} \\ \sigma_1 U_{22} V_{22} - \sigma_2 U_{21} V_{21} \\ \sigma_1 U_{22} V_{32} - \sigma_2 U_{21} V_{31} \\ \sigma_1 U_{32} V_{12} - \sigma_2 U_{31} V_{11} \\ \sigma_1 U_{32} V_{22} - \sigma_2 U_{31} V_{21} \\ \sigma_1 U_{32} V_{32} - \sigma_2 U_{31} V_{31} \end{pmatrix}, \tag{3.37}$$

where U_{ij} and V_{ij} are the (ij) elements of U and V, respectively.

5. Convert the matrix F in Eq. (3.35) to the vector $\boldsymbol{\xi}$ as in Eq. (3.3), and compute the 9×9 matrices

$$\mathsf{M} = \frac{1}{N}\sum_{\alpha=1}^{N} \frac{\boldsymbol{\xi}_\alpha \boldsymbol{\xi}_\alpha^\top}{(\boldsymbol{\theta}, V_0[\boldsymbol{\xi}_\alpha]\boldsymbol{\theta})}, \qquad \mathsf{L} = \frac{1}{N}\sum_{\alpha=1}^{N} \frac{(\boldsymbol{\xi}_\alpha, \boldsymbol{\theta})^2}{(\boldsymbol{\theta}, V_0[\boldsymbol{\xi}_\alpha]\boldsymbol{\theta})^2} V_0[\boldsymbol{\xi}_\alpha]. \tag{3.38}$$

6. Compute the 9×9 matrix

$$\mathsf{X} = \mathsf{M} - \mathsf{L}. \tag{3.39}$$

7. Compute the following first derivatives of J.

$$\nabla_\omega J = 2\mathsf{F}_U^\top \mathsf{X}\boldsymbol{\theta}, \qquad \nabla_{\omega'} J = 2\mathsf{F}_V^\top \mathsf{X}\boldsymbol{\theta}, \qquad \frac{\partial J}{\partial \phi} = 2(\boldsymbol{\theta}_\phi, \mathsf{X}\boldsymbol{\theta}). \tag{3.40}$$

8. Compute the following second derivatives of J.

$$\nabla_{\omega\omega} J = 2\mathsf{F}_U^\top \mathsf{X}\mathsf{F}_U, \qquad \nabla_{\omega'\omega'} J = 2\mathsf{F}_V^\top \mathsf{X}\mathsf{F}_V, \qquad \nabla_{\omega\omega'} J = 2\mathsf{F}_U^\top \mathsf{X}\mathsf{F}_V,$$

$$\frac{\partial J^2}{\partial \phi^2} = 2(\boldsymbol{\theta}_\phi, \mathsf{X}\boldsymbol{\theta}_\phi), \qquad \frac{\partial \nabla_\omega J}{\partial \phi} = 2\mathsf{F}_U^\top \mathsf{X}\boldsymbol{\theta}_\phi, \qquad \frac{\partial \nabla_{\omega'} J}{\partial \phi} = 2\mathsf{F}_V^\top \mathsf{X}\boldsymbol{\theta}_\phi. \tag{3.41}$$

9. Compute the 9×9 matrix

$$\mathsf{H} = \begin{pmatrix} \nabla_{\omega\omega} J & \nabla_{\omega\omega'} J & \partial \nabla_\omega J/\partial \phi \\ (\nabla_{\omega\omega'} J)^\top & \nabla_{\omega'\omega'} J & \partial \nabla_{\omega'} J/\partial \phi \\ (\partial \nabla_\omega J/\partial \phi)^\top & (\partial \nabla_{\omega'} J/\partial \phi)^\top & \partial J^2/\partial \phi^2 \end{pmatrix}. \tag{3.42}$$

10. Solve the 9D linear equation

$$(\mathsf{H} + cD[\mathsf{H}]) \begin{pmatrix} \Delta\boldsymbol{\omega} \\ \Delta\boldsymbol{\omega}' \\ \Delta\phi \end{pmatrix} = - \begin{pmatrix} \nabla_{\boldsymbol{\omega}} J \\ \nabla_{\boldsymbol{\omega}'} J \\ \partial J / \partial \phi \end{pmatrix} \tag{3.43}$$

to determine $\Delta\boldsymbol{\omega}$, $\Delta\boldsymbol{\omega}'$, and $\Delta\phi$, where $D[\,\cdot\,]$ denotes the diagonal matrix obtained by taking out the diagonal elements only.
11. Update U, V, and ϕ to

$$\mathsf{U}' = \mathsf{R}(\Delta\boldsymbol{\omega})\mathsf{U}, \qquad \mathsf{V}' = \mathsf{R}(\Delta\boldsymbol{\omega}')\mathsf{V}, \qquad \phi' = \phi + \Delta\phi, \tag{3.44}$$

where $\mathsf{R}(\mathbf{w})$ denotes the rotation matrix by angle $\|\mathbf{w}\|$ around axis $\mathbf{w}$ screw-wise.
12. Update F to

$$\mathsf{F}' = \mathsf{U}' \begin{pmatrix} \cos\phi' & 0 & 0 \\ 0 & \sin\phi' & 0 \\ 0 & 0 & 0 \end{pmatrix} \mathsf{V}'^{\top}. \tag{3.45}$$

13. Let J' be the value of Eq. (3.23) for F'.
14. Unless $J' < J$ or $J' \approx J$, let $c \leftarrow 10c$ and go back to Step 10.
15. If $\mathsf{F}' \approx \mathsf{F}$, return F' and stop. Else, let $\mathsf{F} \leftarrow \mathsf{F}'$, $\mathsf{U} \leftarrow \mathsf{U}'$, $\mathsf{V} \leftarrow \mathsf{V}'$, $\phi \leftarrow \phi'$, and $c \leftarrow c/10$, and go back to Step 3.

Comments The above method is based on the observation that the differentiation, or gradient, of the Sampson error J with respect to $\boldsymbol{\theta}$ is written in the form (↪ Problem 2.5(1)):

$$\nabla_{\boldsymbol{\theta}} J = \frac{1}{N}\sum_{\alpha=1}^{N} \frac{2(\boldsymbol{\xi}_\alpha, \boldsymbol{\theta})\boldsymbol{\xi}_\alpha}{(\boldsymbol{\theta}, V_0[\boldsymbol{\xi}_\alpha]\boldsymbol{\theta})} - \frac{1}{N}\sum_{\alpha=1}^{N} \frac{2(\boldsymbol{\xi}_\alpha, \boldsymbol{\theta})^2 V_0[\boldsymbol{\xi}_\alpha]\boldsymbol{\theta}}{(\boldsymbol{\theta}, V_0[\boldsymbol{\xi}_\alpha]\boldsymbol{\theta})^2}$$
$$= 2(\mathsf{M} - \mathsf{L})\boldsymbol{\theta} = 2\mathsf{X}\boldsymbol{\theta}. \tag{3.46}$$

Here, M and L are the matrices defined by Eq. (3.38), and X is the matrix in Eq. (3.39). Using Eq. (3.46), we search for the values of U, V, and ϕ in Eq. (3.35) that minimize Eq. (3.23). The main point is the update of U and V. Note that an orthogonal matrix represents a posture, that is, rotation and reflection, of an orthogonal frame and has three degrees of freedom. We may introduce three particular parameters such as the Euler angles, but the resulting expressions are very complicated. An elegant method is to describe, not the three degrees of freedom themselves, but their *infinitesimal changes* using three parameters (↪ Problem 3.6(1)). Specifically, we consider rotations of the frame defined by U around the x-, y-, and z-axes by infinitesimal angles $\Delta\omega_1$, $\Delta\omega_2$, and $\Delta\omega_3$, respectively. Similarly, we consider a rotation of V by infinitesimal angles $\Delta\omega_1'$, $\Delta\omega_2'$, and $\Delta\omega_3'$. We also consider an infinitesimal change of ϕ into $\phi + \Delta\phi$. Then, the resulting infinitesimal change $\Delta\mathsf{F}$ of F in Eq. (3.35) is expressed in $\Delta\omega_1$, $\Delta\omega_2$, $\Delta\omega_3$, $\Delta\omega_1'$, $\Delta\omega_2'$, $\Delta\omega_3'$, and $\Delta\phi$ (↪ Problem 3.6(2)). Using this expression, we choose such $\Delta\omega_1$, $\Delta\omega_2$, $\Delta\omega_3$, $\Delta\omega_1'$, $\Delta\omega_2'$, $\Delta\omega_3'$, and $\Delta\phi$ that decrease the Sampson error J in Eq. (3.23), and increment U, V, and ϕ accordingly. We repeat this.

For numerically minimizing the Sampson error J, we need derivatives of J with respect to the parameters. For U and V, even though they are not explicitly parameterized, their derivatives can be obtained through their infinitesimal variations. Note that when the angle ϕ changes by $\Delta\phi$, the resulting change of J is $(\partial J/\partial\phi)\Delta\phi$ to a first approximation. Similarly, we define the derivative $\partial J/\partial\omega_1$ in such a way that the change of U by $\Delta\omega_1$ results in the change of J by $(\partial J/\partial\omega_1)\Delta\omega_1$ to a first approximation. Derivatives with respect to the other parameters can be similarly defined, and we introduce the vector notation

$$\nabla_\omega J \equiv \begin{pmatrix} \partial J/\partial\omega_1 \\ \partial J/\partial\omega_2 \\ \partial J/\partial\omega_3 \end{pmatrix}, \qquad \nabla_{\omega'} J \equiv \begin{pmatrix} \partial J/\partial\omega'_1 \\ \partial J/\partial\omega'_2 \\ \partial J/\partial\omega'_3 \end{pmatrix}. \tag{3.47}$$

Equation (3.40) is obtained in this way ($\hookrightarrow$ Problem 3.6(3)).

Second derivatives can be defined similarly. Note that the second derivative $\partial^2 J/\partial^2\phi$ is defined in such a way that the change ΔJ of J caused by the change $\Delta\phi$ of ϕ has a quadratic term in $\Delta\phi$ in the form $(1/2)(\partial^2 J/\partial^2\phi)\Delta\phi^2$. Similarly, we define $\partial^2 J/\partial^2\Delta\omega_1$ in such a way that the change ΔJ caused by the change $\Delta\omega_1$ of U has a quadratic term in $\Delta\omega_1$ in the form $(1/2)(\partial^2 J/\partial^2\omega_1)\Delta\omega_1^2$ and define $\partial^2 J/\partial\omega_1\partial\omega_2$ $(= \partial^2 J/\partial\omega_2\partial\omega_1)$ in such a way that the change ΔJ caused by the changes $\Delta\omega_1$ and $\Delta\omega_2$ of U has a quadratic term in $\Delta\omega_1$ and $\Delta\omega_2$ in the form $(\partial^2 J/\partial\omega_1\partial\omega_2)\Delta\omega_1\Delta\omega_2$. Other second-order derivatives are also defined similarly. However, the rigorous analysis is very complicated. Note, on the other hand, that the second derivatives in the optimization process are used for controlling the speed of convergence, and they do not affect the accuracy of the final solution. Hence, we may introduce approximations in evaluating the second derivatives, ignoring high-order small quantities. Then, Eq. (3.41) is obtained ($\hookrightarrow$ Problem 3.6(4)).

Equation (3.52) gives the Hessian of the function J. The Levenberg-Marquardt method is a modification of the well-known Newton iterations, and Eq. (3.43) reduces to the Newton iterations if we let $c = 0$. It is well known that the Newton iterations converge very quickly but may converge to a wrong solution depending on the starting value and the form of the function. For alleviating this, the update process of Eq. (3.43) is slowed down by introducing $c > 0$, checking if J actually decreases at each iteration. If J does not decrease, we increase the value c and recompute the next value (Step 10). If J decreases, the value c is decreased to do a Newton-like update (Step 11). The rotation matrix specified by an axis and angle used in Step 11 has the form:

$$\mathsf{R}(\Omega\mathbf{l}) = \begin{pmatrix} \cos\Omega + l_1^2(1-\cos\Omega) & l_1 l_2(1-\cos\Omega) - l_3\sin\Omega & l_1 l_3(1-\cos\Omega) + l_2\sin\Omega \\ l_2 l_1(1-\cos\Omega) + l_3\sin\Omega & \cos\Omega + l_2^2(1-\cos\Omega) & l_2 l_3(1-\cos\Omega) - l_1\sin\Omega \\ l_3 l_1(1-\cos\Omega) - l_2\sin\Omega & l_3 l_2(1-\cos\Omega) + l_1\sin\Omega & \cos\Omega + l_3^2(1-\cos\Omega) \end{pmatrix}. \tag{3.48}$$

This matrix defines the rotation around axis $\mathbf{l}$ (unit vector) by angle Ω (in radians) screw-wise.

3.7 Extended FNS

We can do iterations directly on $\boldsymbol{\theta}$ without using hidden variables in such a way that $\boldsymbol{\theta}$ approaches its true value at each iteration and at the same time the rank constraint is satisfied when the iterations have converged. This can be done by extending the FNS of Sect. 2.5.2 as follows.

Procedure 3.6 (Extended FNS)

1. Initialize $\boldsymbol{\theta}$.
2. Compute the 9×9 matrices

$$\mathsf{M} = \frac{1}{N}\sum_{\alpha=1}^{N}\frac{\boldsymbol{\xi}_\alpha\boldsymbol{\xi}_\alpha^\top}{(\boldsymbol{\theta}, V_0[\boldsymbol{\xi}_\alpha]\boldsymbol{\theta})}, \qquad \mathsf{L} = \frac{1}{N}\sum_{\alpha=1}^{N}\frac{(\boldsymbol{\xi}_\alpha, \boldsymbol{\theta})^2}{(\boldsymbol{\theta}, V_0[\boldsymbol{\xi}_\alpha]\boldsymbol{\theta})^2}V_0[\boldsymbol{\xi}_\alpha]. \tag{3.49}$$

3. Compute the 9D vector $\boldsymbol{\theta}^\dagger$ of Eq. (3.30), and define the 9×9 matrix

$$\mathsf{P}_{\boldsymbol{\theta}^\dagger} = \mathsf{I} - \frac{\boldsymbol{\theta}^\dagger\boldsymbol{\theta}^{\dagger\top}}{\|\boldsymbol{\theta}^\dagger\|^2}. \tag{3.50}$$

4. Compute the 9×9 matrices

$$\mathsf{X} = \mathsf{M} - \mathsf{L}, \qquad \mathsf{Y} = \mathsf{P}_{\boldsymbol{\theta}^\dagger}\mathsf{X}\mathsf{P}_{\boldsymbol{\theta}^\dagger}. \tag{3.51}$$

5. Let $\mathbf{v}_1$ and $\mathbf{v}_2$ be the unit eigenvectors of Y for the two smallest eigenvalues, and compute the following vector $\hat{\boldsymbol{\theta}}$.

$$\hat{\boldsymbol{\theta}} = (\boldsymbol{\theta}, \mathbf{v}_1)\mathbf{v}_1 + (\boldsymbol{\theta}, \mathbf{v}_2)\mathbf{v}_2. \tag{3.52}$$

6. Compute the vector

$$\boldsymbol{\theta}' = \mathcal{N}[\mathsf{P}_{\boldsymbol{\theta}^\dagger}\hat{\boldsymbol{\theta}}]. \tag{3.53}$$

7. If $\boldsymbol{\theta}' \approx \boldsymbol{\theta}$ up to sign, return $\boldsymbol{\theta}'$ as $\boldsymbol{\theta}$ and stop. Else, let

$$\boldsymbol{\theta} \leftarrow \mathcal{N}[\boldsymbol{\theta} + \boldsymbol{\theta}'], \tag{3.54}$$

and go back to Step 2.

Comments The matrix $\mathsf{P}_{\boldsymbol{\theta}^\dagger}$ in Eq. (3.50) projects vectors onto a plane orthogonal to the unit vector $\boldsymbol{\theta}^\dagger/\|\boldsymbol{\theta}^\dagger\|$ ($\hookrightarrow$ Problem 3.3). From Eq. (3.32), the rank constraint of Eq. (3.24) is written as $(\boldsymbol{\theta}^\dagger, \boldsymbol{\theta}) = 0$. As described in Sect. 2.5.2, the FNS there computes the eigenvector of the matrix X of Eq. (3.51), but here we first modify X to Y by applying the projection operation onto the plane orthogonal to $\boldsymbol{\theta}^\dagger$ and then compute not only the smallest eigenvalue but also the second smallest eigenvalue and their unit eigenvectors $\mathbf{v}_1$ and $\mathbf{v}_2$. Next, we project the current $\boldsymbol{\theta}$ onto the plane spanned by $\mathbf{v}_1$ and $\mathbf{v}_2$ in the form of Eq. (3.52). The resulting vector is further projected onto the plane orthogonal to $\boldsymbol{\theta}^\dagger$ in the form of Eq. (3.53). However, if $\boldsymbol{\theta}$ is updated, the vector $\boldsymbol{\theta}^\dagger$ also changes. Thus $(\boldsymbol{\theta}^\dagger, \boldsymbol{\theta}) = 0$ is not strictly satisfied after the update of $\boldsymbol{\theta}$, therefore we do iterations. Equation (3.54) is a technique for accelerating the convergence: we compute the mean of the values of $\boldsymbol{\theta}$ before and after the update (note that $\mathcal{N}[\boldsymbol{\theta} + \boldsymbol{\theta}'] = \mathcal{N}[(\boldsymbol{\theta} + \boldsymbol{\theta}')/2]$). It can be shown that when the iterations have converged, the rank constraint $(\boldsymbol{\theta}^\dagger, \boldsymbol{\theta}) = 0$ is satisfied and at the same time the two smallest eigenvalues of Y are both 0 ($\hookrightarrow$ Problem 3.7). This implies that the resulting value $\boldsymbol{\theta}$ minimizes the Sampson error J of Eq. (3.23) subject to the rank constraint.

3.8 Geometric Distance Minimization

We can obtain the value of $\boldsymbol{\theta}$ that minimizes the geometric distance by iteratively modifying the Sampson error in the form of Eqs. (2.42) and (2.43), using the value $\boldsymbol{\theta}$ obtained by the above extended FNS. The procedure is as follows.

Procedure 3.7 (Geometric distance minimization)

1. Let $J_0 = \infty$ (a sufficiently large number), $\hat{x}_\alpha = x_\alpha$, $\hat{y}_\alpha = y_\alpha$, $\hat{x}'_\alpha = x'_\alpha$, $\hat{y}'_\alpha = y'_\alpha$, and $\tilde{x}_\alpha = \tilde{y}_\alpha = \tilde{x}'_\alpha = \tilde{y}'_\alpha = 0$, $\alpha = 1, \ldots, N$.
2. Compute the normalized covariance matrix $V_0[\hat{\boldsymbol{\xi}}_\alpha]$ obtained by replacing $\bar{x}_\alpha$, $\bar{y}_\alpha$, $\bar{x}'_\alpha$, and $\bar{y}'_\alpha$ in the definition of $V_0[\boldsymbol{\xi}_\alpha]$ in Eq. (3.12) by $\hat{x}_\alpha$, $\hat{y}_\alpha$, $\hat{x}'_\alpha$, and $\hat{y}'_\alpha$, respectively.
3. Compute the following modified data $\boldsymbol{\xi}^*_\alpha$.

$$\boldsymbol{\xi}^*_\alpha = \begin{pmatrix} \hat{x}_\alpha \hat{x}'_\alpha + \hat{x}'_\alpha \tilde{x}_\alpha + \hat{x}_\alpha \tilde{x}'_\alpha \\ \hat{x}_\alpha \hat{y}'_\alpha + \hat{y}'_\alpha \tilde{x}_\alpha + \hat{x}_\alpha \tilde{y}'_\alpha \\ f_0(\hat{x}_\alpha + \tilde{x}_\alpha) \\ \hat{y}_\alpha \hat{x}'_\alpha + \hat{x}'_\alpha \tilde{y}_\alpha + \hat{y}_\alpha \tilde{x}'_\alpha \\ \hat{y}_\alpha \hat{y}'_\alpha + \hat{y}'_\alpha \tilde{y}_\alpha + \hat{y}_\alpha \tilde{y}'_\alpha \\ f_0(\hat{y}_\alpha + \tilde{y}_\alpha) \\ f_0(\hat{x}'_\alpha + \tilde{x}'_\alpha) \\ f_0(\hat{y}'_\alpha + \tilde{y}'_\alpha) \\ f_0^2 \end{pmatrix}. \tag{3.55}$$

4. Compute the value $\boldsymbol{\theta}$ that minimizes the following *modified Sampson error.*

$$J^* = \frac{1}{N} \sum_{\alpha=1}^{N} \frac{(\boldsymbol{\xi}^*_\alpha, \boldsymbol{\theta})^2}{(\boldsymbol{\theta}, V_0[\hat{\boldsymbol{\xi}}_\alpha]\boldsymbol{\theta})}. \tag{3.56}$$

5. Update $\tilde{x}_\alpha$, $\tilde{y}_\alpha$, $\tilde{x}'_\alpha$, and $\tilde{y}'_\alpha$ as follows.

$$\begin{pmatrix} \tilde{x}_\alpha \\ \tilde{y}_\alpha \end{pmatrix} \leftarrow \frac{(\boldsymbol{\xi}^*_\alpha, \boldsymbol{\theta})}{(\boldsymbol{\theta}, V_0[\hat{\boldsymbol{\xi}}_\alpha]\boldsymbol{\theta})} \begin{pmatrix} \theta_1 & \theta_2 & \theta_3 \\ \theta_4 & \theta_5 & \theta_6 \end{pmatrix} \begin{pmatrix} \hat{x}'_\alpha \\ \hat{y}'_\alpha \\ f_0 \end{pmatrix},$$

$$\begin{pmatrix} \tilde{x}'_\alpha \\ \tilde{y}'_\alpha \end{pmatrix} \leftarrow \frac{(\boldsymbol{\xi}^*_\alpha, \boldsymbol{\theta})}{(\boldsymbol{\theta}, V_0[\hat{\boldsymbol{\xi}}_\alpha]\boldsymbol{\theta})} \begin{pmatrix} \theta_1 & \theta_4 & \theta_7 \\ \theta_2 & \theta_5 & \theta_8 \end{pmatrix} \begin{pmatrix} \hat{x}_\alpha \\ \hat{y}_\alpha \\ f_0 \end{pmatrix}. \tag{3.57}$$

6. Update $\hat{x}_\alpha$, $\hat{y}_\alpha$, $\hat{x}'_\alpha$, and $\hat{y}'_\alpha$ as follows.

$$\hat{x}_\alpha \leftarrow x_\alpha - \tilde{x}_\alpha, \qquad \hat{y}_\alpha \leftarrow y_\alpha - \tilde{y}_\alpha, \qquad \hat{x}'_\alpha \leftarrow x'_\alpha - \tilde{x}'_\alpha, \qquad \hat{y}'_\alpha \leftarrow y'_\alpha - \tilde{y}'_\alpha. \tag{3.58}$$

7. Compute

$$J^* = \frac{1}{N}\sum_{\alpha=1}^{N}(\tilde{x}_\alpha^2 + \tilde{y}_\alpha^2 + \tilde{x}'^2_\alpha + \tilde{y}'^2_\alpha). \tag{3.59}$$

If $J^* \approx J_0$, return $\boldsymbol{\theta}$ and stop. Else, let $J_0 \leftarrow J^*$, and go back to Step 2.

Comments Because $\hat{x}_\alpha = x_\alpha$, $\hat{y}_\alpha = y_\alpha$, $\hat{x}'_\alpha = x'_\alpha$, $\hat{y}'_\alpha = y'_\alpha$, and $\tilde{x}_\alpha = \tilde{y}_\alpha = \tilde{x}'_\alpha = \tilde{y}'_\alpha = 0$ in the initial iteration, the value $\boldsymbol{\xi}^*_\alpha$ of Eq. (3.55) is the same as $\boldsymbol{\xi}_\alpha$. Hence, the modified Sampson error J^* in (3.56) is the same as the function J in Eq. (3.23), and the value $\boldsymbol{\theta}$ that minimizes J is computed initially. We rewrite Eq. (3.15) in the form

$$\begin{aligned}
S &= \frac{1}{N}\sum_{\alpha=1}^{N}\Big((\hat{x}_\alpha + (x_\alpha - \hat{x}_\alpha) - \bar{x}_\alpha)^2 + (\hat{y}_\alpha + (y_\alpha - \hat{y}_\alpha) - \bar{y}_\alpha)^2 \\
&\quad + (\hat{x}'_\alpha + (x'_\alpha - \hat{x}'_\alpha) - \bar{x}'_\alpha)^2 + (\hat{y}'_\alpha + (y'_\alpha - \hat{y}'_\alpha) - \bar{y}'_\alpha)^2\Big) \\
&= \frac{1}{N}\sum_{\alpha=1}^{N}\Big((\hat{x}_\alpha + \tilde{x}_\alpha - \bar{x}_\alpha)^2 + (\hat{y}_\alpha + \tilde{y}_\alpha - \bar{y}_\alpha)^2 + (\hat{x}'_\alpha + \tilde{x}'_\alpha - \bar{x}'_\alpha)^2 + (\hat{y}'_\alpha + \tilde{y}'_\alpha - \bar{y}'_\alpha)^2\Big),
\end{aligned} \tag{3.60}$$

where

$$\tilde{x}_\alpha = x_\alpha - \hat{x}_\alpha, \qquad \tilde{y}_\alpha = y_\alpha - \hat{y}_\alpha, \qquad \tilde{x}'_\alpha = x'_\alpha - \hat{x}'_\alpha, \qquad \tilde{y}'_\alpha = y'_\alpha - \hat{y}'_\alpha. \tag{3.61}$$

In the next iteration step, we regard the corrected values $(\hat{x}_\alpha, \hat{y}_\alpha)$ and $(\hat{x}'_\alpha, \hat{y}'_\alpha)$ as the input data and minimize Eq. (3.60) subject to Eq. (3.16). Let $(\hat{\hat{x}}_\alpha, \hat{\hat{y}}_\alpha)$, $(\hat{\hat{x}}'_\alpha, \hat{\hat{y}}'_\alpha)$ be the solution. If we ignore high-order small terms in $\hat{x}_\alpha - \bar{x}_\alpha$, $\hat{y}_\alpha - \bar{y}_\alpha$, $\hat{x}'_\alpha - \bar{x}'_\alpha$, and $\hat{y}'_\alpha - \bar{y}'_\alpha$ and rewrite Eq. (3.60), we obtain the modified Sampson error J^* in Eq. (3.56) ($\hookrightarrow$ Problem 3.8). We minimize this, regard $(\hat{\hat{x}}_\alpha, \hat{\hat{y}}_\alpha)$ and $(\hat{\hat{x}}'_\alpha, \hat{\hat{y}}'_\alpha)$ as $(\hat{x}_\alpha, \hat{y}_\alpha)$ and $(\hat{x}'_\alpha, \hat{y}'_\alpha)$, and repeat the same process. Because the current $(\hat{x}_\alpha, \hat{y}_\alpha)$ and $(\hat{x}'_\alpha, \hat{y}'_\alpha)$ are the best approximation of $(\bar{x}_\alpha, \bar{y}_\alpha)$ and $(\bar{x}'_\alpha, \bar{y}'_\alpha)$, Equation (3.61) implies that $\tilde{x}_\alpha^2 + \tilde{y}_\alpha^2 + \tilde{x}'^2_\alpha + \tilde{y}'^2_\alpha$ is the corresponding approximation of $(\bar{x}_\alpha - x_\alpha)^2 + (\bar{y}_\alpha - y_\alpha)^2 + (\bar{x}'_\alpha - x'_\alpha)^2 + (\bar{y}'_\alpha - y'_\alpha)^2$. Thus we use Eq. (3.59) to evaluate the geometric distance S. The ignored high-order terms decrease after each iteration, therefore we obtain in the end the value $\boldsymbol{\theta}$ that minimizes the geometric distance S, and Eq. (3.59) coincides with S. According to experiments, however, the correction of $\boldsymbol{\theta}$ by this procedure is very small: the three or four significant digits are unchanged in typical problems. Therefore we can practically identify FNS with a method to minimize the geometric distance.

3.9 Outlier Removal

For extracting point correspondences between two images of the same scene, one needs to find similar-looking points in them and match them, using some similarity measure, and various methods for this have been studied. However, existing methods are not perfect, sometimes matching wrong points. Such false matches are called *outliers*; correct matches are called *inliers*. Inasmuch as fundamental matrix computation has mathematically the same form as ellipse fitting, the outlier detection techniques for ellipse fitting can also be applied to fundamental matrix computation. The underlying idea is, for a given set of point pairs, to find a fundamental matrix such that *the number of point pairs that approximately satisfy the epipolar equation is as large as possible*. The best-known method is the RANSAC described in Chap. 2. The procedure is as follows.

Procedure 3.8 (RANSAC)

1. From among the input point pairs randomly select eight, and let $\boldsymbol{\xi}_1, \ldots, \boldsymbol{\xi}_8$ be their vector representations in the form of $\boldsymbol{\xi}$ of Eq. (3.3).
2. Compute the unit eigenvector $\boldsymbol{\theta}$ of the matrix

$$\mathsf{M}_8 = \sum_{\alpha=1}^{8} \boldsymbol{\xi}_\alpha \boldsymbol{\xi}_\alpha^\top, \tag{3.62}$$

   for the smallest eigenvalue.
3. Store that $\boldsymbol{\theta}$ as a candidate, and count the number n of point pairs in the input that satisfy

$$\frac{(\boldsymbol{\xi}, \boldsymbol{\theta})^2}{(\boldsymbol{\theta}, V_0[\boldsymbol{\xi}]\boldsymbol{\theta})} < 2d^2, \tag{3.63}$$

   where d is a threshold for admissible deviation of each point of a corresponding pair; for example, $d = 2$ (pixels). Store that n.
4. Select a new set of eight point pairs from the input, and repeat Step 3. Repeat this many times, and return from among the stored candidates the one for which n is the largest.

Comments Step 2 computes the fundamental matrix from the eight point pairs selected in Step 1. The fundamental matrix F has nine elements, but it has scale indeterminacy, thus F is determined up to scale by solving the eight epipolar equations obtained from the eight point pairs; it is later normalized in the form of Eq. (3.2). However, the use of least squares as in Procedure 3.1 is easier to program. Here the rank constraint of Eq. (3.24) is not considered, because speed is more important than accuracy for repeated voting. The coefficient $1/N$ in Eq. (3.13) is omitted for M_8; the computed eigenvectors are the same. Step 3 tests to what extent each point pair satisfies the epipolar equation given by the candidate $\boldsymbol{\theta}$. This is measured by Eq. (3.17), which describes the sum of squares of the minimum distances necessary to move the point pair (x, y) and (x', y') to $(\bar{x}, \bar{y})$ and $(\bar{x}', \bar{y}')$ such that the epipolar equation is satisfied. This sum of squares can be approximated by Eq. (3.22). As in ellipse fitting, we can directly go on to the next sampling if the count n is smaller than the stored count, and we can stop if the current value of n is not updated over a fixed number of samplings. In the end, those point pairs that do not satisfy Eq. (3.63) for the chosen $\boldsymbol{\theta}$ are removed as outliers.

3.10 Examples

Figure 3.2 shows simulated images of a curved grid surface in the scene viewed from two different positions. The image size is assumed to be 600×600 pixels. To the x and y coordinates of each grip point is added independent Gaussian noise of mean 0 and standard deviation 1 (pixel). The true value of the fundamental matrix between these two images is

$$\bar{\mathsf{F}} = \begin{pmatrix} 0.07380 & -0.34355 & -0.28357 \\ 0.21858 & 0.41655 & 0.33508 \\ 0.66823 & -0.08789 & -0.09100 \end{pmatrix}. \tag{3.64}$$

Computing the fundamental matrix from the noisy gridpoint correspondences of Fig. 3.2, we obtain

$$\mathsf{F}^{(0)} = \begin{pmatrix} 0.21115 & -0.52234 & -0.38029 \\ 0.32188 & 0.32504 & 0.18557 \\ 0.53935 & 0.05232 & -0.02506 \end{pmatrix}, \quad \mathsf{F}^{(1)} = \begin{pmatrix} 0.09599 & -0.41151 & -0.34263 \\ 0.25978 & 0.36820 & 0.28133 \\ 0.64538 & -0.02586 & -0.06821 \end{pmatrix}, \tag{3.65}$$

where $\mathsf{F}^{(0)}$ is the value obtained by least squares (Procedure 3.1) followed by SVD rank correction (Procedure 3.3), and $\mathsf{F}^{(1)}$ is the value obtained by FNS (Procedure 2.5) followed by SVD rank correction. The value $\mathsf{F}^{(2)}$ obtained by FNS followed by optimal rank correction (Procedure 3.4) and the value $\mathsf{F}^{(3)}$ obtained by hidden variable LM (Procedure 3.5) are

$$\mathsf{F}^{(2)} = \begin{pmatrix} 0.07506 & -0.34616 & -0.27188 \\ 0.21826 & 0.43547 & 0.33471 \\ 0.65834 & -0.09763 & -0.09158 \end{pmatrix}, \quad \mathsf{F}^{(3)} = \begin{pmatrix} 0.09265 & -0.36657 & -0.30765 \\ 0.24157 & 0.40747 & 0.33578 \\ 0.65177 & -0.05101 & -0.07704 \end{pmatrix}. \tag{3.66}$$

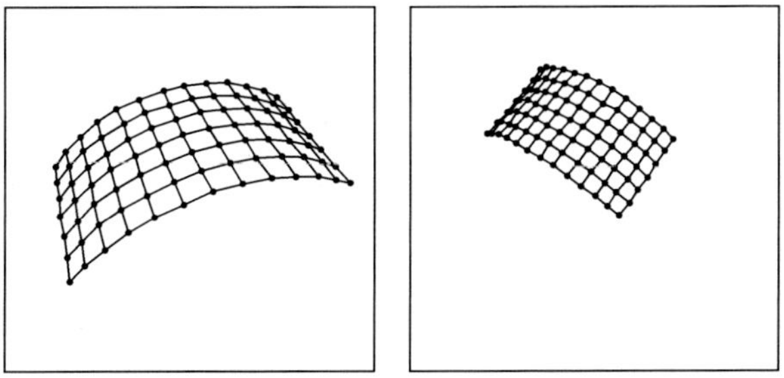

Fig. 3.2 Simulated images of a curved grid surface

Table 3.1 The error of fundamental matrix computation for different methods

Method	E
Least squares + SVD	0.370992
FNS + SVD	0.142874
FNS + optimal correction	0.026385
Hidden variable LM	0.062475
Extended FNS	0.026202
Geometric distance minimization	0.026149

The value $\mathsf{F}^{(4)}$ by extended FNS (Procedure 3.6) and the geometric distance minimization solution $\mathsf{F}^{(5)}$ obtained by iterating the extended FNS (Procedure 3.7) are

$$\mathsf{F}^{(4)} = \begin{pmatrix} 0.06067 & -0.33702 & -0.27208 \\ 0.21213 & 0.42767 & 0.33980 \\ 0.66834 & -0.10005 & -0.09306 \end{pmatrix}, \qquad \mathsf{F}^{(5)} = \begin{pmatrix} 0.06068 & -0.33706 & -0.27210 \\ 0.21215 & 0.42764 & 0.33979 \\ 0.66833 & -0.10002 & -0.09306 \end{pmatrix}. \tag{3.67}$$

Measuring the error of the computed $\mathsf{F} = (F_{ij})$ by

$$E = \sqrt{\sum_{i,j=1}^{3} (F_{ij} - \bar{F}_{ij})^2}, \tag{3.68}$$

we obtain the comparison result in Table 3.1. As we see from this, the accuracy of FNS is higher than least squares, and the accuracy improves more by optimal rank correction than using SVD. We also see that the hidden variable LM and the extended FNS both achieve almost the same accuracy and that the accuracy improvement is very small if the extended FNS is iterated for geometric distance minimization.

3.11 Supplemental Note

The epipolar equation of Eq. (3.1) that constrains two images of the same 3D scene gives one of the most important foundations of computer vision and is explained in many textbooks [3,4,6]. For fixed (x, y), Eq. (3.1) defines a line in the $x'y'$ plane, called the *epipolar line* of point (x, y); for fixed (x', y'), it defines a line in the xy plane, called the epipolar line of point (x', y'). Equation (3.1) implies that one point of a corresponding pair is on the epipolar line of the other point, therefore point correspondence detection reduces to search on the epipolar line. This is the basic principle of correspondence search for stereo vision. Thus, computing the fundamental matrix is the first step of many computer vision applications.

The above consideration leads to various observations about two camera images. Note that the position of the lens center of the first camera is not imaged by the first camera itself but can be imaged by the other camera if the image is sufficiently large. That image position (x'_e, y'_e) in the second image is called the *epipole* of the second camera. Similarly, the image position (x_e, y_e) in the first image of the lens center of the second camera is called the epipole of the first image. If we let $\mathbf{x}'_e$ and $\mathbf{x}_e$ be the vector representation of Eq. (3.19) of these epipoles, we conclude that no vectors $\mathbf{a}$ and $\mathbf{b}$ with the third component being 1 exist that satisfy $(\mathbf{a}, \mathsf{F}\mathbf{x}'_e) = 0$ and $(\mathbf{x}_e, \mathsf{F}\mathbf{b})(= (\mathsf{F}^\top\mathbf{x}_e, \mathbf{b})) = 0$, because although the lens center may be seen in the other camera image, it cannot be seen in its own camera image. This implies that $\mathsf{F}\mathbf{x}'_e = \mathbf{0}$ and $\mathsf{F}^\top\mathbf{x}_e = \mathbf{0}$; that is, $\mathbf{x}'_e$ and $\mathbf{x}_e$ are, respectively, the eigenvectors of F and $\mathsf{F}^\top$ for eigenvalue 0. The rank constraint of Eq. (3.24) expresses this fact. We can also see that the epipolar lines in one image defined by points in the other image all pass through the epipole, defining a *pencil* of lines.

The epipolar equation is written in the form of Eq. (3.1) in the literature in the 1980 s and 1990s. However, in the Hartley and Zisserman [3] textbook published in the 2000s, the x and y in Eq. (3.1) are written as x' and y', respectively, whereas x' and y' in Eq. (3.1) are written as x and y, respectively. This means reversing the order of the first and second images. It is irrelevant which of the two images is called the "first" image and which the "second" image. However, this order change replaces the fundamental matrix F by its transpose $\mathsf{F}^\top$. Because the Hartley and Zisserman textbook was widely read, we find that F and $\mathsf{F}^\top$ are frequently interchanged in the literature before and after 2000; readers must be careful about which definition is used. Also, it should be noted that the numerical values of the elements of the fundamental matrix F depend on the coordinate system used there, for example, whether the origin is at the center of the image or at its upper-left corner and in which direction each axis is directed. They also depend on the scaling constant f_0; most literature uses the default $f_0 = 1$. Hence, we must be careful when comparing the numerical results of different papers.

The fundamental matrix can be determined up to scale by solving eight epipolar equations of the form of Eq. (3.1) obtained from eight corresponding point pairs. The rank constraint can be imposed by SVD. This is the most classical method for fundamental matrix computation and is known as the *8-point algorithm*. When more point pairs are given, we can use least squares for solving linear equations followed by SVD. This is a trivial extension of the 8-point algorithm,thus it is also called the *8-point algorithm*. Because Hartley [2] wrote a paper about this, it is also called *Hartley's 8-point algorithm* and widely used as a recommended method, partly due to Hartley's fame. However, Hartley [2] does not recommend it as a high-accuracy method. His emphasis is on the influence of the choice of the image origin and the scaling of the data over numerical accuracy, recommending that they should be chosen so that all the data have order 1. In this chapter, we take the image origin at the center of the image, rather than at the upper-left corner, and use the scaling constant f_0, and this is based on the same observation. Although there are still many today who think that Hartley's 8-point algorithm computes the fundamental matrix accurately, it is actually of the poorest accuracy among existing methods, as demonstrated in the examples of this chapter.

Because fundamental matrix computation has the same form as ellipse fitting, except for the rank constraint, it has been studied in the same framework of geometric estimation from noisy observation. In fact, both the renormalization of Kanatani [5,6] and the FNS of Chojnacki et al. [1] were originally intended for fundamental matrix computation. The principle of the optimal rank correction of Sect. 3.5 is given in [6] and its application to fundamental matrix computation is described in [9, 11, 12]. Mathematically, it is equivalent to the Newton iterations known in numerical analysis, thus the convergence is quadratic and converges very rapidly; usually, one or two iterations are sufficient in real applications.

Recall the remark in Sect. 3.5 that "$\sigma^2 V_0[\boldsymbol{\theta}]$ coincides with the covariance matrix $V[\boldsymbol{\theta}]$ of an optimally computed $\boldsymbol{\theta}$ except for $O(\sigma^4)$ terms." That covariance matrix $V[\boldsymbol{\theta}]$ is called the *KCR (Kanatani-Cramer-Rao) lower bound*, which gives, under certain general conditions, the lower bound on the covariance matrix of $\boldsymbol{\theta}$ computed by any method (the details are given in Chap. 16). It can be shown that the covariance matrices of $\boldsymbol{\theta}$ obtained by FNS and geometric distance minimization both agree with the KCR lower bound except for $O(\sigma^4)$ terms. In this senes, their accuracy cannot be improved any further. Detailed analyses of variance and bias for different methods along with the derivation of the KCR lower bound are given in [7].

The hidden variable LM of Sect. 3.6 is presented in Sugaya and Kanatani [10, 16]. The technique of optimizing an orthogonal or rotation matrix by considering infinitesimal variations, rather than introducing three parameters, is known as the *Lie algebra method* . Continuous transformations of space such as rotation, translation, affine transformation, and similarity form a group of transformations with respect to the composition operation, known as a *Lie group*. The set of infinitesimal transformations of a Lie group forms an algebra called a *Lie algebra* with respect to addition/subtraction, scalar multiplication, and a product operation called the *commutator production*. Its mathematical foundations and computer vision applications are given in [4]. Optimization of rotation using the Lie algebra approach is also shown in [4]. The remark in the Comments after Procedure 3.5 that "we may introduce approximations in evaluating the second derivatives, ignoring high-order small quantities" actually means that we can ignore terms that contain the inner product $(\boldsymbol{\xi}_\alpha, \boldsymbol{\theta})$ because $(\boldsymbol{\xi}_\alpha, \boldsymbol{\theta}) = 0$ (the epipolar equation) should hold if the data are accurate. This corresponds to what is known in numerical analysis as *Gauss-Newton approximation*.

The general formulation of the extended FNS of Procedure 3.6 is given in Kanatani and Matsunaga [8], and its application to fundamental matrix computation is given in Kanatani and Sugaya [13]. As in the case of ellipse fitting, iterative methods do not necessarily converge in the presence of large noise in the data. In particular, the extended FNS may not converge if the initial value is not close to the true solution. For initialization, the least square of Procedure 3.1 is mostly sufficient, but the use of the Taubin method of Procedure 3.2 improves the convergence.

The meaning of the a posteriori correction, the hidden variable approach, and the extended FNS described in this chapter can be visually understood as follows. We want to compute the solution that minimizes a given cost, for example, the least-square error, the Sampson error, and the geometric distance, subject to the rank constraint $\det \mathsf{F} = 0$. The a posteriori correction means that we first go to the point in the solution space where the cost is minimum, without considering the rank constraint, and then move to the hyper-surface defined by the rank constraint

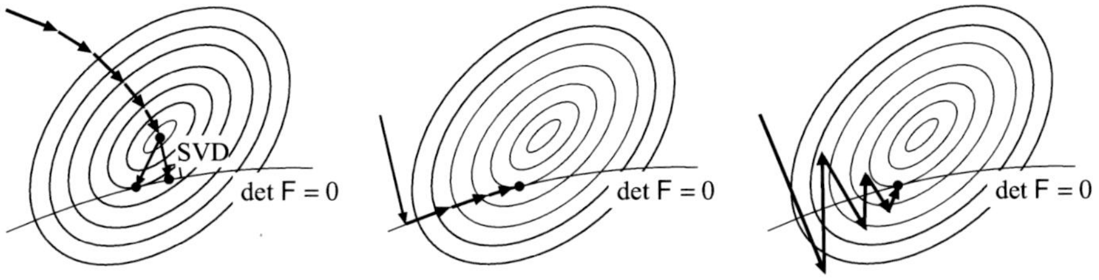

(a) A posteriori correction (b) Hidden variable approach (c) Extended FNS

Fig. 3.3 **a** We first examine the minimum of the cost (the contours are drawn) without considering the rank constraint and then move to the hyper-surface defined by det $\mathsf{F} = 0$. The SVD correction means moving to the hyper-surface orthogonally, and the optimal correction moves in the direction along which the cost least increases. **b** We parameterize the hyper-space of det $\mathsf{F} = 0$ and do a search on it, optimizing the surface parameters. **c** We do iterations in the outside space, moving in the direction of decreasing cost toward the constraint hyper-surface at each step in such a way that in the end the rank constraint is exactly satisfied and the cost cannot decrease any further

det $\mathsf{F} = 0$ (Fig. 3.3a). The SVD correction can be thought of as moving in the shortest distance orthogonally to the hyper-surface, whereas the optimal correction can be interpreted to move in the direction along which the cost least increases. In the hidden variable approach, the hyper-surface of the rank constraint det $\mathsf{F} = 0$ is parameterized, and the search is done on it, optimizing the surface parameters (Fig. 3.3b). In contrast, the extended FNS does iterations in the outside space, moving in the direction of decreasing cost toward the constraint hyper-surface at each step in such a way that in the end the rank constraint is exactly satisfied and the cost cannot decrease any further (Fig. 3.3c). The fact that the geometric distance can be minimized by repeating Sampson error minimization was pointed out by Kanatani and Sugaya [14], and its application to fundamental matrix computation is shown in [13].

There is one thing that needs attention when we do numerical simulation. The fundamental matrix cannot be uniquely determined if the corresponding points are exactly coplanar in the scene or exactly have a special symmetry, for example, at the vertices of a cube, This does not occur for real data with noise. However, we should be careful, inasmuch as we tend to define a simple configuration for simulation in order to save time. For details, see [15]. For outlier removal, many methods have been proposed other than the RANSAC described in Sect. 3.9, including the LMedS and M-estimation.

Problems

3.1 Show that the minimum of Eq. (3.17) subject to the epipolar equation of Eq. (3.16) can be approximated in the form of Eq. (3.18) if high-order terms in Δx_α, Δy_α, $\Delta x'_\alpha$, and $\Delta y'_\alpha$ are ignored.

3.2 If F is expressed in the form of Eq. (3.25), show that $\|\mathsf{F}\|^2 = \sigma_1^2 + \sigma_2^2 + \sigma_2^3$ holds.

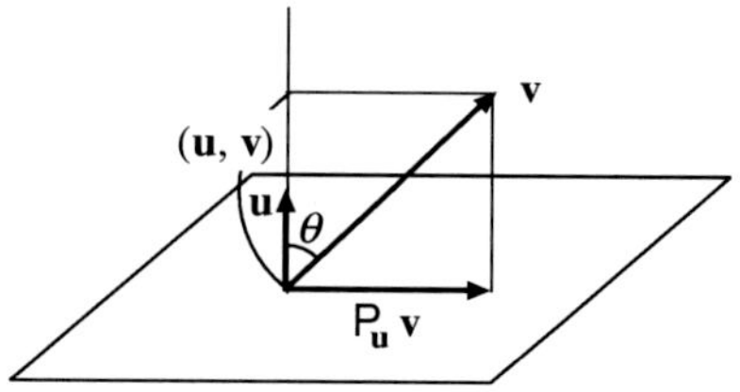

Fig. 3.4 The projection $P_{\mathbf{u}}\mathbf{v}$ of a vector $\mathbf{v}$ onto a plane with unit surface normal $\mathbf{u}$

3.3 (1) Show that the projection of a vector $\mathbf{v}$ onto a plane with unit surface normal $\mathbf{u}$ is given by $P_{\mathbf{u}}\mathbf{v}$ (Fig. 3.4), where $P_{\mathbf{u}}$ is the *projection matrix* given by

$$P_{\mathbf{u}} \equiv I - \mathbf{u}\mathbf{u}^\top. \tag{3.69}$$

(2) The projection matrix $P_{\mathbf{u}}$ of Eq. (3.69) is symmetric by definition. Show that it is also *idempotent*; that is,

$$P_{\mathbf{u}}^2 = P_{\mathbf{u}}, \tag{3.70}$$

which states that once projected, things remain there if projected again, as geometric interpretation implies.

3.4 (1) Let $\bar{\boldsymbol{\xi}}_\alpha$ and $\bar{\boldsymbol{\theta}}$ be the true values (for noiseless data) of $\boldsymbol{\xi}_\alpha$ and $\boldsymbol{\theta}$, respectively, and write $\boldsymbol{\xi}_\alpha = \bar{\boldsymbol{\xi}}_\alpha + \Delta_1\boldsymbol{\xi}_\alpha + \Delta_2\boldsymbol{\xi}_\alpha$ and $\boldsymbol{\theta} = \bar{\boldsymbol{\theta}} + \Delta_1\boldsymbol{\theta} + \Delta_2\boldsymbol{\theta} + \cdots$, where Δ_k denotes terms of order k in σ (the standard deviation of the data noise). Show that if these are substituted into the Sampson error J of Eq. (3.23), we can write it in the form:

$$J = \frac{1}{N}\sum_{\alpha=1}^{N}\frac{(\bar{\boldsymbol{\xi}}_\alpha, \Delta_1\boldsymbol{\theta})^2 + 2(\bar{\boldsymbol{\xi}}_\alpha, \Delta_1\boldsymbol{\theta})(\Delta_1\boldsymbol{\xi}_\alpha, \bar{\boldsymbol{\theta}}) + (\Delta_1\boldsymbol{\xi}_\alpha, \bar{\boldsymbol{\theta}})^2}{(\bar{\boldsymbol{\theta}}, V_0[\boldsymbol{\theta}]\bar{\boldsymbol{\theta}})} + O(\sigma^3). \tag{3.71}$$

(2) Show that the value of $\Delta_1\boldsymbol{\theta}$ that minimizes Eq. (3.71) is given, if terms of $O(\sigma^3)$ are ignored, in the form:

$$\Delta_1\boldsymbol{\theta} = -\bar{M}\Big(\frac{1}{N}\sum_{\alpha=1}^{N}\frac{\bar{\boldsymbol{\xi}}_\alpha\bar{\boldsymbol{\theta}}^\top}{(\bar{\boldsymbol{\theta}}, V_0[\boldsymbol{\theta}]\bar{\boldsymbol{\theta}})}\Delta_1\boldsymbol{\xi}_\alpha\Big), \qquad \bar{M} \equiv \frac{1}{N}\sum_{\alpha=1}^{N}\frac{\bar{\boldsymbol{\xi}}_\alpha\bar{\boldsymbol{\xi}}_\alpha^\top}{(\bar{\boldsymbol{\theta}}, V_0[\boldsymbol{\theta}]\bar{\boldsymbol{\theta}})}. \tag{3.72}$$

(3) Define the covariance matrix $V[\boldsymbol{\theta}]$ of the $\boldsymbol{\theta}$ that minimize the Sampson error J of Eq. (3.23) by $E[\Delta_1\boldsymbol{\theta}\Delta_1\boldsymbol{\theta}^\top]$. If we write $V[\boldsymbol{\theta}] = \sigma^2 V_0[\boldsymbol{\theta}]$, show that the normalized covariance matrix $V_0[\boldsymbol{\theta}]$ is written in the form:

$$V_0[\boldsymbol{\theta}] = \frac{1}{N}\bar{M}^-. \tag{3.73}$$

3.5 (1) Show that the following identity holds for an arbitrary matrix $A = (A_{ij})$ and a small value ε

$$\det[I - \varepsilon A] = 1 - \varepsilon \mathrm{tr}[A] + O(\varepsilon^2), \tag{3.74}$$

where tr denotes the matrix trace. Using this, show that the following identity holds for arbitrary matrices $A = (A_{ij})$ and $B = (B_{ij})$ and a small value ε.

$$\det[A - \varepsilon B] = \det A - \varepsilon \mathrm{tr}[A^\dagger B] + O(\varepsilon^2), \tag{3.75}$$

where $\mathsf{A}^\dagger$ is a cofactor of matrix A.
(2) We want to correct a fundamental matrix F such that $\det \mathsf{F} \approx 0$ by a small quantity $\Delta\mathsf{F}$ such that $\det[\mathsf{F} - \Delta\mathsf{F}] = 0$ holds. Let $\boldsymbol{\theta}$ and $\Delta\boldsymbol{\theta}$ be the 9D vector representations of F and $\Delta\mathsf{F}$, respectively. Show that if we ignore high-order terms, we obtain the relationship:

$$(\boldsymbol{\theta}^\dagger, \Delta\boldsymbol{\theta}) = \frac{1}{3}(\boldsymbol{\theta}^\dagger, \boldsymbol{\theta}). \tag{3.76}$$

(3) Show that the value of $\Delta\boldsymbol{\theta}$ that minimizes $(\Delta\boldsymbol{\theta}, V_0[\boldsymbol{\theta}]^- \Delta\boldsymbol{\theta})$ for the $V_0[\boldsymbol{\theta}]$ of Eq. (3.73) subject to the constraint of Eq. (3.76) is given by

$$\Delta\boldsymbol{\theta} = \frac{(\boldsymbol{\theta}^\dagger, \boldsymbol{\theta}) V_0[\boldsymbol{\theta}]\boldsymbol{\theta}^\dagger}{3(\boldsymbol{\theta}^\dagger, V_0[\boldsymbol{\theta}]\boldsymbol{\theta}^\dagger)}. \tag{3.77}$$

3.6 (1) Let U be an orthogonal matrix, and $\mathsf{U} + \Delta\mathsf{U}$ its infinitesimal variation. Show that there exists an infinitesimal vector $\Delta\boldsymbol{\omega}$ such that the following holds up to high-order infinitesimal terms.

$$\Delta\mathsf{U} = \Delta\boldsymbol{\omega} \times \mathsf{U}. \tag{3.78}$$

Here, the right-hand side denotes the matrix consisting of columns obtained by the vector products of $\Delta\boldsymbol{\omega}$ and the corresponding columns of U (this corresponds to rotating each column of F around $\Delta\boldsymbol{\omega}$ by angle $\|\Delta\boldsymbol{\omega}\|$).
(2) Let $\mathsf{U} + \Delta\mathsf{U}$ be the infinitesimal variation of the orthogonal matrix U in Eq. (3.35) caused by an infinitesimal vector $\Delta\boldsymbol{\omega}$. Similarly, let $\mathsf{V} + \Delta\mathsf{V}$ be the infinitesimal variation of V caused by $\Delta\boldsymbol{\omega}'$, and let $\phi + \Delta\phi$ be the infinitesimal variation of ϕ caused by $\Delta\phi$. Show that if we write F and $\Delta\mathsf{F}$ as 9D vectors $\boldsymbol{\theta}$ and $\Delta\boldsymbol{\theta}$, respectively, we can write $\Delta\boldsymbol{\theta}$ in the form

$$\Delta\boldsymbol{\theta} = \mathsf{F}_U \Delta\boldsymbol{\omega} + \boldsymbol{\theta}_\phi \Delta\phi + \mathsf{F}_V \Delta\boldsymbol{\omega}' + \cdots, \tag{3.79}$$

where F_U and F_V are the matrices defined by Eq. (3.36), $\boldsymbol{\theta}_\phi$ is the vector defined by Eq. (3.37), and $\cdots$ indicates high-order infinitesimal quantities.
(3) Show that the first derivatives of the Sampson error J of Eq. (3.23) with respect to $\boldsymbol{\omega}$, $\boldsymbol{\omega}'$, and ϕ are given in the form of Eq. (3.40).
(4) Show that the second derivatives of the Sampson error J of Eq. (3.23) with respect to $\boldsymbol{\omega}$, $\boldsymbol{\omega}'$, and ϕ are given in the form of Eq. (3.41).

3.7 (1) Show that the value of $\boldsymbol{\theta}$ that minimizes the Sampson error J of Eq. (3.23) subject to the rank constraint of Eq. (3.24) satisfies

$$\mathsf{P}_{\boldsymbol{\theta}^\dagger}\boldsymbol{\theta} = \boldsymbol{\theta}, \qquad \mathsf{Y}\boldsymbol{\theta} = \mathbf{0}, \tag{3.80}$$

where $\mathsf{P}_{\mathbf{u}^\dagger}$ and Y are the matrices defined by Equations (3.50) and (3.51), respectively.
(2) Show that if $\boldsymbol{\theta}$ satisfies the rank constraint, the following identically holds.

$$(\boldsymbol{\theta}, \mathsf{Y}\boldsymbol{\theta}) = 0. \tag{3.81}$$

(3) Show that if the two smallest eigenvalues of Y are both 0 at the time of convergence, the resulting solution $\boldsymbol{\theta}$ satisfies Eq. (3.80).

3.8 (1) Show that the values $(\bar{x}_\alpha, \bar{y}_\alpha)$ and $(\bar{x}'_\alpha, \bar{y}'_\alpha)$ that minimize Eq. (3.60) subject to the epipolar equation of Eq. (3.16) are approximated, if high-order small terms in $\hat{x}_\alpha - \bar{x}_\alpha$, $\hat{y}_\alpha - \bar{y}_\alpha$, $\hat{x}'_\alpha - \bar{x}'_\alpha$, and $\hat{y}'_\alpha - \bar{y}'_\alpha$ are omitted, in the form

$$\begin{pmatrix} \hat{\hat{x}}_\alpha \\ \hat{\hat{y}}_\alpha \end{pmatrix} = \begin{pmatrix} x_\alpha \\ y_\alpha \end{pmatrix} - \frac{(\hat{\mathbf{x}}_\alpha, \mathbf{F}\hat{\mathbf{x}}'_\alpha) + (\mathbf{F}\hat{\mathbf{x}}'_\alpha, \tilde{\mathbf{x}}_\alpha) + (\mathbf{F}^\top \hat{\mathbf{x}}_\alpha, \tilde{\mathbf{x}}'_\alpha)}{(\mathbf{F}\hat{\mathbf{x}}'_\alpha, \mathbf{P}_\mathbf{k}\mathbf{F}\hat{\mathbf{x}}'_\alpha) + (\mathbf{F}^\top \hat{\mathbf{x}}_\alpha, \mathbf{P}_\mathbf{k}\mathbf{F}^\top \hat{\mathbf{x}}_\alpha)} \begin{pmatrix} F_{11} & F_{12} & F_{13} \\ F_{21} & F_{22} & F_{23} \end{pmatrix} \begin{pmatrix} \hat{x}'_\alpha \\ \hat{y}'_\alpha \\ f_0 \end{pmatrix},$$

$$\begin{pmatrix} \hat{\hat{x}}'_\alpha \\ \hat{\hat{y}}'_\alpha \end{pmatrix} = \begin{pmatrix} x'_\alpha \\ y'_\alpha \end{pmatrix} - \frac{(\hat{\mathbf{x}}_\alpha, \mathbf{F}\hat{\mathbf{x}}'_\alpha) + (\mathbf{F}\hat{\mathbf{x}}'_\alpha, \tilde{\mathbf{x}}_\alpha) + (\mathbf{F}^\top \hat{\mathbf{x}}_\alpha, \tilde{\mathbf{x}}'_\alpha)}{(\mathbf{F}\hat{\mathbf{x}}'_\alpha, \mathbf{P}_\mathbf{k}\mathbf{F}\hat{\mathbf{x}}'_\alpha) + (\mathbf{F}^\top \hat{\mathbf{x}}_\alpha, \mathbf{P}_\mathbf{k}\mathbf{F}^\top \hat{\mathbf{x}}_\alpha)} \begin{pmatrix} F_{11} & F_{21} & F_{31} \\ F_{12} & F_{22} & F_{32} \end{pmatrix} \begin{pmatrix} \hat{x}_\alpha \\ \hat{y}_\alpha \\ f_0 \end{pmatrix}, \tag{3.82}$$

where we define

$$\hat{\mathbf{x}}_\alpha = \begin{pmatrix} \hat{x}_\alpha/f_0 \\ \hat{y}_\alpha/f_0 \\ 1 \end{pmatrix}, \quad \hat{\mathbf{x}}'_\alpha = \begin{pmatrix} \hat{x}'_\alpha/f_0 \\ \hat{y}'_\alpha/f_0 \\ 1 \end{pmatrix}, \quad \tilde{\mathbf{x}}_\alpha = \begin{pmatrix} \tilde{x}_\alpha/f_0 \\ \tilde{y}_\alpha/f_0 \\ 0 \end{pmatrix}, \quad \tilde{\mathbf{x}}' = \begin{pmatrix} \tilde{x}'_\alpha/f_0 \\ \tilde{y}'_\alpha/f_0 \\ 0 \end{pmatrix}. \tag{3.83}$$

Here, $\tilde{x}_\alpha$, $\tilde{y}_\alpha$, $\tilde{x}'_\alpha$, and $\tilde{y}'_\alpha$ are defined by Eq. (3.61).
(2) Show that if we define the 9D vector $\boldsymbol{\xi}^*_\alpha$ by Eq. (3.55), Eq. (3.82) is written in the form

$$\begin{pmatrix} \hat{\hat{x}}_\alpha \\ \hat{\hat{y}}_\alpha \end{pmatrix} = \begin{pmatrix} x_\alpha \\ y_\alpha \end{pmatrix} - \frac{(\boldsymbol{\xi}^*_\alpha, \boldsymbol{\theta})}{(\boldsymbol{\theta}, V_0[\hat{\boldsymbol{\xi}}_\alpha]\boldsymbol{\theta})} \begin{pmatrix} \theta_1 & \theta_2 & \theta_3 \\ \theta_4 & \theta_5 & \theta_6 \end{pmatrix} \begin{pmatrix} \hat{x}'_\alpha \\ \hat{y}'_\alpha \\ f_0 \end{pmatrix},$$

$$\begin{pmatrix} \hat{\hat{x}}'_\alpha \\ \hat{\hat{y}}'_\alpha \end{pmatrix} = \begin{pmatrix} x'_\alpha \\ y'_\alpha \end{pmatrix} - \frac{(\boldsymbol{\xi}^*_\alpha, \boldsymbol{\theta})}{(\boldsymbol{\theta}, V_0[\hat{\boldsymbol{\xi}}_\alpha]\boldsymbol{\theta})} \begin{pmatrix} \theta_1 & \theta_4 & \theta_7 \\ \theta_2 & \theta_5 & \theta_8 \end{pmatrix} \begin{pmatrix} \hat{x}_\alpha \\ \hat{y}_\alpha \\ f_0 \end{pmatrix}, \tag{3.84}$$

where $V_0[\hat{\boldsymbol{\xi}}_\alpha]$ is the normalized covariance matrix obtained by replacing $(\bar{x}, \bar{y})$ and $(\bar{x}', \bar{y}')$ in Eq. (3.12) by $(\hat{x}_\alpha, \hat{y}_\alpha)$ and $(\hat{x}'_\alpha, \hat{y}'_\alpha)$, respectively.
(3) Show that the geometric distance S can be written in the form of Eq. (3.56) if $(\bar{x}_\alpha, \bar{y}_\alpha)$ and $(\bar{x}'_\alpha, \bar{y}'_\alpha)$ in Eq. (3.15) are replaced by $(\hat{\hat{x}}_\alpha, \hat{\hat{y}}_\alpha)$ and $(\hat{\hat{x}}'_\alpha, \hat{\hat{y}}'_\alpha)$, respectively.

References

1. W. Chojnacki, M.J. Brooks, A. van den Hengel, D. Gawley, On the fitting of surfaces to data with covariances. IEEE Trans. Pattern Anal. Mach. Intell. **22**(11), 1294–1303 (2000)
2. R. Hartley, In defense of the eight-point algorithm. IEEE Trans. Pattern Anal. Mach. Intell. **19**(6), 580–593 (1997)
3. R. Hartley, A. Zisserman, *Multiple View Geometry in Computer Vision*, 2nd edn. (Cambridge University Press, Cambridge, U.K., 2003)
4. K. Kanatani, *Geometric Computation for Machine Vision* (Oxford University Press, Oxford, U.K., 1993)

5. K. Kanatani, Renormalization for unbiased estimation, in *Proceedings of 4th International Conference on Computer Vision*, Berlin, Germany, pp. 599–606 (1993)
6. K. Kanatani, *Statistical Optimization for Geometric Computation: Theory and Practice, Elsevier, Amsterdam, The Netherlands (1996)* (Reprinted by Dover, New York, U.S., 2005)
7. K. Kanatani, Statistical optimization for geometric fitting: theoretical accuracy bound and high order error analysis. Int. J. Comput. Vision **80**(2), 167–188 (2008)
8. K. Kanatani, C. Matsunaga, Computing internally constrained motion of 3-D sensor data for motion interpretation. Pattern Recogn. **46**(6), 1700–1709 (2013)
9. K. Kanatani, N. Ohta, Comparing optimal three-dimensional reconstruction for finite motion and optical flow. J. Electron. Imaging **12**(3), 478–488 (2003)
10. Y. Sugaya, K. Kanatani, High accuracy computation of rank-constrained fundamental matrix, in *Proceedings of 18th British Machine Vision Conference*, Coventry, U.K., vol. 1, pp. 282–291 (2007)
11. K. Kanatani, Y. Sugaya, High accuracy fundamental matrix computation and its performance evaluation. IEICE Trans. Inf. Syst. **E90-D**(2), 579–585 (2007)
12. K. Kanatani, Y. Sugaya, Performance evaluation of iterative geometric fitting algorithms. Comput. Stat. Data Anal. **52**(2), 1208–1222 (2007)
13. K. Kanatani, Y. Sugaya, Compact fundamental matrix computation. IPSJ Trans. Comput. Vision Appl. **2**, 59–70 (2010)
14. K. Kanatani, Y. Sugaya, Unified computation of strict maximum likelihood for geometric fitting. J. Math. Imaging Vision **38**(1), 1–13 (2010)
15. K. Kanatani, Y. Sugaya, Y. Kanazawa, Latest algorithms for 3-D reconstruction form two view, in *Handbook of Pattern Recognition and Computer Vision*, 4th edn., ed. by C.H. Chen (World Scientific Publishing, Singapore, 2009), pp. 201–234
16. Y. Sugaya, K. Kanatani, Highest accuracy fundamental matrix computation, in *Proceedings of 8th Asian Conference on Computer Vision*, Tokyo, Japan, vol. 2, pp. 311–321 (2007)

Triangulation

4

Abstract

This chapter describes the principles and computational procedures for triangulation that compute the 3D position of the point determined by a given pair of corresponding points over two images, using the knowledge of the positions, orientations, and internal parameters of the two cameras, which are specified by their camera matrices. First, we illustrate the geometry of perspective projection and describe the procedure for optimally correcting the corresponding point pair so that the associated lines of sight intersect in the scene, considering the statistical properties of image noise. This turns out to be closely related to the optimal fundamental matrix computation described in the preceding chapter.

4.1 Perspective Projection

Given two images of the same scene, we can compute the 3D position of the point determined by a corresponding pair of points if the positions, orientations, and internal parameters of the two cameras that took the images are known. This is because we can determine from a point in the image the corresponding 3D direction of the line of sight, or the ray, by using the knowledge of the camera. Hence, we determine from a corresponding point pair the triangle made by two lines of sight and the line connecting the lens centers of the two cameras, called the *baseline*. This is the well-known principle of land measurement by triangulation, and the same term is used for 3D computation from two images, which is also known as *stereo vision*.

For this analysis, we need a mathematical description of camera imaging geometry. Let us fix an XYZ coordinate system in the scene, which we call the *world coordinate system*. We also consider another $X_cY_cZ_c$ coordinate system attached to the camera, which we call the *camera coordinate system*; it is defined in such a way that its origin O_c is at the lens center and the Z_c axis is the *optical axis*, that is, the

K. Kanatani et al., *Guide to 3D Vision Computation*, Advances in Computer Vision and Pattern Recognition, DOI 10.1007/978-3-319-48493-8_4

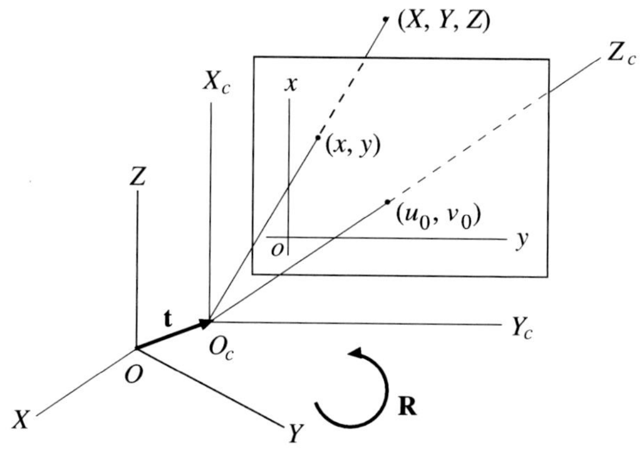

Fig. 4.1 Perspective projection modeling of camera imaging

symmetry axis of the lens. We fix a plane to the camera, called the *image plane*, on which we define an xy coordinate system; we identify with the actual image that we see. Suppose a point (X, Y, Z) in the scene is imaged by the camera to the intersection (x, y) of the image plane with the ray passing through the lens center O and (X, Y, Z) (Fig. 4.1). This imaging model is called *perspective projection*. The lens center O is often called the *viewpoint*. The intersection (u_0, v_0) of the image plane with the optical axis (i.e., the camera Z_c axis) is called the *principal point*. Most commercially available cameras today can be modeled by this perspective projection fairly accurately (more details of the relationship between the image plane and the camera coordinate system are discussed in Sect. 5.1).

Let $\mathbf{t}$ be the position of the camera viewpoint O_c with respect to the world coordinate system, and let R be the relative rotation of the camera $X_cY_cZ_c$ frame with respect to the world coordinate system. The position and orientation of the camera are specified with respect to the world coordinate system by the vector $\mathbf{t}$ and the matrix R; for brevity, we call $\mathbf{t}$ and R, respectively, the *translation* and the *rotation* of the camera, and $\{\mathbf{t}, \mathsf{R}\}$ the *motion parameters* of the camera.

4.2 Camera Matrix and Triangulation

If a point (X, Y, Z) is projected to a point (x, y) on the image by perspective projection, it can be shown they are related by the following fractional equations in the following form (we discuss this further in Chap. 5).

$$x = f_0 \frac{P_{11}X + P_{12}Y + P_{13}Z + P_{14}}{P_{31}X + P_{32}Y + P_{33}Z + P_{34}}, \quad y = f_0 \frac{P_{21}X + P_{22}Y + P_{23}Z + P_{24}}{P_{31}X + P_{32}Y + P_{33}Z + P_{34}}. \tag{4.1}$$

Here, f_0 is a fixed scale constant; it has the same meaning as in Chaps. 2 and 3. The coefficients P_{ij}, $i = 1, 2, 3$, $j = 1, 2, 3, 4$, are determined from the intrinsic camera parameters, such as the focal length, and the motion parameters $\mathbf{t}$, R. We can rewrite Eq. (4.1) in the form

$$\begin{pmatrix} x/f_0 \\ y/f_0 \\ 1 \end{pmatrix} \simeq \begin{pmatrix} P_{11} & P_{12} & P_{13} & P_{14} \\ P_{21} & P_{22} & P_{23} & P_{24} \\ P_{31} & P_{32} & P_{33} & P_{34} \end{pmatrix} \begin{pmatrix} X \\ Y \\ Z \\ 1 \end{pmatrix}, \tag{4.2}$$

where the symbol $\simeq$ means that the left side is a nonzero constant multiple of the right side ($\hookrightarrow$ Problem 4.1). The 3×4 matrix $\mathsf{P} = (P_{ij})$ is called the *camera matrix*. To accomplish triangulation, we need to determine the camera matrix in advance; this process is called *camera calibration*. In this chapter, we assume that the cameras are already calibrated.

Let $\mathsf{P} = (P_{ij})$ and $\mathsf{P}' = (P'_{ij})$ be the camera matrices of two cameras, which, for convenience, we call the "first" and the "second" cameras. Suppose a point (X, Y, Z) in the scene is projected to (x, y) and (x', y') on the images of the first and second camera, respectively. If there is no noise in the observation, we can compute (X, Y, Z) from (x, y) and (x', y') as follows.

Procedure 4.1 (Triangulation with known camera matrices)

1. Compute the following 4×3 matrix T and 4D vector $\mathbf{p}$:

$$\mathsf{T} = \begin{pmatrix} f_0 P_{11} - x P_{31} & f_0 P_{12} - x P_{32} & f_0 P_{13} - x P_{33} \\ f_0 P_{21} - y P_{31} & f_0 P_{22} - y P_{32} & f_0 P_{23} - y P_{33} \\ f_0 P'_{11} - x' P'_{31} & f_0 P'_{12} - x' P'_{32} & f_0 P'_{13} - x' P'_{33} \\ f_0 P'_{21} - y' P'_{31} & f_0 P'_{22} - y' P'_{32} & f_0 P'_{23} - y' P'_{33} \end{pmatrix}, \quad \mathbf{p} = \begin{pmatrix} f_0 P_{14} - x P_{34} \\ f_0 P_{24} - y P_{34} \\ f_0 P'_{14} - x' P'_{34} \\ f_0 P'_{24} - y' P'_{34} \end{pmatrix}. \tag{4.3}$$

2. Determine X, Y, and Z by solving the linear equation:

$$\mathsf{T}^\top \mathsf{T} \begin{pmatrix} X \\ Y \\ Z \end{pmatrix} = -\mathsf{T}^\top \mathbf{p}. \tag{4.4}$$

Comments Canceling the denominators in Eq. (4.1) and doing the same for the second camera as well, we obtain the following simultaneous linear equations in X, Y, and Z.

$$\begin{aligned} (f_0 P_{11} - x P_{31})X + (f_0 P_{12} - x P_{32})Y + (f_0 P_{13} - x P_{33})Z + f_0 P_{14} - x P_{34} &= 0, \\ (f_0 P_{21} - y P_{31})X + (f_0 P_{22} - y P_{32})Y + (f_0 P_{23} - y P_{33})Z + f_0 P_{24} - y P_{34} &= 0, \\ (f_0 P'_{11} - x' P'_{31})X + (f_0 P'_{12} - x' P'_{32})Y + (f_0 P'_{13} - x' P'_{33})Z + f_0 P'_{14} - x' P'_{34} &= 0, \\ (f_0 P'_{21} - y' P'_{31})X + (f_0 P'_{22} - y' P'_{32})Y + (f_0 P'_{23} - y' P'_{33})Z + f_0 P'_{24} - y' P'_{34} &= 0. \end{aligned} \tag{4.5}$$

These are four equations in three unknowns X, Y, and Z, but because there exist (X, Y, Z) that satisfy all, the four equations are linearly dependent; only three are linearly independent. Hence, we can arbitrarily select three of these and solve them;

the remaining equation is automatically satisfied. However, the same solution is obtained using all of them. If we write Eq. (4.5) in the form

$$\mathsf{T}\begin{pmatrix} X \\ Y \\ Z \end{pmatrix} = -\mathbf{p}, \tag{4.6}$$

using the matrix T and the vector $\mathbf{p}$ in Eq. (4.3), and multiply Eq. (4.6) by $\mathsf{T}^\top$ from the left on both sides, we obtain Eq. (4.4), which gives three equations for three unknowns; note that $\mathsf{T}^\top\mathsf{T}$ is a 3×3 matrix, and $\mathsf{T}^\top\mathbf{p}$ is a 3D vector. Solving Eq. (4.6) is nothing but solving Eq. (4.6) by least squares (↪ Problem 4.2). Inasmuch as Eq. (4.6) has a solution that satisfies all the equations, the same solution is obtained whether by solving three selected equations or by using least squares.

4.3 Triangulation from Noisy Correspondence

If the observed points (x, y) and (x', y') are noisy (i.e., not exact due to uncertainty of correspondence detection by image processing), their rays need not intersect in the scene. Traditionally, the desired intersection was approximated by the midpoint of the shortest segment connecting the two rays (Fig. 4.2). For the first camera, the ray is the line passing through the viewpoint O_c and the point (x, y) on the image (Fig. 4.1). This line is determined by solving the first two equations of Eq. (4.5). Because only two equations are available for three unknowns X, Y, and Z, the solution is determined up to one free parameter, say Z, defining a line in the scene. Similarly, the ray for the second camera is obtained by solving the last two equations of Eq. (4.5) up to one free parameter.

However, considering a 3D point "close" to the two rays in the scene does not have not much meaning. The noise occurs in the image, therefore the closeness should be measured on the image plane. The sensible strategy is to modify the corresponding points (x, y) and (x', y') in the shortest distance so that their rays intersect in the scene (Fig. 4.3) and compute the 3D position (X, Y, Z) by Procedure 4.1. The condition for

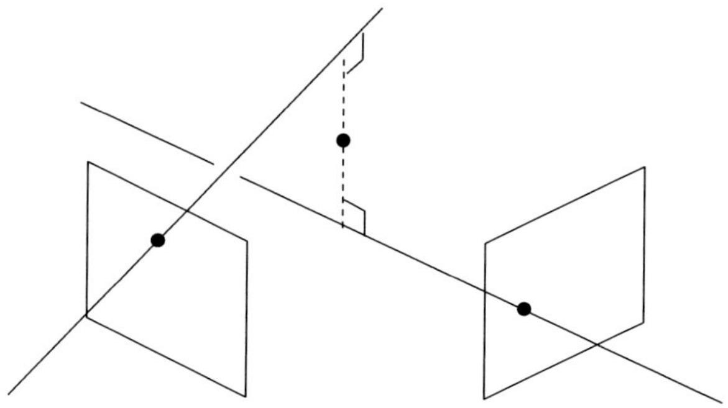

Fig. 4.2 If the two rays do not intersect due to noise, we compute the midpoint of the shortest segment connecting them

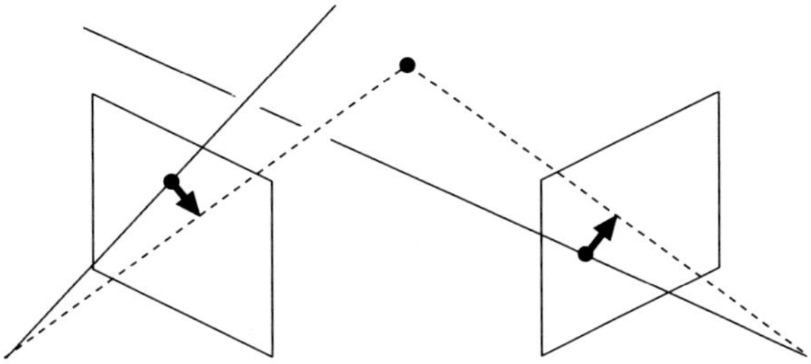

Fig. 4.3 The observed corresponding points are modified in the shortest distance so that their rays intersect

the rays to intersect is that Eq. (4.5) has a unique solution: namely, only three of the four equations are independent, and the resulting solution automatically satisfies the remaining equation. The condition that the four equations of Eq. (4.5) are linearly dependent is written in the form (↪ Problem 4.3):

$$\left(\begin{pmatrix} x/f_0 \\ y/f_0 \\ 1 \end{pmatrix}, \mathsf{F} \begin{pmatrix} x'/f_0 \\ y'/f_0 \\ 1 \end{pmatrix}\right) = 0. \tag{4.7}$$

Here, F is a 3×3 matrix determined from the camera matrices P and P'. Equation (4.7) is nothing but the epipolar equation of Eq. (3.1), and hence F is the fundamental matrix. Thus, the epipolar equation is the necessary and sufficient condition that the two rays intersect, and the fundamental matrix F is determined by the camera matrices P and P'. In this chapter, the cameras are assumed to be calibrated and the camera matrices P and P' are known. Therefore the fundamental matrix F is also known.

Thus the task of triangulation is to modify the observed corresponding points (x, y) and (x', y') in the shortest distance so that the epipolar equation of Eq. (4.7) is satisfied. By saying "in the shortest distance," we mean that we minimize the sum of squares of the displacements

$$S = (x - \bar{x})^2 + (y - \bar{y})^2 + (x' - \bar{x}')^2 + (y' - \bar{y}')^2. \tag{4.8}$$

4.4 Optimal Correction of Correspondences

We represent the known fundamental matrix F as a 3D vector $\boldsymbol{\theta}$ as in Eq. (3.3). The epipolar equation of Eq. (4.7) can be written as $(\boldsymbol{\xi}, \boldsymbol{\theta}) = 0$. The values $(\hat{x}, \hat{y})$ and $(\hat{x}', \hat{y}')$ that minimize Eq. (4.8) subject to Eq. (4.7) are computed by the following procedure.

Procedure 4.2 (Optimal correspondence correction)

1. Let $S_0 = \infty$ (a sufficiently large number), $\hat{x} = x$, $\hat{y} = y$, $\hat{x}' = x'$, $\hat{y}' = y'$, and $\tilde{x} = \tilde{y} = \tilde{x}' = \tilde{y}' = 0$.

2. Compute the normalized covariance matrix $V_0[\hat{\boldsymbol{\xi}}]$ obtained by replacing $\bar{x}$, $\bar{y}$, $\bar{x}'$, and $\bar{y}'$ of $V_0[\boldsymbol{\xi}]$ in Eq. (3.12) by $\hat{x}$, $\hat{y}$, $\hat{x}'$, and $\hat{y}'$, respectively.
3. Compute the following 9D vector $\boldsymbol{\xi}^*$.

$$\boldsymbol{\xi}^* = \begin{pmatrix} \hat{x}\hat{x}' + \hat{x}'\tilde{x} + \hat{x}'\tilde{x}' \\ \hat{x}\hat{y}' + \hat{y}'\tilde{x} + \hat{x}'\tilde{y}' \\ f_0(\hat{x} + \tilde{x}) \\ \hat{y}\hat{x}' + \hat{x}'\tilde{y} + \hat{y}\tilde{x}' \\ \hat{y}\hat{y}' + \hat{y}'\tilde{y} + \hat{y}\tilde{y}' \\ f_0(\hat{y} + \tilde{y}) \\ f_0(\hat{x}' + \tilde{x}') \\ f_0(\hat{y}' + \tilde{y}') \\ f_0^2 \end{pmatrix}. \tag{4.9}$$

4. Update $\tilde{x}$, $\tilde{y}$, $\tilde{x}'$, and $\tilde{y}'$ to

$$\begin{pmatrix} \tilde{x} \\ \tilde{y} \end{pmatrix} \leftarrow \frac{(\boldsymbol{\xi}^*, \boldsymbol{\theta})}{(\boldsymbol{\theta}, V_0[\hat{\boldsymbol{\xi}}]\boldsymbol{\theta})} \begin{pmatrix} \theta_1 & \theta_2 & \theta_3 \\ \theta_4 & \theta_5 & \theta_6 \end{pmatrix} \begin{pmatrix} \hat{x}' \\ \hat{y}' \\ f_0 \end{pmatrix},$$

$$\begin{pmatrix} \tilde{x}' \\ \tilde{y}' \end{pmatrix} \leftarrow \frac{(\boldsymbol{\xi}^*, \boldsymbol{\theta})}{(\boldsymbol{\theta}, V_0[\hat{\boldsymbol{\xi}}]\boldsymbol{\theta})} \begin{pmatrix} \theta_1 & \theta_4 & \theta_7 \\ \theta_2 & \theta_5 & \theta_8 \end{pmatrix} \begin{pmatrix} \hat{x} \\ \hat{y} \\ f_0 \end{pmatrix}. \tag{4.10}$$

5. Update $\hat{x}$, $\hat{y}$, $\hat{x}'$, and $\hat{y}'$ to

$$\hat{x} \leftarrow x - \tilde{x}, \quad \hat{y} \leftarrow y - \tilde{y}, \quad \hat{x}' \leftarrow x' - \tilde{x}', \quad \hat{y}' \leftarrow y' - \tilde{y}'. \tag{4.11}$$

6. Compute the following S.

$$S = (\tilde{x}^2 + \tilde{y}^2 + \tilde{x}'^2 + \tilde{y}'^2). \tag{4.12}$$

If $S \approx S_0$, return $(\hat{x}, \hat{y})$ and $(\hat{x}', \hat{y}')$ and stop. Else, let $S_0 \leftarrow S$, and go back to Step 2.

Comments This procedure is, *except for Step 4, identical* to the Procedure 3.7 for geometric distance minimization for fundamental matrix computation. Thus, we can view Procedure 3.7 as *triangulating and updating* ϑ, which we assume is known in this chapter. The derivation of the above Procedure 4.2 is basically identical to that of Procedure 3.7. After the initial corrections $(\hat{x}, \hat{y})$ and $(\hat{x}', \hat{y}')$ are obtained, we rewrite Eq. (4.8) in the form

$$\begin{aligned} S &= (\hat{x} + (x - \hat{x}) - \bar{x})^2 + (\hat{y} + (y - \hat{y}) - \bar{y})^2 \\ &\quad + (\hat{x}' + (x' - \hat{x}') - \bar{x}')^2 + (\hat{y}' + (y' - \hat{y}') - \bar{y}')^2 \\ &= (\hat{x} + \tilde{x} - \bar{x})^2 + (\hat{y} + \tilde{y} - \bar{y})^2 + (\hat{x}' + \tilde{x}' - \bar{x}')^2 + (\hat{y}' + \tilde{y}' - \bar{y}')^2, \end{aligned} \tag{4.13}$$

where

$$\tilde{x} = x - \hat{x}, \quad \tilde{y} = y - \hat{y}, \quad \tilde{x}' = x' - \hat{x}', \quad \tilde{y}' = y' - \hat{y}', \tag{4.14}$$

which describe the displacement for correction. Then, we regard the corrected positions $(\hat{x}, \hat{y})$ and $(\hat{x}', \hat{y}')$ as the input positions and compute the values $(\hat{\hat{x}}, \hat{\hat{y}})$ and

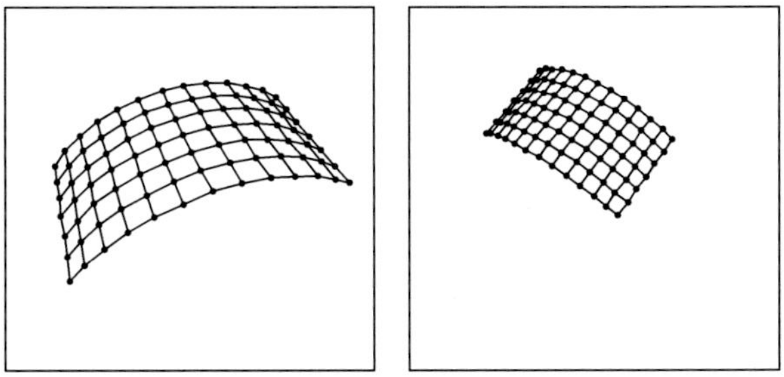

Fig. 4.4 The grid surface images obtained by optimally correcting the grid points in Fig. 3.2 by Procedure 4.2

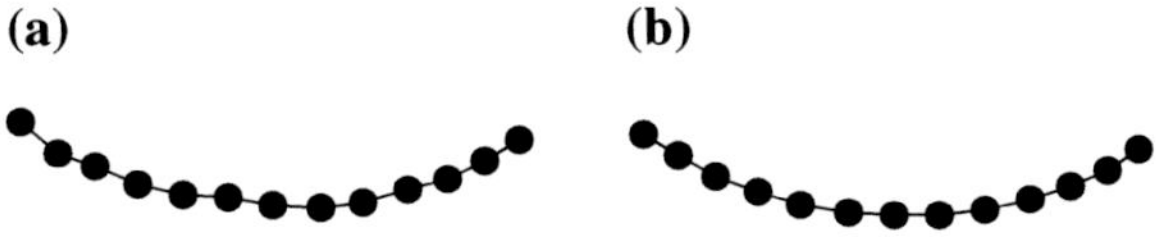

Fig. 4.5 **a** The 3D grid point positions reconstructed from Fig. 4.4 by triangulation, viewed from the direction of the cylindrical surface axis. **b** The true 3D grid point positions viewed from the same direction

$(\hat{\hat{x}}', \hat{\hat{y}}')$ of $(\bar{x}, \bar{y})$ and $(\bar{x}', \bar{y}')$ that minimize S, using Lagrange multipliers and ignoring high-order terms in $\hat{x} - \bar{x}$, $\hat{y} - \bar{y}$, $\hat{x}' - \bar{x}'$, and $\hat{y} - \bar{y}'$ (↪ Problem 4.4(1), (2)). Because $\hat{x} - \bar{x}$, $\hat{y} - \bar{y}$, $\hat{x}' - \bar{x}'$, and $\hat{y} - \bar{y}'$ are higher-order small quantities than $x - \bar{x}$, $y - \bar{y}$, $x' - \bar{x}'$, and $y - \bar{y}'$, the resulting values $(\hat{\hat{x}}, \hat{\hat{y}})$ and $(\hat{\hat{x}}', \hat{\hat{y}}')$ are better approximations of $(\bar{x}, \bar{y})$ and $(\bar{x}', \bar{y}')$ than $(\hat{x}, \hat{y})$ and $(\hat{x}', \hat{y}')$. We identify $(\hat{\hat{x}}, \hat{\hat{y}})$ and $(\hat{\hat{x}}', \hat{\hat{y}}')$ as $(\hat{x}, \hat{y})$ and $(\hat{x}', \hat{y}')$, respectively, and write Eq. (4.8) in the form of Eq. (4.13) (↪ Problem 4.4(3), (4)). By repeating this process, we obtain a better approximation iteration by iteration. We stop if the square sum of the displacements for correction no longer changes. In the end, the value S of Eq. (4.12) coincides with the value S of Eq. (4.8). The amount of correction rapidly decreases after each iteration, and usually one or two iterations are sufficient in practical applications.

4.5 Examples

Figure 4.4 shows the result of the optimal correction of Procedure 4.2 applied to the noisy grid surface images in Fig. 3.2. Figure 4.5a shows the 3D grid positions reconstructed by triangulation, using Procedure 4.1, viewed from the direction of the cylindrical surface axis. For comparison, Fig. 4.5b shows the true 3D grid point positions viewed from the same direction. As we see, the reconstructed shape is somewhat affected by the noise in the images.

4.6 Supplemental Note

The term "triangulation" essentially has the same meaning as "stereo vision," but in order to compute 3D positions from two images, we first need to extract corresponding points between the two images and calibrate the two cameras. The term "stereo vision" puts emphasis on constructing an image processing system for this task, whereas "triangulation" emphasizes actual computational procedures for computing 3D positions and 3D shapes. Detailed discussions on the perspective projection modeling of Eqs. (4.1) and (4.2) are found in many textbooks, such as [1].

The optimal correction of Procedure 4.2 was presented by Kanatani et al. [2], which is an iterative refinement of their earlier version [3]. A direct algebraic method was presented by Hartley and Sturm [4]. Their method reduces the computation to six-degree polynomial solving. It attracted much attention because of their emphasis that it theoretically computes a globally optimal solution. Certainly, gradient-based search "in 3D" may fall into a local minimum if started from a point far from the true value. However, Procedure 4.2 is an iterative correction scheme "in the 2D image domain," starting from the *observed image positions* of the corresponding pair and converging in their vicinity; usually two or three iterations are sufficient. The method of Hartley and Sturm [4] and Procedure 4.2 compute the same solution except when one of the corresponding points is at the epipole (the image of the viewpoint of the other camera), in which case the Hartley and Sturm [4] method fails because the epipole is a singularity in their formulation (no such problem occurs in Procedure 4.2). However, the computational efficiency greatly differs: Procedure 4.2 is far more efficient than the Hartley and Sturm method, which spends a great deal of time on six-degree polynomial solving; an experimental computation time comparison is found in [2]. Thus, there is no reason to use the Hartley and Sturm [4] method in real situations. The fact that optimal triangulation results from optimal fundamental matrix computation by geometric distance minimization if the update step of the fundamental matrix is removed is pointed out by Kanatani and Sugaya [5].

Problems

4.1 Show that Eq. (4.2) describes the same relationship as Eq. (4.1).

4.2 (1) Show that the nD vector $\mathbf{x}$ that minimizes $\|\mathsf{A}\mathbf{x} - \mathbf{b}\|^2$ for an $m \times n$ $(m > n)$ matrix A is obtained by solving the following *normal equation*.

$$\mathsf{A}^\top \mathsf{A}\mathbf{x} = \mathsf{A}^\top \mathbf{b}. \tag{4.15}$$

(2) We define the *pseudoinverse* of an $n \times m$ $(m > n)$ A by $\mathsf{A}^- = (\mathsf{A}^\top \mathsf{A})^{-1}\mathsf{A}^\top$ when $\mathsf{A}^\top \mathsf{A}$ is nonsingular. Show that the solution of the normal equation of Eq. (4.15) is written:

$$\mathbf{x} = \mathsf{A}^- \mathbf{b}. \tag{4.16}$$

4.3 (1) Show that the necessary and sufficient condition that the four equations of Eq. (4.5) are linearly dependent is given by

$$\begin{vmatrix} P_{11} & P_{12} & P_{13} & P_{14} & x/f_0 & 0 \\ P_{21} & P_{22} & P_{23} & P_{24} & y/f_0 & 0 \\ P_{31} & P_{32} & P_{33} & P_{34} & 1 & 0 \\ P'_{11} & P'_{12} & P'_{13} & P'_{14} & 0 & x'/f_0 \\ P'_{21} & P'_{22} & P'_{23} & P'_{24} & 0 & y'/f_0 \\ P'_{31} & P'_{32} & P'_{33} & P'_{34} & 0 & 1 \end{vmatrix} = 0. \tag{4.17}$$

(2) Show that Eq. (4.17) can be rewritten in the form of Eq. (4.7) for some matrix F. Also, express each element of F in terms of P_{ij} and P'_{ij}, $i = 1, 2, 3$, $j = 1, 2, 3, 4$.

4.4 (1) Let $\hat{x}$, $\hat{y}$, $\hat{x}'$, and $\hat{y}'$ be the values of $\bar{x}$, $\bar{y}$, $\bar{x}'$, and $\bar{y}'$, respectively, that minimize Eq. (4.8) subject to the constraint of Eq. (4.7). Show that if we ignore high-order small terms in $x - \bar{x}$, $y - \bar{y}$, $x' - \bar{x}'$, and $y - \bar{y}'$, we obtain the expression

$$\begin{pmatrix} \hat{x} \\ \hat{y} \end{pmatrix} = \begin{pmatrix} x \\ y \end{pmatrix} - \frac{(\mathbf{x}, \mathsf{F}\mathbf{x}')}{(\mathsf{F}\mathbf{x}', \mathsf{P}_\mathbf{k}\mathsf{F}\mathbf{x}') + (\mathsf{F}^\top\mathbf{x}, \mathsf{P}_\mathbf{k}\mathsf{F}^\top\mathbf{x})} \begin{pmatrix} F_{11} & F_{12} & F_{13} \\ F_{21} & F_{22} & F_{23} \end{pmatrix} \begin{pmatrix} x' \\ y' \\ f_0 \end{pmatrix},$$

$$\begin{pmatrix} \hat{x}' \\ \hat{y}' \end{pmatrix} = \begin{pmatrix} x' \\ y' \end{pmatrix} - \frac{(\mathbf{x}, \mathsf{F}\mathbf{x}')}{(\mathsf{F}\mathbf{x}', \mathsf{P}_\mathbf{k}\mathsf{F}\mathbf{x}') + (\mathsf{F}^\top\mathbf{x}, \mathsf{P}_\mathbf{k}\mathsf{F}^\top\mathbf{x})} \begin{pmatrix} F_{11} & F_{21} & F_{31} \\ F_{12} & F_{22} & F_{32} \end{pmatrix} \begin{pmatrix} x \\ y \\ f_0 \end{pmatrix}, \tag{4.18}$$

where

$$\mathbf{x} = \begin{pmatrix} x/f_0 \\ y/f_0 \\ 1 \end{pmatrix}, \quad \mathbf{x}' = \begin{pmatrix} x'/f_0 \\ y'/f_0 \\ 1 \end{pmatrix}, \quad \mathsf{P}_\mathbf{k} = \begin{pmatrix} 1 & 0 & 0 \\ 0 & 1 & 0 \\ 0 & 0 & 0 \end{pmatrix}. \tag{4.19}$$

(2) Write F as a 9D vector $\boldsymbol{\theta}$ as in Eq. (3.3), and define the 9D vector $\boldsymbol{\xi}$ as in Eq. (3.3). Show that Eq. (4.18) is written in the form

$$\begin{pmatrix} \hat{x} \\ \hat{y} \end{pmatrix} = \begin{pmatrix} x \\ y \end{pmatrix} - \frac{(\boldsymbol{\xi}, \boldsymbol{\theta})}{(\boldsymbol{\theta}, V_0[\boldsymbol{\xi}]\boldsymbol{\theta})} \begin{pmatrix} \theta_1 & \theta_2 & \theta_3 \\ \theta_4 & \theta_5 & \theta_6 \end{pmatrix} \begin{pmatrix} x' \\ y' \\ f_0 \end{pmatrix},$$

$$\begin{pmatrix} \hat{x}' \\ \hat{y}' \end{pmatrix} = \begin{pmatrix} x' \\ y' \end{pmatrix} - \frac{(\boldsymbol{\xi}, \boldsymbol{\theta})}{(\boldsymbol{\theta}, V_0[\boldsymbol{\xi}]\boldsymbol{\theta})} \begin{pmatrix} \theta_1 & \theta_4 & \theta_7 \\ \theta_2 & \theta_5 & \theta_8 \end{pmatrix} \begin{pmatrix} x \\ y \\ f_0 \end{pmatrix}, \tag{4.20}$$

where $V_0[\boldsymbol{\xi}]$ is the normalized covariance matrix obtained from Eq. (3.12) by replacing $\bar{x}$, $\bar{y}$, $\bar{x}'$, and $\bar{y}'$ by x, y, x', and y', respectively.

(3) Let $\hat{\hat{x}}$, $\hat{\hat{y}}$, $\hat{\hat{x}}'$, and $\hat{\hat{y}}'$ be the values of $\bar{x}$, $\bar{y}$, $\bar{x}'$, and $\bar{y}'$, respectively, that minimize Eq. (4.13), which is obtained from Eq. (4.8), subject to the constraint of Eq. (4.7). Show that if we ignore high-order small terms in $\hat{x} - \bar{x}$, $\hat{y} - \bar{y}$, $\hat{x}' - \bar{x}'$, and $\hat{y} - \bar{y}'$,

we obtain the expression

$$\begin{pmatrix} \hat{\hat{x}} \\ \hat{\hat{y}} \end{pmatrix} = \begin{pmatrix} x \\ y \end{pmatrix} - \frac{(\hat{\mathbf{x}}, \mathsf{F}\hat{\mathbf{x}}') + (\mathsf{F}\hat{\mathbf{x}}', \tilde{\mathbf{x}}) + (\mathsf{F}^\top\hat{\mathbf{x}}, \tilde{\mathbf{x}}')}{(\mathsf{F}\hat{\mathbf{x}}', \mathsf{P}_\mathbf{k}\mathsf{F}\hat{\mathbf{x}}') + (\mathsf{F}^\top\hat{\mathbf{x}}, \mathsf{P}_\mathbf{k}\mathsf{F}^\top\hat{\mathbf{x}})} \begin{pmatrix} F_{11} & F_{12} & F_{13} \\ F_{21} & F_{22} & F_{23} \end{pmatrix} \begin{pmatrix} \hat{x}' \\ \hat{y}' \\ f_0 \end{pmatrix},$$

$$\begin{pmatrix} \hat{\hat{x}}' \\ \hat{\hat{y}}' \end{pmatrix} = \begin{pmatrix} x' \\ y' \end{pmatrix} - \frac{(\hat{\mathbf{x}}, \mathsf{F}\hat{\mathbf{x}}') + (\mathsf{F}\hat{\mathbf{x}}', \tilde{\mathbf{x}}) + (\mathsf{F}^\top\hat{\mathbf{x}}, \tilde{\mathbf{x}}')}{(\mathsf{F}\hat{\mathbf{x}}', \mathsf{P}_\mathbf{k}\mathsf{F}\hat{\mathbf{x}}') + (\mathsf{F}^\top\hat{\mathbf{x}}, \mathsf{P}_\mathbf{k}\mathsf{F}^\top\hat{\mathbf{x}})} \begin{pmatrix} F_{11} & F_{21} & F_{31} \\ F_{12} & F_{22} & F_{32} \end{pmatrix} \begin{pmatrix} \hat{x} \\ \hat{y} \\ f_0 \end{pmatrix}, \tag{4.21}$$

where

$$\hat{\mathbf{x}} = \begin{pmatrix} \hat{x}/f_0 \\ \hat{y}/f_0 \\ 1 \end{pmatrix}, \quad \hat{\mathbf{x}}' = \begin{pmatrix} \hat{x}'/f_0 \\ \hat{y}'/f_0 \\ 1 \end{pmatrix}, \quad \tilde{\mathbf{x}} = \begin{pmatrix} \tilde{x}/f_0 \\ \tilde{y}/f_0 \\ 0 \end{pmatrix}, \quad \tilde{\mathbf{x}}' = \begin{pmatrix} \tilde{x}'/f_0 \\ \tilde{y}'/f_0 \\ 0 \end{pmatrix}. \tag{4.22}$$

Note that $\tilde{x}$, $\tilde{y}$, $\tilde{x}'$, and $\tilde{y}'$ are defined by Eq. (4.14).

(4) Show that if the 9D vector $\boldsymbol{\xi}^*$ is defined by Eq. (4.9), Eq. (4.21) can be written in the form

$$\begin{pmatrix} \hat{\hat{x}} \\ \hat{\hat{y}} \end{pmatrix} = \begin{pmatrix} x \\ y \end{pmatrix} - \frac{(\boldsymbol{\xi}^*, \boldsymbol{\theta})}{(\boldsymbol{\theta}, V_0[\hat{\boldsymbol{\xi}}]\boldsymbol{\theta})} \begin{pmatrix} \theta_1 & \theta_2 & \theta_3 \\ \theta_4 & \theta_5 & \theta_6 \end{pmatrix} \begin{pmatrix} \hat{x}' \\ \hat{y}' \\ f_0 \end{pmatrix},$$

$$\begin{pmatrix} \hat{\hat{x}}' \\ \hat{\hat{y}}' \end{pmatrix} = \begin{pmatrix} x' \\ y' \end{pmatrix} - \frac{(\boldsymbol{\xi}^*, \boldsymbol{\theta})}{(\boldsymbol{\theta}, V_0[\hat{\boldsymbol{\xi}}]\boldsymbol{\theta})} \begin{pmatrix} \theta_1 & \theta_4 & \theta_7 \\ \theta_2 & \theta_5 & \theta_8 \end{pmatrix} \begin{pmatrix} \hat{x} \\ \hat{y} \\ f_0 \end{pmatrix}, \tag{4.23}$$

where $V_0[\hat{\boldsymbol{\xi}}]$ is the normalized covariance matrix obtained from Eq. (3.12) by replacing $\bar{x}$, $\bar{y}$, $\bar{x}'$, and $\bar{y}'$ by $\hat{x}$, $\hat{y}$, $\hat{x}'$, and $\hat{y}'$, respectively.

References

1. R. Hartley, A. Zisserman, *Multiple View Geometry in Computer Vision*, 2nd edn. (Cambridge University Press, Cambridge, 2003)
2. K. Kanatani, Y. Sugaya, H. Niitsuma, Triangulation from two views revisited: Hartley-Sturm vs. optimal correction, in *Proceedings of 19th British Machine Vision Conference*, Leeds, UK (2008), pp. 173–182
3. Y. Kanazawa, K. Kanatani, Reliability of 3-D reconstruction by stereo vision. IEICE Trans. Inf. Syst. **E78-D**(10), 1301–1306 (1995)
4. R. Hartley, P. Sturm, Triangulation. Comput. Vis. Image Understand. **68**(2), 146–157 (1997)
5. K. Kanatani, Y. Sugaya, Unified computation of strict maximum likelihood for geometric fitting. J. Math. Imaging Vis. **38**(1), 1–13 (2010)

3D Reconstruction from Two Views 5

Abstract

This chapter describes a method for 3D reconstruction from two views, that is, computing the 3D positions of corresponding point pairs. To do this, we need to know the camera matrices that specify the positions, orientations, and internal parameters, such as focal lengths, of the two cameras. We estimate them from the fundamental matrix computed from the two images; this process is called self-calibration. We first express the fundamental matrix in terms of the positions, orientations, and focal lengths of the two cameras. Then we show that the focal lengths of the two cameras are computed from the fundamental matrix by an analytical formula. Using them, we can compute the positions and orientations of the two cameras, which determine their camera matrices. Once the camera matrices are determined, the 3D positions of corresponding point pairs are computed by the triangulation computation described in the preceding chapter.

5.1 Camera Modeling and Self-calibration

To do the triangulation computation described in the preceding chapter, we need to know the camera matrices P and P' of the two cameras. They can be obtained by prior camera calibration, using special devices in laboratories, but we can directly determine P and P' from the fundamental matrix F computed from the two images we observe. Directly determining the camera matrices from the images without prior calibration is called *self-calibration*. However, some conditions are necessary for this to be possible. The fundamental matrix F is determined only up to scale, and the rank constraint $\det \mathsf{F} = 0$ is imposed, so the number of independent elements, or the degree of freedom, is seven. The 3×4 camera matrix also has scale indeterminacy, as seen from Eq. (4.2), so it has 11 degrees of freedom. Hence, P and P' have in total

K. Kanatani et al., *Guide to 3D Vision Computation*, Advances in Computer Vision and Pattern Recognition, DOI 10.1007/978-3-319-48493-8_5

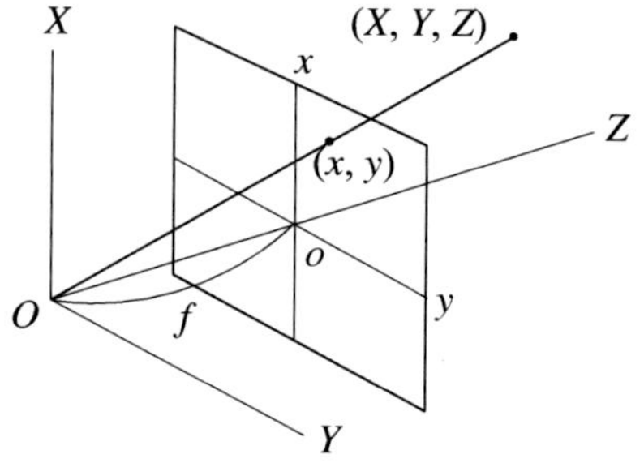

Fig. 5.1 Idealized perspective projection with the camera coordinate system identified with the world coordinate system

22 degrees of freedom. This means that in order for self-calibration to be possible, the camera parameters P and P′ need to be expressed in seven parameters.

Here, we identify the first camera coordinate system with the world coordinate system. Namely, we regard the viewpoint O_c of the first camera as the world origin, and the $X_cY_cZ_c$ camera coordinate axes (the Z_c-axis is the optical axis) as the world coordinate axes (Fig. 5.1). We assume that the principal point (the optical axis position in the image plane) is known (for most cameras, it is at the center of the image) and define an image xy coordinate system with origin o at the principal point. We further assume that there is no image distortion, that the image x- and y-axes are parallel to the X- and Y-axes, respectively, and that the image plane is orthogonal to the optical axis (= the Z-axis; Fig. 5.1).

When we define an xy coordinate system on an image, it is natural to imagine that the optical axis, that is, the Z-axis, extends away from the viewer, just as if we were viewing the scene through a glass window. This implies, as we can see from Fig. 5.1, that the relative orientation of the x- and y-axes in the image must be *reversed* as compared with the usual mathematical convention. For example, we take the x-axis in the upward direction and the y-axis in the rightward direction. Alternatively, we may direct the x-axis rightward and the y-axis downward.

If all the above conditions are satisfied, a point (X, Y, Z) in the scene is projected to a position (x, y) in the image given by

$$x = f\frac{X}{Z}, \qquad y = f\frac{Y}{Z}, \tag{5.1}$$

where f is the distance of the image plane from the XY plane, commonly known as the *focal length* (Fig. 5.1). Equation (5.1) can be written in the form:

$$\begin{pmatrix} x/f_0 \\ y/f_0 \\ 1 \end{pmatrix} \simeq \begin{pmatrix} f & 0 & 0 \\ 0 & f & 0 \\ 0 & 0 & f_0 \end{pmatrix} \begin{pmatrix} X \\ Y \\ Z \end{pmatrix}. \tag{5.2}$$

Similarly, a point (X'_c, Y'_c, Z'_c) in the scene with respect to the $X'_cY'_cZ'_c$ camera coordinate system of the second camera is projected to a point (x', y') in the second image such that

$$\begin{pmatrix} x'/f_0 \\ y'/f_0 \\ 1 \end{pmatrix} \simeq \begin{pmatrix} f' & 0 & 0 \\ 0 & f' & 0 \\ 0 & 0 & f_0 \end{pmatrix} \begin{pmatrix} X'_c \\ Y'_c \\ Z'_c \end{pmatrix}, \tag{5.3}$$

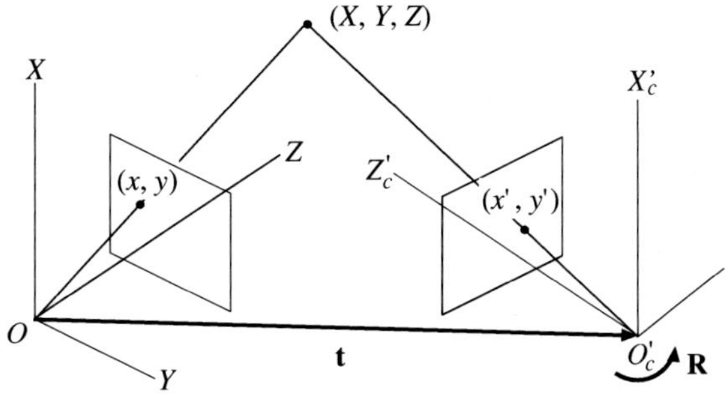

Fig. 5.2 Relationship of two camera coordinate systems

where f' is the focal length of the second camera. Because the $X'_cY'_cZ'_c$ coordinate system is rotated by rotation matrix R relative to the XYZ world coordinate system (= the first camera coordinate system), the direction of the point (X'_c, Y'_c, Z'_c) from O'_c is $\mathsf{R}(X'_c, Y'_c, Z'_c)^\top$ with respect to the XYZ world coordinate system. In addition, the origin O'_c of the $X'_cY'_cZ'_c$ coordinate system is at $\mathbf{t}$ with respect to the XYZ world coordinate system (Fig. 5.2). Therefore the position of the point (X'_c, Y'_c, Z'_c) with respect to the XYZ world coordinate system is given by

$$\begin{pmatrix} X \\ Y \\ Z \end{pmatrix} = \mathsf{R}\begin{pmatrix} X'_c \\ Y'_c \\ Z'_c \end{pmatrix} + \mathbf{t}. \tag{5.4}$$

Noting that the inverse of the rotation matrix R is $\mathsf{R}^{-1} = \mathsf{R}^\top$, we can combine Eq. (5.4) with Equation (5.3) to obtain

$$\begin{pmatrix} x'/f_0 \\ y'/f_0 \\ 1 \end{pmatrix} \simeq \begin{pmatrix} f' & 0 & 0 \\ 0 & f' & 0 \\ 0 & 0 & f_0 \end{pmatrix} \mathsf{R}^\top\left(\begin{pmatrix} X \\ Y \\ Z \end{pmatrix} - \mathbf{t}\right) = \begin{pmatrix} f' & 0 & 0 \\ 0 & f' & 0 \\ 0 & 0 & f_0 \end{pmatrix} \begin{pmatrix} \mathsf{R}^\top & -\mathsf{R}^\top\mathbf{t} \end{pmatrix} \begin{pmatrix} X \\ Y \\ Z \\ 1 \end{pmatrix}. \tag{5.5}$$

If we write Eqs. (5.2) and (5.5) in the form of Eq. (4.2) as

$$\begin{pmatrix} x/f_0 \\ y/f_0 \\ 1 \end{pmatrix} \simeq \mathsf{P}\begin{pmatrix} X \\ Y \\ Z \\ 1 \end{pmatrix}, \qquad \begin{pmatrix} x'/f_0 \\ y'/f_0 \\ 1 \end{pmatrix} \simeq \mathsf{P}'\begin{pmatrix} X \\ Y \\ Z \\ 1 \end{pmatrix}, \tag{5.6}$$

we obtain the following expressions of the camera matrices P and P′.

$$\mathsf{P} = \begin{pmatrix} f & 0 & 0 \\ 0 & f & 0 \\ 0 & 0 & f_0 \end{pmatrix} \begin{pmatrix} \mathsf{I} & \mathbf{0} \end{pmatrix}, \qquad \mathsf{P}' = \begin{pmatrix} f' & 0 & 0 \\ 0 & f' & 0 \\ 0 & 0 & f_0 \end{pmatrix} \begin{pmatrix} \mathsf{R}^\top & -\mathsf{R}^\top\mathbf{t} \end{pmatrix}. \tag{5.7}$$

Thus the two camera matrices P and P′ can be determined if we compute f, f', R, and $\mathbf{t}$ from the two images. However, we cannot determine the absolute magnitude of the translation $\mathbf{t}$, because images have no depth information such that we cannot distinguish a large camera motion relative to a far away scene from a small camera motion relative to a nearby scene. In order to remove this scale indeterminacy, we normalize the camera displacement to unit distance: $\|\mathbf{t}\| = 1$.

In this setting, the unknowns to be computed are the focal lengths f and f' (two degrees of freedom), the rotation R (three degrees of freedom), and the translation $\mathbf{t}$ (two degrees of freedom), having seven degrees of freedom in total. The fundamental matrix F has seven degrees of freedom, therefore we can in principle compute f, f', R, and $\mathbf{t}$ from F.

5.2 Expression of the Fundamental Matrix

If the camera matrices $\mathsf{P} = (P_{ij})$ and $\mathsf{P}' = (P'_{ij})$ are given by Eq. (5.7), the necessary and sufficient conditions for the four equations of Eq. (4.5) to be linearly dependent are

$$\left| \begin{pmatrix} x \\ y \\ f \end{pmatrix}, \mathbf{t}, \mathsf{R} \begin{pmatrix} x' \\ y' \\ f' \end{pmatrix} \right| = 0, \tag{5.8}$$

where $|\mathbf{a}, \mathbf{b}, \mathbf{c}|$ denotes the scalar triple product of vectors $\mathbf{a}$, $\mathbf{b}$, and $\mathbf{c}$. This is geometrically interpreted as follows. As seen from Fig. 5.1, the vector starting from the viewpoint O of the first camera and pointing to the point (x, y) on the image plane is $(x, y, f)^\top$. Similarly, the vector starting from the viewpoint O'_c of the second camera and pointing to the point (x', y') on the image plane is $(x', y', f)^\top$. However, the $X'_cY'_cZ'_c$ coordinate system is rotated relative to the XYZ coordinate system by R, thus the viewing direction of (x', y') is $\mathsf{R}(x', y', f')^\top$ with respect to the world coordinate system. Equation (5.8) implies that these two viewing directions and the camera translation $\mathbf{t}$ are coplanar (Fig. 5.2). Evidently, this is the condition for the two lines of sight to intersect.

As is well known in vector calculus, Eq. (5.8) can be written in terms of the inner product and the vector product in the form

$$\left(\begin{pmatrix} x \\ y \\ f \end{pmatrix}, \mathbf{t} \times \mathsf{R} \begin{pmatrix} x' \\ y' \\ f' \end{pmatrix} \right) = 0. \tag{5.9}$$

Here, we introduce a new notation. For a vector $\mathbf{a} = (a_i)$, we define the "matrix" $\mathbf{a}\times$ by

$$\mathbf{a}\times = \begin{pmatrix} 0 & -a_3 & a_2 \\ a_3 & 0 & -a_1 \\ -a_2 & a_1 & 0 \end{pmatrix}. \tag{5.10}$$

As a result, we see that for vectors $\mathbf{a}$ and $\mathbf{b}$, the product $\mathbf{a} \times \mathbf{b}$ can be interpreted to be either the vector product of the vectors $\mathbf{a}$ and $\mathbf{b}$ or the product of the "matrix" $\mathbf{a}\times$ and the vector $\mathbf{b}$. Using this notation, we can rewrite the left side of Eq. (5.9) in the form

$$
(\begin{pmatrix} f_0 & 0 & 0 \\ 0 & f_0 & 0 \\ 0 & 0 & f \end{pmatrix}\begin{pmatrix} x/f_0 \\ y/f_0 \\ 1 \end{pmatrix}, (\mathbf{t}\times\mathsf{R})\begin{pmatrix} f_0 & 0 & 0 \\ 0 & f_0 & 0 \\ 0 & 0 & f' \end{pmatrix}\begin{pmatrix} x'/f_0 \\ y'/f_0 \\ 1 \end{pmatrix})
$$

$$
= (\begin{pmatrix} x/f_0 \\ y/f_0 \\ 1 \end{pmatrix}, \begin{pmatrix} f_0 & 0 & 0 \\ 0 & f_0 & 0 \\ 0 & 0 & f \end{pmatrix}(\mathbf{t}\times\mathsf{R})\begin{pmatrix} f_0 & 0 & 0 \\ 0 & f_0 & 0 \\ 0 & 0 & f' \end{pmatrix}\begin{pmatrix} x'/f_0 \\ y'/f_0 \\ 1 \end{pmatrix}). \tag{5.11}
$$

Comparing this with Eq. (4.7), we see that the fundamental matrix F is expressed in the form

$$
\mathsf{F} \simeq \begin{pmatrix} f_0 & 0 & 0 \\ 0 & f_0 & 0 \\ 0 & 0 & f \end{pmatrix}(\mathbf{t}\times\mathsf{R})\begin{pmatrix} f_0 & 0 & 0 \\ 0 & f_0 & 0 \\ 0 & 0 & f' \end{pmatrix}, \tag{5.12}
$$

where $\mathbf{t}\times\mathsf{R}$ is the product of the "matrix" $\mathbf{t}\times$ and the matrix R. If we let $\mathbf{r}_1$, $\mathbf{r}_2$, and $\mathbf{r}_3$ be the columns of R, this product equals the matrix consisting of the columns $\mathbf{t}\times\mathbf{r}_1$, $\mathbf{t}\times\mathbf{r}_2$, and $\mathbf{t}\times\mathbf{r}_3$. From the above result, the self-calibration of the camera matrices P and P' reduces to the computation of f, f', $\mathbf{t}$, and R that satisfy Eq. (5.12) for a given fundamental matrix F.

5.3 Focal Length Computation

Given a fundamental matrix F, the focal lengths f and f' that satisfy Eq. (5.12) are computed by the following procedure (↪ Problem 5.1).

Procedure 5.1 (Focal length computation)

1. Compute the unit eigenvectors $\mathbf{e}$ and $\mathbf{e}'$ of the matrices $\mathsf{F}\mathsf{F}^\top$ and $\mathsf{F}^\top\mathsf{F}$, respectively, for the smallest eigenvalues.
2. Compute ξ and η given by

$$
\xi = \frac{\|\mathsf{F}\mathbf{k}\|^2 - (\mathbf{k}, \mathsf{F}\mathsf{F}^\top\mathsf{F}\mathbf{k})\|\mathbf{e}'\times\mathbf{k}\|^2/(\mathbf{k}, \mathsf{F}\mathbf{k})}{\|\mathbf{e}'\times\mathbf{k}\|^2\|\mathsf{F}^\top\mathbf{k}\|^2 - (\mathbf{k}, \mathsf{F}\mathbf{k})^2},
$$

$$
\eta = \frac{\|\mathsf{F}^\top\mathbf{k}\|^2 - (\mathbf{k}, \mathsf{F}\mathsf{F}^\top\mathsf{F}\mathbf{k})\|\mathbf{e}\times\mathbf{k}\|^2/(\mathbf{k}, \mathsf{F}\mathbf{k})}{\|\mathbf{e}\times\mathbf{k}\|^2\|\mathsf{F}\mathbf{k}\|^2 - (\mathbf{k}, \mathsf{F}\mathbf{k})^2}, \tag{5.13}
$$

   where $\mathbf{k} = (0, 0, 1)^\top$.
3. Compute f and f' as follows.

$$
f = \frac{f_0}{\sqrt{1+\xi}}, \qquad f' = \frac{f_0}{\sqrt{1+\eta}}. \tag{5.14}
$$

Comments The vectors $\mathbf{e}$ and $\mathbf{e}'$ computed in Step 1 are actually the unit eigenvectors of $\mathsf{F}^\top$ and F, respectively, for eigenvalue 0 (note that $\det\mathsf{F} = 0$ implies the existence

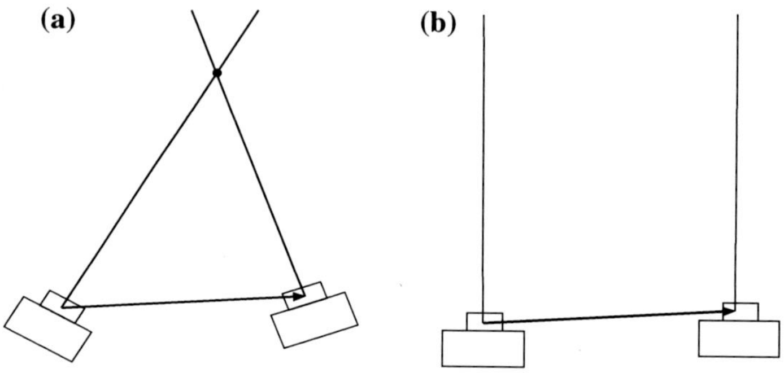

Fig. 5.3 Fixating configuration of two cameras. **a** The two optical axes intersect. **b** The two optical axes are parallel, intersecting at infinity

of the 0 eigenvalue). The reason that we do not solve $\mathsf{F}^\top \mathbf{e} = \mathbf{0}$ and $\mathsf{F}\mathbf{e}' = \mathbf{0}$ is that most numerical tools for eigenvalue computation require a symmetric matrix as the input. Therefore we solve $\mathsf{F}\mathsf{F}^\top \mathbf{e} = \mathbf{0}$ and $\mathsf{F}^\top\mathsf{F}\mathbf{e}' = \mathbf{0}$ for the symmetric matrices $\mathsf{F}\mathsf{F}^\top$ and $\mathsf{F}^\top\mathsf{F}$. Because these are positive semi-definite in general ($\hookrightarrow$ Problem 5.2), the eigenvalue 0 is also the "smallest" eigenvalue, for example, 10^{-10} numerically.

The vector $\mathbf{e}$ points to the viewpoint O' of the second camera seen from the viewpoint O of the first camera, and its image position is called the *epipole* of the second camera. Similarly, $\mathbf{e}'$ points to the viewpoint O of the first camera seen from the viewpoint O' of the second camera; its image position is the epipole of the first camera. The computation of Eq. (5.13) fails if $(\mathbf{k}, \mathsf{F}\mathbf{k})$ $(= F_{33})$ in the denominators vanishes. The vector $\mathbf{k} = (0, 0, 1)^\top$ can be thought of as the values of the vectors $(x/f_0, y/f_0, 1)^\top$ and $(x'/f_0, y'/f_0, 1)^\top$ in Eq. (4.7) for $(x, y) = (0, 0)$ and (x', y') $= (0, 0)$. Hence, $(\mathbf{k}, \mathsf{F}\mathbf{k}) = 0$ indicates that the origins (= principal points) of the two images correspond to each other. This means that the optical axes of the two cameras intersect in the scene (Fig. 5.3a), the special case being parallel optical axes with their intersection at infinity (Fig. 5.3b). Such a camera setting is said to be a *fixating configuration*. This is the most natural way for many to take two images of the same object, but this must be avoided for self-calibration. This is one of the main restrictions of doing self-calibrating using only two images.

Another potential problem is that one or both of the values of ξ and η computed by Eq. (5.13) can be less than -1, in which case no real values of f and f' are computed by Eq. (5.14). This is called the *imaginary focal length problem*. This can occur when some of the detected point correspondences have large errors or the assumed principal point position is very different from its true position. In such a case, we need to use some estimates of f and f', such as the electronic log data of the camera or the values obtained from other images taken by the same camera.

5.4 Motion Parameter Computation

After the focal lengths f and f' have been determined, the motion parameters, that is, the relative translation $\mathbf{t}$ and rotation R of the two cameras, are computed by the following procedure, which requires not only the fundamental matrix F but also the corresponding point pairs (x_α, y_α) and (x'_α, y'_α), $\alpha = 1, ..., N$, from which F is computed.

Procedure 5.2 (Motion parameter computation)

1. Compute the following matrix E:

$$\mathsf{E} = \begin{pmatrix} 1/f_0 & 0 & 0 \\ 0 & 1/f_0 & 0 \\ 0 & 0 & 1/f \end{pmatrix} \mathsf{F} \begin{pmatrix} 1/f_0 & 0 & 0 \\ 0 & 1/f_0 & 0 \\ 0 & 0 & 1/f' \end{pmatrix}. \tag{5.15}$$

2. Compute the unit eigenvector $\mathbf{t}$ of $\mathsf{E}\mathsf{E}^\top$ for the smallest eigenvalue.
3. Represent the corresponding point pairs (x_α, y_α) and (x'_α, y'_α), $\alpha = 1, ..., N$, by the following vectors:

$$\mathbf{x}_\alpha = \begin{pmatrix} x_\alpha/f \\ y_\alpha/f \\ 1 \end{pmatrix}, \qquad \mathbf{x}'_\alpha = \begin{pmatrix} x'_\alpha/f' \\ y'_\alpha/f' \\ 1 \end{pmatrix}. \tag{5.16}$$

4. If

$$\sum_{\alpha=1}^{N} |\mathbf{t}, \mathbf{x}_\alpha, \mathsf{E}\mathbf{x}'_\alpha| > 0 \tag{5.17}$$

does not hold, reverse the sign of $\mathbf{t}$ ($|\mathbf{a}, \mathbf{b}, \mathbf{c}|$ is the scalar triple product of vectors $\mathbf{a}$, $\mathbf{b}$, and $\mathbf{c}$).
5. Let

$$\mathsf{K} = -\mathbf{t} \times \mathsf{E}, \tag{5.18}$$

where $\mathbf{t} \times \mathsf{E}$ is the product of the "matrix" $\mathbf{t}\times$ and the matrix E, that is, the matrix consisting of the vector product of $\mathbf{t}$ and each column of E as columns. Then, compute the SVD (singular value decomposition) of K:

$$\mathsf{K} = \mathsf{U}\boldsymbol{\Lambda}\mathsf{V}^\top, \qquad \boldsymbol{\Lambda} = \begin{pmatrix} \sigma_1 & 0 & 0 \\ 0 & \sigma_2 & 0 \\ 0 & 0 & \sigma_3 \end{pmatrix}, \qquad \sigma_1 \geq \sigma_2 \geq \sigma_3 (= 0). \tag{5.19}$$

6. Compute the rotation R by

$$\mathsf{R} = \mathsf{U} \begin{pmatrix} 1 & 0 & 0 \\ 0 & 1 & 0 \\ 0 & 0 & \det(\mathsf{U}\mathsf{V}^\top) \end{pmatrix} \mathsf{V}^\top. \tag{5.20}$$

Comments The matrix E defined by Eq. (5.15) is called the *essential matrix*. From Eq. (5.12), we see that

$$\mathsf{E} \simeq \mathbf{t} \times \mathsf{R}. \tag{5.21}$$

If we define $\mathbf{x}_\alpha$ and $\mathbf{x}'_\alpha$ by Eq. (5.16), the epipolar equation of Eq. (5.9) can be written as

$$(\mathbf{x}_\alpha, \mathsf{E}\mathbf{x}'_\alpha) = 0, \tag{5.22}$$

which holds in the absence of noise, that is, if (x_α, y_α) and (x'_α, y'_α) exactly correspond. The computation of $\mathbf{t}$ in Step 2 is based on the observation that from Eq. (5.21), $\mathsf{E}^\top \mathbf{t} = \mathbf{0}$ and hence $\mathsf{E}\mathsf{E}^\top \mathbf{t} = \mathbf{0}$ in the absence of noise. The sign judgment of Step 4 is necessary, because the translation $\mathbf{t}$ is obtained in Step 2 as an eigenvector, which has sign indeterminacy. The condition of Eq. (5.17) states that all the observed points in the scene are on the same side of both cameras (↪ Problem 5.3). The rotation matrix R is determined from Eq. (5.21) in such a way that

$$\|c\mathsf{E} - \mathbf{t} \times \mathsf{R}\|^2 \tag{5.23}$$

is minimized for some constant c, where $\|\cdot\|$ denotes the matrix norm defined by $\|\mathsf{A}\|^2 = \sum_{i,j=1}^3 A_{ij}^2$ $(= \mathrm{tr}[\mathsf{A}^\top \mathsf{A}])$ (↪ Eq. (3.2)). Here we assume that the sign of E is correctly chosen such that $c > 0$ (we discuss this in the next section). Then it can be shown that for the matrix K of Eq. (5.18), the rotation matrix R that minimizes Eq. (5.23) maximizes $\mathrm{tr}[\mathsf{K}^\top \mathsf{R}]$ (↪ Problem 5.4(1)); the solution is obtained in the form of Eq. (5.20) using the SVD of K (↪ Problem 5.4(3)). Note that the fundamental matrix F and hence the essential matrix E have determinant 0 ($\det \mathsf{F} = \det \mathsf{E} = 0$). Therefore $\det \mathsf{K} = \det(-\mathbf{t}\times) \det \mathsf{E} = 0$, and the smallest singular value σ_3 of K is 0.

5.5 3D Shape Computation

After the focal lengths f and f' and the motion parameters $\{\mathbf{t}, \mathsf{R}\}$ have been determined, the 3D positions of the corresponding point pairs (x_α, y_α) and (x'_α, y'_α), $\alpha = 1, \ldots, N$, are computed and scaled so that $\|\mathbf{t}\| = 1$, by the following procedure.

Procedure 5.3 (3D shape computation)

1. Compute the camera matrices P and P' by Eq. (5.7).
2. Optimally correct the corresponding points (x_α, y_α) and (x'_α, y'_α) to $(\hat{x}_\alpha, \hat{y}_\alpha)$ and $(\hat{x}'_\alpha, \hat{y}'_\alpha)$, using Procedure 4.2.
3. Compute the 3D positions $(X_\alpha, Y_\alpha, Z_\alpha)$ from the corrected corresponding point pairs, using Procedure 4.1.
4. If the computed Z_α do not satisfy

$$\sum_{\alpha=1}^{N} \mathrm{sgn}(Z_\alpha) > 0, \tag{5.24}$$

reverse the sign of all $(X_\alpha, Y_\alpha, Z_\alpha)$, where $\mathrm{sgn}(x)$ is the signature function, returning 1, 0, and -1 for $x > 0$, $x = 0$, and $x < 0$, respectively.

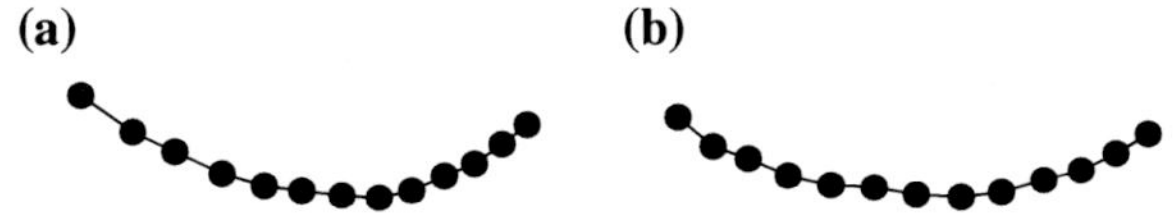

Fig. 5.4 **a** The fundamental matrix is computed from the grid point correspondence between the images in Fig. 3.2, and the 3D positions of the grid points are computed by the self-calibration method of this chapter. This figure shows the reconstructed grid points viewed from the direction of the cylindrical surface axis. **b** The reconstructed grid points by triangulation using the true camera matrices viewed from the same direction (the same as Fig. 4.5a)

Comments The sign judgment of Step 4 is to select the correct sign of the fundamental matrix F. As shown in Chap. 3, the fundamental matrix F is defined up to sign, and if it is determined by solving an eigenvalue problem or a generalized eigenvalue for the 9D vector $\boldsymbol{\theta}$ that represents the fundamental matrix F, its sign is arbitrarily chosen. We can see that for the focal length computation by Procedure 5.1, the values computed by Eq. (5.13) are not affected by the sign reversal of F. In Procedure 5.2, however, reversing the sign of F in Eq. (5.15) changes the sign of the computed essential matrix E. Then, the sign of $\mathbf{t}$ chosen by Eq. (5.17) is also reversed. However, the simultaneous sign change of E and $\mathbf{t}$ does not affect the value of K of Eq. (5.18) or the value of Eq. (5.23), and the rotation R is correctly computed. As a result, the triangulation using the camera matrices P and P' determined from the computed $\mathbf{t}$ and R reconstructs a mirror image, that is, a reversed 3D shape behind the two cameras. This is because the mathematical perspective projection modeling holds irrespective of whether the scene is in front of or behind the camera. Note that Eq. (5.17) in Procedure 5.2 is the condition that the scene is on the "same side" of the two cameras; it does not exclude the case of the scene being behind both cameras. Therefore we check by Eq. (5.24) if most of Z_α are positive. Theoretically, all Z_α should be positive, but if the lines of sight from the two cameras are nearly parallel, its intersection could be computed far behind the cameras due to noise. The requirement of Eq. (5.24) takes this into account, using the signature function $\mathrm{sgn}(x)$ instead of simply computing $\sum_{\alpha=1}^{N} Z_\alpha$.

5.6 Examples

We computed the fundamental matrix from the grid point correspondences between the noisy grid surface images in Fig. 3.2; we minimized the geometric distance, using Procedure 3.7. Then, the camera matrices were self-calibrated, using the method shown in this chapter, and the 3D shape was reconstructed. Figure 5.4a shows the computed 3D positions of the grid points viewed from the direction of the cylindrical surface axis. For comparison, Fig. 5.4b shows the 3D grid positions reconstructed by triangulation using the true camera matrices, assuming that they are known, viewed from the same direction; this is the same as Fig. 4.5. However, self-calibration using the fundamental matrix reconstructs the 3D shape only up to the scale of $\|\mathbf{t}\| = 1$. The shape in Fig. 5.4a is scaled to the true size. We can see that due to the error in the fundamental matrix computed from noisy correspondences, the self-calibrated camera matrices are not exact, and hence the reconstructed shape is slightly distorted.

5.7 Supplemental Note

The study of 3D shape reconstruction from two images was started by photogrammetrists, mainly in Germany, far before the advent of computer vision research in the United States in the 1980s, and its mathematical foundation was established by German mathematicians including Erwin Kruppa (1885–1967). From the computer vision viewpoint, on the other hand, the English physicist/cognitive psychologist H.C. Longuet-Higgins (1923–2004) showed in his 1981 paper [7] that the essential matrix can be determined from eight point correspondences and that the 3D shape and the camera motion parameters are analytically obtained from it. Similar results were also presented by Tsai and Huang [8] in the United States. In these analyses, the camera focal length is assumed to be known. It was Bougnoux [1] who derived an analytical formula for computing the focal lengths f and f' from the fundamental matrix F. Equation (5.13) is obtained by modifying his formulation; the derivation is shown in Kanatani et al. [4,5], and an alternative formation is given in [3]. The analyses in those papers point out that the computation of f and f' fails if the two cameras are in a fixating configuration. It is also pointed out that if the focal length is fixed, that is, if $f = f'$, the computation is possible in fixating configurations except in the case where the two optical axes and the two camera viewpoints make an isosceles triangle.

The problem of fixating configurations and imaginary focal lengths is the main limitation of self-calibration using only two images. In order to alleviate this, Kanazawa et al. [6] proposed the use of three cameras, decomposing the three fundamental matrices obtained from three image pairs to compute the focal lengths and motion parameters of the three cameras. It is shown that the computation does not fail to fixate configurations and that the occurrence of imaginary focal lengths decreases. Once the focal lengths are determined, the essential matrix E of Eq. (5.15) is obtained from the fundamental matrix F. The technique for computing $\mathbf{t}$ and R from E such that Eq. (5.21) is satisfied has been known since the 1980s and discussed in many textbooks (e.g., see [2]). This chapter is based on the authors' article [5].

Problems

5.1 Derive the Procedure 5.1 for computing the focal lengths in the order:

(1) For vectors $\mathbf{a}$ and $\mathbf{b}$ and a nonsingular matrix A, let $\mathbf{a}' = \mathsf{A}\mathbf{a}$ and $\mathbf{b}' = \mathsf{A}\mathbf{b}$. Show that the relation

$$\mathbf{a}' \times \mathbf{b}' = |\mathsf{A}|(\mathsf{A}^{-1})^\top(\mathbf{a} \times \mathbf{b}) \tag{5.25}$$

holds, where $|\mathsf{A}|$ denotes the determinant of A. Using this, show that the following identity holds for an arbitrary nonsingular matrix T.

$$\mathsf{T}(\mathbf{a}\times) = |\mathsf{T}|((\mathsf{T}^{-1})^\top\mathbf{a}) \times (\mathsf{T}^{-1})^\top. \tag{5.26}$$

(2) For the fundamental matrix F of Eq. (5.12), show that the eigenvectors $\mathbf{e}$ and $\mathbf{e}'$ of $\mathsf{F}^\top$ and F, respectively, for eigenvalue 0 are given as follows.

$$\mathbf{e} \simeq \begin{pmatrix} 1/f_0 & 0 & 0 \\ 0 & 1/f_0 & 0 \\ 0 & 0 & 1/f \end{pmatrix}\mathbf{t}, \qquad \mathbf{e}' \simeq \begin{pmatrix} 1/f_0 & 0 & 0 \\ 0 & 1/f_0 & 0 \\ 0 & 0 & 1/f' \end{pmatrix}\mathsf{R}^\top\mathbf{t}. \tag{5.27}$$

(3) Show the relationships:

$$\mathsf{F} \simeq \mathbf{e} \times \begin{pmatrix} 1/f_0 & 0 & 0 \\ 0 & 1/f_0 & 0 \\ 0 & 0 & 1/f \end{pmatrix}\mathsf{R}\begin{pmatrix} f_0 & 0 & 0 \\ 0 & f_0 & 0 \\ 0 & 0 & f' \end{pmatrix}, \tag{5.28}$$

$$\mathsf{F}^\top \simeq \mathbf{e}' \times \begin{pmatrix} 1/f_0 & 0 & 0 \\ 0 & 1/f_0 & 0 \\ 0 & 0 & 1/f' \end{pmatrix}\mathsf{R}^\top\begin{pmatrix} f_0 & 0 & 0 \\ 0 & f_0 & 0 \\ 0 & 0 & f \end{pmatrix}. \tag{5.29}$$

(4) Show that the following matrix form of the *Kruppa equation* holds.

$$\mathsf{F}\begin{pmatrix} 1 & 0 & 0 \\ 0 & 1 & 0 \\ 0 & 0 & f_0^2/f'^2 \end{pmatrix}\mathsf{F}^\top \simeq \mathbf{e} \times \begin{pmatrix} 1 & 0 & 0 \\ 0 & 1 & 0 \\ 0 & 0 & f_0^2/f^2 \end{pmatrix} \times \mathbf{e}, \tag{5.30}$$

$$\mathsf{F}^\top\begin{pmatrix} 1 & 0 & 0 \\ 0 & 1 & 0 \\ 0 & 0 & f_0^2/f^2 \end{pmatrix}\mathsf{F} \simeq \mathbf{e}' \times \begin{pmatrix} 1 & 0 & 0 \\ 0 & 1 & 0 \\ 0 & 0 & f_0^2/f'^2 \end{pmatrix} \times \mathbf{e}'. \tag{5.31}$$

where $\times\mathbf{a}$ denotes $(\mathbf{a}\times)^\top$ $(= -\mathbf{a}\times)$ for a vector $\mathbf{a}$.

(5) If ξ and η are defined by

$$\xi = \left(\frac{f_0}{f}\right)^2 - 1, \qquad \eta = \left(\frac{f_0}{f'}\right)^2 - 1, \tag{5.32}$$

show that Eqs. (5.30) and (5.31) are written, respectively, in the form

$$\mathsf{F}\mathsf{F}^\top\mathbf{k} + \eta(\mathbf{k}, \mathsf{F}\mathbf{k})\mathsf{F}\mathbf{k} = c\mathsf{P}_\mathbf{e}\mathbf{k}, \tag{5.33}$$

$$\mathsf{F}^\top\mathsf{F}\mathbf{k} + \xi(\mathbf{k}, \mathsf{F}\mathbf{k})\mathsf{F}^\top\mathbf{k} = c'\mathsf{P}_{\mathbf{e}'}\mathbf{k}, \tag{5.34}$$

for some constants c and c', where the matrices $\mathsf{P}_\mathbf{e}$ and $\mathsf{P}_{\mathbf{e}'}$ are defined by

$$\mathsf{P}_\mathbf{e} = \mathsf{I} - \mathbf{e}\mathbf{e}^\top, \qquad \mathsf{P}_{\mathbf{e}'} = \mathsf{I} - \mathbf{e}'\mathbf{e}'^\top. \tag{5.35}$$

(6) Show that the focal lengths f and f' are computed in the form of Eqs. (5.13) and (5.14).

5.2 For an arbitrary matrix A, show that the eigenvalues of $\mathsf{A}^\top\mathsf{A}$ and $\mathsf{A}\mathsf{A}^\top$ are all nonnegative.

5.3 Show that the point P_α in the scene determined by Eq. (5.16) is on the same side of the two cameras if and only if

$$|\mathbf{t}, \mathbf{x}_\alpha, \mathsf{E}\mathbf{x}'_\alpha| > 0. \tag{5.36}$$

5.4 Show that the rotation R is obtained by Eqs. (5.19) and (5.20) in the following order.

(1) If we define the matrix K by Eq. (5.18), show that the rotation matrix R that minimizes Eq. (5.23) maximizes $\mathrm{tr}[\mathsf{K}^\top\mathsf{R}]$.
(2) For the diagonal matrix $\boldsymbol{\Lambda}$ of Eq. (5.19), show that the orthogonal matrix T that maximizes $\mathrm{tr}[\mathsf{T}\boldsymbol{\Lambda}]$ is given either by the identity I or by $\mathrm{diag}(1, 1, -1)$, the diagonal matrix with diagonal elements $1, 1, -1$ in that order.
(3) If the SVD of the matrix K is given in the form of Eq. (5.19), show that the rotation matrix R that maximizes $\mathrm{tr}[\mathsf{K}^\top\mathsf{R}]$ is given by Eq. (5.20).

References

1. S. Bougnoux, From projective to Euclidean space under any practical situation, a criticism of self calibration, in *Proceedings of 6th International Conference on Computer Vision*, Bombay, India (1998), pp. 790–796
2. K. Kanatani, *Geometric Computation for Machine Vision* (Oxford University Press, Oxford, 1993)
3. K. Kanatani, Statistical optimization for geometric fitting: theoretical accuracy bound and high order error analysis. Int. J. Comput. Vis. **80**(2), 167–188 (2008)
4. K. Kanatani, A. Nakatsuji, Y. Sugaya, Stabilizing the focal length computation for 3-D reconstruction from two uncalibrated views. Int. J. Comput. Vis. **66**(2), 109–122 (2006)
5. K. Kanatani, Y. Sugaya, Y. Kanazawa, Latest algorithms for 3-D reconstruction form two view, in *Handbook of Pattern Recognition and Computer Vision*, 4th edn., ed. by C.H. Chen (World Scientific Publishing, Singapore, 2009), pp. 201–234
6. Y. Kanazawa, Y. Sugaya, K. Kanatani, Decomposing three fundamental matrices for initializing 3-D reconstruction from three views. IPSJ Trans. Comput. Vis. Appl. **6** (2014)
7. H.C. Longuet-Higgins, A computer algorithm for reconstructing a scene from two projections. Nature **293**(10), 133–135 (1981)
8. R.Y. Tsai, T.S. Huang, Uniqueness and estimation of three-dimensional motion parameters of rigid objects with curved surfaces. IEEE Trans. Pattern Anal. Mach. Intell. **6**(1), 13–27 (1984)

Homography Computation

6

Abstract

If we take two images of a planar surface from two different places, the two images are related by a mapping called a homography. Computing a homography from point correspondences over two images is one of the most fundamental processes of computer vision. This is because, among other things, the 3D positions of the planar surface we are viewing and the two cameras that took the images can be computed from the computed homography. Such applications are discussed in Chaps. 7 and 8. This chapter describes the principles and typical computational procedures for accurately computing the homography by considering the statistical properties of the noise involved in correspondence detection. As in ellipse fitting and fundamental matrix computation, the methods are classified into algebraic (least squares, iterative reweight, the Taubin method, renormalization, HyperLS, and hyper-renormalization) and geometric (FNS, geometric distance minimization, and hyperaccurate correction). We also describe the RANSAC procedure for removing wrong correspondences (outliers).

6.1 Homographies

Consider taking images of a planar surface by two cameras. Suppose point (x, y) in one image corresponds to point (x', y') in the other. It is known that they are related by the following equations (the details are given in Chaps. 7 and 8).

$$x' = f_0 \frac{H_{11}x + H_{12}y + H_{13}f_0}{H_{31}x + H_{32}y + H_{33}f_0}, \quad y' = f_0 \frac{H_{21}x + H_{22}y + H_{23}f_0}{H_{31}x + H_{32}y + H_{33}f_0}. \tag{6.1}$$

As in ellipse fitting and fundamental matrix computation, f_0 is a scaling constant; theoretically, it can be 1, but considering finite length numerical computation, we choose it to make x/f_0 and y/f_0 approximately 1, for example, $f_0 = 600$ for an

K. Kanatani et al., *Guide to 3D Vision Computation*, Advances in Computer Vision and Pattern Recognition, DOI 10.1007/978-3-319-48493-8_6

approximately 600 × 600 pixel planar region. We also take the image origin of the xy coordinate system at the center of the image.

Equation (6.1) can be equivalently written as

$$\begin{pmatrix} x'/f_0 \\ y'/f_0 \\ 1 \end{pmatrix} \simeq \begin{pmatrix} H_{11} & H_{12} & H_{13} \\ H_{21} & H_{22} & H_{23} \\ H_{31} & H_{32} & H_{33} \end{pmatrix} \begin{pmatrix} x/f_0 \\ y/f_0 \\ 1 \end{pmatrix}, \tag{6.2}$$

where the symbol $\simeq$ denotes that the left-hand side is a nonzero multiple of the right-hand side. This mapping from (x, y) to (x', y') is called a *homography* or *projective transformation* when the matrix $\mathsf{H} = (H_{ij})$ is nonsingular. Because Eq. (6.2) implies the left and right sides are parallel vectors, we may rewrite it as

$$\begin{pmatrix} x'/f_0 \\ y'/f_0 \\ 1 \end{pmatrix} \times \begin{pmatrix} H_{11} & H_{12} & H_{13} \\ H_{21} & H_{22} & H_{23} \\ H_{31} & H_{32} & H_{33} \end{pmatrix} \begin{pmatrix} x/f_0 \\ y/f_0 \\ 1 \end{pmatrix} = \begin{pmatrix} 0 \\ 0 \\ 0 \end{pmatrix}. \tag{6.3}$$

The nonsingular matrix $\mathsf{H} = (H_{ij})$ is determined by the relative positions of the two cameras and their intrinsic parameters, such as the focal length, and the position and orientation of the planar scene (we discuss this in Chap. 8); it is called the *homography* (or *projective transformation*) *matrix*. Inasmuch as it represents the same homography if multiplied by a nonzero constant, we normalize it to

$$\|\mathsf{H}\| \left(\equiv \sqrt{\sum_{i,j=1}^{3} H_{ij}^2} \right) = 1. \tag{6.4}$$

If we define the 9D vectors

$$\boldsymbol{\theta} = \begin{pmatrix} H_{11} \\ H_{12} \\ H_{13} \\ H_{21} \\ H_{22} \\ H_{23} \\ H_{31} \\ H_{32} \\ H_{33} \end{pmatrix}, \quad \boldsymbol{\xi}^{(1)} = \begin{pmatrix} 0 \\ 0 \\ 0 \\ -f_0 x \\ -f_0 y \\ -f_0^2 \\ xy' \\ yy' \\ f_0 y' \end{pmatrix}, \quad \boldsymbol{\xi}^{(2)} = \begin{pmatrix} f_0 x \\ f_0 y \\ f_0^2 \\ 0 \\ 0 \\ 0 \\ -xx' \\ -yx' \\ -f_0 x' \end{pmatrix}, \quad \boldsymbol{\xi}^{(3)} = \begin{pmatrix} -xy' \\ -yy' \\ -f_0 y' \\ xx' \\ yx' \\ f_0 x' \\ 0 \\ 0 \\ 0 \end{pmatrix}, \tag{6.5}$$

the vector equation of Eq. (6.3) has the following three components.

$$(\boldsymbol{\xi}^{(1)}, \boldsymbol{\theta}) = 0, \qquad (\boldsymbol{\xi}^{(2)}, \boldsymbol{\theta}) = 0, \qquad (\boldsymbol{\xi}^{(3)}, \boldsymbol{\theta}) = 0. \tag{6.6}$$

The vector $\boldsymbol{\theta}$ in these equations has scale indeterminacy. The normalization of Eq. (6.4) is equivalent to the normalization of $\boldsymbol{\theta}$ to unit norm: $\|\boldsymbol{\theta}\| = 1$.

6.2 Noise and Covariance Matrices

Computing the homography matrix H from point correspondences of two planar scene images (Fig. 6.1) is the first step of many computer vision applications. Mathematically, computing the matrix H that satisfies Eq. (6.1) for corresponding points (x_α, y_α) and (x'_α, y'_α), $\alpha = 1, ..., N$, in the presence of noise is equivalent to computing a unit vector $\boldsymbol{\theta}$ such that

$$(\boldsymbol{\xi}_\alpha^{(1)}, \boldsymbol{\theta}) \approx 0, \quad (\boldsymbol{\xi}_\alpha^{(2)}, \boldsymbol{\theta}) \approx 0, \quad (\boldsymbol{\xi}_\alpha^{(3)}, \boldsymbol{\theta}) \approx 0, \quad \alpha = 1, ..., N, \tag{6.7}$$

where $\boldsymbol{\xi}_\alpha^{(k)}$ is the value of $\boldsymbol{\xi}^{(k)}$ for $x = x_\alpha$, $y = y_\alpha$, $x' = x'_\alpha$, and $y' = y'_\alpha$. We regard the data x_α, x_α, y_α, x'_α, and y'_α for x, y_α, x'_α, and y'_α as deviations of their true values $\bar{x}_\alpha$, $\bar{y}_\alpha$, and $\bar{x}'_\alpha$, by noise terms Δx_α, Δy_{S_α}, $\Delta x'_\alpha$, and $\Delta y'_\alpha$, respectively, and write

$$x_\alpha = \bar{x}_\alpha + \Delta x_\alpha, \qquad y_\alpha = \bar{y}_\alpha + \Delta y_\alpha, \qquad x'_\alpha = \bar{x}'_\alpha + \Delta x'_\alpha, \qquad y'_\alpha = \bar{y}'_\alpha + \Delta y'_\alpha. \tag{6.8}$$

Substituting these into $\boldsymbol{\xi}_\alpha^{(k)}$, we can write

$$\boldsymbol{\xi}_\alpha^{(k)} = \bar{\boldsymbol{\xi}}_\alpha^{(k)} + \Delta_1 \boldsymbol{\xi}_\alpha^{(k)} + \Delta_2 \boldsymbol{\xi}_\alpha^{(k)}, \tag{6.9}$$

where $\bar{\boldsymbol{\xi}}_\alpha^{(k)}$ is the value of $\boldsymbol{\xi}_\alpha^{(k)}$ for $x_\alpha = \bar{x}_\alpha$, $y_\alpha = \bar{y}_\alpha$, $x'_\alpha = \bar{x}'_\alpha$, and $y'_\alpha = \bar{y}'_\alpha$, and $\Delta_1 \boldsymbol{\xi}_\alpha^{(k)}$ and $\Delta_2 \boldsymbol{\xi}_\alpha^{(k)}$ are, respectively, the first and the second-order noise terms. The first-order noise term is

$$\Delta_1 \boldsymbol{\xi}_\alpha^{(k)} = \mathsf{T}_\alpha^{(k)} \begin{pmatrix} \Delta x_\alpha \\ \Delta y_\alpha \\ \Delta x'_\alpha \\ \Delta y'_\alpha \end{pmatrix}, \tag{6.10}$$

where $\mathsf{T}_\alpha^{(k)}$ is the matrix consisting of columns obtained by differentiating $\boldsymbol{\xi}_\alpha^{(k)}$ with respect to x_α, y_α, x'_α, and y'_α, respectively:

$$\mathsf{T}_\alpha^{(1)} = \begin{pmatrix} 0 & 0 & 0 & 0 \\ 0 & 0 & 0 & 0 \\ 0 & 0 & 0 & 0 \\ -f_0 & 0 & 0 & 0 \\ 0 & -f_0 & 0 & 0 \\ 0 & 0 & 0 & 0 \\ \bar{y}'_\alpha & 0 & 0 & \bar{x}_\alpha \\ 0 & \bar{y}'_\alpha & 0 & \bar{y}_\alpha \\ 0 & 0 & 0 & f_0 \end{pmatrix}, \qquad \mathsf{T}_\alpha^{(2)} = \begin{pmatrix} f_0 & 0 & 0 & 0 \\ 0 & f_0 & 0 & 0 \\ 0 & 0 & 0 & 0 \\ 0 & 0 & 0 & 0 \\ 0 & 0 & 0 & 0 \\ 0 & 0 & 0 & 0 \\ -\bar{x}'_\alpha & 0 & -\bar{x}_\alpha & 0 \\ 0 & -\bar{x}'_\alpha & -\bar{y}_\alpha & 0 \\ 0 & 0 & -f_0 & 0 \end{pmatrix},$$

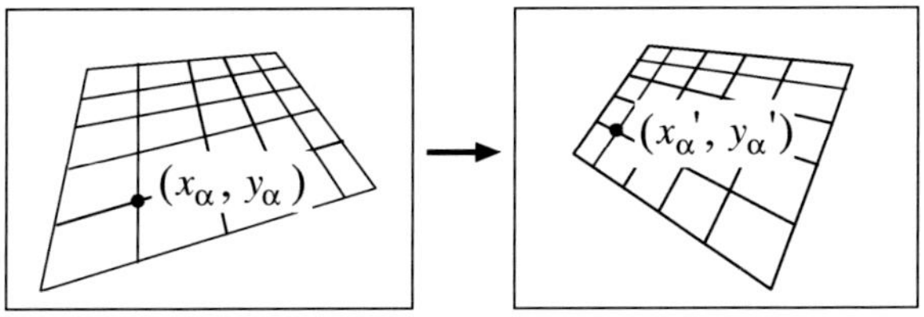

Fig. 6.1 Computing a homography from noisy point correspondences

$$\mathsf{T}_\alpha^{(3)} = \begin{pmatrix} -\bar{y}'_\alpha & 0 & 0 & -\bar{x}_\alpha \\ 0 & -\bar{y}'_\alpha & 0 & -\bar{y}_\alpha \\ 0 & 0 & 0 & -f_0 \\ \bar{x}'_\alpha & 0 & \bar{x}_\alpha & 0 \\ 0 & \bar{x}'_\alpha & \bar{y}_\alpha & 0 \\ 0 & 0 & f_0 & 0 \\ 0 & 0 & 0 & 0 \\ 0 & 0 & 0 & 0 \\ 0 & 0 & 0 & 0 \end{pmatrix}. \tag{6.11}$$

The second-order noise terms are

$$\Delta_2\boldsymbol{\xi}_\alpha^{(1)} = \begin{pmatrix} 0 \\ 0 \\ 0 \\ 0 \\ 0 \\ 0 \\ \Delta x_\alpha \Delta y'_\alpha \\ \Delta y_\alpha \Delta y'_\alpha \\ 0 \end{pmatrix}, \quad \Delta_2\boldsymbol{\xi}_\alpha^{(2)} = \begin{pmatrix} 0 \\ 0 \\ 0 \\ 0 \\ 0 \\ 0 \\ -\Delta x'_\alpha \Delta x_\alpha \\ -\Delta x'_\alpha \Delta y_\alpha \\ 0 \end{pmatrix}, \quad \Delta_2\boldsymbol{\xi}_\alpha^{(3)} = \begin{pmatrix} -\Delta y'_\alpha \Delta x_\alpha \\ -\Delta y'_\alpha \Delta y_\alpha \\ 0 \\ \Delta x'_\alpha \Delta x_\alpha \\ \Delta x'_\alpha \Delta y_\alpha \\ 0 \\ 0 \\ 0 \\ 0 \end{pmatrix}. \tag{6.12}$$

Regarding the noise terms Δx_α, Δy_α, $\Delta x'_\alpha$, and Δy_α as random variables, we define the *covariance matrices* of $\boldsymbol{\xi}_\alpha^{(k)}$ and $\boldsymbol{\xi}_\alpha^{(l)}$ by

$$V^{(kl)}[\boldsymbol{\xi}_\alpha] = E[\Delta_1\boldsymbol{\xi}_\alpha^{(k)}\Delta_1\boldsymbol{\xi}_\alpha^{(l)\top}], \tag{6.13}$$

where $E[\,\cdot\,]$ denotes expectation for their probability distribution. If Δx_α, Δy_α, $\Delta x'_\alpha$, and $\Delta y'_\alpha$ are independent Gaussian variables of mean 0 and standard deviation σ, we have

$$E[\Delta x_\alpha] = E[\Delta y_\alpha] = E[\Delta x'_\alpha] = E[\Delta y'_\alpha] = 0,$$

$$E[\Delta x_\alpha^2] = E[\Delta y_\alpha^2] = E[\Delta x'^2_\alpha] = E[\Delta y'^2_\alpha] = \sigma^2,$$

$$E[\Delta x_\alpha \Delta y_\alpha] = E[\Delta x'_\alpha \Delta y'_\alpha] = E[\Delta x_\alpha \Delta y'_\alpha] = E[\Delta x'_\alpha \Delta y_\alpha] = 0. \tag{6.14}$$

Substituting Eq. (6.10) into Eq. (6.13), we can express the covariance matrix $V^{(kl)}[\boldsymbol{\xi}_\alpha]$ in the form

$$V^{(kl)}[\boldsymbol{\xi}_\alpha] = \sigma^2 V_0^{(kl)}[\boldsymbol{\xi}_\alpha], \qquad V_0^{(kl)}[\boldsymbol{\xi}_\alpha] = \mathsf{T}_\alpha^{(k)}\mathsf{T}_\alpha^{(l)\top}. \tag{6.15}$$

We call $V_0^{(kl)}[\boldsymbol{\xi}_\alpha]$ the *normalized covariance matrix*. As in the case of ellipse fitting and fundamental matrix computation, we need not consider $\Delta_2\boldsymbol{\xi}_\alpha^{(k)}$. Also, the true values $\bar{x}_\alpha$, $\bar{y}_\alpha$, $\bar{x}'_\alpha$, and $\bar{y}'_\alpha$ in Eq. (6.11) are, respectively, replaced by observed values x_α, y_α, x'_α, and y'_α in actual computation.

6.3 Algebraic Methods

Computing the homography matrix H from noisy point correspondences means computing the unit vector $\boldsymbol{\theta}$ that satisfies Eq. (6.7) by considering the noise properties described by the covariance matrices $V^{(kl)}[\boldsymbol{\xi}_\alpha]$. This is the same task as ellipse fitting and fundamental matrix computation except that the number of equations is three, rather than one. Therefore we can compute the solution by appropriately modifying the algebraic methods, such as least squares, the Taubin method, HyperLS, iterative reweight, renormalization, and hyper-renormalization. The procedure of iterative reweight is given as follows.

Procedure 6.1 (Iterative reweight)

1. Let $\boldsymbol{\theta}_0 = \mathbf{0}$ and $W_\alpha^{(kl)} = \delta_{kl}$, $\alpha = 1, ..., N$, $k, l = 1, 2, 3$, where δ_{kl} is the Kronecker delta, taking the value 1 for $k = l$ and 0 otherwise.
2. Compute the 9×9 matrices

$$\mathsf{M} = \frac{1}{N}\sum_{\alpha=1}^{N}\sum_{k,l=1}^{3} W_\alpha^{(kl)} \boldsymbol{\xi}_\alpha^{(k)} \boldsymbol{\xi}_\alpha^{(l)\top}. \tag{6.16}$$

3. Solve the eigenvalue problem

$$\mathsf{M}\boldsymbol{\theta} = \lambda\boldsymbol{\theta}, \tag{6.17}$$

and compute the unit eigenvector $\boldsymbol{\theta}$ for the smallest eigenvalue λ.

4. If $\boldsymbol{\theta} \approx \boldsymbol{\theta}_0$ up to sign, return $\boldsymbol{\theta}$ and stop. Else, update

$$W_\alpha^{(kl)} \leftarrow \Big((\boldsymbol{\theta}, V_0^{(kl)}[\boldsymbol{\theta}_\alpha]\boldsymbol{\theta})\Big)_2^{-}, \qquad \boldsymbol{\theta}_0 \leftarrow \boldsymbol{\theta}, \tag{6.18}$$

and go back to Step 2.

Comments. The expression $\Big((\boldsymbol{\theta}, V_0^{(kl)}[\boldsymbol{\theta}_\alpha]\boldsymbol{\theta})\Big)_2^{-}$ in Eq. (6.18) means the (kl) element of the pseudoinverse with truncated rank 2 of the matrix whose (kl) element is $(\boldsymbol{\theta}, V_0^{(kl)}[\boldsymbol{\theta}_\alpha]\boldsymbol{\theta})$; it is an abbreviation of the following matrix.

$$\begin{pmatrix} (\boldsymbol{\theta}, V_0^{(11)}[\boldsymbol{\xi}_\alpha]\boldsymbol{\theta}) & (\boldsymbol{\theta}, V_0^{(12)}[\boldsymbol{\xi}_\alpha]\boldsymbol{\theta}) & (\boldsymbol{\theta}, V_0^{(13)}[\boldsymbol{\xi}_\alpha]\boldsymbol{\theta}) \\ (\boldsymbol{\theta}, V_0^{(21)}[\boldsymbol{\xi}_\alpha]\boldsymbol{\theta}) & (\boldsymbol{\theta}, V_0^{(22)}[\boldsymbol{\xi}_\alpha]\boldsymbol{\theta}) & (\boldsymbol{\theta}, V_0^{(23)}[\boldsymbol{\xi}_\alpha]\boldsymbol{\theta}) \\ (\boldsymbol{\theta}, V_0^{(31)}[\boldsymbol{\xi}_\alpha]\boldsymbol{\theta}) & (\boldsymbol{\theta}, V_0^{(32)}[\boldsymbol{\xi}_\alpha]\boldsymbol{\theta}) & (\boldsymbol{\theta}, V_0^{(33)}[\boldsymbol{\xi}_\alpha]\boldsymbol{\theta}) \end{pmatrix}_2^{-}. \tag{6.19}$$

For a symmetric matrix A with eigenvalues $\lambda_1 \geq \lambda_2 \geq \cdots$ and corresponding orthonormal system of eigenvectors $\mathbf{u}_1, \mathbf{u}_2, ...,$ its pseudoinverse A_r^- with truncated rank r is given by $\mathsf{A}_r^- = \mathbf{u}_1\mathbf{u}_1^\top/\lambda_1 + \cdots + \mathbf{u}_r\mathbf{u}_r^\top/\lambda_r$ ($\hookrightarrow$ Eqs. (2.30) and (3.29)). As in ellipse fitting, computing the unit vector $\boldsymbol{\theta}$ of the symmetric matrix M for the smallest eigenvalue means computing the unit eigenvector $\boldsymbol{\theta}$ that minimizes the quadratic form $(\boldsymbol{\theta}, \mathsf{M}\boldsymbol{\theta})$. Because

$$(\boldsymbol{\theta}, \mathsf{M}\boldsymbol{\theta}) = (\boldsymbol{\theta}, \Big(\frac{1}{N}\sum_{\alpha=1}^{N}\sum_{k,l=1}^{3} W_\alpha^{(kl)} \boldsymbol{\xi}_\alpha^{(k)} \boldsymbol{\xi}_\alpha^{(k)\top}\Big)\boldsymbol{\theta}) = \frac{1}{N}\sum_{\alpha=1}^{N}\sum_{k,l=1}^{3} W_\alpha^{(kl)} (\boldsymbol{\theta}, \boldsymbol{\xi}_\alpha^{(k)} \boldsymbol{\xi}_\alpha^{(k)\top}\boldsymbol{\theta})$$
$$= \frac{1}{N}\sum_{\alpha=1}^{N}\sum_{k,l=1}^{3} W_\alpha^{(kl)} (\boldsymbol{\theta}, \boldsymbol{\xi}_\alpha^{(k)})(\boldsymbol{\theta}, \boldsymbol{\xi}_\alpha^{(l)}), \tag{6.20}$$

we are computing the unit vector $\boldsymbol{\theta}$ that minimizes the weighted sum of squares $\sum_{\alpha=1}^{N}\sum_{k,l=1}^{3} W_\alpha^{(kl)}(\boldsymbol{\theta}, \boldsymbol{\xi}_\alpha^{(k)})(\boldsymbol{\theta}, \boldsymbol{\xi}_\alpha^{(l)})$. This is commonly known as *weighted least squares*. According to statistics, the weights are optimal if they are proportional to the inverse of the covariance matrices, being large for uncertain terms and small for uncertain terms. Because $(\boldsymbol{\xi}_\alpha^{(k)}, \boldsymbol{\theta}) = (\bar{\boldsymbol{\xi}}_\alpha^{(k)}, \boldsymbol{\theta}) + (\Delta_1\boldsymbol{\xi}_\alpha^{(k)}, \boldsymbol{\theta}) + (\Delta_2\boldsymbol{\xi}_\alpha^{(k)}, \boldsymbol{\theta})$ and $(\bar{\boldsymbol{\xi}}_\alpha^{(k)}, \boldsymbol{\theta}) = 0$, the covariance matrices are given from Eqs. (6.13) and (6.15) by

$$E[(\Delta_1\boldsymbol{\xi}_\alpha^{(k)}, \boldsymbol{\theta})(\Delta_1\boldsymbol{\xi}_\alpha^{(l)}, \boldsymbol{\theta})] = (\boldsymbol{\theta}, E[\Delta_1\boldsymbol{\xi}_\alpha^{(k)}\Delta_1\boldsymbol{\xi}_\alpha^{(l)\top}]\boldsymbol{\theta}) = \sigma^2(\boldsymbol{\theta}, V_0^{(kl)}[\boldsymbol{\xi}_\alpha]\boldsymbol{\theta}). \tag{6.21}$$

Thus the weight $W_\alpha^{(kl)}$ should be the (kl) element of the inverse of the matrix $\big((\boldsymbol{\theta}, V_0^{(kl)}[\boldsymbol{\xi}_\alpha]\boldsymbol{\theta})\big)$. However, the determinant of $\big((\boldsymbol{\theta}, V_0^{(kl)}[\boldsymbol{\xi}_\alpha]\boldsymbol{\theta})\big)$ is 0, and hence its inverse does not exist. This is because the three equations of Eq. (6.6) are linearly dependent. Indeed, we see from Eq. (6.5) that $x'\boldsymbol{\xi}^{(1)} - y'\boldsymbol{\xi}^{(2)} = \boldsymbol{\xi}^{(3)}$. Hence, the three equations of Eq. (6.6) are redundant; if the first and second equations are satisfied, for example, the third one is automatically satisfied. However, the selection of two of them is not unique. To resolve this, we use the pseudoinverse of truncated rank 2, which corresponds to using two equations of Eq. (6.6) without specifying which two to use. However, $\boldsymbol{\theta}$ is not known yet, therefore we instead use the weights $W_\alpha^{(kl)}$ determined in the preceding iteration to compute $\boldsymbol{\theta}$ and update the weights as in Eq. (6.18). Initially, we set $W_\alpha^{(kl)} = \delta_{kl}$, thus Eq. (6.20) implies that we first minimize the sum of squares $\sum_{\alpha=1}^{N}\sum_{k=1}^{3}(\boldsymbol{\theta}, \boldsymbol{\xi}_\alpha^{(k)})^2$. This means solving Eq. (6.7) by *least squares*(LS) ($\hookrightarrow$ Problem 6.1). The phrase "up to sign" in Step 4 reflects the fact that eigenvectors have sign indeterminacy. Therefore we align $\boldsymbol{\theta}$ and $\boldsymbol{\theta}_0$ by reversing the sign $\boldsymbol{\theta} \leftarrow -\boldsymbol{\theta}$ when $(\boldsymbol{\theta}, \boldsymbol{\theta}_0) < 0$.

The procedure of renormalization is given as follows.

Procedure 6.2 (Renormalization)

1. Let $\boldsymbol{\theta}_0 = \mathbf{0}$ and $W_\alpha^{(kl)} = \delta_{kl}$, $\alpha = 1, ..., N$, $k, l = 1, 2, 3$.
2. Compute the 9×9 matrices

$$\mathsf{M} = \frac{1}{N}\sum_{\alpha=1}^{N}\sum_{k,l=1}^{3} W_\alpha^{(kl)} \boldsymbol{\xi}_\alpha^{(k)} \boldsymbol{\xi}_\alpha^{(l)\top}, \quad \mathsf{N} = \frac{1}{N}\sum_{\alpha=1}^{N}\sum_{k,l=1}^{3} W_\alpha^{(kl)} V_0^{(kl)}[\boldsymbol{\xi}_\alpha]. \tag{6.22}$$

3. Solve the generalized eigenvalue problem

$$\mathsf{M}\boldsymbol{\theta} = \lambda \mathsf{N}\boldsymbol{\theta}, \tag{6.23}$$

and compute the unit generalized eigenvector $\boldsymbol{\theta}$ for the generalized eigenvalue λ of the smallest absolute value.
4. If $\boldsymbol{\theta} \approx \boldsymbol{\theta}_0$ up to sign, return $\boldsymbol{\theta}$ and stop. Else, update

$$W_\alpha^{(kl)} \leftarrow \Big((\boldsymbol{\theta}, V_0^{(kl)}[\boldsymbol{\xi}_\alpha]\boldsymbol{\theta})\Big)_2^-, \qquad \boldsymbol{\theta}_0 \leftarrow \boldsymbol{\theta}, \tag{6.24}$$

and go back to Step 2.

Comments. As in ellipse fitting, solving the generalized eigenvalue problem of Eq. (6.23) is equivalent to minimizing the quadratic form $(\boldsymbol{\theta}, \mathsf{M}\boldsymbol{\theta})$ subject to the constraint $(\boldsymbol{\theta}, \mathsf{N}\boldsymbol{\theta})$ = constant. Initially $W_\alpha^{(kl)} = \delta_{kl}$, therefore the first iteration minimizes the sum of the squares $\sum_{\alpha=1}^N \sum_{k=1}^3 (\boldsymbol{\xi}_\alpha^{(k)}, \boldsymbol{\theta})^2$ subject to the constraint $(\boldsymbol{\theta}, \big(\sum_{\alpha=1}^N \sum_{k=1}^3 V_0^{(kk)}[\boldsymbol{\xi}_\alpha]\big)\boldsymbol{\theta})$ = constant. This is a variant of the Taubin method for ellipse fitting, thus we also call this the *Taubin method* (↪ Problem 6.2). It is experimentally confirmed that it has higher accuracy than least squares or iterative reweight and that renormalization outperforms both. The symmetric matrix N of Eq. (6.22) is not necessarily positive definite, therefore we rewrite Eq. (6.23) in the form of Eq. (2.24) and solve it using a standard numerical tool.

The procedure of hyper-renormalization is described as follows.

Procedure 6.3 (Hyper-renormalization)

1. Let $\boldsymbol{\theta}_0 = \mathbf{0}$ and $W_\alpha^{(kl)} = \delta_{kl}$, $\alpha = 1, ..., N$, $k, l = 1, 2, 3$.
2. Compute the 9×9

$$\mathsf{M} = \frac{1}{N}\sum_{\alpha=1}^N \sum_{k,l=1}^3 W_\alpha^{(kl)} \boldsymbol{\xi}_\alpha^{(k)} \boldsymbol{\xi}_\alpha^{(l)\top},$$

$$\begin{aligned}\mathsf{N} =& \frac{1}{N}\sum_{\alpha=1}^N \sum_{k,l=1}^3 W_\alpha^{(kl)} V_0^{(kl)}[\boldsymbol{\xi}_\alpha] \\ &- \frac{1}{N^2}\sum_{\alpha=1}^N \sum_{k,l,m,n=1}^3 W_\alpha^{(kl)} W_\alpha^{(mn)} \Big((\boldsymbol{\xi}_\alpha^{(k)}, \mathsf{M}_8^- \boldsymbol{\xi}_\alpha^{(m)}) V_0^{(ln)}[\boldsymbol{\xi}_\alpha] \\ &+ 2\mathscr{S}[V_0^{(km)}[\boldsymbol{\xi}_\alpha]\mathsf{M}_8^- \boldsymbol{\xi}_\alpha^{(l)} \boldsymbol{\xi}_\alpha^{(n)\top}]\Big).\end{aligned} \tag{6.25}$$

3. Solve the generalized eigenvalue problem

$$\mathsf{M}\boldsymbol{\theta} = \lambda \mathsf{N}\boldsymbol{\theta}, \tag{6.26}$$

and compute the unit generalized eigenvector $\boldsymbol{\theta}$ for the generalized eigenvalue λ of the smallest absolute value.

4. If $\boldsymbol{\theta} \approx \boldsymbol{\theta}_0$ up to sign, return $\boldsymbol{\theta}$ and stop. Else, update

$$W_\alpha^{(kl)} \leftarrow \Big((\boldsymbol{\theta}, V_0^{(kl)}[\boldsymbol{\xi}_\alpha]\boldsymbol{\theta})\Big)_2^-, \qquad \boldsymbol{\theta}_0 \leftarrow \boldsymbol{\theta}, \tag{6.27}$$

and go back to Step 2.

Comments As in ellipse fitting, the operator $\mathscr{S}[\,\cdot\,]$ in Eq. (6.25) denotes symmetrization ($\mathscr{S}[\mathsf{A}] = (\mathsf{A} + \mathsf{A}^\top)/2$), and M_8^- is the pseudoinverse of truncated rank 8. The derivation of Eq. (6.25) is given in Chap. 14. The initial solution obtained in the first iteration minimizes the sum of squares $\sum_{\alpha=1}^N \sum_{k=1}^3 (\boldsymbol{\xi}_\alpha^{(k)}, \boldsymbol{\theta})^2$ subject to the constraint $(\boldsymbol{\theta}, \mathsf{N}\boldsymbol{\theta})$ = constant for the matrix N of Eq. (6.25) with $W_\alpha^{(kl)} = \delta_{kl}$. As in ellipse fitting, this scheme is called *HyperLS* (↪ Problem 6.3). The symmetric matrix N of Eq. (6.22) is not necessarily positive definite, therefore we rewrite Eq. (6.26) in the form of Eq. (2.24) and solve it using a standard numerical tool. For homography computation, however, it has been known that the accuracy of HyperLS is almost the same as the Taubin method and that hyper-renormalization does not improve the accuracy very much as compared with renormalization (see the numerical examples in Sect. 6.9 below).

6.4 Geometric Distance and Sampson Error

The geometric approach for homography computation is to displace optimally the observed corresponding points (x_α, y_α) and (x'_α, y'_α) to new positions $(\bar{x}_\alpha, \bar{y}_\alpha)$ and $(\bar{x}'_\alpha, \bar{y}'_\alpha)$ that satisfy the homography of Eq. (6.1) for some $\mathsf{H} = (H_{ij})$. This displacement is done in such a way that the average square distance

$$S = \frac{1}{N}\sum_{\alpha=1}^{N}\Big((x_\alpha - \bar{x}_\alpha)^2 + (y_\alpha - \bar{y}_\alpha)^2 + (x'_\alpha - \bar{x}'_\alpha)^2 + (y'_\alpha - \bar{y}'_\alpha)^2\Big) \tag{6.28}$$

is minimized. In other words, if $(\bar{x}_\alpha, \bar{y}_\alpha)$ and $(\bar{x}'_\alpha, \bar{y}'_\alpha)$ are expressed as vectors $\bar{\boldsymbol{\xi}}_\alpha^{(k)}$ in the form of Eq. (6.5), we minimize Eq. (6.28) subject to the constraint $(\bar{\boldsymbol{\xi}}_\alpha^{(k)}, \boldsymbol{\theta}) = 0$, $\alpha = 1, ..., N$, $k = 1, 2, 3$. If we identify $(\bar{x}_\alpha, \bar{y}_\alpha)$ and $(\bar{x}'_\alpha, \bar{y}'_\alpha)$ with the true positions of the observed positions (x_α, y_α) and (x'_α, y'_α), we see from Eq. (6.8) that Eq. (6.28) can be written as $S = (1/N)\sum_{\alpha=1}^N (\Delta x_\alpha^2 + \Delta y_\alpha^2 + \Delta x_\alpha'^2 + \Delta y_\alpha'^2)$, where we interpret Δx_α^2, and so on to be unknown variables rather than stochastic quantities. Equation (6.28) is commonly called the *geometric distance* (although it has the dimension of square length) or *reprojection error* of homography computation.

The minimum of

$$S_\alpha = \Delta x_\alpha^2 + \Delta y_\alpha^2 + \Delta x_\alpha'^2 + \Delta y_\alpha'^2 \tag{6.29}$$

with respect to $(\bar{x}_\alpha, \bar{y}_\alpha)$ and $(\bar{x}'_\alpha, \bar{y}'_\alpha)$ subject to the constraint $(\bar{\boldsymbol{\xi}}_\alpha^{(k)}, \boldsymbol{\theta}) = 0$, $k = 1, 2, 3$, can be approximated, if high-order terms in Δx_α, Δy_α, $\Delta x'_\alpha$, and $\Delta y'_\alpha$ are ignored, in the form (↪ Problem 6.4):

$$S_\alpha \approx \sum_{k,l=1}^{3} W_\alpha^{(kl)}(\boldsymbol{\xi}_\alpha^{(k)}, \boldsymbol{\theta})(\boldsymbol{\xi}_\alpha^{(l)}, \boldsymbol{\theta}), \quad W_\alpha^{(kl)} = \Big((\boldsymbol{\theta}, V_0^{(kl)}[\boldsymbol{\xi}_\alpha]\boldsymbol{\theta})\Big)_2^{-}. \tag{6.30}$$

If we use this, the geometric distance S of Eq. (6.28) is approximated by

$$J = \frac{1}{N}\sum_{\alpha=1}^{N}\sum_{k,l=1}^{3} W_\alpha^{(kl)}(\boldsymbol{\xi}_\alpha^{(k)}, \boldsymbol{\theta})(\boldsymbol{\xi}_\alpha^{(l)}, \boldsymbol{\theta}), \quad W_\alpha^{(kl)} = \Big((\boldsymbol{\theta}, V_0^{(kl)}[\boldsymbol{\xi}_\alpha]\boldsymbol{\theta})\Big)_2^{-}, \tag{6.31}$$

which is called the *Sampson error* of homography computation. This is known to be a very good approximation of the geometric distance S of Eq. (6.28).

6.5 FNS

The Sampson error of Eq. (6.31) can be minimized by modifying the FNS for ellipse fitting. We also call this modification *FNS* the (*fundamental numerical scheme*).

Procedure 6.4 (FNS)

1. Let $\boldsymbol{\theta} = \boldsymbol{\theta}_0 = \mathbf{0}$ and $W_\alpha^{(kl)} = \delta_{kl}$, $\alpha = 1, \ldots, N$, $k, l = 1, 2, 3$.
2. Compute the 9×9 matrices

$$\mathsf{M} = \frac{1}{N}\sum_{\alpha=1}^{N}\sum_{k,l=1}^{3} W_\alpha^{(kl)}\boldsymbol{\xi}_\alpha^{(k)}\boldsymbol{\xi}_\alpha^{(l)\top}, \quad \mathsf{L} = \frac{1}{N}\sum_{\alpha=1}^{N}\sum_{k,l=1}^{3} v_\alpha^{(k)} v_\alpha^{(l)} V_0^{(kl)}[\boldsymbol{\xi}_\alpha], \tag{6.32}$$

   where

$$v_\alpha^{(k)} = \sum_{l=1}^{3} W_\alpha^{(kl)}(\boldsymbol{\xi}_\alpha^{(l)}, \boldsymbol{\theta}). \tag{6.33}$$

3. Compute the 9×9 matrix

$$\mathsf{X} = \mathsf{M} - \mathsf{L}. \tag{6.34}$$

4. Solve the eigenvalue problem

$$\mathsf{X}\boldsymbol{\theta} = \lambda\boldsymbol{\theta}, \tag{6.35}$$

   and compute the unit eigenvector $\boldsymbol{\theta}$ for the smallest eigenvalue λ.
5. If $\boldsymbol{\theta} \approx \boldsymbol{\theta}_0$ up to sign, return $\boldsymbol{\theta}$ and stop. Else, update

$$W_\alpha^{(kl)} \leftarrow \Big((\boldsymbol{\theta}, V_0^{(kl)}[\boldsymbol{\xi}_\alpha]\boldsymbol{\theta})\Big)_2^{-}, \qquad \boldsymbol{\theta}_0 \leftarrow \boldsymbol{\theta}, \tag{6.36}$$

   and go back to Step 2.

Comments. This method computes the value $\boldsymbol{\theta}$ that satisfies $\nabla_{\boldsymbol{\theta}} J = \mathbf{0}$ for the Sampson error J of Eq. (6.31). Differentiating Eq. (6.31) we obtain the expression ($\hookrightarrow$ Problem 6.5(1),(2))

$$\nabla_{\boldsymbol{\theta}} J = 2(\mathsf{M} - \mathsf{L})\boldsymbol{\theta} = 2\mathsf{X}\boldsymbol{\theta}, \tag{6.37}$$

where M, L, and X are the matrices defined in Eqs. (6.32) and (6.34). It can be shown that when the FNS iterations have converged, the eigenvalue λ in Eq. (6.35) is 0 ($\hookrightarrow$ Problem 6.5(3)) and hence $\nabla_{\boldsymbol{\theta}} J = \mathbf{0}$ is satisfied. Initially, we set $\boldsymbol{\theta} = \mathbf{0}$, so $v_\alpha^{(k)}$ in Eq. (6.33) is 0, and hence $\mathsf{L} = \mathsf{O}$, which reduces Eq. (6.35) to $\mathsf{M}\boldsymbol{\theta} = \lambda\boldsymbol{\theta}$. Thus, we are starting the FNS iterations from the least squares solution.

6.6 Geometric Distance Minimization

We can compute the solution $\boldsymbol{\theta}$ that strictly minimizes the geometric distance S of Eq. (6.28) by iterating FNS: we modify the data $\boldsymbol{\xi}_\alpha^{(k)}$ using the $\boldsymbol{\theta}$ computed by FNS, minimize the Sampson error for the modified data, modify the data using that solution, and repeat this form several times. The procedure is as follows.

Procedure 6.5 (**Geometric distance minimization**)

1. Let $J_0^* = \infty$ (a sufficiently large number), $\hat{x}_\alpha = x_\alpha$, $\hat{y}_\alpha = y_\alpha$, $\hat{x}'_\alpha = x'_\alpha$, $\hat{y}'_\alpha = y'_\alpha$, and $\tilde{x}_\alpha = \tilde{y}_\alpha = \tilde{x}'_\alpha = \tilde{y}'_\alpha = 0$, $\alpha = 1, ..., N$, and define the following 4D vectors:

$$\mathbf{p}_\alpha = \begin{pmatrix} x_\alpha \\ y_\alpha \\ x'_\alpha \\ y'_\alpha \end{pmatrix}, \qquad \hat{\mathbf{p}}_\alpha = \begin{pmatrix} \hat{x}_\alpha \\ \hat{y}_\alpha \\ \hat{x}'_\alpha \\ \hat{y}'_\alpha \end{pmatrix}, \qquad \tilde{\mathbf{p}}_\alpha = \begin{pmatrix} \tilde{x}_\alpha \\ \tilde{y}_\alpha \\ \tilde{x}'_\alpha \\ \tilde{y}'_\alpha \end{pmatrix}. \tag{6.38}$$

2. Compute the matrices $\hat{\mathsf{T}}_\alpha^{(k)}$ from the matrices $\mathsf{T}_\alpha^{(k)}$ in Eq. (6.11) by replacing $\bar{x}_\alpha$, $\bar{y}_\alpha$, $\bar{x}'_\alpha$, and $\bar{y}'_\alpha$ by $\hat{x}_\alpha$, $\hat{y}_\alpha$, $\hat{x}'_\alpha$, and $\hat{y}'_\alpha$, respectively, and compute the normalized covariance matrices $V_0^{(kl)}[\hat{\boldsymbol{\xi}}_\alpha]$, $k, l = 1, 2, 3$, as follows.

$$V_0^{(kl)}[\hat{\boldsymbol{\xi}}_\alpha] = \hat{\mathsf{T}}_\alpha^{(k)}\hat{\mathsf{T}}_\alpha^{(l)\top}. \tag{6.39}$$

3. Compute the following $\hat{W}_\alpha^{(kl)}$, $k, l = 1, 2, 3$.

$$\hat{W}_\alpha^{(kl)} = \Big((\boldsymbol{\theta}, V_0^{(kl)}[\hat{\boldsymbol{\xi}}_\alpha]\boldsymbol{\theta})\Big)_2^-. \tag{6.40}$$

4. Compute the following 9D vectors $\boldsymbol{\xi}_\alpha^{(1)*}$, $\boldsymbol{\xi}_\alpha^{(2)*}$, and $\boldsymbol{\xi}_\alpha^{(3)*}$.

$$\boldsymbol{\xi}_\alpha^{(1)*} = \begin{pmatrix} 0 \\ 0 \\ 0 \\ -f_0\hat{x}_\alpha \\ -f_0\hat{y}_\alpha \\ -f_0^2 \\ \hat{x}_\alpha\hat{y}'_\alpha \\ \hat{y}_\alpha\hat{y}'_\alpha \\ f_0\hat{y}'_\alpha \end{pmatrix} + \hat{\mathsf{T}}_\alpha^{(1)}\tilde{\mathbf{p}}_\alpha, \qquad \boldsymbol{\xi}_\alpha^{(2)*} = \begin{pmatrix} f_0\hat{x}_\alpha \\ f_0\hat{y}_\alpha \\ f_0^2 \\ 0 \\ 0 \\ 0 \\ -\hat{x}_\alpha\hat{x}'_\alpha \\ -\hat{y}_\alpha\hat{x}'_\alpha \\ -f_0\hat{x}'_\alpha \end{pmatrix} + \hat{\mathsf{T}}_\alpha^{(2)}\tilde{\mathbf{p}}_\alpha,$$

$$\boldsymbol{\xi}_\alpha^{(3)*} = \begin{pmatrix} -\hat{x}_\alpha\hat{y}'_\alpha \\ -\hat{y}_\alpha\hat{y}'_\alpha \\ -f_0\hat{y}'_\alpha \\ \hat{x}_\alpha\hat{x}'_\alpha \\ \hat{y}_\alpha\hat{x}'_\alpha \\ f_0\hat{x}'_\alpha \\ 0 \\ 0 \\ 0 \end{pmatrix} + \hat{\mathsf{T}}_\alpha^{(3)}\tilde{\mathbf{p}}_\alpha. \tag{6.41}$$

5. Compute the value $\boldsymbol{\theta}$ that minimizes the following *modified Sampson error*.

$$J^* = \frac{1}{N}\sum_{\alpha=1}^{N}\sum_{k,l=1}^{3} \hat{W}_\alpha^{(kl)}(\boldsymbol{\xi}_\alpha^{(k)*}, \boldsymbol{\theta})(\boldsymbol{\xi}_\alpha^{(l)*}, \boldsymbol{\theta}). \tag{6.42}$$

6. Update $\tilde{\mathbf{p}}_\alpha$ and $\hat{\mathbf{p}}_\alpha$ to

$$\tilde{\mathbf{p}}_\alpha \leftarrow \sum_{k,l=1}^{3} \hat{W}_\alpha^{(kl)}(\boldsymbol{\xi}_\alpha^{(k)*}, \boldsymbol{\theta})\hat{\mathsf{T}}_\alpha^{(l)\top}\boldsymbol{\theta}, \qquad \hat{\mathbf{p}}_\alpha \leftarrow \mathbf{p}_\alpha - \tilde{\mathbf{p}}_\alpha. \tag{6.43}$$

7. Compute

$$J^* = \frac{1}{N}\sum_{\alpha=1}^{N} \|\tilde{\mathbf{p}}_\alpha\|^2. \tag{6.44}$$

If $J^* \approx J_0$, return $\boldsymbol{\theta}$ and stop. Else, let $J_0 \leftarrow J^*$, and go back to Step 2.

Comments. Because $\hat{x}_\alpha = x_\alpha$, $\hat{y}_\alpha = y_\alpha$, $\hat{x}'_\alpha = x'_\alpha$, $\hat{y}'_\alpha = y'_\alpha$, $\tilde{x}_\alpha = \tilde{y}_\alpha = \tilde{x}'_\alpha = \tilde{y}'_\alpha = 0$ in the initial iteration, the value $\boldsymbol{\xi}_\alpha^{(k)*}$ of Eq. (6.41) is the same as $\boldsymbol{\xi}_\alpha^{(k)}$. Hence, the modified Sampson error J^* in Eq. (6.42) is the same as the value J in Eq. (6.31), and the value $\boldsymbol{\theta}$ that minimizes J is first computed. We can write Eq. (6.28) as $S = (1/N)\sum_{\alpha=1}^{N} \|\mathbf{p}_\alpha - \bar{\mathbf{p}}_\alpha\|^2$, where $\bar{\mathbf{p}}_\alpha$ is the value of $\mathbf{p}_\alpha$ in Eq. (6.38) for $x_\alpha = \bar{x}_\alpha$, $y_\alpha = \bar{y}_\alpha$, $x'_\alpha = \bar{x}'_\alpha$, and $y'_\alpha = \bar{y}'_\alpha$. This is rewritten in the form

$$S = \frac{1}{N}\sum_{\alpha=1}^{N} \|\hat{\mathbf{p}}_\alpha + (\mathbf{p}_\alpha - \hat{\mathbf{p}}_\alpha) - \bar{\mathbf{p}}_\alpha\|^2 = \frac{1}{N}\sum_{\alpha=1}^{N} \|\hat{\mathbf{p}}_\alpha + \tilde{\mathbf{p}}_\alpha - \bar{\mathbf{p}}_\alpha\|^2, \tag{6.45}$$

where $\tilde{\mathbf{p}}_\alpha = \mathbf{p}_\alpha - \hat{\mathbf{p}}_\alpha$.

In the next iteration step, we regard the vector $\hat{\mathbf{p}}_\alpha$ for the corrected values $(\hat{x}_\alpha, \hat{y}_\alpha)$, $(\hat{x}'_\alpha, \hat{y}'_\alpha)$ as the input data and minimize Eq. (6.45) with respect to $\bar{\mathbf{p}}_\alpha$. Let $\hat{\hat{\mathbf{p}}}_\alpha$ be the solution. If we ignore high-order small terms in $\hat{\mathbf{p}}_\alpha - \bar{\mathbf{p}}_\alpha$ and rewrite Eq. (6.45), we obtain the modified Sampson error J^* in Eq. (6.42) (↪ Problem 6.6). We minimize this, regard $\hat{\hat{\mathbf{p}}}_\alpha$ as $\hat{\mathbf{p}}_\alpha$, and repeat the same process. Inasmuch as the current $\hat{\mathbf{p}}_\alpha$ is the best approximation of $\bar{\mathbf{p}}_\alpha$, the second equation of Eq. (6.43) implies that $\|\tilde{\mathbf{p}}_\alpha\|^2$ is the corresponding approximation of $\|\bar{\mathbf{p}}_\alpha - \mathbf{p}_\alpha\|^2$. Therefore we use Eq. (6.44) to evaluate the geometric distance S. Because the ignored high-order terms decrease after each iteration, we obtain in the end the value $\boldsymbol{\theta}$ that minimizes the geometric distance S, and Eq. (6.44) coincides with S. According to experiments, however, the correction of $\boldsymbol{\theta}$ by this procedure is very small: the three or four significant digits are unchanged in typical problems. Hence, we can practically identify FNS with a method to minimize the geometric distance.

6.7 Hyperaccurate Correction

As in ellipse fitting, the solution of FNS or geometric distance minimization theoretically has statistical bias, and we can improve the accuracy by subtracting it. However, this is in theory; in practice, not much difference results, as shown in the examples below. The procedure is as follows.

Procedure 6.6 (Hyperaccurate correction)

1. Compute $\boldsymbol{\theta}$ by FNS (Procedure 6.4).
2. Estimate σ^2 in the form
$$\hat{\sigma}^2 = \frac{(\boldsymbol{\theta}, \mathsf{M}\boldsymbol{\theta})}{2(1-4/N)}, \tag{6.46}$$
where M is the value of the matrix M in Eq. (6.32) after the FNS iterations have converged.
3. Compute the correction term
$$\Delta_c\boldsymbol{\theta} = \frac{\hat{\sigma}^2}{N^2}\mathsf{M}_8^- \sum_{\alpha=1}^{N}\sum_{k,l=1}^{3} W_\alpha^{(km)} W_\alpha^{(ln)} (\boldsymbol{\xi}_\alpha^{(l)}, \mathsf{M}_8^- V_0^{(mn)}[\boldsymbol{\xi}_\alpha]\boldsymbol{\theta})\boldsymbol{\xi}_\alpha^{(k)}, \tag{6.47}$$
where $W_\alpha^{(kl)}$ is the value of $W_\alpha^{(kl)}$ in Eq. (6.31) after the FNS iterations have converged.
4. Correct $\boldsymbol{\theta}$ to
$$\boldsymbol{\theta} \leftarrow \mathcal{N}[\boldsymbol{\theta} - \Delta_c\boldsymbol{\theta}], \tag{6.48}$$
where $\mathcal{N}[\,\cdot\,]$ denotes normalization to the unit norm.

Comments. The derivation of Eqs. (6.46) and (6.47) is given in Chap. 15. In practice, however, not much improvement is made by this correction, as shown by the numerical examples in Sect. 6.9 below.

6.8 Outlier Removal

Homography computation requires point correspondences between two images, and various techniques have been presented for correspondence detection. As in fundamental matrix computation, however, wrong matches (i.e., outliers) may be included. In Sect. 3.9, we determined the fundamental matrix such that the number of point pairs that approximately satisfy the resulting epipolar equation is as large as possible. Likewise, we can robustly determine the homography by choosing a solution such that the number of point pairs that approximately satisfy the resulting homography relationship is as large as possible. The best-known technique is RANSAC (Random Sample Consensus). The actual computation is as follows.

Procedure 6.7 (RANSAC)

1. From among the input point pairs randomly select four, and let $\boldsymbol{\xi}_1^{(k)}, \ldots, \boldsymbol{\xi}_4^{(k)}$, $k = 1, 2, 3$, be their vector representations in the form of $\boldsymbol{\xi}^{(k)}$ in Eq. (6.5).
2. Compute the eigenvector $\boldsymbol{\theta}$ of the matrix

$$\mathsf{M}_4 = \sum_{\alpha=1}^{4} \sum_{k=1}^{3} \boldsymbol{\xi}_\alpha^{(k)} \boldsymbol{\xi}_\alpha^{(k)\top}, \tag{6.49}$$

   for the smallest eigenvalue.
3. Store that $\boldsymbol{\theta}$ as a candidate, and count the number n of point pairs in the input that satisfy

$$\sum_{k,l=1}^{3} W^{(kl)} (\boldsymbol{\xi}^{(k)}, \boldsymbol{\theta})(\boldsymbol{\xi}^{(l)}, \boldsymbol{\theta}) < 2d^2, \quad W^{(kl)} = \Big((\boldsymbol{\theta}, V_0^{(kl)}[\boldsymbol{\xi}]\boldsymbol{\theta}) \Big)_2^-, \tag{6.50}$$

   where d is a threshold for admissible deviation of each point of a corresponding pair, for example, $d = 2$ (pixels). Store that n.
4. Select a new set of four point pairs from the input, and do the same. Repeat this many times, and return from among the stored candidates the one for which n is the largest.

Comments Step 2 computes the homography from the four point pairs selected in Step 1. The homography matrix H has nine elements, but it has scale indeterminacy, therefore the degree of freedom is eight. The homography equation of Eq. (6.1) for one point pair consists of two equations; if they are written in the form of Eq. (6.6), only two are independent. Therefore we can determine H by solving simultaneous equations in the form of Eqs. (6.1) or (6.6) for four point pairs. However, the use of least squares (↪ Problem 6.1) is easier to program; the coefficient 1/4 is omitted for M_4, inasmuch as the computed eigenvectors are the same. Step 3 tests to what extent each point pair satisfies the homography given by the candidate $\boldsymbol{\theta}$. This is measured by Eq. (6.29), which describes the sum of squares of the minimum distances necessary to move the point pair (x, y) and (x', y') to $(\bar{x}, \bar{y})$ and $(\bar{x}', \bar{y}')$ such that the homography equation is satisfied. This sum of squares can be approximated by Eq. (6.30). As in ellipse fitting and fundamental matrix computation, we can directly go on to the next sampling if the count n is smaller than the stored count, and we can stop if the current value of n is not updated over a fixed number of samplings. In the end, those point pairs that do not satisfy Eq. (6.50) for the chosen $\boldsymbol{\theta}$ are removed as outliers.

6.9 Examples

Figure 6.2 shows simulated images of a planar grid surface in the scene viewed from two different positions. The image size is assumed to be 500 × 500 pixels. To the x and y coordinates of each grid point is added independent Gaussian noise of mean 0 and standard deviation 1 (pixel). The true value of the homography matrix between these two images is

$$\bar{\mathsf{H}} = \begin{pmatrix} 0.57773 & 0.00000 & 0.00000 \\ 0.00000 & 0.47171 & 0.00000 \\ 0.00000 & -0.31587 & 0.57773 \end{pmatrix}. \tag{6.51}$$

Computing the homography matrix from the noisy grid point correspondences of Fig. 6.2, we obtain

$$\mathsf{H}^{(0)} = \begin{pmatrix} 0.57483 & -0.00026 & -0.00018 \\ 0.00153 & 0.47107 & -0.00010 \\ -0.00706 & -0.33990 & 0.57626 \end{pmatrix}, \quad \mathsf{H}^{(1)} = \begin{pmatrix} 0.57690 & -0.00023 & -0.00018 \\ 0.00155 & 0.47284 & 0.00001 \\ -0.00679 & -0.33143 & 0.57768 \end{pmatrix}, \tag{6.52}$$

where $\mathsf{H}^{(0)}$ is the value obtained by least squares (↪ Problem 6.1), and $\mathsf{H}^{(1)}$ is the value obtained by hyper-renormalization (Procedure 6.3). If we use FNS (Procedure 6.4) and geometric distance minimization (Procedure 6.5), we obtain, respectively,

$$\mathsf{H}^{(2)} = \begin{pmatrix} 0.57694 & -0.00020 & -0.00018 \\ 0.00158 & 0.47282 & 0.00001 \\ -0.00671 & -0.33138 & 0.57769 \end{pmatrix}, \quad \mathsf{H}^{(3)} = \begin{pmatrix} 0.57695 & -0.00020 & -0.00018 \\ 0.00158 & 0.47282 & 0.00001 \\ -0.00571 & -0.33135 & 0.57769 \end{pmatrix}. \tag{6.53}$$

If we measure the error of the computed $\mathsf{H} = (H_{ij})$ by

$$E = \sqrt{\sum_{i,j=1}^{3} (H_{ij} - \bar{H}_{ij})^2}, \tag{6.54}$$

we obtain the result in Table 6.1, where some other methods are also included. As we can see from this, the accuracy of least squares and iterative reweight is low,

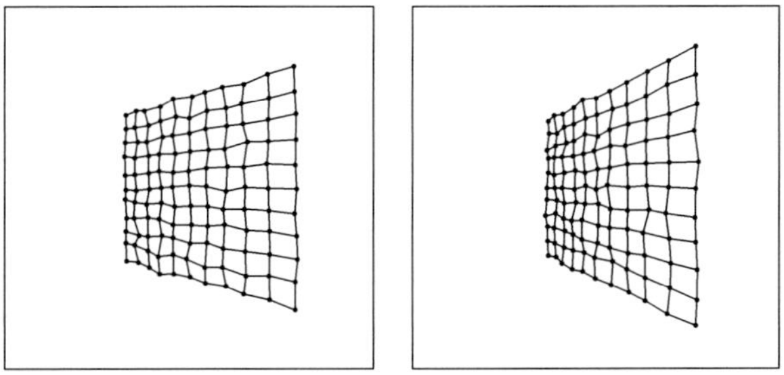

Fig. 6.2 Simulated images of a planar grid surface

Table 6.1 The error of homography computation for different methods

Method	E
Least squares	1.15042×10^{-2}
Iterative reweight	1.07295×10^{-2}
Taubin	0.73568×10^{-2}
Renormalization	0.71149×10^{-2}
HyperLS	0.73513×10^{-2}
Hyper-renormalization	0.71154×10^{-2}
FNS	0.70337×10^{-2}
Geometric distance minimization	0.70304×10^{-2}
FNS + hyper-accurate correction	0.70296×10^{-2}

but all the other methods exhibit almost the same high accuracy. For this particular example, the hyper-accurate correction of FNS exhibits the smallest error, but the difference from hyper-renormalization and geometric distance minimization is very small.

6.10 Supplemental Note

The homography of Eq. (6.1) between two images of a planar scene is one of the most fundamental principles of computer vision and is discussed in many textbooks [1–3]. From the camera, a distant scene can be regarded as a picture painted on a planar surface placed far away, therefore the homography of Eq. (6.1) holds very well for whatever scene if the distance is sufficiently large, say, more than 10 m.

As mentioned in Chaps. 2 and 3, the method of renormalization was first considered for ellipse fitting and fundamental matrix computation. It was later extended to homography computation in Kanatani et al. [4], but the computational procedure is slightly different from Procedure 6.2. The form of Procedure 6.2 was presented in Kanatani et al. [5]. The fact that the Taubin method for ellipse fitting and fundamental matrix computation can be extended to homography computation was pointed out by

Rangarajan and Papamichalis [6] (↪ Problem 6.2), and the scheme was extended to HyperLS (↪ Problem 6.3) by Kanatani et al. [7] and to hyper-renormalization (Procedure 6.3) by the same authors [5].

Scoleri et al. [8] pointed out that the FNS for ellipse fitting and fundamental matrix computation can also be used for homography computation. However, their description was formal and symbolic in a high-dimensional space, and actual computation required a software tool. Kanatani and Niitsuma [9] reduced it to the directly computable numerical form of Procedure 6.4. The geometric distance minimization of Procedure 6.5 was obtained by Kanatani and Niitsuma [9] by extending the geometric distance minimization for ellipse fitting and fundamental matrix computation. The hyper-accurate correction of Procedure 6.6 was obtained by Kanatani and Sugaya [10] by extending the hyper-accurate correction for ellipse fitting and fundamental matrix computation.

As we can see from the numerical examples of Sect. 6.9, all the methods other than least squares and iterative reweight have almost the same accuracy. This is because, as experimentally confirmed, the bias of these solutions is very small. It can be inferred that this is perhaps due to the fact that although the ellipse equation contains square terms x^2 and y^2, the homography equation is, as seen from Eq. (6.5), a *bilinear form*, consisting of terms such as xx' and xy' and that noise in the different variables is assumed to be statistically independent between these variables. In particular, the vector $\mathbf{e}$ defined via $E[\Delta_2 \boldsymbol{\xi}_\alpha] = \sigma^2 \mathbf{e}$ that appears in ellipse fitting does not appear for homography computation due to Eq. (6.12). This also applies to the fundamental matrix computation and, in general, all geometric computations from multiple images, where the noise in each image is assumed to be independent of the noise in other images (this is discussed in more detail in Chaps. 14 and 15).

Problems

6.1 Specifically write down the procedure for least squares for computing homography as the computation of the initial solution of Procedure 6.1.

6.2 Specifically write down the procedure for the Taubin method as the computation of the initial solution of Procedure 6.2.

6.3 Specifically write down the HyperLS procedure as the computation of the initial solution of Procedure 6.3.

6.4 Show that the minimum of Eq. (6.29) subject to the constraint $(\bar{\boldsymbol{\xi}}_\alpha^{(k)}, \boldsymbol{\theta}) = 0$, $k, l = 1, 2, 3$, is approximated by (Eq. 6.30) if high-order terms in Δx_α, Δy_α, $\Delta x'_\alpha$, and $\Delta y'_\alpha$ are omitted.

6.5 (1) Show that the derivative of $W_\alpha^{(kl)}$ of Eq. (6.31) is written in the form:

$$\nabla_{\boldsymbol{\theta}} W_\alpha^{(kl)} = -2 \sum_{m,n=1}^{3} W_\alpha^{(km)} W_\alpha^{(ln)} V_0^{mn}[\boldsymbol{\xi}_\alpha]\boldsymbol{\theta}. \tag{6.55}$$

(2) Show that the derivative of J in Eq. (6.31) is written in the form of Eq. (6.37).
(3) Show that when the iterations of FNS have converged, the value λ in Eq. (6.35) is 0.

6.6 Show that the minimum of S in Eq. (6.45) is approximated by the modified Sampson error J^* in Eq. (6.42) if high-order terms in $\hat{\mathbf{p}}_\alpha - \bar{\mathbf{p}}_\alpha$ are omitted. Hint: Let $\Delta\hat{\mathbf{p}}_\alpha = \hat{\mathbf{p}}_\alpha - \bar{\mathbf{p}}_\alpha$, and rewrite Eq. (6.45) as $S = (1/N)\sum_{\alpha=1}^{N} \|\tilde{\mathbf{p}}_\alpha + \Delta\hat{\mathbf{p}}_\alpha\|^2$. Then, compute the value of $\Delta\hat{\mathbf{p}}_\alpha$ that minimizes S.

References

1. R. Hartley, A. Zisserman, *Multiple View Geometry in Computer Vision*, 2nd edn. (Cambridge University Press, Cambridge, U.K., 2003)
2. K. Kanatani, *Geometric Computation for Machine Vision* (Oxford University Press, Oxford, U.K., 1993)
3. K. Kanatani, *Statistical Optimization for Geometric Computation: Theory and Practice* (Elsevier, Amsterdam, The Netherlands, 1996). Reprinted by Dover, New York, U.S. (2005)
4. K. Kanatani, N. Ohta, Y. Kanazawa, Optimal homography computation with a reliability measure, in *IEICE Transactions on Information and Systems*, vol. E83-D, No. 7, pp. 1369–1374 (2000)
5. K. Kanatani, A. Al-Sharadqah, N. Chernov, Y. Sugaya, Hyper-renormalization: non-minimization approach for geometric estimation. IPSJ Trans. Comput. Vis. Appl. **6**, 143–159 (2014)
6. P. Rangarajan, P. Papamichalis, Estimating homographies without normalization, in *Proceedings of International Conference on Image Processing*, Cairo, Egypt, pp. 3517–3520 (2009)
7. K. Kanatani, P. Rangarajan, Y. Sugaya, H. Niitsuma, HyperLS and its applications. IPSJ Trans. Comput. Vis. Appl. **3**, 80–94 (2011)
8. T. Scoleri, W. Chojnacki, M.J. Brooks, A multi-objective parameter estimation for image mosaicing, in *Proceedings of 8th International Symposium on Signal Processing and its Applications*, Sydney, Australia, vol. 2, pp. 551–554 (2005)
9. K. Kanatani, H. Niitsuma, Optimal two-view planar triangulation. IPSJ Trans. Comput. Vis. Appl. **3**, 67–79 (2011)
10. K. Kanatani, Y. Sugaya, Hyperaccurate correction of maximum likelihood for geometric estimation. IPSJ Trans. Comput. Vis. Appl. **5**, 19–29 (2013)

Planar Triangulation

7

Abstract

This chapter describes the principles and procedure for computing the 3D position of a corresponding point pair between two images of a known planar surface by assuming knowledge of the camera matrices of the two cameras. This process is called planar triangulation. We first show that the homography between the two images is determined from the equation of the plane and the camera matrices. The principle of planar triangulation is to correct the corresponding point pair optimally such that the associated lines of sight intersect precisely at a point on the assumed plane, using knowledge of the statistical properties of image noise. It turns out that the procedure is closely related to the optimal homography computation described in the preceding chapter (Chap. 6).

7.1 Perspective Projection of a Plane

Suppose a planar surface in the scene is imaged by two cameras having camera matrices P and P'. If a point (X, Y, Z) on the plane is projected to (x, y) and (x', y') on the images, Eq. (4.2) implies the relations:

$$\begin{pmatrix} x/f_0 \\ y/f_0 \\ 1 \end{pmatrix} \simeq \mathsf{P} \begin{pmatrix} X \\ Y \\ Z \\ 1 \end{pmatrix}, \qquad \begin{pmatrix} x'/f_0 \\ y'/f_0 \\ 1 \end{pmatrix} \simeq \mathsf{P}' \begin{pmatrix} X \\ Y \\ Z \\ 1 \end{pmatrix}. \tag{7.1}$$

Let $Z = aX + bY + c$ be the equation of the plane; this does not describe a plane parallel to the Z-axis, in which case we let $X = aY + bZ + c$ or $Y = aX + bZ + c$,

K. Kanatani et al., *Guide to 3D Vision Computation*, Advances in Computer Vision and Pattern Recognition, DOI 10.1007/978-3-319-48493-8_7

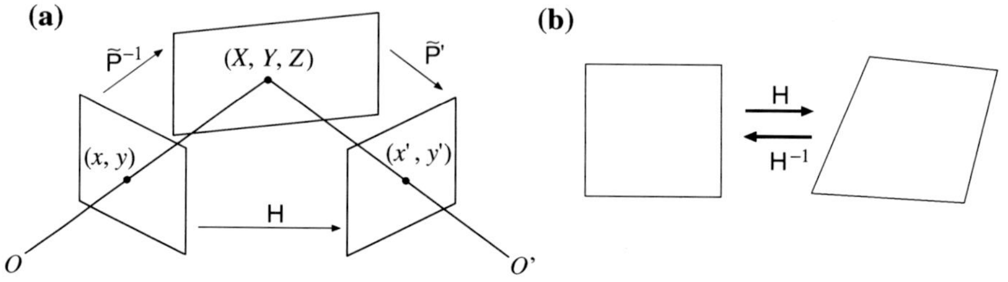

Fig. 7.1 **a** The perspective projection of a plane onto the image plane and its inverse are homographies. The homography between the two images is their composition. **b** A square is mapped to a general quadrilateral by a homography

and the subsequent discussions are the same. Inasmuch as

$$\begin{pmatrix} X \\ Y \\ Z \\ 1 \end{pmatrix} = \mathsf{C} \begin{pmatrix} X \\ Y \\ 1 \end{pmatrix}, \qquad \mathsf{C} \equiv \begin{pmatrix} 1 & 0 & 0 \\ 0 & 1 & 0 \\ a & b & c \\ 0 & 0 & 1 \end{pmatrix}, \tag{7.2}$$

Equation (7.1) is written in the form

$$\begin{pmatrix} x/f_0 \\ y/f_0 \\ 1 \end{pmatrix} \simeq \tilde{\mathsf{P}} \begin{pmatrix} X \\ Y \\ 1 \end{pmatrix}, \quad \begin{pmatrix} x'/f_0 \\ y'/f_0 \\ 1 \end{pmatrix} \simeq \tilde{\mathsf{P}}' \begin{pmatrix} X \\ Y \\ 1 \end{pmatrix}, \quad \tilde{\mathsf{P}} \equiv \mathsf{PC}, \quad \tilde{\mathsf{P}}' \equiv \mathsf{P}'\mathsf{C}. \tag{7.3}$$

A point on the plane and its projection in the image are in one-to-one correspondence, thus the 3×3 matrices $\tilde{\mathsf{P}}$ and $\tilde{\mathsf{P}}'$ are both nonsingular. This means that *the perspective projection of a plane onto the image and its inverse are both homographies*. The correspondence between the two images is described by the composition Fig. (7.1a):

$$\begin{pmatrix} x'/f_0 \\ y'/f_0 \\ 1 \end{pmatrix} \simeq \mathsf{H} \begin{pmatrix} x/f_0 \\ y/f_0 \\ 1 \end{pmatrix}, \qquad \mathsf{H} \equiv \tilde{\mathsf{P}}'\tilde{\mathsf{P}}^{-1}. \tag{7.4}$$

This has the form of Eq. (6.2), and because $\mathsf{H} = \tilde{\mathsf{P}}'\tilde{\mathsf{P}}^{-1}$ is nonsingular, this is a homography. The set of all homographies constitutes a group with respect to composition, called the *homography group*, or the *group of projective transformations* ($\hookrightarrow$ Problems 7.1). A homography maps a line to a line ($\hookrightarrow$ Problem 7.2). Thus a quadrilateral is mapped to a quadrilateral. However, the length, the angle, and the ratio are not preserved, therefore a square is mapped to a general quadrilateral (Fig. 7.1b). An ellipse is mapped to an ellipse, and a circle is mapped to a general ellipse. As can be seen from Eq. (7.4), the homography matrix H between two planar surface images is determined by the equation of the plane and the two camera matrices P and P'.

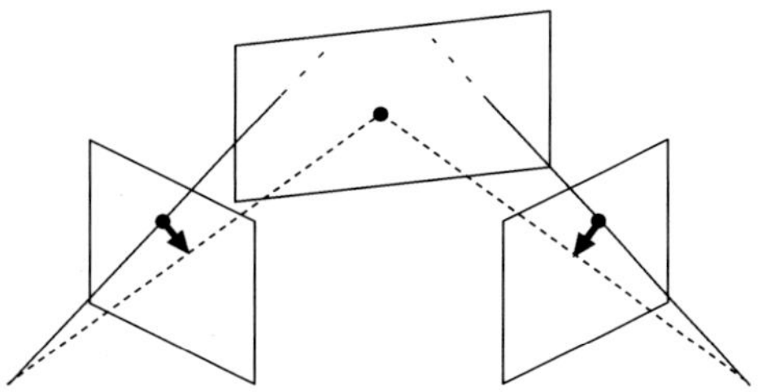

Fig. 7.2 Correcting observed corresponding points in the shortest distance such that their lines of sight intersect precisely at a point on the specified plane

7.2 Planar Triangulation

In the following, we assume that the equation, the plane, and the camera matrices P and P', are known. Therefore the homography matrix H is also known. Note that given corresponding points (x, y) and (x', y') between the two images, their 3D position can be computed by the triangulation procedure described in Chap. 4. However, the computed position (X, Y, Z) may not be on the assumed plane. We call the procedure of mapping a corresponding point pair onto a specified plane *planar triangulation*. A naive idea is first to compute the 3D point by the triangulation procedure of Chap. 4 and then to replace it by the closest point on the plane, that is, the foot of the perpendicular line from it. However, noise occurs in images therefore the "closeness" in the scene has no special logical meaning.

In Chap. 4, we modified noisy corresponding points (x, y) and (x', y') in the shortest distance such that *their lines of sight intersect in the scene*, where "in the shortest distance" means that the sum of squares of the displacement distances in the images is minimized. If we want to reconstruct a point on a known plane, it is reasonable to modify (x, y) and (x', y') to $(\bar{x}, \bar{y})$ and $(\bar{x}', \bar{y}')$ in the shortest distance such that *their lines of sight intersect precisely at a point on that plane* (Fig. 7.2). Mathematically, we minimize the sum of squares

$$S = (x - \bar{x})^2 + (y - \bar{y})^2 + (x' - \bar{x}')^2 + (y - \bar{y}')^2, \tag{7.5}$$

subject to the constraint that $(\bar{x}, \bar{y})$ and $(\bar{x}', \bar{y}')$ satisfy the homography of Eq. (7.4).

7.3 Procedure of Planar Triangulation

We write the homography matrix H, which we assume to be known, as the 9D vector $\boldsymbol{\theta}$ in Eq. (6.5). If we define the 9D vector $\boldsymbol{\xi}^{(k)}$ as in Eq. (6.5), the homography of Eq. (7.4) is written as $(\boldsymbol{\xi}^{(k)}, \boldsymbol{\theta}) = 0$, $k = 1, 2, 3$. The values of $(\hat{x}, \hat{y})$ and $(\hat{x}', \hat{y}')$ that minimize Eq. (7.5) subject to Eq. (7.4) are computed by the following procedure.

Procedure 7.1 (Optimal correspondence correction)

1. Let $S = \infty$ (a sufficiently large number), $\hat{x} = x$, $\hat{y} = y$, $\hat{x}' = x'$, $\hat{y}' = y'$, and $\tilde{x} = \tilde{y} = \tilde{x}' = \tilde{y}' = 0$, and define the 4D vectors:

$$\mathbf{p} = \begin{pmatrix} x \\ y \\ x' \\ y' \end{pmatrix}, \qquad \hat{\mathbf{p}} = \begin{pmatrix} \hat{x} \\ \hat{y} \\ \hat{x}' \\ \hat{y}' \end{pmatrix}, \qquad \tilde{\mathbf{p}} = \begin{pmatrix} \tilde{x} \\ \tilde{y} \\ \tilde{x}' \\ \tilde{y}' \end{pmatrix}. \tag{7.6}$$

2. Define the following 9×4 matrices $\mathsf{T}^{(k)}$, $k = 1, 2, 3$.

$$\mathsf{T}^{(1)} = \begin{pmatrix} 0 & 0 & 0 & 0 \\ 0 & 0 & 0 & 0 \\ 0 & 0 & 0 & 0 \\ -f_0 & 0 & 0 & 0 \\ 0 & -f_0 & 0 & 0 \\ 0 & 0 & 0 & 0 \\ \hat{y}' & 0 & 0 & \hat{x} \\ 0 & \hat{y}' & 0 & \hat{y} \\ 0 & 0 & 0 & f_0 \end{pmatrix}, \qquad \mathsf{T}^{(2)} = \begin{pmatrix} f_0 & 0 & 0 & 0 \\ 0 & f_0 & 0 & 0 \\ 0 & 0 & 0 & 0 \\ 0 & 0 & 0 & 0 \\ 0 & 0 & 0 & 0 \\ 0 & 0 & 0 & 0 \\ -\hat{x}' & 0 & -\hat{x} & 0 \\ 0 & -\hat{x}' & -\hat{y} & 0 \\ 0 & 0 & -f_0 & 0 \end{pmatrix},$$

$$\mathsf{T}^{(3)} = \begin{pmatrix} -\hat{y}' & 0 & 0 & -\hat{x} \\ 0 & -\hat{y}' & 0 & -\hat{y} \\ 0 & 0 & 0 & -f_0 \\ \hat{x}' & 0 & \hat{x} & 0 \\ 0 & \hat{x}' & \hat{y} & 0 \\ 0 & 0 & f_0 & 0 \\ 0 & 0 & 0 & 0 \\ 0 & 0 & 0 & 0 \\ 0 & 0 & 0 & 0 \end{pmatrix}. \tag{7.7}$$

3. Compute the following 9×9 matrices $V_0^{(kl)}[\hat{\boldsymbol{\xi}}]$ and the coefficients $\hat{W}^{(kl)}$, $k, l = 1, 2, 3$.

$$V_0^{(kl)}[\hat{\boldsymbol{\xi}}] = \hat{\mathsf{T}}^{(k)}\hat{\mathsf{T}}^{(l)}, \qquad \hat{W}^{(kl)} = \Big((\boldsymbol{\theta}, V_0^{(kl)}[\hat{\boldsymbol{\xi}}]\boldsymbol{\theta})\Big)_2^{-}. \tag{7.8}$$

4. Compute the following 9D vectors $\boldsymbol{\xi}^{(1)*}$, $\boldsymbol{\xi}^{(2)*}$, and $\boldsymbol{\xi}^{(3)*}$.

$$\boldsymbol{\xi}^{(1)*} = \begin{pmatrix} 0 \\ 0 \\ 0 \\ -f_0\hat{x} \\ -f_0\hat{y} \\ -f_0^2 \\ \hat{x}\hat{y}' \\ \hat{y}\hat{y}' \\ f_0\hat{y}' \end{pmatrix} + \hat{\mathsf{T}}^{(1)}\tilde{\mathbf{p}}, \qquad \boldsymbol{\xi}^{(2)*} = \begin{pmatrix} f_0\hat{x} \\ f_0\hat{y} \\ f_0^2 \\ 0 \\ 0 \\ 0 \\ -\hat{x}\hat{x}' \\ -\hat{y}\hat{x}' \\ -f_0\hat{x}' \end{pmatrix} + \hat{\mathsf{T}}^{(2)}\tilde{\mathbf{p}},$$

$$\boldsymbol{\xi}^{(3)*} = \begin{pmatrix} -\hat{x}\hat{y}' \\ -\hat{y}\hat{y}' \\ -f_0\hat{y}' \\ \hat{x}\hat{x}' \\ \hat{y}\hat{x}' \\ f_0\hat{x}' \\ 0 \\ 0 \\ 0 \end{pmatrix} + \hat{\mathsf{T}}^{(3)}\tilde{\mathbf{p}}. \tag{7.9}$$

5. Update $\tilde{\mathbf{p}}$ and $\hat{\mathbf{p}}$ to

$$\tilde{\mathbf{p}} \leftarrow \sum_{k,l=1}^{3} \hat{W}^{(kl)}(\boldsymbol{\xi}^{(k)*}, \boldsymbol{\theta})\hat{\mathsf{T}}^{(l)\top}\boldsymbol{\theta}, \qquad \hat{\mathbf{p}} \leftarrow \mathbf{p} - \tilde{\mathbf{p}}. \tag{7.10}$$

6. If $\|\tilde{\mathbf{p}}\|^2 \approx S$, return $(\hat{x}, \hat{y})$ and $(\hat{x}', \hat{y}')$ and stop. Else, let $S \leftarrow \|\tilde{\mathbf{p}}\|^2$, and go back to Step 2.

Comments. This procedure is *identical except for Step 5* to Procedure 6.5 for geometric distance minimization for homography computation. Thus, we can view Procedure 6.5 as *doing planar triangulation and updating* $\boldsymbol{\theta}$, which we assume is known in this chapter. The derivation of the above Procedure 7.1 is basically identical to Procedure 6.5. If we let $\bar{\mathbf{p}}$ be the value of $\mathbf{p}$ of Eq. (7.6) for $x = \bar{x}$, $y = \bar{y}$, $x' = \bar{x}'$, $y' = \bar{y}'$, Eq. (7.5) can be written as $S = \|\mathbf{p} - \bar{\mathbf{p}}\|^2$. After the initial correction $\hat{\mathbf{p}}$ is obtained, we rewrite this in the form

$$S = \|\hat{\mathbf{p}} + (\mathbf{p} - \hat{\mathbf{p}}) - \bar{\mathbf{p}}\|^2 = \|\hat{\mathbf{p}} + \tilde{\mathbf{p}} - \bar{\mathbf{p}}\|^2, \tag{7.11}$$

where we put $\tilde{\mathbf{p}} = \mathbf{p} - \hat{\mathbf{p}}$, which describes the displacement for correction. Then, we regard the corrected value $\hat{\mathbf{p}}$ as the input value and compute the value $\hat{\hat{\mathbf{p}}}$ of $\bar{\mathbf{p}}$ that minimizes S, using Lagrange multipliers and ignoring high-order terms in $\hat{\mathbf{p}} - \bar{\mathbf{p}}$ ($\hookrightarrow$ Problems 7.3). Because $\hat{\mathbf{p}} - \bar{\mathbf{p}}$ is a smaller-order quantity than $\mathbf{p} - \bar{\mathbf{p}}$, the resulting value $\hat{\hat{\mathbf{p}}}$ is a better approximation of $\bar{\mathbf{p}}$ than $\hat{\mathbf{p}}$. We identify $\hat{\hat{\mathbf{p}}}$ as $\hat{\mathbf{p}}$ and write S in the form of Eq. (7.11). Repeating this process, we obtain a better approximation iteration by iteration. We stop if the displacement for correction no longer changes. In the end, the value of $\|\tilde{\mathbf{p}}\|^2$ coincides with the value of S in Eq. (7.5).

The positions $(\hat{x}, \hat{y})$ and $(\hat{x}', \hat{y}')$ thus corrected strictly satisfy the specified homography (and thus the epipolar equation also holds $\hookrightarrow$ Problem 7.4). Therefore the 3D position (X, Y, Z) computed by Procedure 4.1 is guaranteed to be on the specified plane.

7.4 Examples

Figure 7.3 shows the grid positions obtained by optimally correcting the noisy grid points in Fig. 6.2 by Procedure 7.1, using the knowledge of the planar surface equation. As compared with Fig. 6.2, we can see that the noise is somewhat reduced by

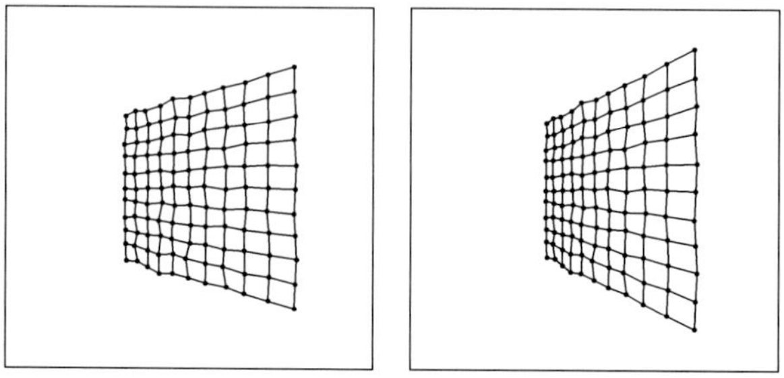

Fig. 7.3 The grid-point positions obtained by optimally correcting the noisy grid points in Fig. 6.2 by Procedure 7.1

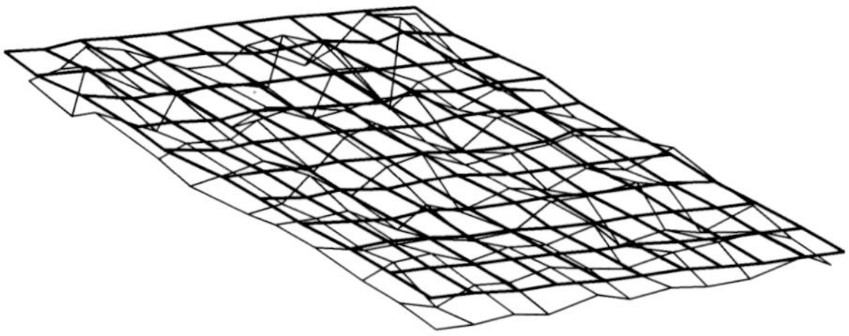

Fig. 7.4 The thick lines connect the 3D positions of the grid points reconstructed from the two images in Fig. 7.3 by Procedure 4.1. The thin lines connect the 3D positions reconstructed without considering the planarity of the grid points, correcting the grid points in Fig. 6.2 by Procedure 4.2 and computing the 3D positions by Procedure 4.1. Both are viewed from the same direction

considering the planarity of the points. The thick lines in Fig. 7.4 connect the 3D positions of the grid points reconstructed from the two images in Fig. 7.3 by Procedure 4.1. The thin lines connect the 3D positions reconstructed without considering the planarity of the points. Both are viewed from the same direction. We can see that the reconstructed 3D points after the correction by Procedure 7.1 all lie on a plane. We also see that the distortion is smaller to some extent than the case of not considering the planarity of the points.

7.5 Supplemental Note

The technique for computing the 3D positions of corresponding points in two images on a specified plane was presented by Kanazawa and Kanatani [4] in 1995. The computed solution corresponds to the solution in the initial iteration of Procedure 7.1. The technique of iterating it to compute the exact solution was presented in Kanatani and Niitsuma [3], where it was pointed out that the procedure results from the geometric distance minimization for homography computation by removing the updating step of the homography.

On the other hand, Chum et al. [1] presented an algebraic method, reducing the computation to solving an eight-degree polynomial. This is, like the triangulation of Hartley and Sturm [2] without assuming planarity, motivated by a theoretical interest of computing a globally optimal solution. This, however, does not have much practical meaning, because Procedure 7.1 is not a gradient-based search in 3D, which may fall into a local minimum; it is an iterative correction scheme "in the 2D image domain" starting from the *observed image positions* of the corresponding pair and converging in their vicinity. The computation is very efficient; usually, two or three iterations are sufficient.

Problems

7.1 Show that homographies between images form a group with respect to composition, that is, that the associativity is satisfied, and that the identity and the inverse exist.

7.2 To what line does the homography with matrix H map a line $ax + by + cf_0 = 0$?

7.3 Show that, given an approximation $\hat{\mathbf{p}}$ of $\mathbf{p}$, the value $\bar{\mathbf{p}}$ that minimizes Eq. (7.11) is obtained, if high-order terms in $\hat{\mathbf{p}} - \bar{\mathbf{p}}$ are ignored, in the form:

$$\hat{\hat{\mathbf{p}}} = \mathbf{p} - \sum_{k=1}^{3}\sum_{l=1}^{3} \hat{W}^{(kl)}(\boldsymbol{\xi}^{*(k)}, \boldsymbol{\theta})\hat{\mathsf{T}}^{(l)\top}\boldsymbol{\theta}. \tag{7.12}$$

Hint: Let $\Delta\hat{\mathbf{p}} = \hat{\mathbf{p}} - \bar{\mathbf{p}}$, and write Eq. (7.11) as $S = \|\tilde{\mathbf{p}} + \Delta\hat{\mathbf{p}}\|^2$. Then, compute the value $\Delta\hat{\mathbf{p}}$ that minimizes S.

7.4 Show that if two images are related by a homography with matrix H, the fundamental matrix F between them satisfies the identity:

$$\mathsf{FH} + \mathsf{H}^\top\mathsf{F}^\top = \mathsf{O}. \tag{7.13}$$

References

1. O. Chum, T. Pajdla, P. Sturm, The geometric error for homographies. Comput. Vis. Image Unders. **97**(1), 86–102 (2002)
2. R. Hartley, P. Sturm, Triangulation. Comput. Vis. Image Unders. **68**(2), 146–157 (1997)
3. K. Kanatani, H. Niitsuma, Optimal two-view planar triangulation. IPSJ Trans. Comput. Vis. Appl. **3**, 67–79 (2011)
4. Y. Kanazawa, K. Kanatani, Direct reconstruction of planar surface by stereo vision, in *IEICE Transactions on Information and Systems*, Vol. E78-D, No. 10, pp. 917–922 (1995)

3D Reconstruction of a Plane

8

Abstract

This chapter describes the procedure for computing, from corresponding points between two images of a planar scene, the 3D position of that plane and the camera matrices of the two cameras that took those images. First, we express the matrix of the homography between the two images in terms of the 3D position of the plane and the two camera matrices. We then show how to decompose the homography matrix into the 3D position of the plane and the two camera matrices in an analytical form. The solution is not unique; we describe the procedure for selecting the correct one. Once the camera matrices are obtained, the 3D positions of the corresponding point pairs are computed by the planar triangulation procedure of the preceding chapter.

8.1 Self-calibration with a Plane

In order to do the planar triangulation of the preceding chapter, we need to know the camera matrices P and P' and the equation of the plane. We now consider a self-calibration procedure for computing them only from the homography matrix H between two images. Because the homography matrix has scale indeterminacy, it has eight degrees of freedom. A plane in space is specified by three parameters. For self-calibration to be possible, therefore, the camera matrices P and P' must be specified by five parameters. To reduce the degrees of freedom, we let the world coordinate system coincide with the first camera coordinate system, as we did in Chap. 5. We also assume that the principal point is known and that no image distortions exist. Still, even if we take into account the fact that the relative camera translation is determined only up to scale, the unknowns have seven degrees of freedom: the focal lengths f and f' of the two cameras, the relative camera translation $\mathbf{t}$ (two degrees of freedom), and their relative rotation R (three degrees of freedom). In order to reduce

K. Kanatani et al., *Guide to 3D Vision Computation*, Advances in Computer Vision and Pattern Recognition, DOI 10.1007/978-3-319-48493-8_8

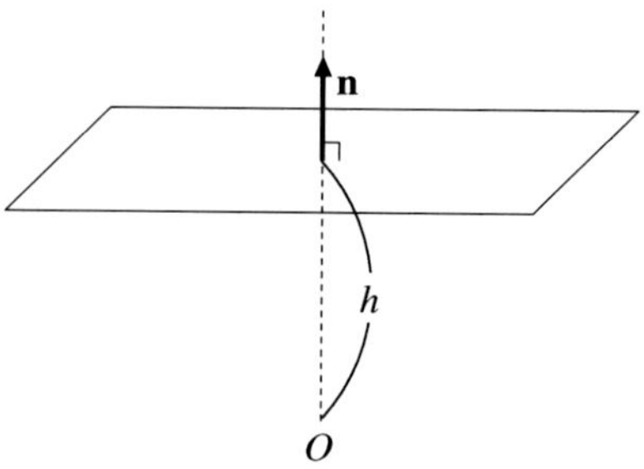

Fig. 8.1 A plane space is specified by the unit surface normal **n** and the distance h from the origin

the degrees of freedom further, we assume that the focal lengths f and f' are known. They do not depend on the object we see, thus we can determine them in advance, using a reference board, or we can read them out from the electronic data provided by today's digital cameras.

A plane is specified by an equation in the form

$$n_1X + n_2Y + n_3Z = h. \tag{8.1}$$

Inasmuch as multiplication by any nonzero constant on both sides does not change the represented plane, we normalize the coefficient so that $\mathbf{n} = (n_1, n_2, n_3)^\top$ is a unit vector. Then, $\mathbf{n}$ describes the unit surface normal, and h is its signed distance (positive in the direction of $\mathbf{n}$) from the origin (Fig. 8.1); we call $\mathbf{n}$ and h the *surface parameters*. We now consider a procedure for computing the motion parameters $\{\mathbf{t}, \mathsf{R}\}$ and the surface parameters $\{\mathbf{n}, h\}$ from the homography matrix H. The absolute scale of the scene cannot be reconstructed from images alone, as pointed out in Chap. 5, therefore we assume that $\mathbf{t}$ is not $\mathbf{0}$ and normalize it to the unit norm, as we did in Chap. 5.

8.2 Computation of Surface Parameters and Motion Parameters

If the camera focal lengths f and f' are known, the motion parameters $\mathbf{t}$ (unit vector) and R (rotation matrix) have five degrees of freedom, and the surface parameters $\mathbf{n}$ (unit vector) and h have three degrees of freedom. Because the homography matrix H has eight degrees of freedom, we can in principle compute all the unknowns. In order to restrict the solution, however, we further assume the following.

- The camera translation is not $\mathbf{0}$; that is, we are viewing the plane from different positions.
- The plane does not pass through the viewpoint of either camera; that is, we are not seeing the plane as a "line."
- The cameras are on the same sides of the plane; that is, we are seeing only one side of the plane.

- The distance h to the plane is positive; that is, the surface unit normal is defined in the direction away from the cameras.

Then, we obtain four sets of solutions as follows (↪ Problems 8.2–8.5).

Procedure 8.1 (Computing the plane and the motion)

1. Transform the input homography matrix H into

$$\tilde{\mathsf{H}} = \begin{pmatrix} f_0 & 0 & 0 \\ 0 & f_0 & 0 \\ 0 & 0 & f' \end{pmatrix} \mathsf{H} \begin{pmatrix} 1/f_0 & 0 & 0 \\ 0 & 1/f_0 & 0 \\ 0 & 0 & 1/f \end{pmatrix}. \tag{8.2}$$

2. Normalize $\tilde{\mathsf{H}}$ to determinant 1:

$$\tilde{\mathsf{H}} \leftarrow \frac{\tilde{\mathsf{H}}}{\sqrt[3]{|\tilde{\mathsf{H}}|}}. \tag{8.3}$$

3. Compute the SVD (singular value decomposition) of $\tilde{\mathsf{H}}$,

$$\tilde{\mathsf{H}} = \mathsf{U} \begin{pmatrix} \sigma_1 & 0 & 0 \\ 0 & \sigma_2 & 0 \\ 0 & 0 & \sigma_3 \end{pmatrix} \mathsf{V}^\top, \qquad \sigma_1 \geq \sigma_2 \geq \sigma_3 > 0, \tag{8.4}$$

where U and V are orthogonal matrices.
4. Let $\mathbf{v}_1$, $\mathbf{v}_2$, and $\mathbf{v}_3$ be the columns of V. Compute the surface parameters $\{\mathbf{n}, h\}$ in the form

$$\mathbf{n} = \mathcal{N}[\sqrt{\sigma_1^2 - \sigma_2^2}\mathbf{v}_1 \pm \sqrt{\sigma_2^2 - \sigma_3^2}\mathbf{v}_3], \qquad h = \frac{\sigma_2}{\sigma_1 - \sigma_3}, \tag{8.5}$$

where $\mathcal{N}[\,\cdot\,]$ designates normalization to the unit norm.
5. Compute the motion parameters $\{\mathbf{t}, \mathsf{R}\}$ in the form

$$\mathbf{t} = \mathcal{N}[-\sigma_3\sqrt{\sigma_1^2 - \sigma_2^2}\mathbf{v}_1 \pm \sigma_1\sqrt{\sigma_2^2 - \sigma_3^2}\mathbf{v}_3], \qquad \mathsf{R} = \frac{1}{\sigma_2}\Big(\mathsf{I} + \frac{\sigma_2^3\mathbf{n}\mathbf{t}^\top}{h}\Big)\tilde{\mathsf{H}}^\top, \tag{8.6}$$

where the double sign $\pm$ corresponds to that in Eq. (8.5).
6. Return the resulting solutions along with additional solutions obtained by simultaneously changing the signs of $\mathbf{n}$ and $\mathbf{t}$.

8.3 Selection of the Solution

The reason that we obtain multiple solutions by Procedure 8.1 originates from the perspective projection modeling of the camera imaging geometry (Fig. 4.1), where a 3D point is imaged to the intersection of the image plane with the ray passing through it and the viewpoint, without considering where the 3D point is on the ray.

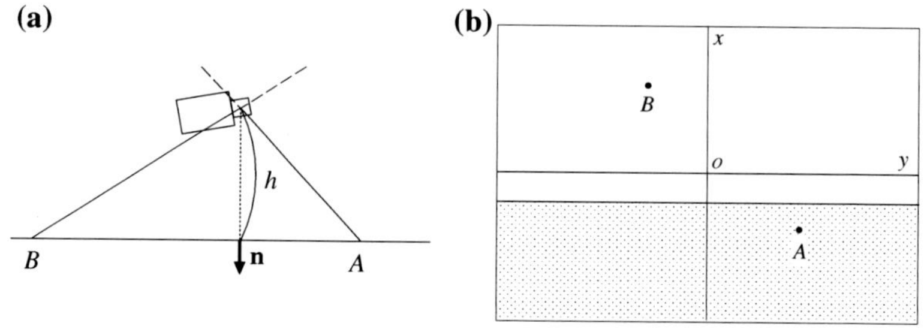

Fig. 8.2 **a** The mathematical model of perspective projection allows the scene behind the camera to be seen. **b** The projection of an infinitely extending plane corresponds to one side of the vanishing line in the image. The other side corresponds to the part behind the camera. Point A is the actual image of the point A in (**a**) and point B is the "mathematical" image of the point B in (**a**)

Thus points "behind" the camera are also imaged by this model (Fig. 8.2a). This is also the reason that the 3D reconstruction from two views in Chap. 5 computes the true shape in front of the cameras and its mirror image behind them; we removed the mirror image solution by judging the position of the computed shape relative to the camera.

For a planar scene, however, the following situation occurs. If we take an image of an infinitely extended plane, its projection is only a part of the image plane; its boundary is a line called the *vanishing line*. In Fig. 8.2b, the part below the vanishing line is the actual projection of the plane, but mathematically the upper part is the projection of the scene "behind" the camera. Therefore if an image such as Fig. 8.2b is given as a projection of a planar surface without the knowledge of where the vanishing line lies and which part is the actually visible part, points A and B are both interpreted to be projections of points on the plane, as shown in Fig. 8.2a.

In a real situation, visible planar surfaces are all finite and projected onto some finite regions in the image. We assume below that we can judge which of the following situations is true.

Case 1: The principal point is in the region of the plane (or its extension).
Case 2: The principal point is outside the region of the plane (or its extension).

For example, suppose we can tell, based on other available information, that the image of Fig. 8.2b is in Case 2; that is, the principal point o is outside the region of the plane. If we have such knowledge, we can narrow down the four solutions of Procedure 8.1 by the following procedure.

Procedure 8.2 (Selection of solution 1)

1. If the first image is in Case 1, choose the solutions for which the Z component of $\mathbf{n}$ is positive; otherwise, choose the solutions for which it is negative.

2. If the second image is in Case 1, choose the solutions for which the Z component of $\mathsf{R}^\top \mathbf{n}$ is positive; otherwise, choose the solutions for which it is negative.

Comments In this chapter, we regard the viewpoint of the first camera as the world coordinate origin O and regard the optical axis as the Z axis. We also adopt the convention that the distance h to the plane is positive and that the unit surface normal $\mathbf{n}$ to the plane extends in the direction away from O. Hence, if the first image is in Case 1, the Z component of $\mathbf{n}$ should be positive. If it is in Case 2, $\mathbf{n}$ points toward the camera, therefore its Z component is negative (Fig. 8.2a). Inasmuch as the second camera is rotated by R relative to the first camera, the unit surface normal is, seen from the second camera, inversely rotated by $\mathsf{R}^\top$. Therefore if the second camera is in Case 1, the Z component of $\mathsf{R}^\top \mathbf{n}$ is positive, and negative for Case 2.

In most practical cases, we can choose one solution from the four candidates by Procedure 8.2, but mathematically there exist possibilities that two or more solutions remain. In that case, and in situations where the visible planar part is too small to judge if the principal point is included in the extension of the planar region or outside it, we compute the 3D positions of all the corresponding point pairs (x_α, x_α) and (x'_α, x'_α), $\alpha = 1, \ldots, N$, that we used for homography computation and check if all of them are in front of both cameras. The actual procedure is as follows.

Procedure 8.3 (Selection of solution 2)

1. Compute the 3D positions $(X_\alpha, Y_\alpha, Z_\alpha)$ of all the corresponding point pairs (x_α, y_α) and (x'_α, y'_α), $\alpha = 1, \ldots, N$, by planar triangulation, and choose the solutions for which $Z_\alpha > 0$ for all α.
2. From among them, choose the solution for which the Z component of $\mathsf{R}^\top ((X_\alpha, Y_\alpha, Z_\alpha)^\top - \mathbf{t})$ is positive for all α.

Comments For doing planar triangulation, we first need to correct optimally the corresponding pairs (x_α, y_α) and (x'_α, y'_α) to $(\hat{x}_\alpha, \hat{y}_\alpha)$ and $(\hat{x}'_\alpha, \hat{y}'_\alpha)$ by Procedure 7.1, using the input homography matrix H. However, if we compute the homography matrix H from these corresponding pairs by geometric distance minimization of Procedure 6.5, the corrected positions $(\hat{x}_\alpha, \hat{y}_\alpha)$ and $(\hat{x}'_\alpha, \hat{y}'_\alpha)$ are already obtained as a result of that procedure. Then, we compute the camera matrices P and P' by Eq. (5.7), using the computed $\mathbf{t}$ and R and the known values of f' and f'. Finally, we compute the 3D positions $(X_\alpha, Y_\alpha, Z_\alpha)$ by Procedure 4.1.

Mathematically, we are assuming that the image plane is infinitely large and that the corresponding points can be anywhere in it. Therefore there is no theoretical guarantee that we obtain a unique solution by Procedure 8.3. However, such an anomaly is a very pathological case. In real situations, we can usually obtain a unique solution, using only Procedure 8.2.

8.4 Examples

Figure 8.3 shows optimally corrected grid positions obtained by applying Procedure 7.1 to Fig. 6.2, using the homography matrix H estimated from Fig. 6.2. Note that Fig. 7.4 was obtained by using the knowledge of the plane equation and the camera matrices. Here, we used the estimated homography matrix; we used geometric distance minimization. We computed the surface and motion parameters by Procedure 8.1. Assuming Case 1, we applied Procedure 8.2, which chose a unique solution.

The thick lines in Fig. 8.4 connect the 3D positions of the grid points reconstructed using the resulting camera matrices. The thin lines show the true grid plane viewed from the same direction. The shape reconstructed from images is scale indeterminate, therefore we adjusted the scale such that the camera relative translation $\mathbf{t}$ is $\|\mathbf{t}\| = 1$. We can see that the estimated plane is slightly deviated from the true position, because the homography computation from noisy images involves some errors.

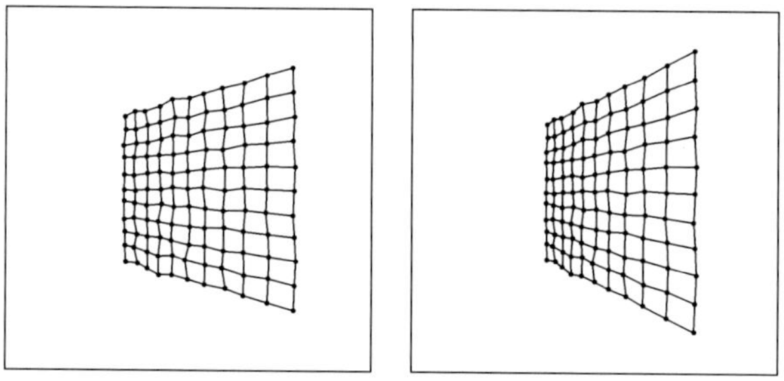

Fig. 8.3 The grid positions obtained by optimally correcting the grid positions in Fig. 6.2 by Procedure 7.1, using the homography matrix H computed from Fig. 6.2

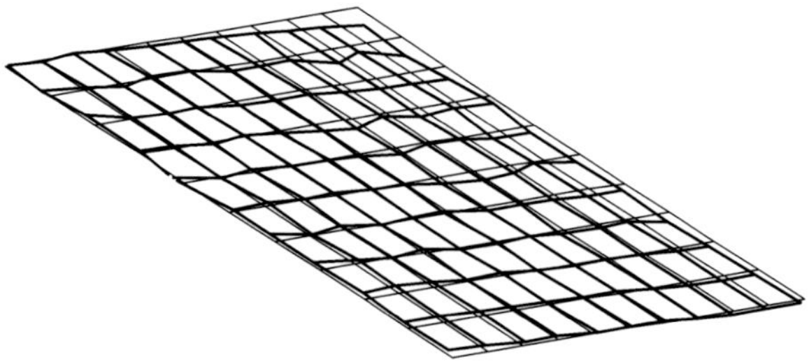

Fig. 8.4 The thick lines connect the 3D positions of the grid points reconstructed from the two images in Fig. 8.3 by Procedure 8.1, using the homography matrix H computed from Fig. 8.3. The thin lines show the true grid plane viewed from the same direction

8.5 Supplemental Note

Given two images of a planar surface, how can we tell the 3D geometric relationship of the plane and the viewpoint motion? This is a very old question and was studied far before computer vision research started in the 1980s. This was a topic of cognitive psychology studied by many cognitive psychologists, among whom J.J. Gibson (1904–1979) is well known. They were interested, in particular, in the *optical flow*, that is, a small displacement field in the image caused by a small motion of the viewpoint relative to a planar surface. It was already known that multiple configurations exist for the plane position and the viewpoint motion that cause an identical optical flow. Cognitive psychologists including Gibson inferred that migratory birds might perceive their motion from the visual changes of the ground and that the nonuniqueness of the 3D interpretation might be the reason that birds sometimes confuse their course of motion. For finite motions, a mathematical analysis was published by another cognitive psychologist, J.C. Hay [1], in 1966. Later, the same conclusion was obtained from the standpoint of computer vision by Tsai and Huang [6,7] and Tsai et al. [8], and a refined mathematical formulation was given by the English physicist/cognitive psychologist H.C. Longuet-Higgins (1923–2004). Procedure 8.1 is based on his 1986 paper [4] and the textbooks [2,3] that introduce his theory.

In this chapter, we analyzed two images of a planar surface. For a plane with a known texture, however, we only need one image, because we can artificially generate a front view of the texture and regard it as the "first image." As an application of this technique, for instance, we take video images of a known textured plane while moving the camera arbitrarily. Then we can compute the position, orientation, and intrinsic parameters of the camera at each instance by comparing the image at that instance and the known reference image (see, e.g., Matsunaga and Kanatani [5]).

Problems

8.1 (1) Show that the following identity holds for arbitrary vectors **a** and **b**.

$$|\mathsf{I} + \mathbf{a}\mathbf{b}^\top| = 1 + (\mathbf{a}, \mathbf{b}). \tag{8.7}$$

(2) Show that the following identity holds for arbitrary vectors **a** and **b**.

$$(\mathsf{I} + \mathbf{a}\mathbf{b}^\top)^{-1} = \mathsf{I} - \frac{\mathbf{a}\mathbf{b}^\top}{1 + (\mathbf{a}, \mathbf{b})}. \tag{8.8}$$

(3) Show that if $\mathbf{a} \neq \mathbf{0}$ and $\mathbf{b} \neq \mathbf{0}$, the matrix $\mathsf{I} + \mathbf{a}\mathbf{b}^\top$ is not an orthogonal matrix.

8.2 (1) If we represent corresponding points (x, y) and (x', y') by the vectors

$$\mathbf{x} = \begin{pmatrix} x \\ y \\ f \end{pmatrix}, \qquad \mathbf{x}' = \begin{pmatrix} x' \\ y' \\ f' \end{pmatrix}, \tag{8.9}$$

show that the mapping from (x, y) to (x', y') can be written in terms of the matrix $\tilde{\mathsf{H}}$ in Eq. (8.2) in the form:

$$\mathbf{x}' \simeq \tilde{\mathsf{H}}\mathbf{x}. \tag{8.10}$$

(2) Show that the matrix $\tilde{\mathsf{H}}$ obtained by the normalization of Eq. (8.3) can be expressed in the form:

$$\tilde{\mathsf{H}} = \frac{1}{k}\mathsf{R}^\top(\mathsf{I} - \frac{\mathbf{t}\mathbf{n}^\top}{h}), \qquad k = \sqrt[3]{1 - \frac{(\mathbf{n}, \mathbf{t})}{h}}. \tag{8.11}$$

(3) Show that $(\mathbf{n}, \mathbf{t}) < h$ for Eq. (8.11) and hence $k > 0$.

8.3 Show that for Eq. (8.4) the equality $\sigma_1\sigma_2\sigma_3 = 1$ holds but not $\sigma_1 = \sigma_2 = \sigma_3 = 1$.

8.4 Using the matrix V in Eq. (8.2), define vectors $\boldsymbol{\tau} = (\tau_i)$ and $\boldsymbol{\nu} = (\nu_i)$ by

$$\boldsymbol{\tau} = \mathsf{V}^\top\mathbf{t}, \qquad \boldsymbol{\nu} = \frac{\mathsf{V}^\top\mathbf{n}}{h}. \tag{8.12}$$

Then show that we obtain from Eqs. (8.2) and (8.11) the equality:

$$k^2\begin{pmatrix} \sigma_1^2 & 0 & 0 \\ 0 & \sigma_2^2 & 0 \\ 0 & 0 & \sigma_3^2 \end{pmatrix}$$
$$= \begin{pmatrix} 1 - -2\nu_1\tau_1 + \nu_1^2 & -\tau_1\nu_2 - \nu_1\tau_2 + \nu_1\nu_2 & -\tau_1\nu_3 - \nu_1\tau_3 + \nu_1\nu_3 \\ -\tau_2\nu_1 - \nu_2\tau_1 + \nu_2\nu_1 & 1 - -2\nu_2\tau_2 + \nu_2^2 & -\tau_2\nu_3 - \nu_2\tau_3 + \nu_2\nu_3 \\ -\tau_3\nu_1 - \nu_3\tau_1 + \nu_3\nu_1 & -\tau_3\nu_2 - \nu_3\tau_2 + \nu_3\nu_2 & 1 - -2\nu_3\tau_3 + \nu_3^2 \end{pmatrix}. \tag{8.13}$$

8.5 (1) Show that if $\nu_1 \neq 0$, $\nu_2 = 0$, $\nu_3 \neq 0$, Eq. (8.13) implies that ν_i are τ_i, $i = 1$, 2, 3, are expressed either in the form

$$\nu_1 = \frac{1}{\sigma_2}\sqrt{\frac{\sigma_1 - \sigma_3}{\sigma_1 + \sigma_3}}\sqrt{\sigma_1^2 - \sigma_2^2}, \qquad \nu_2 = 0,$$
$$\nu_3 = \pm\frac{1}{\sigma_2}\sqrt{\frac{\sigma_1 - \sigma_3}{\sigma_1 + \sigma_3}}\sqrt{\sigma_2^2 - \sigma_3^2}, \tag{8.14}$$

$$\tau_1 = -\frac{\sigma_3}{\sigma_2}\sqrt{\frac{\sigma_1^2 - \sigma_2^2}{\sigma_1^2 - \sigma_3^2}}, \qquad \tau_2 = 0, \qquad \tau_3 = \pm\frac{\sigma_1}{\sigma_2}\sqrt{\frac{\sigma_2^2 - \sigma_3^2}{\sigma_1^2 - \sigma_3^2}}, \tag{8.15}$$

or their sign reversals.
(2) Show that the solution $\nu_2 \neq 0$ that satisfies Eq. (8.13) does not exist.
(3) Show that if $\nu_2 = 0$, we obtain the solution of Eqs. (8.14) and (8.15) even if $\nu_1 = 0$ or $\nu_3 = 0$.
(4) Show that from Eqs. (8.14) and (8.15) the solutions $\mathbf{n}$ and $\mathbf{t}$ are given by Eqs. (8.5) and (8.6) and their sign reversals.

(5) Show that the distance h to the plane is given by Eq. (8.5).
(6) Show that the rotation R is given by Eq. (8.6).

References

1. J.C. Hay, Optical motions and space perception: an extension of Gibson's analysis. Psychol. Rev. **73**(6), 550–656 (1966)
2. K. Kanatani, *Geometric Computation for Machine Vision* (Oxford University Press, Oxford, UK, 1993)
3. K. Kanatani, *Statistical Optimization for Geometric Computation: Theory and Practice* (Elsevier, Amsterdam, The Netherlands, 1996) (Reprinted by Dover, New York, US, 2005)
4. H.C. Longuet-Higgins, The reconstruction of a plane surface from two perspective views. Proc. Royal Soc. Lond. Ser. B **227**, 399–410 (1986)
5. C. Matsunaga, K. Kanatani, Calibration of a moving camera using a planer pattern: optimal computation, reliability evaluation and stabilization by model selection, in *Proceedings of 6th European Conference on Computer Vision*, Dublin, Ireland, Vol. 2 (2000), pp. 595–609
6. R.Y. Tsai, T.S. Huang, Estimating three-dimensional motion parameters of a rigid planar patch. IEEE Trans. Acoust. Speech Signal Process. **29**(6), 1147–1152 (1981)
7. R.Y. Tsai, T.S. Hunag, Estimating 3-D motion parameters of a rigid planar patch III: finite point correspondences and the three-view problem. IEEE Trans. Acoust. Speech Signal Process. **32**(2), 213–220 (1984)
8. R.Y. Tsai, T.S. Huang, W.-L. Zhu, Estimating three-dimensional motion parameters of a rigid planar patch II: singular value decomposition. IEEE Trans. Acoust. Speech Signal Process. **30**(4), 525–534 (1982)

Ellipse Analysis and 3D Computation of Circles

9

Abstract

Circular objects in the scene are projected onto the image plane as ellipses. By fitting ellipse equations to them as shown in Chap. 2, we can compute the 3D properties of the circular objects. This chapter describes typical procedures for such computations. First, we show how we can compute various attributes of ellipses such as intersections, centers, tangents, and perpendiculars. Then we describe the procedure for computing from an ellipse image of a circle the position and orientation of the circle in the scene and the 3D position of its center. These computations allow us to generate an image of the circle seen from the front. The basic principle underlying these computations is the analysis of homographies induced by hypothetical camera rotation around the viewpoint, where projective geometry plays a central role.

9.1 Intersections of Ellipses

The ellipse equation

$$Ax^2 + 2Bxy + Cy^2 + 2f_0(Dx + Ey) + f_0^2 F = 0 \tag{9.1}$$

can be written in terms of the vector $\mathbf{x}$ and the matrix Q in Eq. (2.1) in the form:

$$(\mathbf{x}, \mathsf{Q}\mathbf{x}) = 0, \quad \mathbf{x} = \begin{pmatrix} x/f_0 \\ y/f_0 \\ 1 \end{pmatrix}, \quad \mathsf{Q} = \begin{pmatrix} A & B & D \\ B & C & E \\ D & E & F \end{pmatrix}. \tag{9.2}$$

However, Eq. (9.1) may not always represent an ellipse. It defines a curve only when the matrix Q is nonsingular, that is, $|\mathsf{Q}| \neq 0$, where $|\mathsf{Q}|$ denotes the determinant of Q; if $|\mathsf{Q}| = 0$, two real or imaginary lines result. For $|\mathsf{Q}| \neq 0$, Eq. (9.1) describes an ellipse (possibly an imaginary ellipse such as $x^2 + y^2 = -1$), a parabola, or a hyperbola (↪ Problem 9.1). In general terms, these curves are called *conics*.

K. Kanatani et al., *Guide to 3D Vision Computation*, Advances in Computer Vision and Pattern Recognition, DOI 10.1007/978-3-319-48493-8_9

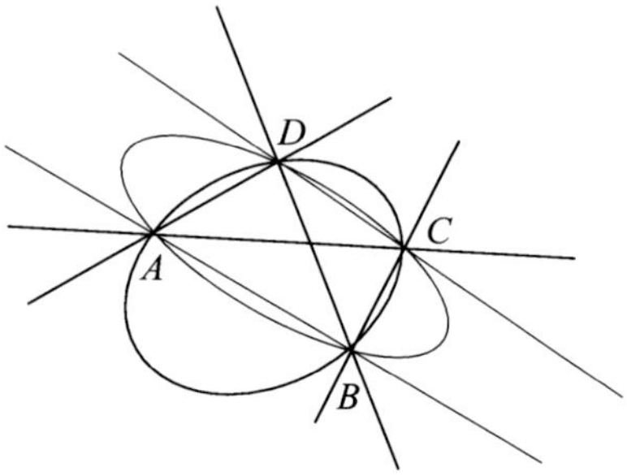

Fig. 9.1 The three pairs of lines $\{AC, BD\}$, $\{AB, CD\}$, and $\{AD, BC\}$ passing through the four intersections A, B, C, and D of two ellipses

Suppose two ellipses $(\mathbf{x}, \mathsf{Q}_1\mathbf{x}) = 0$ and $(\mathbf{x}, \mathsf{Q}_2\mathbf{x}) = 0$ intersect. Here, we consider only real intersections; in the following, we do not consider imaginary points or imaginary curves/lines. The intersections of two ellipses are computed as follows.

Procedure 9.1 (Ellipse intersections)

1. Compute one solution of the following cubic equation in λ.
$$|\lambda \mathsf{Q}_1 + \mathsf{Q}_2| = 0. \tag{9.3}$$
2. Compute the two lines represented by the quadratic equation in x and y for that λ:
$$(\mathbf{x}, (\lambda \mathsf{Q}_1 + \mathsf{Q}_2)\mathbf{x}) = 0, \tag{9.4}$$
3. Return the intersection of each of the lines with the ellipse $(\mathbf{x}, \mathsf{Q}_1\mathbf{x}) = 0$ (or $(\mathbf{x}, \mathsf{Q}_2\mathbf{x}) = 0$).

Comments If two ellipses $(\mathbf{x}, \mathsf{Q}_1\mathbf{x}) = 0$ and $(\mathbf{x}, \mathsf{Q}_2\mathbf{x}) = 0$ intersect at $\mathbf{x}$, it satisfies $\lambda(\mathbf{x}, \mathsf{Q}_1\mathbf{x}) + (\mathbf{x}, \mathsf{Q}_2\mathbf{x}) = 0$ for an arbitrary λ. This is a quadratic equation in $\mathbf{x}$ in the form of Eq. (9.4), which describes a curve or a line pair (possibly imaginary) that passes through all the intersections of the two ellipses. If we choose the λ so that Eq. (9.3) is satisfied, Eq. (9.4) represents two real or imaginary lines ($\hookrightarrow$ Problem 9.2). Inasmuch as we are assuming that real intersections exist, we obtain two lines by factorizing Eq. (9.4) ($\hookrightarrow$ Problem 9.3). Hence, we can locate the intersections by computing the intersections of each line with either of the ellipses ($\hookrightarrow$ Problem 9.4).

If the two ellipses intersect at four points (Fig. 9.1), the cubic equation of Eq. (9.3) has three real roots, each of which defines a pair of lines. Thus, we obtain a pair of lines intersecting inside the ellipses and a pair of lines intersecting outside the ellipses (or being parallel to each other). If the two ellipses intersect at two points, at least one real line among the six computed lines passes through them. Evidently, the above procedure can be applied if one (or both) of $(\mathbf{x}, \mathsf{Q}_1\mathbf{x}) = 0$ and $(\mathbf{x}, \mathsf{Q}_2\mathbf{x}) = 0$ is a hyperbola or a parabola, computing their intersections if they exist.

9.2 Ellipse Centers, Tangents, and Perpendiculars

The center (x_c, y_c) of the ellipse of Eq. (9.1) has the coordinates:

$$x_c = f_0\frac{-CD+BE}{AC-B^2}, \qquad y_c = f_0\frac{BD-AE}{AC-B^2}. \tag{9.5}$$

These are obtained by differentiating the ellipse equation $F(x, y) = 0$ of Eq. (9.1) and solving $\partial F/\partial x = 0$ and $\partial F/\partial y = 0$; that is,

$$Ax + By + f_0D = 0, \qquad Bx + Cy + f_0E = 0. \tag{9.6}$$

Using Eq. (9.5), we can write Eq. (9.1) in the form

$$A(x-x_c)^2 + 2B(x-x_c)(y-y_c) + C(y-y_c)^2 = Ax_c^2 + 2Bx_cy_c + Cy_c^2 - f_0F. \tag{9.7}$$

The tangent line $n_1x + n_2y + n_3f_0 = 0$ to the ellipse of Eq. (9.1) at point (x_0, y_0) on it is given as follows (Fig. 9.2).

$$\begin{aligned} n_1 &= Ax_0 + By_0 + Df_0, \\ n_2 &= Bx_0 + Cy_0 + Ef_0, \\ n_3 &= Dx_0 + Ey_0 + Ff_0. \end{aligned} \tag{9.8}$$

Let $\mathbf{x}_0$ be the value of the vector $\mathbf{x}$ of Eq. (9.2) for $x = x_0$ and $y = y_0$, and let $\mathbf{n}$ be the vector with components n_1, n_2, and n_3. We can see that Eq. (9.8) implies $\mathbf{n} \simeq \mathbf{Q}\mathbf{x}_0$; note that the line equation represents the same line if multiplied by an arbitrary nonzero constant. The expressions in Eq. (9.8) are easily obtained by differentiating equation (9.1) ($\hookrightarrow$ Problem 9.5).

The foot of the perpendicular line from point (a, b) to the ellipse of Eq. (9.1) (i.e., the closest point on it) is computed as follows (Fig. 9.2).

Procedure 9.2 (Perpendicular to Ellipse)

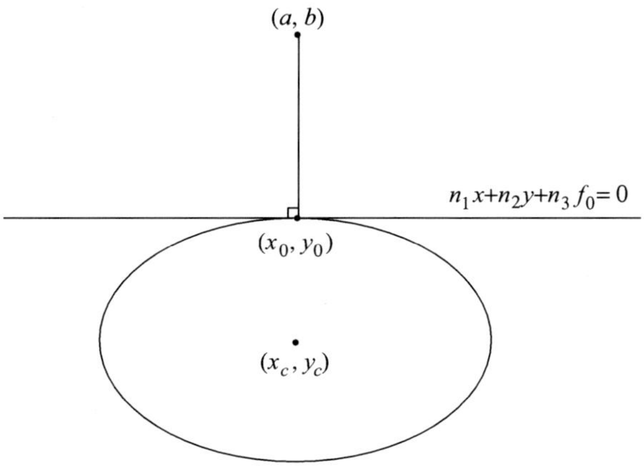

Fig. 9.2 The center (x_c, y_c) of the ellipse, the tangent line $n_1x + n_2y + n_3f_0 = 0$ at point (x_0, y_0) on the ellipse, and the perpendicular line from point (a, b) to point (x_0, y_0)

1. Define the matrix

$$\mathsf{D} = \begin{pmatrix} B & (C-A)/2 & (Ab-Ba+Ef_0)/2f_0 \\ (C-A)/2 & -B & (Bb-Ca-Df_0)/2f_0 \\ (Ab-Ba+Ef_0)/2f_0 & (Bb-Ca-Df_0)/2f_0 & (Db-Ea)/f_0 \end{pmatrix}. \quad (9.9)$$

2. Compute the intersections (x_0, y_0) of the ellipse $(\mathbf{x}, \mathsf{Q}\mathbf{x}) = 0$ and the quadratic curve $(\mathbf{x}, \mathsf{D}\mathbf{x}) = 0$ by Procedure 9.1.
3. From among the computed intersections, return the point (x_0, y_0) for which $(a - x_0)^2 + (b - y_0)^2$ is the smallest.

Comments Let (x_0, y_0) be the foot of the perpendicular line from (a, b) to the ellipse. The line passing through (a, b) and (x_0, y_0) is

$$(b - y_0)x + (x_0 - a)y + (ay_0 - bx_0) = 0. \quad (9.10)$$

Let $n_1x + n_2y + n_3f_0 = 0$ be the tangent line to the ellipse at (x_0, y_0) on it. The condition that this tangent line is orthogonal to the above line is

$$(b - y_0)n_1 + (x_0 - a)n_2 = 0. \quad (9.11)$$

Substituting Eq. (9.8), this condition is rearranged in the form

$$Bx_0^2+(C-A)x_0y_0-By_0^2+(Ab-Ba+Ef_0)x_0+(Bb-Ca-Df_0)y_0+(Db-Ea)f_0 = 0. \quad (9.12)$$

This means that if we define the matrix D of Eqs. (9.9), (9.12) expresses the fact that the point (x_0, y_0) is on the quadratic curve $(\mathbf{x}, \mathsf{D}\mathbf{x}) = 0$. Because (x_0, y_0) is a point on the ellipse $(\mathbf{x}, \mathsf{Q}\mathbf{x}) = 0$, we can compute it as the intersection of the two quadratic curves by Procedure 9.1, which applies if $(\mathbf{x}, \mathsf{D}\mathbf{x}) = 0$ does not represent an ellipse. The above procedure is an analytical computation, but we can also compute the same result iteratively by modifying the geometric distance minimization procedure for ellipse fitting described in Chap. 2 (↪ Problem 9.6).

9.3 Projection of Circles and 3D Reconstruction

A circle in the scene is usually projected onto the image plane as an ellipse (it can be a parabola or a hyperbola, depending on the camera position). This is because, as mentioned in Chap. 7, the perspective projection of a plane in the scene to the image plane is a homography, which maps a conic to a conic. Specifically, a conic $(\mathbf{x}, \mathsf{Q}\mathbf{x}) = 0$ is mapped by a homography of the form $\mathbf{x}' \simeq \mathsf{H}\mathbf{x}$ to a conic $(\mathbf{x}', \mathsf{Q}'\mathbf{x}') = 0$ with

$$\mathsf{Q}' \simeq \mathsf{H}^{-\top}\mathsf{Q}\mathsf{H}^{-1}, \quad (9.13)$$

where $\mathsf{H}^{-\top}$ denotes $(\mathsf{H}^{-1})^\top$ $(= (\mathsf{H}^\top)^{-1})$ (↪ Problem 9.7). In fact, because $\mathbf{x} \simeq \mathsf{H}^{-1}\mathbf{x}'$, the equality $(\mathbf{x}, \mathsf{Q}\mathbf{x}) = 0$ implies $(\mathsf{H}^{-1}\mathbf{x}', \mathsf{Q}\mathsf{H}^{-1}\mathbf{x}') = (\mathbf{x}', \mathsf{H}^{-\top}\mathsf{Q}\mathsf{H}^{-1}\mathbf{x}') = 0$. Because H is nonsingular, we see that $|\mathsf{Q}'| \neq 0$ if $|\mathsf{Q}| \neq 0$.

An important homography is the image transformation induced by camera rotation. If a camera is rotated around its viewpoint, the resulting image change is a

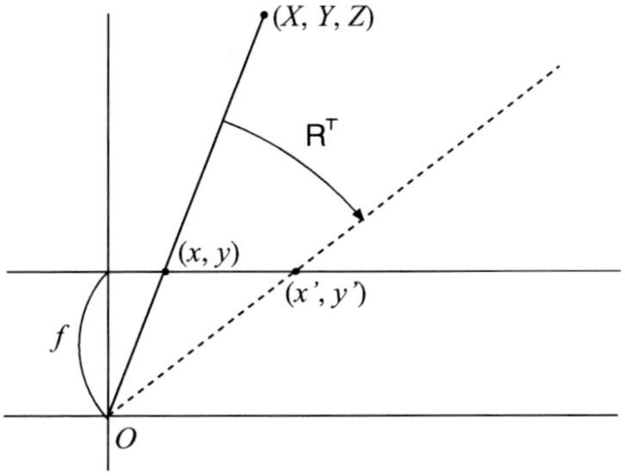

Fig. 9.3 If the camera is rotated by R around its viewpoint O, the scene appears to rotate relative to the camera by R^{-1} $(= \mathsf{R}^\top)$. As a result, the ray direction of (x, y) is rotated to the ray direction of (x', y')

homography, for which the scene may not be a planar surface. In the following analysis, we identify the camera viewpoint with the world coordinate origin O and the optical axis with the Z-axis, assuming that there is no image distortion and that the image coordinate origin o is placed at the principal point, as we did in Chaps. 5 and 8. Then, if the camera is rotated around O by R, the following homography $\mathbf{x}' \simeq \mathsf{H}\mathbf{x}$ is induced.

$$\mathsf{H} = \begin{pmatrix} 1/f_0 & 0 & 0 \\ 0 & 1/f_0 & 0 \\ 0 & 0 & 1/f \end{pmatrix} \mathsf{R}^\top \begin{pmatrix} f_0 & 0 & 0 \\ 0 & f_0 & 0 \\ 0 & 0 & f \end{pmatrix}. \tag{9.14}$$

This result is obtained by the following observation. As shown in Fig. 5.1, the vector $(x, y, f)^\top$ points to the point (X, Y, Z) in the scene that we are viewing. If the camera is rotated by R, the scene appears to rotate relative to the camera in the opposite sense by R^{-1} $(= \mathsf{R}^\top)$. Thus the direction of the vector $(x, y, f)^\top$ rotates to the direction of $\mathsf{R}^\top(x, y, f)^\top$ (Fig. 9.3), which is the ray direction $(x', y', f)^\top$ of the point (x', y') that we see in the image after the rotation. Hence, the following holds.

$$\begin{aligned} \begin{pmatrix} x'/f_0 \\ y'/f_0 \\ 1 \end{pmatrix} &= \begin{pmatrix} 1/f_0 & 0 & 0 \\ 0 & 1/f_0 & 0 \\ 0 & 0 & 1/f \end{pmatrix} \begin{pmatrix} x' \\ y' \\ f \end{pmatrix} \simeq \begin{pmatrix} 1/f_0 & 0 & 0 \\ 0 & 1/f_0 & 0 \\ 0 & 0 & 1/f \end{pmatrix} \mathsf{R}^\top \begin{pmatrix} x \\ y \\ f \end{pmatrix} \\ &= \begin{pmatrix} 1/f_0 & 0 & 0 \\ 0 & 1/f_0 & 0 \\ 0 & 0 & 1/f \end{pmatrix} \mathsf{R}^\top \begin{pmatrix} f_0 & 0 & 0 \\ 0 & f_0 & 0 \\ 0 & 0 & f \end{pmatrix} \begin{pmatrix} x/f_0 \\ y/f_0 \\ 1 \end{pmatrix}. \end{aligned} \tag{9.15}$$

Using this relationship, we can compute, from an ellipse image of a circular object in the scene, its 3D position. We call the plane in the scene on which the circle lies the *supporting plane* of the circle. Let $\mathbf{n}$ be its unit surface normal, and h the distance from the origin O. If we know the radius r of the circle, we can compute $\mathbf{n}$ and h as follows.

Procedure 9.3 (Supporting plane reconstruction)

1. Transform the coefficient matrix Q of the observed ellipse $(\mathbf{x}, \mathsf{Q}\mathbf{x}) = 0$ as follows.

$$\bar{\mathsf{Q}} = \begin{pmatrix} 1/f_0 & 0 & 0 \\ 0 & 1/f_0 & 0 \\ 0 & 0 & 1/f \end{pmatrix} \mathsf{Q} \begin{pmatrix} 1/f_0 & 0 & 0 \\ 0 & 1/f_0 & 0 \\ 0 & 0 & 1/f \end{pmatrix}. \tag{9.16}$$

Normalize the matrix $\bar{\mathsf{Q}}$ to determinant -1:

$$\bar{\mathsf{Q}} \leftarrow \frac{\bar{\mathsf{Q}}}{\sqrt[3]{-|\bar{\mathsf{Q}}|}}. \tag{9.17}$$

2. Compute the eigenvalues λ_1, λ_2, and λ_3 of $\bar{\mathsf{Q}}$ and arrange them in the order $\lambda_2 \geq \lambda_1 > 0 > \lambda_3$. Let $\mathbf{u}_1$, $\mathbf{u}_2$, and $\mathbf{u}_3$ be the corresponding unit eigenvectors.
3. Compute the unit surface normal $\mathbf{n}$ to the supporting plane in the form

$$\mathbf{n} = \mathcal{N}[\sqrt{\lambda_2 - \lambda_1}\mathbf{u}_2 + \sqrt{\lambda_1 - \lambda_3}\mathbf{u}_3]. \tag{9.18}$$

4. Compute the distance h to the supporting plane in the form

$$h = \lambda_1^{3/2} r. \tag{9.19}$$

Comments If the matrix $\bar{\mathsf{Q}}$ is defined by Eq. (9.16), we see from Eq. (9.2) that

$$\bar{\mathsf{Q}} = \begin{pmatrix} A/f_0^2 & B/f_0^2 & D/f_0 f \\ B/f_0^2 & C/f_0^2 & E/f_0 f \\ D/f_0 f & E/f_0 f & F/f^2 \end{pmatrix} \simeq \begin{pmatrix} A & B & (f_0/f)D \\ B & C & (f_0/f)E \\ (f_0/f)D & (f_0/f)E & (f_0/f)^2 F \end{pmatrix}. \tag{9.20}$$

If we write the elements of $\bar{\mathsf{Q}}$ as $\bar{A}$, $\bar{B}$, ..., it is easy to see that the equation

$$\bar{A}x^2 + 2\bar{B}xy + \bar{C}y^2 + 2f(\bar{D}x + \bar{E}y) + f^2\bar{F} = 0 \tag{9.21}$$

is identical to Eq. (9.1). Namely, using the matrix $\bar{\mathsf{Q}}$ instead of Q means replacing the constant f_0 with the focal length f. Because Eq. (9.14) is rewritten as

$$\mathsf{H}^{-1} = \begin{pmatrix} 1/f_0 & 0 & 0 \\ 0 & 1/f_0 & 0 \\ 0 & 0 & 1/f \end{pmatrix} \mathsf{R} \begin{pmatrix} f_0 & 0 & 0 \\ 0 & f_0 & 0 \\ 0 & 0 & f \end{pmatrix}, \tag{9.22}$$

Equation (9.13) implies that the transformation of the ellipse by camera rotation given by

$$\begin{aligned} \mathsf{Q}' &\simeq \begin{pmatrix} f_0 & 0 & 0 \\ 0 & f_0 & 0 \\ 0 & 0 & f \end{pmatrix} \mathsf{R}^\top \begin{pmatrix} 1/f_0 & 0 & 0 \\ 0 & 1/f_0 & 0 \\ 0 & 0 & 1/f \end{pmatrix} \mathsf{Q} \begin{pmatrix} 1/f_0 & 0 & 0 \\ 0 & 1/f_0 & 0 \\ 0 & 0 & 1/f \end{pmatrix} \mathsf{R} \begin{pmatrix} f_0 & 0 & 0 \\ 0 & f_0 & 0 \\ 0 & 0 & f \end{pmatrix} \\ &= \begin{pmatrix} f_0 & 0 & 0 \\ 0 & f_0 & 0 \\ 0 & 0 & f \end{pmatrix} \mathsf{R}^\top \bar{\mathsf{Q}} \mathsf{R} \begin{pmatrix} f_0 & 0 & 0 \\ 0 & f_0 & 0 \\ 0 & 0 & f \end{pmatrix}. \end{aligned} \tag{9.23}$$

In other words, $\bar{\mathsf{Q}}' \simeq \mathsf{R}^\top \bar{\mathsf{Q}} \mathsf{R}$. However, $\bar{\mathsf{Q}}$ and $\bar{\mathsf{Q}}'$ are both normalized to determinant -1, and the determinant is unchanged by multiplication of rotation matrices R and $\mathsf{R}^\top$. Thus we have

$$\bar{\mathsf{Q}}' = \mathsf{R}^\top \bar{\mathsf{Q}} \mathsf{R}. \tag{9.24}$$

Using this, we first consider the case where the ellipse in the image is in canonical form, that is, centered on the image origin o having the major and minor axes in the $x-$ and y-directions (↪ Problem 9.8(1)). In the general case, we rotate the camera so that the ellipse has the canonical form. This is done by first rotating the camera such that the optical axis passes through the center of the ellipse and then rotating the camera around the optical axis such that the major and minor axes are in the x- and y-directions. Applying the result for the canonical form to this, we obtain Procedure 9.3 (↪Problem 9.8(2)).

The distance to the supporting plane is determined by using the radius r of the circle we observe. If r is unknown, the supporting plane is determined up to the distance from the origin O. The 3D position of the circle can be computed by *back-projection* of each point on the ellipse, that is, by computing the intersection of its ray with the supporting plane (↪ Problem 9.9).

9.4 Center of Circle

If we know the unit surface normal $\mathbf{n}$ to the supporting plane, we can compute the image position of the center of the circle, which does not necessarily coincide with the center of the ellipse. The procedure is as follows.

Procedure 9.4 (Center of circle)

1. Compute the following vector $\mathbf{m} = (m_i)$.

$$\mathbf{m} = \begin{pmatrix} f_0 & 0 & 0 \\ 0 & f_0 & 0 \\ 0 & 0 & f \end{pmatrix} \mathsf{Q}^{-1} \begin{pmatrix} f_0 & 0 & 0 \\ 0 & f_0 & 0 \\ 0 & 0 & f \end{pmatrix} \begin{pmatrix} n_1 \\ n_2 \\ n_3 \end{pmatrix}. \tag{9.25}$$

2. Return the following (x_C, y_C).

$$x_C = f\frac{m_1}{m_3}, \qquad y_C = f\frac{m_2}{m_3}. \tag{9.26}$$

Comments In terms of the matrix $\bar{\mathsf{Q}}$ in Eq. (9.16), the right-hand side of Eq. (9.25) can be written as $\bar{\mathsf{Q}}^{-1}\mathbf{n}$. Hence, from Eq. (9.26) it means the relationship:

$$\mathbf{n} \simeq \bar{\mathsf{Q}} \begin{pmatrix} x_C \\ y_C \\ f \end{pmatrix} \tag{9.27}$$

Evidently, this holds if the camera is rotated such that the optical axis is perpendicular to the supporting plane, as shown in Fig. 9.4 (↪ Problem 9.10(1)). We can prove that this relationship holds if the camera is rotated arbitrarily (↪ Problem 9.10(2)). Therefore it holds for an arbitrary camera position.

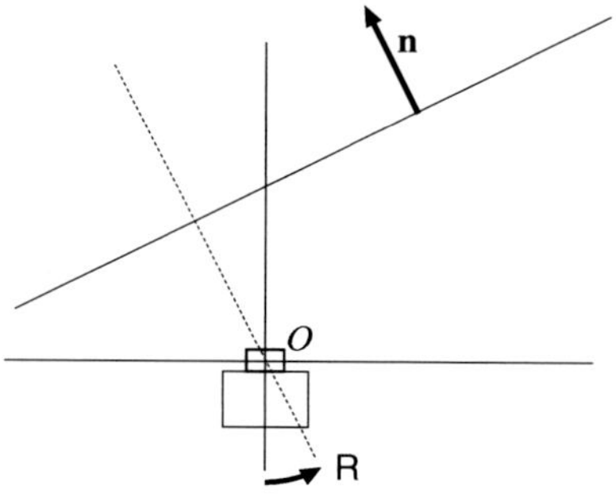

Fig. 9.4 If we rotate the camera such that the optical axis is perpendicular to the supporting plane, the image of the circle becomes a circle

9.5 Front Image of the Circle

Given an ellipse image of a circle in the scene, we can compute the supporting plane of the circle by Procedure 9.3. Then, we can generate its image as if seen from the front in such a way that the center of the circle is at the image origin. Note that translation of the image is also a homography. In fact, translation $x' = x + a$ and $y' = y + b$ by a and b in the $x-$ and y- directions, respectively, is the homography:

$$\begin{pmatrix} x'/f_0 \\ y'/f_0 \\ 1 \end{pmatrix} = \begin{pmatrix} (x+a)/f_0 \\ (y+b)/f_0 \\ 1 \end{pmatrix} = \begin{pmatrix} 1 & 0 & a/f_0 \\ 0 & 1 & b/f_0 \\ 0 & 0 & 1 \end{pmatrix} \begin{pmatrix} x/f_0 \\ y/f_0 \\ 1 \end{pmatrix}. \tag{9.28}$$

Its inverse is obtained by replacing a and b by $-a$ and $-b$, respectively. The hypothetical front image of the circle is computed by the following procedure.

Procedure 9.5 (Front image of the circle)

1. Compute the projected position (x_C, y_C) of the center of the circle by Procedure 9.4.
2. Compute a camera rotation R by which the unit surface normal to the supporting plane comes to the optical axis direction, and determine the corresponding homography matrix H by Eq. (9.14).
3. Compute the projected position (x'_C, y'_C) of the center of the circle after that camera rotation as follows.

$$\begin{pmatrix} x'_C/f_0 \\ y'_C/f_0 \\ 1 \end{pmatrix} \simeq \mathsf{H} \begin{pmatrix} x_C/f_0 \\ y_C/f_0 \\ 1 \end{pmatrix}. \tag{9.29}$$

4. Compute the following homography matrix that translates the point (x'_C, y'_C) to the image origin.

$$\mathsf{H}_0 = \begin{pmatrix} 1 & 0 & -x_C/f_0 \\ 0 & 1 & -y_C/f_0 \\ 0 & 0 & 1 \end{pmatrix}. \tag{9.30}$$

5. Define a new image buffer frame, and for each pixel (x, y) of it compute the point $(\bar{x}, \bar{y})$ that satisfies

$$\begin{pmatrix} \bar{x}/f_0 \\ \bar{y}/f_0 \\ 1 \end{pmatrix} \simeq \mathsf{H}^{-1}\mathsf{H}_0^{-1} \begin{pmatrix} x/f_0 \\ y/f_0 \\ 1 \end{pmatrix}. \tag{9.31}$$

   Then copy the image value of the pixel $(\bar{x}, \bar{y})$ of the input image to the pixel (x, y) of the buffer frame. If $(\bar{x}, \bar{y})$ are not integers, the image value is interpolated from the values of surrounding pixels.

Comments The camera rotation that rotates the unit vector $\mathbf{n}$ to the optical axis direction is not unique, because it has indeterminacy of rotations around the optical axis (Fig. 9.4). We can choose any of such rotations. The simplest one is the rotation around an axis perpendicular to both $\mathbf{n}$ and the optical axis direction (=the Z-axis) by the angle $\boldsymbol{\Omega}$ made by $\mathbf{n}$ and the Z-axis ($\hookrightarrow$ Problem 9.11). Combining it with the translation of Eq. (9.30), we can obtain the homography matrix $\mathsf{H}_0\mathsf{H}$, which maps the ellipse to a circle around the image origin as if we were viewing the circle from the front.

As is well known, in order to generate a new image from a given image by image processing, the transformation equation that maps an old image to a new image is not sufficient. For digital image processing, we first define a new image buffer in which the generated image is to be stored and then give a computational procedure to define the image value at each pixel of this buffer. This is done by first computing the "inverse" of the image generating transformation and then copying to the buffer the image value of the inversely transformed pixel position. In the above procedure, the image generating transformation is the composite homography $\mathsf{H}_0\mathsf{H}$, whose inverse is $\mathsf{H}^{-1}\mathsf{H}_0^{-1}$. The inverse H^{-1} is given by Eq. (9.22), and the inverse H_0^{-1} is obtained from Eq. (9.30) by changing the signs of x_C and y_C. When the computed image coordinates are not integers, we may simply round them to integers, but a more accurate way is to use bilinear interpolation, that is, proportional allocation in the x- and y-directions, of surrounding pixel values ($\hookrightarrow$ Problem 9.12).

9.6 Examples

Figure 9.5a is an image of a circle drawn on a planar surface. Detecting the circle boundary by edge detection and fitting an ellipse to it, we can compute the position and orientation of the surface relative to the camera by Procedure 9.3. Using that knowledge, Fig. 9.5b displays a 3D graphics object shown as if it were placed on the surface Fig. 9.5b.

Figure 9.6a is an outdoor scene of a planar board, on which a circular mark (encircled by a white line in the figure) is painted. Detecting its boundary by edge detection and fitting an ellipse to it, we can compute the position and orientation of the board relative to the camera by Procedure 9.3. Using that knowledge, we can generate its front image by virtual camera rotation, using Procedure 9.5; for visual ease, the image is appropriately translated.

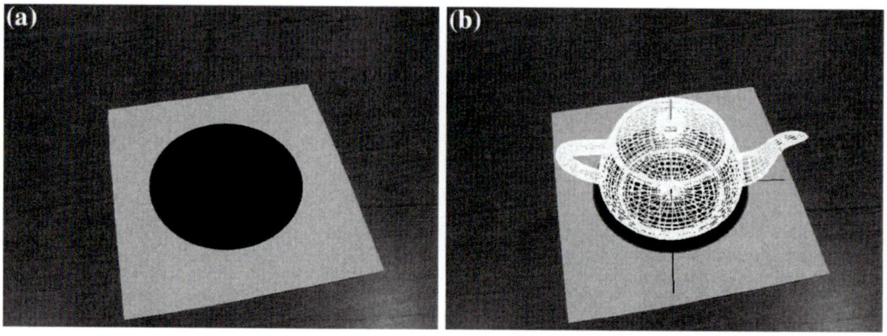

Fig. 9.5 **a** An image of a circle on a planar surface. **b** A 3D graphics object is displayed in a position compatible with the surface in (**a**); its position and orientation relative to the camera are computed from the circle image using Procedure 9.3

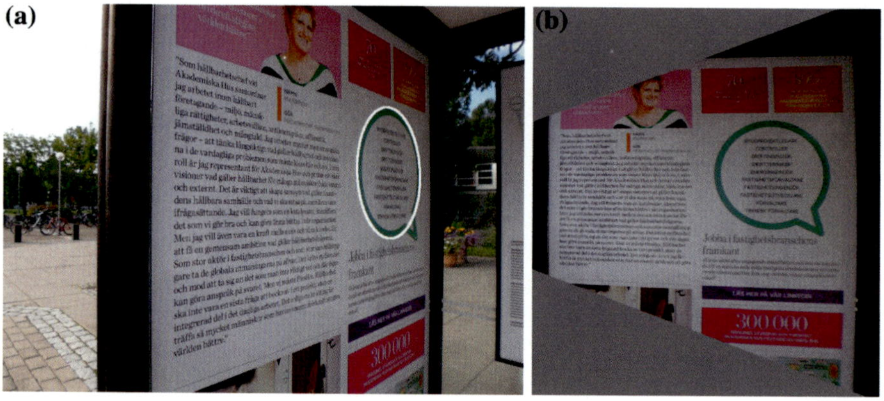

Fig. 9.6 **a** An image of a planar board, on which a circular mark (encircled by a white line) is painted. **b** A front image of the board generated by computing its relative orientation to the camera by Procedure 9.3, using the circular mark, and virtually rotating the camera by Procedure 9.5

9.7 Supplemental Note

As described in most geometry textbooks, the curves described by Eq. (9.1) include a real quadratic curve (ellipse, hyperbola, or parabola), an imaginary ellipse, two real lines (intersecting or parallel), a real line, and two imaginary lines (intersecting at a real point or parallel). From the standpoint of projective geometry, two ellipses always intersect at four points, which may be real or imaginary. Similarly, an ellipse and a line always intersect at two points; a tangent point is regarded as a degenerate double intersection, and no intersection is interpreted to be intersecting at an imaginary point. A well-known textbook on classical projective geometry is Semple and Kneebone [6]. For the analysis of general algebraic curves extended to the complex number domain, called *algebraic geometry*, see the classical textbook of Semple and Roth [5].

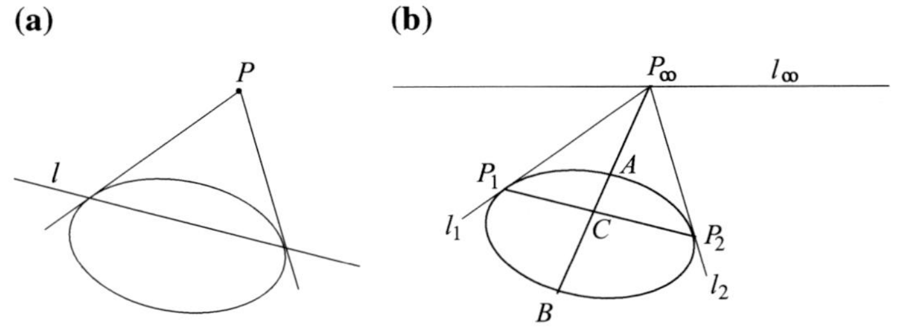

Fig. 9.7 **a** The pole P and the polar l of an ellipse. **b** The ellipse center, tangent lines, vanishing point, and vanishing line

The expression of the tangent line in Eq. (9.8) holds if Eq. (9.1) represents a parabola or a hyperbola. If the point (x_0, y_0) is not on the conic of Eqs. (9.1), (9.8), or $\mathbf{n} \simeq \mathbf{Q}\mathbf{x}_0$ defines a line, called the *polar* of the conic of Eq. (9.1) at (x_0, y_0); the point (x_0, y_0) is called the *pole* of the line $n_1x + n_2y + n_1f_0 = 0$. For a point (x_0, y_0) outside an ellipse, the line passing through the intersections of the two tangent lines from it to the ellipse is the polar of (x_0, y_0) (Fig. 9.7a). The condition that the pole is on the polar is that the pole is on the conic and the polar is the tangent line there. The tangent line (and the polar in general) is obtained by the *polar decomposition* of the conic. This is the process of replacing x^2, xy, y^2, x, and y of the conic equation by x_0x, $(x_0y + xy_0)/2$, y_0y, $(x + x_0)/2$, and $(y + y_0)/2$, respectively.

The poles and polars of an ellipse are closely related to the 3D interpretation of the ellipse. Suppose an ellipse in the image is an oblique view of an ellipse (or a circle) in a scene. Let C be the true center of the ellipse, where by "true" we mean the properties when seen from the front. Consider an arbitrary chord P_1P_2 passing through C (Fig. 9.7b). Let l_1 and l_2 be the tangent line to the ellipse at P_1 and P_2, respectively. These are parallel tangent lines to the "true" ellipse. Their intersection P_∞ in the image is the *vanishing point*, which is infinitely far away on the "true" supporting plane. The chord P_1P_2 is arbitrary, therefore each choice of it defines its vanishing point. All such vanishing points are on a line called the *vanishing line*, which is the "horizon" or the infinitely far away boundary of the supporting plane. In the image, the vanishing line l_∞ is the polar of the center C, which is the pole of the vanishing line l_∞. Procedure 9.4 is a consequence of this interpretation.

Let A and B be the intersection of the ellipse with the line passing through the center C and the vanishing point P_∞. Then, the set $[P_\infty, A, C, B]$ is a *harmonic range*, meaning that it has *cross-ratio* -1. A similar relation holds for the vanishing line l_∞, the tangent lines l_1 and l_2, and the axis $P_\infty C$, and the set $[l_\infty, l_1, P_\infty C, l_2]$ is a *harmonic pencil*, meaning that it has *cross-ratio* -1. See [3] for computational details.

Procedure 9.3 for 3D reconstruction of a circle was first presented by Forsyth et al. [1]. The result was extended to 3D reconstruction of an ellipse by Kanatani and Liu [4]. These analyses are based on the image transformation induced by camera rotation, for which see [2,3] for the details.

Problems

9.1 Show that for $|\mathsf{Q}| \neq 0$, the equation $(\mathbf{x}, \mathsf{Q}\mathbf{x}) = 0$ represents an ellipse (including an imaginary ellipse) if and only if

$$AC - B^2 > 0. \tag{9.32}$$

Also show that $AC - B^2 = 0$ and $AC - B^2 < 0$ correspond to a parabola and a hyperbola, respectively.

9.2 (1) Show that for an arbitrary square matrix A, the identity

$$|\lambda \mathsf{I} + \mathsf{A}| = \lambda^3 + \lambda^2 \mathrm{tr}[\mathsf{A}] + \lambda \mathrm{tr}[\mathsf{A}^\dagger] + |\mathsf{A}| \tag{9.33}$$

holds, where $\mathsf{A}^\dagger$ is the cofactor matrix of A (its (i,j) equals the determinant of the matrix obtained from A by removing the row and the column containing A_{ji} and multiplying it by $(-1)^{i+j}$):

$$\mathsf{A}^\dagger = \begin{pmatrix} A_{22}A_{33} - A_{32}A_{23} & A_{32}A_{13} - A_{12}A_{33} & A_{12}A_{23} - A_{22}A_{13} \\ A_{31}A_{23} - A_{21}A_{33} & A_{11}A_{33} - A_{31}A_{13} & A_{12}A_{23} - A_{21}A_{13} \\ A_{21}A_{32} - A_{31}A_{22} & A_{31}A_{12} - A_{11}A_{32} & A_{11}A_{22} - A_{21}A_{12} \end{pmatrix}. \tag{9.34}$$

(2) Show that the identity for arbitrary square matrices A and B:

$$|\lambda \mathsf{A} + \mathsf{B}| = \lambda^3 |\mathsf{A}| + \lambda^2 \mathrm{tr}[\mathsf{A}^\dagger \mathsf{B}] + \lambda \mathrm{tr}[\mathsf{A}\mathsf{B}^\dagger] + |\mathsf{B}| \tag{9.35}$$

9.3 Show that if $|\mathsf{Q}| = 0$ and $B^2 - AC > 0$, Eq. (9.1) defines the two lines:

$$Ax + (B - \sqrt{B^2 - AC})y + \left(D - \frac{BD - AE}{\sqrt{B^2 - AC}}\right)f_0 = 0,$$
$$Ax + (B + \sqrt{B^2 - AC})y + \left(D + \frac{BD - AE}{\sqrt{B^2 - AC}}\right)f_0 = 0. \tag{9.36}$$

9.4 Describe the procedure for computing the two intersections (x_1, y_1) and (x_2, y_2) of the ellipse of Eq. (9.1) with the line $n_1 x + n_2 y + n_3 f_0 = 1$.

9.5 Show that the tangent line $n_1 x + n_2 y + n_3 f_0 = 0$ to the quadratic curve of Eq. (9.1) at (x_0, y_0) is given by Eq. (9.8).

9.6 Describe the procedure for iteratively computing the foot of the perpendicular to an ellipse drawn from a given point by modifying the ellipse fitting procedure by geometric distance minimization given in Chap. 2.

9.7 Show that $(\mathsf{A}^{-1})^\top = (\mathsf{A}^\top)^{-1}$ holds for an arbitrary nonsingular matrix A.

9.8 (1) Show that when the projected circle in the image is an ellipse in the canonical form of

$$x^2 + \alpha y^2 = \gamma, \qquad \alpha \geq 1, \qquad \gamma > 0, \tag{9.37}$$

the inclination angle θ of the supporting plane is given by

$$\sin\theta = \pm\sqrt{\frac{\alpha - 1}{\alpha + \gamma/f^2}}, \qquad \cos\theta = \sqrt{\frac{1 + \gamma/f^2}{\alpha + \gamma/f^2}}, \tag{9.38}$$

and the distance h to the supporting plane is

$$h = \frac{fr}{\sqrt{\alpha\gamma}}. \tag{9.39}$$

(2) Show that Eqs. (9.18) and (9.19) hold in the general case.

9.9 Show that the 3D point (X, Y, Z) obtained by back-projecting a point (x, y) in the image onto a plane with unit surface normal $\mathbf{n}$ and distance h from the origin O is given by

$$\begin{pmatrix} X \\ Y \\ Z \end{pmatrix} = \frac{h}{n_1 x + n_2 y + n_3 f} \begin{pmatrix} x \\ y \\ f \end{pmatrix}. \tag{9.40}$$

9.10 (1) Show that Eq. (9.27) holds when the supporting plane of the circle is perpendicular to the optical axis of the camera.
(2) Show that Eq. (9.27) holds if the camera is arbitrarily rotated around the viewpoint.

9.11 Find a rotation matrix R that rotates a unit vector $\mathbf{n}$ into the direction of the Z-axis.

9.12 Show how the value of the pixel with noninteger image coordinates (x, y) is determined from surrounding pixels by bilinear interpolation.

References

1. D. Forsyth, J.L. Mundy, A. Zisserman, C. Coelho, A. Heller, C. Rothwell, Invariant descriptors for3-D object recognition and pose. IEEE Trans. Pattern Anal. Mach. Intell. **13**(10), 971–991 (1991)
2. K. Kanatani, *Group-Theoretical Methods in Image Understanding* (Springer, Berlin, 1990)
3. K. Kanatani, *Geometric Computation for Machine Vision* (Oxford University Press, Oxford, UK, 1993)
4. K. Kanatani, W. Liu, 3D interpretation of conics and orthogonality. CVIGP: Image Underst. **58**(3), 286–301 (1993)
5. J.G. Semple, G.T. Kneebone, *Algebraic Projective Geometry* (Oxford University Press, Oxford, UK, 1952)
6. J.G. Semple, L. Roth, *Introduction to Algebraic Geometry* (Oxford University Press, Oxford, UK, 1949)

Part II
Multiview 3D Reconstruction Techniques

Multiview Triangulation

10

Abstract

In Chap. 4, we showed how we can reconstruct the 3D point positions from their two-view images using knowledge of the camera matrices. Here, we extend it, reconstructing the 3D point positions from multiple images. The basic principle is the same as the two-view case: we optimally correct the observed point positions such that the lines of sight they define intersect at a single point in the scene. We begin with the three-view case and describe the optimal triangulation procedure based on the fact that three rays intersect in the scene if and only if the trilinear constraint is satisfied, just in the same way that two rays intersect if and only if the epipolar constraint is satisfied. We then extend this to general M views, imposing the trilinear constraint on all three consecutive images.

10.1 Trilinear Constraint

Suppose a point in the scene is imaged in three images, which we hereafter call the 0th, the first, and the second images, at points (x_0, y_0), (x_1, y_1), and (x_2, y_2). We represent them as vectors

$$\mathbf{x}_\kappa = \begin{pmatrix} x_\kappa/f_0 \\ y_\kappa/f_0 \\ 1 \end{pmatrix}, \qquad \kappa = 0, 1, 2, \tag{10.1}$$

where f_0 is the scaling constant we are using throughout this book. For brevity, we call the point represented by vector $\mathbf{x}_\kappa$ simply "point $\mathbf{x}_\kappa$." The three rays defined by $\mathbf{x}_0$, $\mathbf{x}_1$, and $\mathbf{x}_2$ intersect at one point in the scene, as shown in Fig. 10.1, if and only if the following condition is satisfied (see Sect. 10.2.4 for the details).

$$\sum_{i,j,k,l,m=1}^{3} \varepsilon_{ljp}\varepsilon_{mkq}T_i^{lm}x_{0(i)}x_{1(j)}x_{2(k)} = 0. \tag{10.2}$$

K. Kanatani et al., *Guide to 3D Vision Computation*, Advances in Computer Vision and Pattern Recognition, DOI 10.1007/978-3-319-48493-8_10

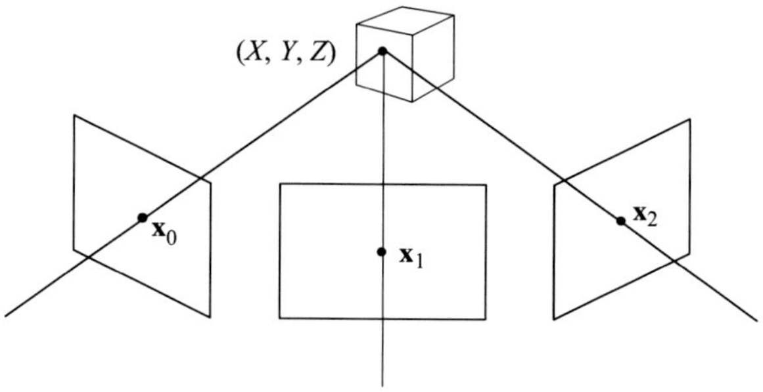

Fig. 10.1 For three views, the rays of points $\mathbf{x}_0$, $\mathbf{x}_1$, and $\mathbf{x}_2$ intersect at one point in the scene if and only if the trilinear constraint of Eq. (10.2) holds

Here, ε_{ijk} is the permutation symbol, taking 1 if (i, j, k) is an even permutation of $(1, 2, 3)$, -1 if it is an odd permutation, and 0 otherwise. We also write the ith component of vector $\mathbf{x}_\kappa$ as $x_{\kappa(i)}$. The coefficients T_i^{lm}, called the *trifocal tensor*, are determined by the relative configuration of the three cameras. Given the camera matrices P_0, P_1, and P_2 of the three cameras, the trifocal tensor is determined as follows (see Sect. 10.2.4 for the details).

$$T_1^{jk} = \begin{vmatrix} P_{0(21)} & P_{0(22)} & P_{0(23)} & P_{0(24)} \\ P_{0(31)} & P_{0(32)} & P_{0(33)} & P_{0(34)} \\ P_{1(j1)} & P_{1(j2)} & P_{1(j3)} & P_{1(j4)} \\ P_{2(k1)} & P_{2(k2)} & P_{2(k3)} & P_{2(k4)} \end{vmatrix}, \quad T_2^{jk} = \begin{vmatrix} P_{0(31)} & P_{0(32)} & P_{0(33)} & P_{0(34)} \\ P_{0(11)} & P_{0(12)} & P_{0(13)} & P_{0(14)} \\ P_{1(j1)} & P_{1(j2)} & P_{1(j3)} & P_{1(j4)} \\ P_{2(k1)} & P_{2(k2)} & P_{2(k3)} & P_{2(k4)} \end{vmatrix},$$
$$T_3^{jk} = \begin{vmatrix} P_{0(11)} & P_{0(12)} & P_{0(13)} & P_{0(14)} \\ P_{0(21)} & P_{0(22)} & P_{0(23)} & P_{0(24)} \\ P_{1(j1)} & P_{1(j2)} & P_{1(j3)} & P_{1(j4)} \\ P_{2(k1)} & P_{2(k2)} & P_{2(k3)} & P_{2(k4)} \end{vmatrix}. \tag{10.3}$$

Here, we write the (ij) element of P_κ as $P_{\kappa(ij)}$. Equation (10.2) is called the *trilinear constraint* and is an extension of the epipolar constraint for two views to three views; the trifocal tensor plays the same role as the fundamental matrix for two views. However, the nine equations of Eq. (10.2) for $p, q = 1, 2, 3$ are not linearly independent; only four are independent ($\hookrightarrow$ Problem 10.1).

10.2 Triangulation from Three Views

We describe the procedure for computing the 3D position in the scene from noisy point correspondences over three images. The computation is divided into several stages.

10.2.1 Optimal Correspondence Correction

We assume in this chapter that the camera matrices P_κ of the three cameras are known. Hence, the trifocal tensor T_i^{lm} is also known. For optimally computing the

3D position from noisy observed positions $\mathbf{x}_0$, $\mathbf{x}_1$, and $\mathbf{x}_2$, we need to correct them optimally to $\bar{\mathbf{x}}_0$, $\bar{\mathbf{x}}_1$, and $\bar{\mathbf{x}}_2$ such that they satisfy the trilinear constraint of Eq. (10.2), just in the same way as the two-view triangulation procedures of Chaps. 4 and 7, where "optimally correcting" means the sum of the square distances for correction, or the *reprojection error*,

$$E = \sum_{\kappa=0}^{2} \|\mathbf{x}_\kappa - \bar{\mathbf{x}}_\kappa\|^2 \tag{10.4}$$

is minimized. This is done by the following procedure.

Procedure 10.1 (Optimal three-view correspondence correction)

1. Let $E_0 = \infty$ (a sufficiently large number), $\hat{\mathbf{x}}_\kappa = \mathbf{x}_\kappa$, and $\tilde{\mathbf{x}}_\kappa = \mathbf{0}$, $\kappa = 0, 1, 2$.
2. Compute the following P_{pqs}, Q_{pqs}, and R_{pqs}, $p, q, s = 1, 2, 3$.

$$P_{pqs} = \sum_{i,j,k,l,m=1}^{3} \varepsilon_{ljp}\varepsilon_{mkq}T_i^{lm} P_{\mathbf{k}(si)}\hat{x}_{1(j)}\hat{x}_{2(k)},$$

$$Q_{pqs} = \sum_{i,j,k,l,m=1}^{3} \varepsilon_{ljp}\varepsilon_{mkq}T_i^{lm} \hat{x}_{0(i)}P_{\mathbf{k}(sj)}\hat{x}_{2(k)},$$

$$R_{pqs} = \sum_{i,j,k,l,m=1}^{3} \varepsilon_{ljp}\varepsilon_{mkq}T_i^{lm} \hat{x}_{0(i)}\hat{x}_{1(j)}P_{\mathbf{k}(sk)}. \tag{10.5}$$

   Here, $P_{\mathbf{k}(ij)}$ is the (i, j) element of the matrix

$$\mathsf{P}_\mathbf{k} = \begin{pmatrix} 1 & 0 & 0 \\ 0 & 1 & 0 \\ 0 & 0 & 0 \end{pmatrix}. \tag{10.6}$$

3. Compute the following C_{pqrs} and F_{pq}.

$$C_{pqrs} = \sum_{i,j,k,l,m=1}^{3} \varepsilon_{ljp}\varepsilon_{mkq}T_i^{lm}\Big(P_{rsi}\hat{x}_{1(j)}\hat{x}_{2(k)} + \hat{x}_{0(i)}Q_{rsj}\hat{x}_{2(k)} + \hat{x}_{0(i)}\hat{x}_{1(j)}R_{rsk}\Big),$$

$$F_{pq} = \sum_{i,j,k,l,m=1}^{3} \varepsilon_{ljp}\varepsilon_{mkq}T_i^{lm}\Big(\hat{x}_{0(i)}\hat{x}_{1(j)}\hat{x}_{2(k)} + \tilde{x}_{0(i)}\hat{x}_{1(j)}\hat{x}_{2(k)} + \hat{x}_{0(i)}\tilde{x}_{1(j)}\hat{x}_{2(k)} + \hat{x}_{0(i)}\hat{x}_{1(j)}\tilde{x}_{2(k)}\Big). \tag{10.7}$$

4. Solve the following linear equation, using the pseudoinverse of truncated rank 3, to compute λ_{pq}.

$$\sum_{r,s=1}^{3} C_{pqrs}\lambda_{rs} = F_{pq}. \tag{10.8}$$

5. Update $\tilde{\mathbf{x}}_\kappa$ and $\hat{\mathbf{x}}_\kappa$, $\kappa = 0, 1, 2$, to

$$\tilde{x}_{0(i)} \leftarrow \sum_{p,q=1}^{3} P_{pqi}\lambda_{pq}, \quad \tilde{x}_{1(i)} \leftarrow \sum_{p,q=1}^{3} Q_{pqi}\lambda_{pq}, \quad \tilde{x}_{2(i)} \leftarrow \sum_{p,q=1}^{3} R_{pqi}\lambda_{pq},$$

$$\hat{\mathbf{x}}_\kappa \leftarrow \mathbf{x}_\kappa - \tilde{\mathbf{x}}_\kappa. \tag{10.9}$$

6. Compute the following reprojection error E.

$$E = \sum_{\kappa=0}^{2} \|\tilde{\mathbf{x}}_\kappa\|^2. \tag{10.10}$$

7. If $E \approx E_0$, return E and $\hat{\mathbf{x}}_\kappa$, $\kappa = 0, 1, 2$, and stop. Else, let $E_0 \leftarrow E$ and go back to Step 2.

Comments The principle of the above procedure is the same as Procedure 4.2 for optimally imposing the epipolar constraint on a two-view correspondence. We write the positions $\bar{\mathbf{x}}_0$, $\bar{\mathbf{x}}_1$, and $\bar{\mathbf{x}}_2$ to which the three points are to be corrected in the form

$$\bar{\mathbf{x}}_\kappa = \mathbf{x}_\kappa - \Delta\mathbf{x}_\kappa, \qquad \kappa = 0, 1, 2, \tag{10.11}$$

and substitute this into the trilinear constraint. We expand it with respect to $\Delta\mathbf{x}_\kappa$ and compute the position at which the reprojection error E of Eq. (10.4) is minimized by ignoring high-order terms in $\Delta\mathbf{x}_\kappa$. We write its difference from $\bar{\mathbf{x}}_\kappa$ as $\Delta\hat{\mathbf{x}}_\kappa$ and expand Eq. (10.2) with respect to $\Delta\hat{\mathbf{x}}_\kappa$. Then we minimize Eq. (10.4) by ignoring high-order terms in $\Delta\hat{\mathbf{x}}_\kappa$. We repeat this until the reproduction error E no longer changes ($\hookrightarrow$ Problem 10.2).

10.2.2 Solving Linear Equations

By "solve Eq. (10.8) using the pseudoinverse of truncated rank 3" in Procedure 10.1, we mean the following. The unknowns λ_{pq} of Eq. (10.8) are the Lagrange multipliers for minimizing Eq. (10.10). In matrix form, we can write (10.8) as

$$\begin{pmatrix} C_{1111} & C_{1112} & C_{1113} & C_{1121} & C_{1122} & C_{1123} & C_{1131} & C_{1132} & C_{1133} \\ C_{1211} & C_{1212} & C_{1213} & C_{1221} & C_{1222} & C_{1223} & C_{1231} & C_{1232} & C_{1233} \\ C_{1311} & C_{1312} & C_{1313} & C_{1321} & C_{1322} & C_{1323} & C_{1331} & C_{1332} & C_{1333} \\ C_{2111} & C_{2112} & C_{2113} & C_{2121} & C_{2122} & C_{2123} & C_{2131} & C_{2132} & C_{2133} \\ C_{2211} & C_{2212} & C_{2213} & C_{2221} & C_{2222} & C_{2223} & C_{2231} & C_{2232} & C_{2233} \\ C_{2311} & C_{2312} & C_{2313} & C_{2321} & C_{2322} & C_{2323} & C_{2331} & C_{2332} & C_{2333} \\ C_{3111} & C_{3112} & C_{3113} & C_{3121} & C_{3122} & C_{3123} & C_{3131} & C_{3132} & C_{3133} \\ C_{3211} & C_{3212} & C_{3213} & C_{3221} & C_{3222} & C_{3223} & C_{3231} & C_{3232} & C_{3233} \\ C_{3311} & C_{3312} & C_{3313} & C_{3321} & C_{3322} & C_{3323} & C_{3331} & C_{3332} & C_{3333} \end{pmatrix} \begin{pmatrix} \lambda_{11} \\ \lambda_{12} \\ \lambda_{13} \\ \lambda_{21} \\ \lambda_{22} \\ \lambda_{23} \\ \lambda_{31} \\ \lambda_{32} \\ \lambda_{33} \end{pmatrix} = \begin{pmatrix} F_{11} \\ F_{12} \\ F_{13} \\ F_{21} \\ F_{22} \\ F_{23} \\ F_{31} \\ F_{32} \\ F_{33} \end{pmatrix}. \tag{10.12}$$

However, *this coefficient matrix has rank 6*, only six columns and six rows being independent, therefore the solution is not unique. The reason for this rank deficiency is that we only need to correct six components Δx_0, Δy_0, Δx_1, Δy_1, Δx_2, and Δx_2 and there exist nine equations. This is because we regard all nine components of the vectors $\Delta \mathbf{x}_0$, $\Delta \mathbf{x}_1$, and $\Delta \mathbf{x}_2$ as unknowns whereas the third components of $\mathbf{x}_\kappa$ are all constant 1 and hence the third components of $\Delta \mathbf{x}_\kappa$ are all 0. In principle, we could obtain a unique solution if we select six from the nine equations of Eq. (10.12).

However, there exists an additional problem. If vectors $\mathbf{x}_\kappa$ exactly satisfy Eq. (10.2), *the rank decreases to 3*. This is because although Eq. (10.2) consists of nine equations for $p, q = 1, 2, 3$, *only three of them are independent*. This is understood as follows. Each equation of Eq. (10.2) defines a cubic polynomial hyper-surface in the 6D space of x_0, y_0, x_1, y_1, x_2, and x_2, and the solution of Eq. (10.2) is the intersection of the resulting nine Hyper-surfaces. This intersection should have dimension 3, because the solution must correspond to positions where the three rays meet at a point. In other words, points in 3D and triplets of their image points over three views must correspond one to one. In 6D, the intersection of three hyper-surfaces generally has dimension 3. Hence, if nine hyper-surfaces intersect, only three of them are independent and the rest are redundant. Of course, the observed values and intermediate values in the correction computation do not exactly satisfy Eq. (10.2), therefore the matrix in Eq. (10.12) generally has rank 6. If we select six equations from Eq. (10.12), they degenerate to rank 3 in the limit of convergence, thus the numerical computation becomes unstable (i.e., rounding errors are magnified) as it approaches convergence. This can be handled by selecting from Eq. (10.12) the three equations that are the "most independent." This is equivalent to applying the SVD (singular value decomposition) to the matrix and solving the equation with respect to the largest three singular values, that is, using the pseudoinverse of truncated rank 3. Specifically, we write Eq. (10.12) as

$$\mathsf{C}\boldsymbol{\lambda} = \mathbf{f}, \tag{10.13}$$

and compute the SVD of C in the form

$$\mathsf{C} = \mathsf{U}\begin{pmatrix} \sigma_1 & \cdots & 0 \\ \vdots & \ddots & \vdots \\ 0 & \cdots & \sigma_9 \end{pmatrix}\mathsf{V}^\top. \tag{10.14}$$

Then we compute its pseudoinverse of truncated rank 3

$$\mathsf{C}_3^- = \mathsf{V}\begin{pmatrix} 1/\sigma_1 & 0 & 0 & \\ 0 & 1/\sigma_2 & 0 & \mathsf{O} \\ 0 & 0 & 1/\sigma_3 & \\ & \mathsf{O} & & \mathsf{O} \end{pmatrix}\mathsf{U}^\top \tag{10.15}$$

and compute $\boldsymbol{\lambda}$ by

$$\boldsymbol{\lambda} = \mathsf{C}_3^-\mathbf{f}. \tag{10.16}$$

10.2.3 Efficiency of Computation

Procedure 10.1 requires repeated evaluation of expressions of the form $\mathsf{T}(\mathbf{x}, \mathbf{y}, \mathbf{z})$, which returns for three input vectors $\mathbf{x} = (x_i)$, $\mathbf{y} = (y_i)$, and $\mathbf{z} = (z_i)$ the following matrix $\mathsf{T} = (T_{pq})$.

$$T_{pq} = \sum_{i,j,k,l,m=1}^{3} \varepsilon_{ljp}\varepsilon_{mkq}T_i^{lm}x_i y_j z_k. \tag{10.17}$$

The right-hand side is the sum of $3^5 = 243$ terms, which is computed nine times for $p, q = 1, 2, 3$, requiring 2187 summations in total. This is a time-consuming process but can be made efficient by rewriting Eq. (10.17) in the form

$$T_{pq} = \frac{1}{4}\sum_{i,j,k,l,m=1}^{3} \varepsilon_{ljp}\varepsilon_{mkq}x_i\Big(T_i^{lm}y_j z_k - T_i^{jm}y_l z_k - T_i^{lk}y_j z_m + T_i^{jk}y_l z_m\Big). \tag{10.18}$$

From the properties of the permutation symbol ε_{ijk}, we can easily confirm that each term in the summand of $\sum_{i,j,k,l=1}^{3}$ equals Eq. (10.17). We can also see that the expression is symmetric with respect to l and j; for example, the term for $l = 1$ and $j = 2$ is equal to the term for $l = 2$ and $j = 1$. Therefore we only need to compute either of them and multiply it by 2. Because the terms are multiplied by ε_{ljp} and summed, we need not consider terms for which l or j equals p. In other words, we only need to sum over l and j not equal to p such that $\varepsilon_{ljp} = 1$ holds. Similarly, we only need to sum over m and k not equal to q such that $\varepsilon_{mkq} = 1$ holds. This is systematically done by defining "addition $\oplus$ modulo 3": $1 \oplus 1 = 2$, $1 \oplus 2 = 3$, $3 \oplus 1 = 1$, and so on. Inasmuch as $\varepsilon_{p\oplus1,p\oplus2,p} = 1$ and $\varepsilon_{q\oplus1,q\oplus2,q} = 1$, Eq. (10.18) is equivalently written as

$$T_{pq} = \sum_{i=1}^{3} x_i\Big(T_i^{p\oplus1,q\oplus1}y_{p\oplus2}z_{q\oplus2} - T_i^{p\oplus2,q\oplus1}y_{p\oplus1}z_{q\oplus2} - T_i^{p\oplus1,q\oplus2}y_{p\oplus2}z_{q\oplus1} + T_i^{p\oplus2,q\oplus2}y_{p\oplus1}z_{q\oplus1}\Big). \tag{10.19}$$

In this form, only 12 terms are summed, reducing the number of additions to about 1/20 as compared with Eq. (10.17). The operation $\oplus$ can be defined as an inline function.

10.2.4 3D Position Computation

Once the observed three points have been corrected by Procedure 10.1 such that their rays meet at a common intersection in the scene, its 3D position (X, Y, Z) is computed in the same way as Procedure 4.1 for two-view triangulation. Let (x_0, y_0), (x_1, y_1), and (x_2, y_2) be the corrected positions in the three images, and P_0, P_1, and P_2 be their camera matrices. Procedure 4.1 is modified to three views as follows.

Procedure 10.2 (Triangulation with known camera matrices)

1. Compute the following 6×3 matrix T and 6D vector $\mathbf{p}$.

$$\mathsf{T} = \begin{pmatrix} f_0 P_{0(11)} - x_0 P_{0(31)} & f_0 P_{0(12)} - x_0 P_{0(32)} & f_0 P_{0(13)} - x_0 P_{0(33)} \\ f_0 P_{0(21)} - y_0 P_{0(31)} & f_0 P_{0(22)} - y_0 P_{0(32)} & f_0 P_{0(23)} - y_0 P_{0(33)} \\ f_0 P_{1(11)} - x_1 P_{1(31)} & f_0 P_{1(12)} - x_1 P_{1(32)} & f_0 P_{1(13)} - x_1 P_{1(33)} \\ f_0 P_{1(21)} - y_1 P_{1(31)} & f_0 P_{1(22)} - y_1 P_{1(32)} & f_0 P_{1(23)} - y_1 P_{1(33)} \\ f_0 P_{2(11)} - x_2 P_{2(31)} & f_0 P_{2(12)} - x_2 P_{2(32)} & f_0 P_{2(13)} - x_2 P_{2(33)} \\ f_0 P_{2(21)} - y_2 P_{2(31)} & f_0 P_{2(22)} - y_2 P_{2(32)} & f_0 P_{2(23)} - y_2 P_{2(33)} \end{pmatrix}, \tag{10.20}$$

$$\mathbf{p} = \begin{pmatrix} f_0 P_{0(14)} - x_0 P_{0(34)} \\ f_0 P_{0(24)} - y_0 P_{0(34)} \\ f_0 P_{1(14)} - x_1 P_{1(34)} \\ f_0 P_{1(24)} - y_1 P_{1(34)} \\ f_0 P_{2(14)} - x_2 P_{2(34)} \\ f_0 P_{2(24)} - y_2 P_{2(34)} \end{pmatrix}. \tag{10.21}$$

2. Determine X, Y, and Z by solving the following linear equation.

$$\mathsf{T}^\top \mathsf{T} \begin{pmatrix} X \\ Y \\ Z \end{pmatrix} = -\mathsf{T}^\top \mathbf{p}. \tag{10.22}$$

Comments The matrix in Eq. (10.20) has rank 3, because there are only three unknowns X, Y, and Z. In principle, we can select an arbitrary three rows of T in Eq. (10.20) and the corresponding three components of $\mathbf{p}$ in Eq. (10.21) and solve

$$\mathsf{T} \begin{pmatrix} X \\ Y \\ Z \end{pmatrix} = -\mathbf{p}. \tag{10.23}$$

Instead of selecting three rows, we can apply least squares to all the rows and solve Eq. (10.22) to obtain the same solution (↪ Problem 4.2). This form has the advantage that we can obtain an approximation of X, Y, and Z, even if the three rays do not meet at a point.

The rays of three corresponding points meet at a point if and only if Eq. (10.23) has a unique solution. As is well known in linear algebra, this is the case when all 4×4 minors of the 6×4 matrix $\begin{pmatrix} \mathsf{T} & \mathbf{p} \end{pmatrix}$ vanish. Applying cofactor expansion to this, we obtain the trilinear constraint of Eq. (10.2) using the definition of the trifocal tensor T_i^{jk} of Eq. (10.3) (↪ Problem 10.3). The principle is the same as obtaining the epipolar equation and the fundamental matrix expression from the matrix T and vector $\mathbf{p}$ in Eq. (4.3) as the condition for the linear equation of Eq. (4.6) to have a solution (↪ Problem 4.3).

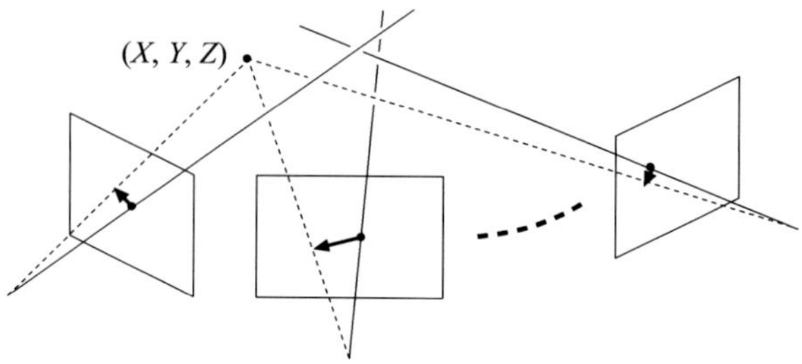

Fig. 10.2 Correcting observed points such that their rays meet in the scene in such a way that the reprojection error (= the sum of square distances of displacements) is minimized

10.3 Triangulation from Multiple Views

Consider the general case, where corresponding points $\mathbf{x}_\kappa$, $\kappa = 0, ..., M-1$, are given over M images. We assume that their camera matrices P_κ are given. As in the two-view and three-view cases, we correct the M points $\mathbf{x}_\kappa$ to $\bar{\mathbf{x}}_\kappa$ by minimizing the reprojection error

$$E = \sum_{\kappa=0}^{M-1} \|\mathbf{x}_\kappa - \bar{\mathbf{x}}_\kappa\|^2, \tag{10.24}$$

subject to the constraint that all the rays meet at a single point in the scene (Fig. 10.2). For this condition, we require the trilinear constraints over all three consecutive images

$$\sum_{i,j,k,l,m=1}^{3} \varepsilon_{ljp}\varepsilon_{mkq} T_{(\kappa)i}^{lm} x_{\kappa(i)} x_{\kappa+1(j)} x_{\kappa+2(k)} = 0, \ \kappa = 0, ..., M-3, \tag{10.25}$$

where $T_{(\kappa)i}^{jk}$ is the trifocal tensor for the κth, $\kappa+1$st, and $\kappa+2$rd images, and $x_{\kappa(i)}$ is the ith component of $\mathbf{x}_\kappa$. Procedure 10.1 for three views is generalized to M views as follows.

Procedure 10.3 (Optimal M-view correspondence correction)

1. Let $E_0 = \infty$ (a sufficiently large number), $\hat{\mathbf{x}}_\kappa = \mathbf{x}_\kappa$, and $\tilde{\mathbf{x}}_\kappa = \mathbf{0}$, $\kappa = 0, ..., M-1$.
2. Compute the following $P_{\kappa(pqs)}$, $Q_{\kappa(pqs)}$, and $R_{\kappa pqs}$, $\kappa = 0, ..., M-1$, $p, q, s = 1, 2, 3$.

$$P_{\kappa(pqs)} = \sum_{i,j,k,l,m=1}^{3} \varepsilon_{ljp}\varepsilon_{mkq} T_{(\kappa)i}^{lm} P_{\mathbf{k}(si)} \hat{x}_{\kappa(j)} \hat{x}_{\kappa+1(k)},$$

$$Q_{\kappa(pqs)} = \sum_{i,j,k,l,m=1}^{3} \varepsilon_{ljp}\varepsilon_{mkq} T_{(\kappa-1)i}^{lm} \hat{x}_{\kappa-1(i)} P_{\mathbf{k}(sj)} \hat{x}_{\kappa+1(k)},$$

$$R_{\kappa(pqs)} = \sum_{i,j,k,l,m=1}^{3} \varepsilon_{ljp}\varepsilon_{mkq} T_{(\kappa-2)i}^{lm} \hat{x}_{\kappa-2(i)} \hat{x}_{\kappa-1(j)} P_{\mathbf{k}(sk)}. \tag{10.26}$$

3. Compute the following. $A_{\kappa(pqrs)}$, $B_{\kappa(pqrs)}$, $C_{\kappa(pqrs)}$, $D_{\kappa(pqrs)}$, $E_{\kappa(pqrs)}$, and $F_{\kappa(pqrs)}$:

$$
\begin{aligned}
A_{\kappa(pqrs)} &= \sum_{i,j,k,l,m=1}^{3} \varepsilon_{ljp}\varepsilon_{mkq}T_{(\kappa)i}^{lm} R_{\kappa(rsi)}\hat{x}_{\kappa+1(j)}\hat{x}_{\kappa+2(k)}, \\
B_{\kappa(pqrs)} &= \sum_{i,j,k,l,m=1}^{3} \varepsilon_{ljp}\varepsilon_{mkq}T_{(\kappa)i}^{lm} \Big(Q_{\kappa(rsi)}\hat{x}_{\kappa+1(j)}\hat{x}_{\kappa+2(k)} + \hat{x}_{\kappa(i)}R_{\kappa+1(rsj)}\hat{x}_{\kappa+2(k)}\Big), \\
C_{\kappa(pqrs)} &= \sum_{i,j,k,l,m=1}^{3} \varepsilon_{ljp}\varepsilon_{mkq}T_{\kappa(i)}^{lm} \Big(P_{\kappa(rsi)}\hat{x}_{\kappa+1(j)}\hat{x}_{\kappa+2(k)} + \hat{x}_{\kappa(i)}Q_{\kappa+1(rsj)}\hat{x}_{\kappa+2(k)} \\
&\quad + \hat{x}_{\kappa(i)}\hat{x}_{\kappa+1(j)}R_{\kappa+2(rsk)}\Big), \\
D_{\kappa(pqrs)} &= \sum_{i,j,k,l,m=1}^{3} \varepsilon_{ljp}\varepsilon_{mkq}T_{(\kappa)i}^{lm} \Big(\hat{x}_{\kappa(i)}\hat{P}_{\kappa+1(rsj)}\hat{x}_{\kappa+2(k)} + \hat{x}_{\kappa(i)}\hat{x}_{\kappa+1(j)}Q_{\kappa+2(rsk)}\Big), \\
E_{\kappa(pqrs)} &= \sum_{i,j,k,l,m=1}^{3} \varepsilon_{ljp}\varepsilon_{mkq}T_{(\kappa)i}^{lm} \hat{x}_{\kappa(i)}\hat{x}_{\kappa+1(j)}P_{\kappa+2(rsk)}, \\
F_{\kappa(pq)} &= \sum_{i,j,k,l,m=1}^{3} \varepsilon_{ljp}\varepsilon_{mkq}T_{(\kappa)i}^{lm} \Big(\hat{x}_{\kappa(i)}\hat{x}_{\kappa+1(j)}\hat{x}_{\kappa+2(k)} + \tilde{x}_{\kappa(i)}\hat{x}_{\kappa+1(j)}\hat{x}_{\kappa+2(k)} \\
&\quad + \hat{x}_{\kappa(i)}\tilde{x}_{\kappa+1(j)}\hat{x}_{\kappa+2(k)} + \hat{x}_{\kappa(i)}\hat{x}_{\kappa+1(j)}\tilde{x}_{\kappa+2(k)}\Big).
\end{aligned}
\tag{10.27}
$$

4. Solve the following linear equation, using the pseudoinverse of truncated rank $2M - 3$, to compute $\lambda_{\kappa(pq)}$, $\kappa = 0, ..., M - 1$, $p, q = 1, 2, 3$.

$$
\begin{aligned}
&\sum_{r,s=1}^{3} A_{\kappa(pqrs)}\lambda_{\kappa-2(rs)} + \sum_{r,s=1}^{3} B_{\kappa(pqrs)}\lambda_{\kappa-1(rs)} + \sum_{r,s=1}^{3} C_{\kappa(pqrs)}\lambda_{\kappa(rs)} \\
&+ \sum_{r,s=1}^{3} D_{\kappa(pqrs)}\lambda_{\kappa+1(rs)} + \sum_{r,s=1}^{3} E_{\kappa(pqrs)}\lambda_{\kappa+2(rs)} = F_{\kappa(pq)}.
\end{aligned}
\tag{10.28}
$$

5. Update $\tilde{\mathbf{x}}_\kappa$ and $\hat{\mathbf{x}}_\kappa$, $\kappa = 0, ..., M - 1$, to

$$
\tilde{x}_{\kappa(i)} \leftarrow \sum_{p,q=1}^{3} P_{\kappa(pqi)}\lambda_{\kappa(pq)} + \sum_{p,q=1}^{3} Q_{\kappa(pqi)}\lambda_{\kappa-1(pq)} + \sum_{p,q=1}^{3} R_{\kappa(pqi)}\lambda_{\kappa-2(pq)},
\tag{10.29}
$$

$$
\hat{\mathbf{x}}_{(\kappa)} \leftarrow \mathbf{x}_\kappa - \tilde{\mathbf{x}}_\kappa.
\tag{10.30}
$$

6. Compute the following reprojection error E.

$$
E = \sum_{\kappa=0}^{M-1} \|\tilde{\mathbf{x}}_\kappa\|^2.
\tag{10.31}
$$

7. If $E \approx E_0$, return E and $\hat{\mathbf{x}}_\kappa$, $\kappa = 0, ..., M - 1$, and stop. Else, let $E_0 \leftarrow E$ and go back to Step 2.

Comments This procedure is based on the same principle as Procedure 10.1. Namely, we write $\bar{\mathbf{x}}_\kappa$ as

$$\bar{\mathbf{x}}_\kappa = \mathbf{x}_\kappa - \Delta\mathbf{x}_\kappa, \qquad \kappa = 0, ..., M-1, \tag{10.32}$$

and substitute this into Eq. (10.25), which we expand with respect to $\Delta\mathbf{x}_\kappa$ and ignore high-order terms. Then we compute the position $\hat{\mathbf{x}}_\kappa$ of $\bar{\mathbf{x}}_\kappa$ that minimizes the reprojection error of Eq. (10.24). We let $\Delta\hat{\mathbf{x}}_\kappa$ be its discrepancy from $\bar{\mathbf{x}}_\kappa$ and expand Eq. (10.25) with respect to $\Delta\hat{\mathbf{x}}_\kappa$ Ignoring its high-order terms, we compute the value that minimizes Eq. (10.24), and repeat this procedure until the reprojection error E is no longer altered ($\hookrightarrow$ Problem 10.4).

Note that not all terms in Eqs. (10.26) and (10.27) can be evaluated. For example, $T^{lm}_{(\kappa-1)i}$ and $T^{lm}_{(\kappa-2)i}$ are not defined for $\kappa = 0$. Such undefined terms do not appear in the final result, therefore we just skip undefined terms in Eqs. (10.26) and (10.27) or assign arbitrary values.

By "solve Eq. (10.28) using the pseudoinverse of truncated rank $2M - 3$," we mean the following. The unknowns $\lambda_{\kappa(pq)}$ of the linear equation of Eq. (10.28) are the Lagrange multipliers for minimizing E of Eq. (10.24). Assigning sequential numbers $\alpha = 1, ..., 9$ to the nine pairs of indices $(p, q) = (1, 1), (1, 2), ..., (3, 3)$ and sequential numbers $\beta = 1, ..., 9$ to the nine pairs of indices $(r, s) = (1, 1), (1, 2), ..., (3, 3)$, we regard $A_{\kappa(pqrs)}$ as a 9×9 matrix $\mathsf{A}_\kappa = (A_{\kappa(\alpha\beta)})$, $\alpha, \beta = 1, ..., 9$. Similarly, we regard $B_{\kappa(pqrs)}$, $C_{\kappa(pqrs)}$, $D_{\kappa(pqrs)}$, and $E_{\kappa(pqrs)}$ as 9×9 matrices B_κ, C_κ, D_κ, and E_κ, respectively. Furthermore, we assign sequential numbers to (p, q) and regard $F_{\kappa(pq)}$ and $\lambda_{\kappa(pq)}$, respectively, as 9D vectors $\mathbf{f}_\kappa$ and $\boldsymbol{\lambda}_\kappa$. Then Eq. (10.28) can be regarded as vector equations

$$\mathsf{A}_\kappa\boldsymbol{\lambda}_{\kappa-2} + \mathsf{B}_\kappa\boldsymbol{\lambda}_{\kappa-1} + \mathsf{C}_\kappa\boldsymbol{\lambda}_\kappa + \mathsf{D}_\kappa\boldsymbol{\lambda}_{\kappa+1} + \mathsf{E}_\kappa\boldsymbol{\lambda}_{\kappa+2} = \mathbf{f}_\kappa, \quad \kappa = 0, ..., M-3. \tag{10.33}$$

These are $9(M-2)$ simultaneous linear equations in $9(M-2)$ unknowns $\boldsymbol{\lambda}_0, ..., \boldsymbol{\lambda}_{M-1}$. However, the coefficient matrix has only rank $2M$ for the same reason as in the three-view case: we are correcting only $2M$ components Δx_0, Δy_0, ..., Δx_{M-1}, and Δy_{M-1}, whereas $9(M-2)$ equations exist, because we are treating $\Delta\mathbf{x}_0$, ..., $\Delta\mathbf{x}_M$ as unknowns although their third components are identically 0. Therefore we can arbitrarily select $2M$ from among the $9(M-2)$ equations of Eq. (10.33).

However, by the same argument as in the three-view case, the rank reduces to $2M - 3$ when all $\mathbf{x}_\kappa$ exactly satisfy Equation (10.25). This is because the set of all $\mathbf{x}_\kappa$ that satisfy Eq. (10.25) is the set of all $\mathbf{x}_\kappa$ for which their rays intersect in the scene. This should correspond one to one to the set of intersections and thus have dimension 3, for the intersection can be anywhere in 3D. Equation (10.25) consists of $9(M-2)$ equations for $\kappa = 0, ..., M-1$ and $p, q = 1, 2, 3$, and each equation defines a 3D cubic polynomial hyper-surface in the $2M$D space of x_0, y_0, ..., x_{M-1}, and y_{M-1}. Because the intersection of $2M-3$ such hyper-surface defines a 3D space (one hyper-surface is $(2M-1)$D, two hyper-surfaces define a $(2M-2)$D intersection, three hyper-surfaces define a $(2M-3)$D intersection, ...), the remaining $7M - 15$ hyper-surfaces are redundant, meeting with the same 3D space. This means we can arbitrarily select $2M - 3$ out of the $9(M-2)$ equations of Eq. (10.33). However, observed values and intermediate values in the correction computation do not exactly satisfy Eq. (10.25), such that the number of independent equations of Eq. (10.33) is generally $2M$, which drops to $2M - 3$ at the time of convergence, causing numerical instability as the computation proceeds. This can be handled, as in the three-view case, by selecting from among the $9(M - 2)$ equations of Eq. (10.33) the most independent $2M - 3$. This is equivalent to using the pseudoinverse of truncated rank $2M - 3$ and solving

$$\begin{pmatrix} \boldsymbol{\lambda}_0 \\ \boldsymbol{\lambda}_1 \\ \boldsymbol{\lambda}_2 \\ \boldsymbol{\lambda}_3 \\ \vdots \\ \boldsymbol{\lambda}_{M-6} \\ \boldsymbol{\lambda}_{M-5} \\ \boldsymbol{\lambda}_{M-4} \\ \boldsymbol{\lambda}_{M-3} \end{pmatrix} = \begin{pmatrix} \mathsf{C}_0 & \mathsf{D}_0 & \mathsf{E}_0 & & & & & & \\ \mathsf{B}_1 & \mathsf{C}_1 & \mathsf{D}_1 & \mathsf{E}_1 & & & & & \\ \mathsf{A}_2 & \mathsf{B}_2 & \mathsf{C}_2 & \mathsf{D}_2 & \mathsf{E}_2 & & & & \\ & \mathsf{A}_3 & \mathsf{B}_3 & \mathsf{C}_3 & \mathsf{D}_3 & & & & \\ & & \ddots & \ddots & \ddots & \ddots & \ddots & & \\ & & & \mathsf{A}_{M-6} & \mathsf{B}_{M-6} & \mathsf{C}_{M-6} & \mathsf{D}_{M-6} & \mathsf{E}_{M-6} & \\ & & & & \mathsf{A}_{M-5} & \mathsf{B}_{M-5} & \mathsf{C}_{M-5} & \mathsf{D}_{M-5} & \mathsf{E}_{M-5} \\ & & & & & \mathsf{A}_{M-4} & \mathsf{B}_{M-4} & \mathsf{C}_{M-4} & \mathsf{D}_{M-4} \\ & & & & & & \mathsf{A}_{M-3} & \mathsf{B}_{M-3} & \mathsf{C}_{M-3} \end{pmatrix}^{-}_{2M-3} \begin{pmatrix} \mathbf{f}_0 \\ \mathbf{f}_1 \\ \mathbf{f}_2 \\ \mathbf{f}_3 \\ \vdots \\ \mathbf{f}_{M-6} \\ \mathbf{f}_{M-5} \\ \mathbf{f}_{M-4} \\ \mathbf{f}_{M-3} \end{pmatrix}, \tag{10.34}$$

where $(\cdot)^{-}_{2M-3}$ denotes the pseudoinverse of truncated rank $2M - 3$ defined in the same way as Eqs. (10.14) and (10.15); in this case we retain the largest $2M - 3$ singular values. The expressions $\sum_{i,j,k,l,m=1}^{3} \varepsilon_{ljp}\varepsilon_{mkq} T^{lm}_{(\kappa)i}(\cdots)$ that appear in the computation can be efficiently evaluated in the form of Eq. (10.19).

Once all the points have been corrected such that their rays meet, the 3D position (X, Y, Z) is computed from the corrected positions (x_κ, y_κ), $\kappa = 0, ..., M - 1$, and the camera matrices $\mathsf{P}_\kappa = (P_{\kappa(ij)})$, $\kappa = 0, ..., M - 1$, using Proposition 10.2. In this case, Eqs. (10.20) and (10.22) are replaced by the following $2M \times 3$ matrix T and the $2M$D vector $\mathbf{p}$.

$$\mathsf{T} = \begin{pmatrix} f_0 P_{0(11)} - x_0 P_{0(31)} & f_0 P_{0(12)} - x_0 P_{0(32)} & f_0 P_{0(13)} - x_0 P_{0(33)} \\ f_0 P_{0(21)} - y_0 P_{0(31)} & f_0 P_{0(22)} - y_0 P_{0(32)} & f_0 P_{0(23)} - y_0 P_{0(33)} \\ \vdots & \vdots & \vdots \\ f_0 P_{M-1(21)} - x_{M-1} P_{M-1(31)} & f_0 P_{M-1(22)} - x_{M-1} P_{M-1(32)} & f_0 P_{M-1(23)} - x_{M-1} P_{M-1(33)} \\ f_0 P_{M-1(21)} - y_{M-1} P_{M-1(31)} & f_0 P_{M-1(22)} - y_{M-1} P_{M-1(32)} & f_0 P_{M-1(23)} - y_{M-1} P_{M-1(33)} \end{pmatrix}, \tag{10.35}$$

$$\mathbf{p} = \begin{pmatrix} f_0 P_{0(14)} - x_0 P_{0(34)} \\ f_0 P_{0(24)} - y_0 P_{0(34)} \\ \vdots \\ f_0 P_{M-1(14)} - x_{M-1} P_{M-1(34)} \\ f_0 P_{M-1(24)} - y_{M-1} P_{M-1(34)} \end{pmatrix}. \tag{10.36}$$

Then, X, Y, and Z are determined by solving Eq. (10.22).

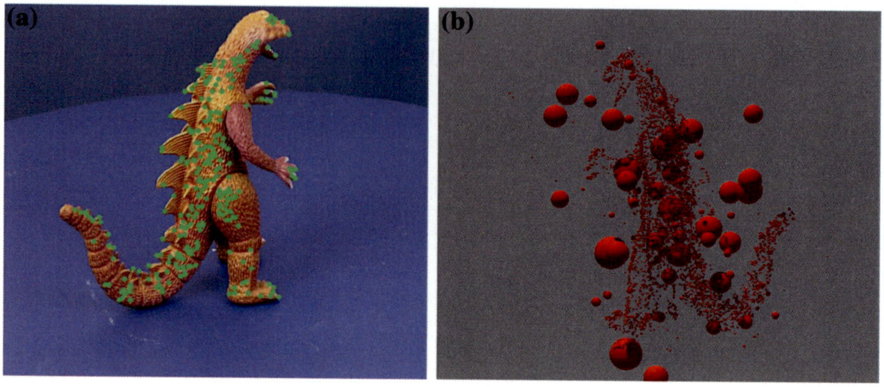

Fig. 10.3 **a** One of the 36 input images. **b** The reconstructed 3D positions. The size of the point represents the degree of its uncertainty

10.4 Examples

Figure 10.3a shows one frame of the image sequence provided by the University of Oxford.[1] It consists of pictures of a toy dinosaur on a turntable viewed from 36 directions, from which 4983 feature points are extracted. Each feature point is tracked over $M = 2 - 21$ frames, and the camera matrix P_κ of each frame is estimated.

Figure 10.3b shows the reconstructed 3D positions of the feature points viewed from one direction. The size (the radius) of each point is proportional to $\sqrt{E/M}$, which describes the average image noise magnitude estimated from the final reprojection E of Eq. (10.31); it represents the uncertainty of the computed 3D position of each point. Actually, there are many points whose uncertainty is too small to be displayed; they are displayed with a fixed size in order for us to be able to discern them. We see a large variation of reliability. Very large uncertainty is perhaps due to mismatching of feature points in the course of similarity-based image matching processing.

10.5 Supplemental Note

Multiview triangulation has been studied by many researchers in the past, but the motivation has mostly been theoretical interest. This is because this is a typical task of global optimization. The standard formulation in the past is as follows. Assuming that the camera matrices P_κ of the M images that we observe are known, we let (X, Y, Z) be the 3D position that we want to compute. Then we can predict the image positions $\bar{\mathbf{x}}_\kappa$ seen by the M cameras with known camera matrices P_κ as functions

[1] http://www.robots.ox.ac.uk/~vgg/data.html.

of (X, Y, Z). If the reprojection error E is defined by Eq. (10.24), that is, as the sum of square distances between predicted and observed positions, it is also a function of (X, Y, Z). Thus the task is to minimize the function $E(X, Y, Z)$. It is known that this function has a lot of local minima, therefore gradient-based search, such as the Levenberg-Marquardt method (LM), may not reach a global minimum. How to compute the global minimum has been the main concern of triangulation research in the past. The two-view triangulation method of Hartley and Sturm [5] for solving a six-degree polynomial mentioned in Chap. 4 and the planar triangulation of Chum et al. [2] for solving an eight-degree polynomial were obtained from this standpoint. Existing approaches for global optimization are roughly classified as follows (see the review paper of Hartley and Kahl [3]).

Algebraic approach We let the derivatives of E with respect to X, Y, and Z be zero and solve the resulting algebraic equations. By using the Gröbner basis, this reduces to solving a single high-degree polynomial equation. The degree is for 3, 4, 5, 6, and 7 images, for example, 47, 148, 336, 638, and 1081, respectively. The computation of the Gröbner basis is known to be numerically unstable, so various stabilization techniques have been studied [1].

Branch-and-bound approach Introducing a function that locally estimates the lower bound of E, we divide the search region into cells and remove those cells in which the estimated lower bond is larger than already achieved values. The remaining cells are recursively subdivided, and the same process is repeated [6,9]. However, lower bound estimation is complicated, requiring a large amount of computational time.

Matrix inequality approach Changing variables, we reduce the problem to minimization of a matrix polynomial subject to a matrix inequality (matrix inequality $\mathsf{A} \succ \mathsf{B}$ means that $\mathsf{A} - \mathsf{B}$ is positive semi-definite) [7]. This is what is called a *semi-definite problem* (*SDP*), for which a MATLAB software tool called GloptiPoly is publicly available. The resulting solution is only an approximation, but the accuracy increases as the initial change of variables is refined. Theoretically, the solution converges to the true value in the limit. However, the procedure is very complicated.

L_∞ **Optimization** Minimization of the sum of square distances between reprojected and observed positions (L_2-norm) makes sense statistically because it can be interpreted as maximum likelihood when the noise is Gaussian, but the global minimization is difficult to compute. To cope with this, the sum of square distances is replaced by the maximum of distances (L_∞-norm). Then, the cost function is shown to be a *quasi-convex function*, for which no local minimums exist. The local minimum is searched by setting a threshold and checking if the cost function takes a value above it. This is a *second-order conic program* (*SOCP*), for which a MATLAB software tool called SeDuMi is publicly available. Then, the threshold is gradually increased from 0, or binary search is used, for finding the absolute minimum.

All of these approaches require complicated procedures and a large amount of computation. They are very interesting from a theoretical point of view but are not very useful in practice. As an alternative, some researchers are focusing on methods for testing whether a solution obtained by gradient-based search, such as the Levenberg-Marquardt method, is indeed a global minimum or merely a local solution [4]. In practice, however, we need not use these theoretically motivated methods, because the fact that the function $E(X, Y, Z)$ has many local minima into which gradient-base search may fall is when the search is started from an *arbitrary* position in the 3D *XYZ* space. If an approximation of the solution is known, a few iterations lead to the correct solution. In fact, no real or synthetic examples are ever known such that a local solution results from gradient-based search starting from a least squares solution.

On the other hand, the procedure described in this chapter does *not* do any search in the *XYZ* space; it *corrects*, as in Chaps. 4 and 7, the *observed image points* in the *2D image region* such that their rays meet at a single point in the scene. Therefore the correction converges to positions *near the observed points*. The procedure of this chapter was presented by the authors in [8], where it is shown that an optimal solution is obtained after a few iterations and the computation time is far shorter than any known global optimization methods. Specifically, the computation time for all the trifocal tensors and all the 3D positions was 2.22 s (0.000446 s per point), using the C++ language on Windows Vista with Intel CPU Core2Duo E6850 (3.0 GHz, main memory 4 GB). When the MATLAB code[2] of Kahl et al. [6] was used, it took 5030 s (1.01 s per point) on a WindowsXP machine with Intel CPU Core2Duo E6400 (2.13 GHz, main memory 2 GB). It was found that 98 % of the computation was spent on the MATLAB optimization tool SeDuMi written partly in the C language. Theoretically, the resulting solution should be identical to ours, but our solution has a lower reprojection error. This is perhaps due to the convergence criterion of the SeDuMi iterations. In another example using different data, it was reported that the C++ implementation of branch-and-bound by Lu and Hartley [9] took 0.02 s per point, for which the pure MATLAB implementation of the method of Kahl et al. took 6 s per point. Thus, although global optimization is theoretically very interesting, it is not useful in practical applications for which the method of this chapter is the most suited.

Problems

10.1 The trilinear constraint of Eq. (10.2) consists of nine equations for $p, q = 1, 2, 3$. Show that only four of them are linearly independent, that is, the remaining ones are expressed as their linear combinations.

10.2 Show that Procedure 10.1 computes the optimal correction of three-view correspondence.

---

[2] http://www.cs.washington.edu/homes/sagarwal/code.html.

10.3 Show that the trilinear constraint of Eq. (10.2) and the trifocal tensor T_i^{jk} of Eq. (10.3) are obtained from the condition that Eq. (10.23) has a unique solution.

10.4 Show that Procedure 10.3 computes the optimal correction of M-view correspondence.

References

1. M. Byröd, K. Josephson, K. Åström, Fast optimal three view triangulation, in *Proceedings of the 8th Asian Conference on Computer Vision*, Tokyo, Japan, vol. 2 (2007), pp. 549–559
2. O. Chum, T. Pajdla, P. Sturm, The geometric error for homographies. Comput. Vis. Image Underst. **97**(1), 86–102 (2002)
3. R. Hartley, F. Kahl, Optimal algorithms in multiview geometry, in *Proceedings of the 8th Asian Conference on Computer Vision*, Tokyo, Japan, vol. 1 (2007), pp. 13–34
4. R. Hartley, Y. Seo, Verifying global minima for L_2 minimization problems, in *Proceedings of the IEEE Conference on Computer Vision and Pattern Recognition*, Anchorage, AK, U.S. (2008)
5. R. Hartley, P. Sturm, Triangulation. Comput. Vis. Image Underst. **68**(2), 146–157 (1997)
6. F. Kahl, S. Agarwal, M.K. Chandraker, D. Kriegman, S. Belongie, Practical global optimization for multiview geometry. Int. J. Comput. Vis. **79**(3), 271–284 (2008)
7. F. Kahl, D. Henrion, Global optimal estimates for geometric reconstruction problems. Int. J. Comput. Vis. **74**(1), 3–15 (2007)
8. K. Kanatani, Y. Sugaya, H. Niitsuma, Optimization without search: constraint satisfaction by orthogonal projection with applications to multiview triangulation. IEICE Trans. Inf. Syst. **E93-D**(10), 2386–2845 (2010)
9. F. Lu, R. Hartley, A fast optimal algorithm for L_2 triangulation, in *Proceedings of the 8th Asian Conference on Computer Vision*, Tokyo, Japan, vol. 2 (2007), pp. 279–288

Bundle Adjustment

11

Abstract

In the preceding chapter, we used knowledge of the camera matrices for reconstructing the 3D from multiple views. In this chapter, we describe a procedure, called bundle adjustment, for computing from multiple views not only the 3D shape but also the positions, orientations, and intrinsic parameters of all the cameras simultaneously. Specifically, starting from given initial values for all the unknowns, we iteratively update them such that the reconstructed shape and observed images better satisfy the perspective projection relationship. Mathematically, this is nothing but minimization of a multivariable function, but the number of unknowns is very large. Hence, we need to devise an ingenious scheme for efficiently storing intermediate values and avoiding unnecessary computations. Here, we describe a typical programming technique for such implementation.

11.1 Principle of Bundle Adjustment

Bundle adjustment is a process for computing from multiple camera images of a 3D scene its 3D shape and the positions and intrinsic parameters of all the cameras simultaneously so that the perspective projection relationship holds. We fix a world coordinate and let $\mathbf{t}_\kappa$ be the viewpoint of the κth camera, $\kappa = 1, ..., M$. We assume that its camera coordinate system is rotated by R_κ relative to the world coordinate system. Let f_κ be its focal length, and $(u_{0\kappa}, v_{0\kappa})$ its principal point.

Let $(x_{\alpha\kappa}, y_{\alpha\kappa})$ be the projection of the αth point $(X_\alpha, Y_\alpha, Z_\alpha)$, $\alpha = 1, ..., N$, in the κth image (Fig. 11.1). From Eq. (4.2), the perspective projection relationship is

© Springer International Publishing AG 2016   149
K. Kanatani et al., *Guide to 3D Vision Computation*, Advances in Computer
Vision and Pattern Recognition, DOI 10.1007/978-3-319-48493-8_11

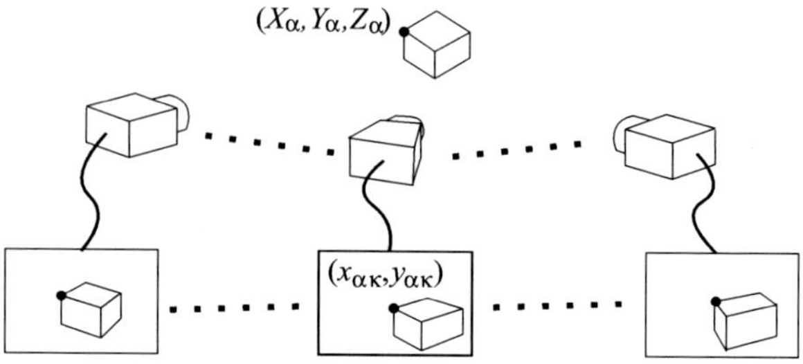

Fig. 11.1 N points in the scene are viewed by M cameras. The αth point $(X_\alpha, Y_\alpha, Z_\alpha)$ is projected to a point $(x_{\alpha\kappa}, y_{\alpha\kappa})$ in the image of the κth camera

written as

$$\begin{pmatrix} x_{\alpha\kappa}/f_0 \\ y_{\alpha\kappa}/f_0 \\ 1 \end{pmatrix} \simeq \mathsf{P}_\kappa \begin{pmatrix} X_\alpha \\ Y_\alpha \\ Z_\alpha \\ 1 \end{pmatrix}. \tag{11.1}$$

From Eq. (5.7), we can write the camera matrix P_κ of the κth camera in the form

$$\mathsf{P}_\kappa = \begin{pmatrix} f_\kappa & 0 & u_{0\kappa} \\ 0 & f_\kappa & v_{0\kappa} \\ 0 & 0 & f_0 \end{pmatrix} \begin{pmatrix} \mathsf{R}_\kappa^\top & -\mathsf{R}_\kappa^\top \mathbf{t}_\kappa \end{pmatrix}. \tag{11.2}$$

Note that Eq. (5.7) was obtained by assuming that the principal point is at the image origin, but here it is assumed to be at $(u_{0\kappa}, v_{0\kappa})$ ($\hookrightarrow$ Problem 11.1). We can rewrite Eq. (11.2) as

$$\mathsf{P}_\kappa = \mathsf{K}_\kappa \mathsf{R}_\kappa^\top \begin{pmatrix} \mathsf{I} & -\mathbf{t}_\kappa \end{pmatrix}, \qquad \mathsf{K}_\kappa \equiv \begin{pmatrix} f_\kappa & 0 & u_{0\kappa} \\ 0 & f_\kappa & v_{0\kappa} \\ 0 & 0 & f_0 \end{pmatrix}. \tag{11.3}$$

The matrix K_κ is called the *intrinsic parameter matrix* of the κth camera.

From Eq. (11.1), we can express $x_{\alpha\kappa}$ and $y_{\alpha\kappa}$ in the form of Eq. (4.1):

$$\begin{aligned} x_{\alpha\kappa} &= f_0 \frac{P_{\kappa(11)}X_\alpha + P_{\kappa(12)}Y_\alpha + P_{\kappa(13)}Z_\alpha + P_{\kappa(14)}}{P_{\kappa(31)}X_\alpha + P_{\kappa(32)}Y_\alpha + P_{\kappa(33)}Z_\alpha + P_{\kappa(34)}}, \\ y_{\alpha\kappa} &= f_0 \frac{P_{\kappa(21)}X_\alpha + P_{\kappa(22)}Y_\alpha + P_{\kappa(23)}Z_\alpha + P_{\kappa(24)}}{P_{\kappa(31)}X_\alpha + P_{\kappa(32)}Y_\alpha + P_{\kappa(33)}Z_\alpha + P_{\kappa(34)}}. \end{aligned} \tag{11.4}$$

Here, we write $P_{\kappa(ij)}$ for the (i, j) element of P_κ. Bundle adjustment is to estimate the 3D positions $(X_\alpha, Y_\alpha, Z_\alpha)$ and the camera matrices from observed $(x_{\alpha\kappa}, y_{\alpha\kappa})$, $\alpha = 1, ..., N$, $\kappa = 1, ..., M$, in such a way that Eq. (11.4) is satisfied as much as possible in the sense that

$$\begin{aligned} E = \sum_{\alpha=1}^{N} \sum_{\kappa=1}^{M} I_{\alpha\kappa} \Bigg(&\Big(\frac{x_{\alpha\kappa}}{f_0} - \frac{P_{\kappa(11)}X_\alpha + P_{\kappa(12)}Y_\alpha + P_{\kappa(13)}Z_\alpha + P_{\kappa(14)}}{P_{\kappa(31)}X_\alpha + P_{\kappa(32)}Y_\alpha + P_{\kappa(33)}Z_\alpha + P_{\kappa(34)}} \Big)^2 \\ &+ \Big(\frac{y_{\alpha\kappa}}{f_0} - \frac{P_{\kappa(21)}X_\alpha + P_{\kappa(22)}Y_\alpha + P_{\kappa(23)}Z_\alpha + P_{\kappa(24)}}{P_{\kappa(31)}X_\alpha + P_{\kappa(32)}Y_\alpha + P_{\kappa(33)}Z_\alpha + P_{\kappa(34)}} \Big)^2 \Bigg) \end{aligned} \tag{11.5}$$

is minimized, where $I_{\alpha\kappa}$ is the *visibility index* that takes 1 if the αth point is viewed in the κth image and 0 otherwise. Because Eq. (11.5) measures the sum of the distances between the perspectively projected points and actually observed positions (in unit f_0), it is called the *reprojection error*.

11.2 Bundle Adjustment Algorithm

The reprojection error E of Eq. (11.5) is iteratively minimized by giving initial values to all unknowns and updating them so that E decreases at each step. The unknowns are the 3D positions $\mathsf{X}_\alpha = (X_\alpha, Y_\alpha, Z_\alpha)^\top$, $\alpha = 1, ..., N$, of all the points and the focal lengths f_κ, the principal points $(u_{0\kappa}, v_{0\kappa})$, translations $\mathbf{t}_\kappa$, and rotations R_κ, $\kappa = 1, ..., M$, of all the cameras. For updating the rotation R_κ, we adopt the method of Lie algebra that we used for fundamental matrix computation in Sect. 3.6. Namely, we rotate the camera infinitesimally around each coordinate axis by $\Delta\omega_{\kappa 1}$, $\Delta\omega_{\kappa 2}$, and $\Delta\omega_{\kappa 3}$, where we write $\Delta\boldsymbol{\omega}_\kappa$ as a vector, and evaluate the resulting infinitesimal variation in E. As in Sect. 3.6, we write the corresponding rate of change in E as $\partial E/\partial\omega_{\kappa 1}$, $\partial E/\partial\omega_{\kappa 2}$, and $\partial E/\partial\omega_{\kappa 3}$.

We need to evaluate $3N + 9M$ update terms $\Delta\mathsf{X}_\alpha$, Δf_κ, $(\Delta u_{0\kappa}, \Delta v_{0\kappa})$, $\Delta\mathbf{t}_\kappa$, and $\Delta\boldsymbol{\omega}_\kappa$, $\alpha = 1, ..., N$, $\kappa = 1, ..., M$. However, we cannot determine all of these, because the camera positions are defined relative to the 3D shape such that their absolute positions are indeterminate. Also, we cannot determine the absolute scale. This is because, as we pointed out for two-view reconstruction in Chap. 5, images do not have depth information, thus we cannot tell if the observed M images are taken by moving a camera by a large amount relative to a distance scene or by moving the camera slightly relative to a nearby scene. In order to fix this ambiguity, we adopt the following normalization.

$$\mathsf{R}_1 = \mathsf{I}, \qquad \mathbf{t}_1 = \mathbf{0}, \qquad t_{22} = 1. \tag{11.6}$$

This means that we adopt the world coordinate system fixed to the first camera as we did in Chap. 5. As for scale, we take the relative displacement of the second camera in the Y-direction as the unit of length. This is because the choice $\|\mathbf{t}_2\| = 1$ that we used in Chap. 5 would be difficult to handle due to nonlinearity. We are assuming that the second camera has a displacement component in the Y-direction, but if we know that it is displaced in the X- or Y-direction, we can let $t_{21} = 1$ or $t_{23} = 1$ instead. The normalization of Eq. (11.6) reduces the number of unknowns to $3N + 9M - 7$.

For convenience, we arrange all the $3M + 9N - 7$ variations $\Delta\mathsf{X}_\alpha$, Δf_κ, $(\Delta u_{0\kappa}, \Delta v_{0\kappa})$, $\Delta\mathbf{t}_\kappa$, and $\Delta\boldsymbol{\omega}_\kappa$ sequentially as $\Delta\xi_1, \Delta\xi_2, ..., \Delta\xi_{3N+9M-7}$. Then, we use the Levenberg-Marquardt Method (LM) in the same way as the hidden variable LM in Sect. 3.6. The procedure is described as follows.

Procedure 11.1 (Bundle adjustment)

1. Assume initial values for X_α, f_κ, $(u_{0\kappa}, v_{0\kappa})$, $\mathbf{t}_\kappa$, and R_κ, and compute the reprojection error E. Let $c = 0.0001$.
2. Evaluate the first and second derivatives $\partial E/\partial\xi_k$ and $\partial^2 E/\partial\xi_k\partial\xi_l$, $k, l = 1, ...,$ $3N + 9M - 7$.

3. Solve the following linear equation to determine $\Delta\xi_k$, $k = 1, ..., 3N + 9M - 7$.

$$\begin{pmatrix} (1+c)\partial^2 E/\partial\xi_1^2 & \partial^2 E/\partial\xi_1\partial\xi_2 & \partial^2 E/\partial\xi_1\partial\xi_3 & \cdots \\ \partial^2 E/\partial\xi_2\partial\xi_1 & (1+c)\partial^2 E/\partial\xi_2^2 & \partial^2 E/\partial\xi_2\partial\xi_3 & \cdots \\ \partial^2 E/\partial\xi_3\partial\xi_1 & \partial^2 E/\partial\xi_3\partial\xi_2 & (1+c)\partial^2 E/\partial\xi_3^2 & \cdots \\ \vdots & \vdots & \vdots & \ddots \end{pmatrix} \begin{pmatrix} \Delta\xi_1 \\ \Delta\xi_2 \\ \Delta\xi_3 \\ \vdots \end{pmatrix} = -\begin{pmatrix} \partial E/\partial\xi_1 \\ \partial E/\partial\xi_2 \\ \partial E/\partial\xi_3 \\ \vdots \end{pmatrix}. \tag{11.7}$$

4. Update X_α, f_κ, $(u_{0\kappa}, v_{0\kappa})$, $\mathbf{t}_\kappa$, and R_κ to

$$\tilde{\mathsf{X}}_\alpha \leftarrow \mathsf{X}_\alpha + \Delta\mathsf{X}_\alpha, \quad \tilde{f}_\kappa \leftarrow f_\kappa + \Delta f_\kappa, \quad (\tilde{u}_{0\kappa}, \tilde{v}_{0\kappa}) \leftarrow (u_{0\kappa}, v_{0\kappa}) \tag{11.8}$$

$$\tilde{\mathbf{t}}_\kappa \leftarrow \mathbf{t}_\kappa + \Delta\mathbf{t}_\kappa, \qquad \tilde{\mathsf{R}}_\kappa \leftarrow \mathsf{R}(\Delta\boldsymbol{\omega}_\kappa)\mathsf{R}_\kappa, \tag{11.9}$$

where $\mathsf{R}(\Delta\boldsymbol{\omega}_\kappa)$ denotes the matrix of rotation around $\Delta\boldsymbol{\omega}_\kappa$ by $\|\Delta\boldsymbol{\omega}_\kappa\|$ ($\hookrightarrow$ Eq. (4.48)).

5. Compute the reprojection error $\tilde{E}$ for $\tilde{\mathsf{X}}_\alpha$, $\tilde{f}_\kappa$, $(\tilde{u}_{0\kappa}, \tilde{v}_{0\kappa})$, $\tilde{\mathbf{t}}_\kappa$, and $\tilde{\mathsf{R}}_\kappa$. If $\tilde{E} > E$, let $c \leftarrow 10c$, and go back to Step 3.
6. Update all the unknowns to

$$\mathsf{X}_\alpha \leftarrow \tilde{\mathsf{X}}_\alpha, \quad f_\kappa \leftarrow \tilde{f}, \quad (u_{0\kappa}, v_{0\kappa}) \leftarrow (\tilde{u}_{0\kappa}, \tilde{v}_{0\kappa}), \quad \mathbf{t}_\kappa \leftarrow \tilde{\mathbf{t}}_\kappa, \quad \mathsf{R}_\kappa \leftarrow \tilde{\mathsf{R}}_\kappa. \tag{11.10}$$

Stop if $|\tilde{E} - E| \leq \delta$ for a pre-fixed small constant δ. Else, let $E \leftarrow \tilde{E}$ and $c \leftarrow c/10$, and go back to Step 2.

Comments The initial values in Step 1 are assumed to be normalized in the form of Eq. (11.6). Otherwise, we rotate the world coordinate system so that R_1 becomes I, translate it such that $\mathbf{t}_1$ comes to the origin, and scale X_α and $\mathbf{t}_\kappa$ such that $t_{22} = 1$. Specifically, the original X_α, $\mathbf{t}_\kappa$, and R_κ are modified to

$$\mathsf{X}'_\alpha = \frac{1}{s}\mathsf{R}_1^\top\left(\mathsf{X}_\alpha - \mathbf{t}_1\right), \quad \mathsf{R}'_\kappa = \mathsf{R}_1^\top\mathsf{R}_\kappa, \quad \mathbf{t}'_\kappa = \frac{1}{s}\mathsf{R}_1^\top(\mathbf{t}_\kappa - \mathbf{t}_1), \tag{11.11}$$

where $s = (\mathbf{j}, \mathsf{R}_1^\top(\mathbf{t}_2 - \mathbf{t}_1))$ and $\mathbf{j} = (0, 1, 0)^\top$. The derivative computation in Step 2 is described in the next section.

In usual numerical iterations, the variables are successively updated until they no longer change. However, the number of unknowns for bundle adjustment is thousands or even tens of thousands, and an impractically long computation time would be necessary if all variables were required to converge over significant digits. However, the purpose of bundle adjustment is to find a solution that minimizes the reprojection error. Hence, it is a practical compromise to stop if the reprojection error almost ceases to decrease, as we described in the above procedure. If we are to stop if the update of the reprojection error is less than ε pixels per point, the constant δ above is $n\varepsilon^2/f_0^2$, where $n = \sum_{\alpha=1}^{N}\sum_{\kappa=1}^{M} I_{\alpha\kappa}$ (=the number of visible points). In real applications, it is reasonable to stop at around $\varepsilon = 0.01$ pixels.

11.3 Derivative Computation

For the Levenberg-Marquardt method of Procedure 11.1, we need first and second derivatives of the reprojection error E of Eq. (11.5). We now show the principle and actual process of their evaluation.

11.3.1 Gauss-Newton Approximation

We write

$$\begin{aligned} p_{\alpha\kappa} &= P_{\kappa(11)}X_\alpha + P_{\kappa(12)}Y_\alpha + P_{\kappa(13)}Z_\alpha + P_{\kappa(14)}, \\ q_{\alpha\kappa} &= P_{\kappa(21)}X_\alpha + P_{\kappa(22)}Y_\alpha + P_{\kappa(23)}Z_\alpha + P_{\kappa(24)}, \\ r_{\alpha\kappa} &= P_{\kappa(31)}X_\alpha + P_{\kappa(32)}Y_\alpha + P_{\kappa(33)}Z_\alpha + P_{\kappa(34)}, \end{aligned} \tag{11.12}$$

and express Eq. (11.5) in the following form.

$$E = \sum_{\alpha=1}^{N}\sum_{\kappa=1}^{M} I_{\alpha\kappa}\left(\left(\frac{p_{\alpha\kappa}}{r_{\alpha\kappa}} - \frac{x_{\alpha\kappa}}{f_0}\right)^2 + \left(\frac{q_{\alpha\kappa}}{r_{\alpha\kappa}} - \frac{y_{\alpha\kappa}}{f_0}\right)^2\right). \tag{11.13}$$

The first derivative of this is

$$\begin{aligned} \frac{\partial E}{\partial \xi_k} = 2\sum_{\alpha=1}^{N}\sum_{\kappa=1}^{M} \frac{I_{\alpha\kappa}}{r_{\alpha\kappa}^2}\Bigg(\left(\frac{p_{\alpha\kappa}}{r_{\alpha\kappa}} - \frac{x_{\alpha\kappa}}{f_0}\right)\left(r_{\alpha\kappa}\frac{\partial p_{\alpha\kappa}}{\partial \xi_k} - p_{\alpha\kappa}\frac{\partial r_{\alpha\kappa}}{\partial \xi_k}\right) \\ + \left(\frac{q_{\alpha\kappa}}{r_{\alpha\kappa}} - \frac{y_{\alpha\kappa}}{f_0}\right)\left(r_{\alpha\kappa}\frac{\partial q_{\alpha\kappa}}{\partial \xi_k} - q_{\alpha\kappa}\frac{\partial r_{\alpha\kappa}}{\partial \xi_k}\right)\Bigg). \end{aligned} \tag{11.14}$$

Next, we consider the second derivatives. We note that after a sufficient number of iterations Eq. (11.13) becomes very small such that $p_{\alpha\kappa}/r_{\alpha\kappa} - x_{\alpha\kappa}/f_0 \approx 0$ and $q_{\alpha\kappa}/r_{\alpha\kappa} - y_{\alpha\kappa}/f_0 \approx 0$. Therefore we ignore terms that contain $p_{\alpha\kappa}/r_{\alpha\kappa} - x_{\alpha\kappa}/f_0$ and $q_{\alpha\kappa}/r_{\alpha\kappa} - y_{\alpha\kappa}/f_0$. This is known as *Gauss-Newton approximation*. In minimization iterations, the second derivatives affect the speed of convergence, but iterations are repeated until $\partial E/\partial \xi_\kappa$ is close to 0. Hence, the accuracy of the solution is not affected by approximation of second derivatives. The second derivatives of E with Gauss-Newton approximation have the expression:

$$\begin{aligned} \frac{\partial^2 E}{\partial \xi_k \partial \xi_l} = 2\sum_{\alpha=1}^{N}\sum_{\kappa=1}^{M} \frac{I_{\alpha\kappa}}{r_{\alpha\kappa}^4}\Bigg(\left(r_{\alpha\kappa}\frac{\partial p_{\alpha\kappa}}{\partial \xi_k} - p_{\alpha\kappa}\frac{\partial r_{\alpha\kappa}}{\partial \xi_k}\right)\left(r_{\alpha\kappa}\frac{\partial p_{\alpha\kappa}}{\partial \xi_l} - p_{\alpha\kappa}\frac{\partial r_{\alpha\kappa}}{\partial \xi_l}\right) \\ + \left(r_{\alpha\kappa}\frac{\partial q_{\alpha\kappa}}{\partial \xi_k} - q_{\alpha\kappa}\frac{\partial r_{\alpha\kappa}}{\partial \xi_k}\right)\left(r_{\alpha\kappa}\frac{\partial q_{\alpha\kappa}}{\partial \xi_l} - q_{\alpha\kappa}\frac{\partial r_{\alpha\kappa}}{\partial \xi_l}\right)\Bigg). \end{aligned} \tag{11.15}$$

From Eqs. (11.14) and (11.15), we see that for evaluating the first derivatives $\partial E/\partial \xi_k$ and the second derivatives $\partial^2 E/\partial \xi_k \partial \xi_l$ of E, we need to evaluate only the *first derivatives* $\partial p_{\alpha\kappa}/\partial \xi_k$, $\partial q_{\alpha\kappa}/\partial \xi_k$, and $\partial r_{\alpha\kappa}/\partial \xi_k$.

11.3.2 Derivatives with Respect to 3D Positions

We write the derivation with respect to X_β, Y_β, and Z_β as the vector operator $\nabla_{\mathbf{X}_\beta} = (\partial/\partial X_\beta, \partial/\partial Y_\beta, \partial/\partial Z_\beta)^\top$. The derivatives of $p_{\alpha\kappa}$, $q_{\alpha\kappa}$, and $r_{\alpha\kappa}$ with respect to X_β, Y_β, and Z_β are written in the following form ($\hookrightarrow$ Problem 11.2(1)).

$$\nabla_{\mathbf{X}_\beta} p_{\alpha\kappa} = \delta_{\alpha\beta}\begin{pmatrix} P_{\kappa(11)} \\ P_{\kappa(12)} \\ P_{\kappa(13)} \end{pmatrix}, \quad \nabla_{\mathbf{X}_\beta} q_{\alpha\kappa} = \delta_{\alpha\beta}\begin{pmatrix} P_{\kappa(21)} \\ P_{\kappa(22)} \\ P_{\kappa(23)} \end{pmatrix}, \quad \nabla_{\mathbf{X}_\beta} r_{\alpha\kappa} = \delta_{\alpha\beta}\begin{pmatrix} P_{\kappa(31)} \\ P_{\kappa(32)} \\ P_{\kappa(33)} \end{pmatrix}. \tag{11.16}$$

Here, $\delta_{\alpha\beta}$ is the Kronecker delta, taking 1 for $\alpha = \beta$ and 0 otherwise.

11.3.3 Derivatives with Respect to Focal Lengths

The derivatives of $p_{\alpha\kappa}$, $q_{\alpha\kappa}$, and $r_{\alpha\kappa}$ with respect to f_λ are given as follows ($\hookrightarrow$ Problem 11.2(2)).

$$\frac{\partial p_{\alpha\kappa}}{\partial f_\lambda} = \frac{\delta_{\kappa\lambda}}{f_\kappa}\Big(p_{\alpha\kappa} - \frac{u_0}{f_0} r_{\alpha\kappa}\Big), \quad \frac{\partial q_{\alpha\kappa}}{\partial f_\lambda} = \frac{\delta_{\kappa\lambda}}{f_\kappa}\Big(q_{\alpha\kappa} - \frac{v_0}{f_0} r_{\alpha\kappa}\Big), \quad \frac{\partial r_{\alpha\kappa}}{\partial f_\lambda} = 0. \tag{11.17}$$

11.3.4 Derivatives with Respect to Principal Points

We write the derivation with respect to $u_{0\lambda}$ and $v_{0\lambda}$ as the vector operator $\nabla_{\mathbf{u}_{0\lambda}} = (\partial/\partial u_{0\lambda}, \partial/\partial u_{0\lambda})^\top$. The derivatives of $p_{\alpha\kappa}$, $q_{\alpha\kappa}$, and $r_{\alpha\kappa}$ with respect to $u_{0\lambda}$ and $v_{0\lambda}$ are written in the following form ($\hookrightarrow$ Problem 11.2(3)).

$$\nabla_{\mathbf{u}_{0\lambda}} p_{\alpha\kappa} = \begin{pmatrix} \delta_{\kappa\lambda} r_{\alpha\kappa}/f_0 \\ 0 \end{pmatrix}, \quad \nabla_{\mathbf{u}_{0\lambda}} q_{\alpha\kappa} = \begin{pmatrix} 0 \\ \delta_{\kappa\lambda} r_{\alpha\kappa}/f_0 \end{pmatrix}, \quad \nabla_{\mathbf{u}_{0\lambda}} r_{\alpha\kappa} = \begin{pmatrix} 0 \\ 0 \end{pmatrix}. \tag{11.18}$$

11.3.5 Derivatives with Respect to Translations

We write the derivation with respect to $t_{\lambda 1}$, $t_{\lambda 2}$, and $t_{\lambda 3}$ as the vector operator $\nabla_{\mathbf{t}_\lambda} = (\partial/\partial t_{\lambda 1}, \partial/\partial t_{\lambda 2}, \partial/\partial t_{\lambda 3})^\top$. The derivatives of $p_{\alpha\kappa}$, $q_{\alpha\kappa}$, and $r_{\alpha\kappa}$ with respect to $t_{\lambda 1}$, $t_{\lambda 2}$, and $t_{\lambda 3}$ are written in the following form ($\hookrightarrow$ Problem 11.2(4)).

$$\nabla_{\mathbf{t}_\lambda} p_{\alpha\kappa} = -\delta_{\kappa\lambda}(f_\kappa \mathbf{r}_{\kappa 1} + u_0 \mathbf{r}_{\kappa 3}), \quad \nabla_{\mathbf{t}_\lambda} q_{\alpha\kappa} = -\delta_{\kappa\lambda}(f_\kappa \mathbf{r}_{\kappa 2} + v_0 \mathbf{r}_{\kappa 3}),$$

$$\nabla_{\mathbf{t}_\lambda} r_{\alpha\kappa} = -\delta_{\kappa\lambda} f_0 \mathbf{r}_{\kappa 3}. \tag{11.19}$$

Here, $\mathbf{r}_{\kappa 1}$, $\mathbf{r}_{\kappa 2}$, and $\mathbf{r}_{\kappa 3}$ are, respectively, the first, the second, and the third columns of the rotation R_κ; that is,

$$\mathbf{r}_{\kappa 1} = \begin{pmatrix} R_{\kappa(11)} \\ R_{\kappa(21)} \\ R_{\kappa(31)} \end{pmatrix}, \quad \mathbf{r}_{\kappa 2} = \begin{pmatrix} R_{\kappa(12)} \\ R_{\kappa(22)} \\ R_{\kappa(32)} \end{pmatrix}, \quad \mathbf{r}_{\kappa 3} = \begin{pmatrix} R_{\kappa(13)} \\ R_{\kappa(23)} \\ R_{\kappa(33)} \end{pmatrix}, \tag{11.20}$$

where $R_{\kappa(ij)}$ is the (i, j) element of R_κ.

11.3.6 Derivatives with Respect to Rotations

We write the derivations with respect to the rotation R_λ as the vector operator $\nabla_{\omega_\lambda} = (\partial/\partial\omega_{\lambda 1}, \partial/\partial\omega_{\lambda 2}, \partial/\partial\omega_{\lambda 3})^\top$. The derivatives of $p_{\alpha\kappa}$, $q_{\alpha\kappa}$, and $r_{\alpha\kappa}$ with respect to R_λ are given in the following form (↪ Problem 11.2(5)).

$$\begin{aligned}
\nabla_{\omega_\lambda} p_{\alpha\kappa} &= \delta_{\kappa\lambda}(f_\kappa \mathbf{r}_{\kappa 1} + u_{0\kappa}\mathbf{r}_{\kappa 3}) \times (\mathbf{X}_\alpha - \mathbf{t}_\kappa), \\
\nabla_{\omega_\lambda} q_{\alpha\kappa} &= \delta_{\kappa\lambda}(f_\kappa \mathbf{r}_{\kappa 2} + v_{0\kappa}\mathbf{r}_{\kappa 3}) \times (\mathbf{X}_\alpha - \mathbf{t}_\kappa), \\
\nabla_{\omega_\lambda} r_{\alpha\kappa} &= \delta_{\kappa\lambda} f_0 \mathbf{r}_{\kappa 3} \times (\mathbf{X}_\alpha - \mathbf{t}_\kappa).
\end{aligned} \tag{11.21}$$

11.3.7 Efficient Computation and Memory Use

A significant portion of the summand terms of $\sum_{\alpha=1}^N \sum_{\kappa=1}^M$ in Eqs. (11.14) and (11.15) is 0, therefore if they are directly computed, a large amount of time would be spent on meaningless zero evaluation. To avoid this, we pick out nonzero terms. Consider $\partial E/\partial\xi_k$ of Eq. (11.14). If $\Delta\xi_k$ is the update $\Delta\mathbf{X}_\beta$ of the βth point, we need to evaluate in the summation $\sum_{\alpha=1}^N$ only terms for which $\alpha = \beta$ due to the Kronecker delta $\delta_{\alpha\beta}$ in Eq. (11.16). If $\Delta\xi_k$ is the update of f_λ, $(u_{0\lambda}, v_{0\lambda})$, $\mathbf{t}_\lambda$, or R_λ of the λth image, we need to evaluate in the summation $\sum_{\kappa=1}^M$ only terms for which $\kappa = \lambda$ due to the Kronecker delta $\delta_{\kappa\lambda}$ in Eqs. (11.17), (11.18), (11.19), and (11.21). Thus the summation $\sum_{\alpha=1}^N \sum_{\kappa=1}^M$ in Eq. (11.14) need to be summed over either α or κ. Moreover, we need to consider for the αth point only those κth images in which it is seen and for the κth image only those αth points that are seen there.

The same applies to $\partial^2 E/\partial\xi_k\partial\xi_l$. Namely, if $\Delta\xi_k$ and $\Delta\xi_l$ both refer to points, Eq. (11.15) is 0 if they are different points, and if they are the same point, we need to evaluate in $\sum_{\alpha=1}^N$ only the term corresponding to that point. If $\Delta\xi_k$ and $\Delta\xi_l$ both refer to images, Eq. (11.15) is 0 if they are different images, and if they are the same image, we need to evaluate in $\sum_{\kappa=1}^M$ only the term corresponding to that image. If one of $\Delta\xi_k$ and $\Delta\xi_l$ refers to a point and the other to an image, we need to evaluate in $\sum_{\alpha=1}^N \sum_{\kappa=1}^M$ only the term corresponding to that point and that image, provided that point is seen in that image.

In this way, the computation time for $\partial E/\partial\xi_k$ and $\partial^2 E/\partial\xi_k\partial\xi_l$ can be held to the minimum, but the Hessian $H(k,l)$, that is, the matrix that contains the second derivatives $\partial^2 E/\partial\xi_k\partial\xi_l$, is of size $(3N + 9M - 7) \times (3N + 9M - 7)$, which can be too large to store in the computer memory when N and M are large. To reduce the memory size, we define a $3N \times 3$ array E that stores second derivatives with respect to points, such as $\partial^2 E/\partial X_\alpha^2$ and $\partial^2 E/\partial X_\alpha \partial Y_\alpha$, define a $3N \times 9M$ array F that stores the second derivatives with respect to points and images, such as $\partial^2 E/\partial X_\alpha \partial f_\kappa$, $\partial^2 E/\partial X_\alpha \partial u_{0\kappa}$, and define a $9M \times 9$ array G that stores the second derivatives with respect to images, such as $\partial^2 E/\partial f_\kappa^2$, and $\partial^2 E/\partial f_\kappa \partial u_{0\kappa}$. This reduces the required memory size to $27NM + 9N + 81M$. If we remove the redundancies in the second derivative elements, the net array elements are $27NM + 6N + 41M$ in number, among which those of $I_{\alpha\kappa} = 0$ are not necessary. Considering all these, it is better to define the Hessian $H(k,l)$ not as an array but as a function program that returns its (k,l) element.

11.4 Efficient Linear Equation Solving

Equation (11.7) is a set of simultaneous linear equations in $3N+9M-7$ unknowns, and the matrix on the left-hand side has size $(3N+9M-7)\times(3N+9M-7)$. For large N and M, the computational burden is heavy with a large amount of complexity (the number of operations), and the required space for storing data and intermediate values may not be maintained within the memory. There exist techniques for resolving these difficulties and making the computation efficient. Equation (11.7) consists of the part for the 3D coordinates of the points and the part for the camera parameters in the form

$$\begin{pmatrix} \mathsf{E}_1^{(c)} & & & \mathsf{F}_1 \\ & \ddots & & \vdots \\ & & \mathsf{E}_N^{(c)} & \mathsf{F}_N \\ \mathsf{F}_1^\top & \cdots & \mathsf{F}_N^\top & \mathsf{G}^{(c)} \end{pmatrix} \begin{pmatrix} \Delta\boldsymbol{\xi}_P \\ \Delta\boldsymbol{\xi}_F \end{pmatrix} = -\begin{pmatrix} \mathbf{d}_P \\ \mathbf{d}_F \end{pmatrix}, \tag{11.22}$$

where $\Delta\boldsymbol{\xi}_P$ is the $3N$D vector for the 3D positions and $\Delta\boldsymbol{\xi}_F$ is the $(9M-7)$D vector for the camera parameters. Vectors $\mathbf{d}_P$ and $\mathbf{d}_F$ are the corresponding $3N$D and $(9M-7)$D parts of the right side of Eq. (11.7). Each of $\mathsf{E}_\alpha^{(c)}$, $\alpha = 1, ..., N$, is the 3×3 block containing the second derivatives of the reprojection error E with respect to the coordinates $(X_\alpha, Y_\alpha, Z_\alpha)$ of the αth point, and the superscript (c) means that the diagonal elements are multiplied by $(1+c)$. Each F_α is the $3\times(9M-7)$ block containing the second derivatives of E with respect to the coordinates $(X_\alpha, Y_\alpha, Z_\alpha)$ of the αth point and the parameters of the cameras that view it, and $\mathsf{G}^{(c)}$ is the $(9M-7)\times(9M-7)$ block containing the second derivatives of E with respect to the camera parameters with diagonal elements multiplied by $(1+c)$. Equation (11.22) can be solved efficiently by the following procedure.

Procedure 11.2 (Efficient linear equation solving)

1. For each α, let $\nabla_{\mathsf{X}_\alpha} E$ be

$$\nabla_{\mathsf{X}_\alpha} E = \begin{pmatrix} \partial E/\partial X_\alpha \\ \partial E/\partial Y_\alpha \\ \partial E/\partial Z_\alpha \end{pmatrix}. \tag{11.23}$$

2. Solve the following $(9M-7)$D linear equations for $\Delta\boldsymbol{\xi}_F$.

$$\Big(\mathsf{G}^{(c)} - \sum_{\alpha=1}^{N} \mathsf{F}_\alpha^\top \mathsf{E}_\alpha^{(c)-1}\mathsf{F}_\alpha\Big)\Delta\boldsymbol{\xi}_F = \sum_{\alpha=1}^{N} \mathsf{F}_\alpha^\top \mathsf{E}_{\mathsf{X}_\alpha}^{(c)-1}\nabla_{\mathsf{X}_\alpha} E - \mathbf{d}_F. \tag{11.24}$$

3. Compute ΔX_α, ΔY_α, and ΔZ_α for the αth point by

$$\begin{pmatrix} \Delta X_\alpha \\ \Delta Y_\alpha \\ \Delta Z_\alpha \end{pmatrix} = -\mathsf{E}_\alpha^{(c)-1}(\mathsf{F}_\alpha\Delta\boldsymbol{\xi}_F + \nabla_{\mathsf{X}_\alpha} E). \tag{11.25}$$

Comments The standard method for solving simultaneous linear equations is the use of the *Cholesky decomposition* for symmetric coefficient matrices and *LU decomposition* for general coefficient matrices. Whatever method we use, however, the complexity is nearly proportional to the cubes of the number of unknowns. Thus the complexity of directly solving Eq. (11.7) is around $(3N+9M-7)^3$. In the above procedure, in contrast, we only need to solve Eq. (11.24), for which the complexity is around $(9M-7)^3$. This is a significant efficiency improvement when N (= the number of points) is very large and M (= the number of images) is not very large. The memory space required for storing intermediate values is only N arrays $\mathsf{E}_\alpha^{(c)}$ of size 3×3, N arrays F_α of size $3\times(9M-7)$, one array $\mathsf{G}^{(c)}$ of size $(9M-7)\times(9M-7)$, one $3N$D vector $\mathbf{d}_P$, and one $(9M-7)$D vector $\mathbf{d}_F$.

Procedure 11.2 is obtained by the following consideration (↪ Problem 11.3). Equation (11.22) is split into two parts in the form

$$\begin{pmatrix} \mathsf{E}_1^{(c)} & & \\ & \ddots & \\ & & \mathsf{E}_N^{(c)} \end{pmatrix} \Delta\boldsymbol{\xi}_P + \begin{pmatrix} \mathsf{F}_1 \\ \vdots \\ \mathsf{F}_N \end{pmatrix} \Delta\boldsymbol{\xi}_F = -\mathbf{d}_P \tag{11.26}$$

$$\begin{pmatrix} \mathsf{F}_1^\top & \cdots & \mathsf{F}_N^\top \end{pmatrix} \Delta\boldsymbol{\xi}_P + \mathsf{G}^{(c)}\Delta\boldsymbol{\xi}_F = -\mathbf{d}_F. \tag{11.27}$$

We first solve Eq. (11.26) for $\Delta\boldsymbol{\xi}_P$. If we note that $\mathbf{d}_P$ is a vertical array of $\nabla_{\mathsf{X}_\alpha}E$, $\alpha = 1, ..., N$, of Eq. (11.23), we can write the solution in the form

$$\begin{aligned} \Delta\boldsymbol{\xi}_P &= -\begin{pmatrix} \mathsf{E}_1^{(c)-1} & & \\ & \ddots & \\ & & \mathsf{E}_N^{(c)-1} \end{pmatrix} \begin{pmatrix} \mathsf{F}_1 \\ \vdots \\ \mathsf{F}_N \end{pmatrix} \Delta\boldsymbol{\xi}_F - \begin{pmatrix} \mathsf{E}^{(c)-1} & & \\ & \ddots & \\ & & \mathsf{E}^{(c)-1} \end{pmatrix} \mathbf{d}_P \\ &= -\begin{pmatrix} \mathsf{E}_1^{(c)-1}\mathsf{F}_1 \\ \vdots \\ \mathsf{E}_N^{(c)-1}\mathsf{F}_N \end{pmatrix} \Delta\boldsymbol{\xi}_F - \begin{pmatrix} \mathsf{E}_1^{(c)-1}\nabla_{\mathsf{X}_1}E \\ \vdots \\ \mathsf{E}_N^{(c)-1}\nabla_{\mathsf{X}_N}E \end{pmatrix}. \end{aligned} \tag{11.28}$$

Substituting this into Eq. (11.27), we obtain

$$\begin{aligned} &-\begin{pmatrix} \mathsf{F}_1^\top & \cdots & \mathsf{F}_N^\top \end{pmatrix} \begin{pmatrix} \mathsf{E}_1^{(c)-1}\mathsf{F}_1 \\ \vdots \\ \mathsf{E}_N^{(c)-1}\mathsf{F}_N \end{pmatrix} \Delta\boldsymbol{\xi}_F - \begin{pmatrix} \mathsf{F}_1^\top & \cdots & \mathsf{F}_N^\top \end{pmatrix} \begin{pmatrix} \mathsf{E}_1^{(c)-1}\nabla_{\mathsf{X}_1}E \\ \vdots \\ \mathsf{E}_N^{(c)-1}\nabla_{\mathsf{X}_N}E \end{pmatrix} + \mathsf{G}^{(c)}\Delta\boldsymbol{\xi}_F \\ &= -\mathbf{d}_F, \end{aligned} \tag{11.29}$$

from which Eq. (11.24) is obtained. The part of Eq. (11.28) corresponding to X_α is written in the form of Eq. (11.25).

11.5 Examples

Bundle adjustment was applied to the image frames of the Oxford University data used in Fig. 10.3. For all the frames, the camera matrices P_κ are provided in the data, but they may not necessarily be correct. We decomposed each P_κ in the form of Eq. (11.2) and used the estimated f_κ, $(u_{0\kappa}, v_{0\kappa})$, $\mathbf{t}_\kappa$, and R_κ as initial camera parameters (the decomposition of P_κ in the form of Eq. (11.2) is discussed in Chap. 13 ↪ Problem 13.11). For the 3D positions X_α, we used the triangulation result in the preceding chapter as their initial values.

The number of unknowns is 15,266, thus the number of elements of the matrix in Eq. (11.7) is around 200,000,000, which cannot be stored in the computer memory. If we use the technique of Sect. 11.3.7, however, the number of matrix elements is around 400,000 (1/600). Moreover, each point is viewed in only a small number of frames, so most matrix elements are 0. Figure 11.2a shows nonzero elements of the 1008 × 1008 Hessian obtained by randomly selecting 100 points from the original 4983 points; it is displayed as a binary image whose black and white pixels indicate nonzero and zero elements, respectively. We can see that nonzero elements are only 13 %.

The number $n = \sum_{\alpha=1}^{N} \sum_{\kappa=1}^{M} I_{\alpha\kappa}$ of points seen in the images is 16,432. We normalize the reprojection error E to its value per pixel measured in length (the unit is pixel), considering the degree of freedom of the unknowns, in the form

$$e = f_0 \sqrt{\frac{E}{2n - (3N + 9M - 7)}} \tag{11.30}$$

(see Sect. 14.4 for the reasoning). The reprojection error of the initial reconstruction (the triangulation result of the preceding chapter) was $e = 3.27797$ (pixels), which was decreased to $e = 1.625876$ (pixels) after bundle adjustment. The decrease of e for the number of iterations is plotted in Fig. 11.2b. The actual values of e are listed

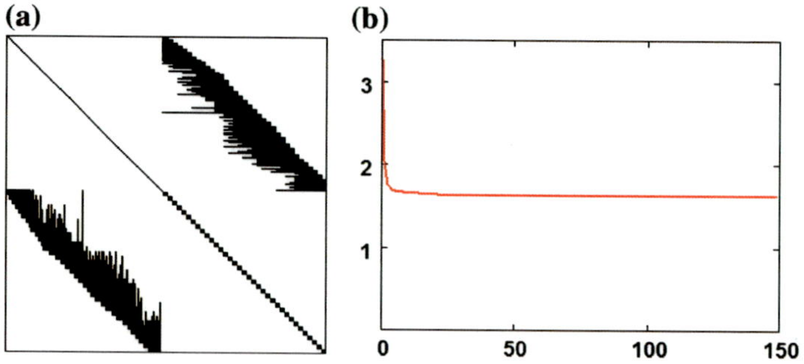

Fig. 11.2 **a** Nonzero elements in the Hessian for 100 points. **b** The decrease of the reprojection error e for the number of iterations

Table 11.1 The values of the reprojection error e for the number of iterations

	Reprojection error		Reprojection error
0	3.277965703463469	⋮	⋮
1	2.037807322757024	140	1.626138870635717
2	1.767180606187605	141	1.626109073343624
3	1.721032319350261	142	1.626079434501709
4	1.698429496315309	143	1.626049951753774
5	1.684614811452468	144	1.626020622805242
6	1.675366012050569	145	1.625991445421568
7	1.668829491793228	146	1.625962417425169
8	1.664028486785132	147	1.625933536694230
9	1.660393246948761	148	1.625904801160639
10	1.657569357560945	149	1.625876208807785

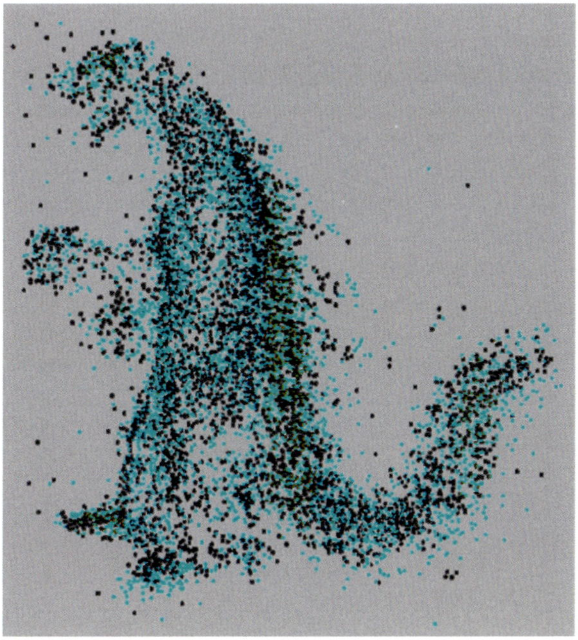

Fig. 11.3 3D reconstruction: initial positions (light color) and final positions (dark color)

in Table 11.1. Figure 11.3 shows the reconstructed shape. The light color points are the initial positions obtained by triangulation, and the dark color points are the final positions. We can see that a large number of initially dispersed points have coalesced into the body part. On the other hand, some points (possibly outliers) have moved away, perhaps because the camera matrices are estimated better.

11.6 Supplemental Note

The term bundle means the set of rays corresponding to image points, and bundle adjustment means adjusting camera positions so that the rays starting from them meet in the scene. This term originates from photogrammetry, and photogrammetrists were doing this for producing maps from multiple aerial photographs before the advent of computer vision research. The technique has made significant progress as the performance of computers has increased. Today, various software tools are provided for bundle adjustment, including the *SBA*[1] of Lourakis and Argyros [2]. Snaverly et al. [3,4] combined it with correspondence extraction and offered a tool called the *bundler*.[2] The procedure of this chapter is based on Kanatani and Sugaya [1].

The biggest problem of bundle adjustment is that the number of unknowns is very large (usually tens of thousands) making the coefficient matrix of the linear equation (see Eqs. (11.7) and (11.22)) to be solved in each iteration become very large. As we see from Fig. 11.2a, however, most of the elements are 0, therefore if they are directly computed, most of the computational time would be wasted for 0 evaluation. Matrices whose elements are mostly 0 are said to be *sparse*, and there have been many studies for avoiding meaningless computation of sparse matrices. For efficient bundle adjustment programming, see Triggs et al. [5]. A typical computational technique for this is to decompose the linear equation in the form of Eq. (11.31) in Problem 11.3 into the form of Eqs. (11.32) and (11.33). However, the degree of efficiency improvement depends on the size and the sparsity of the submatrices A, B, C, and D. For the structure of Eq. (11.22), however, this technique is very effective, as we see from the above example.

The matrix $\mathsf{A} - \mathsf{BD}^{-1}\mathsf{C}$ that appears in Eqs. (11.33) and (11.34) is known as *the Schur complement*. The complexity of solving a linear equation has the same order as inverting the coefficient matrix. As we see from Eqs. (11.32) and (11.33), inversion computation for solving Eq. (11.31) occurs only for the submatrix D and the Schur complement $\mathsf{A} - \mathsf{BD}^{-1}\mathsf{C}$.

Problems

11.1 Show that if the principal point is at $(u_{0\kappa}, v_{0\kappa})$, the camera matrix has the form of Eq. (11.2).

11.2 (1) Show that the derivatives of $p_{\alpha\kappa}$, $q_{\alpha\kappa}$, and $r_{\alpha\kappa}$ with respect to X_β, Y_β, and Z_β are written in the form of Eq. (11.16).
(2) Show that the derivatives of $p_{\alpha\kappa}$, $q_{\alpha\kappa}$, and $r_{\alpha\kappa}$ with respect to f_λ are written in the form of Eq. (11.17).
(3) Show that the derivatives of $p_{\alpha\kappa}$, $q_{\alpha\kappa}$, and $r_{\alpha\kappa}$ with respect to $u_{0\lambda}$ and $v_{0\lambda}$ are written in the form of Eq. (11.18).

---

[1] http://users.ics.forth.gr/~lourakis/sba/.
[2] http://www.cs.cornell.edu/~snavely/bundler/.

(4) Show that the derivatives of $p_{\alpha\kappa}$, $q_{\alpha\kappa}$, and $r_{\alpha\kappa}$ with respect to $t_{\lambda 1}$, $t_{\lambda 2}$, and $t_{\lambda 3}$ are written in the form of Eq. (11.19).
(5) Show that the derivatives of $p_{\alpha\kappa}$, $q_{\alpha\kappa}$, and $r_{\alpha\kappa}$ with respect to R_λ are written in Eq. (11.21).

11.3 (1) Show that the linear equation

$$\begin{pmatrix} \mathsf{A} & \mathsf{B} \\ \mathsf{C} & \mathsf{D} \end{pmatrix} \begin{pmatrix} \mathbf{x} \\ \mathbf{y} \end{pmatrix} = \begin{pmatrix} \mathbf{a} \\ \mathbf{b} \end{pmatrix} \tag{11.31}$$

in $\mathbf{x}$ and $\mathbf{y}$ can be solved in the following two steps.

1. Solve the linear equation
$$(\mathsf{A} - \mathsf{B}\mathsf{D}^{-1}\mathsf{C})\mathbf{x} = \mathbf{a} - \mathsf{B}\mathsf{D}^{-1}\mathbf{b} \tag{11.32}$$
in $\mathbf{x}$.
2. Using the obtained $\mathbf{x}$, compute $\mathbf{y}$ in the form
$$\mathbf{y} = -\mathsf{D}^{-1}\mathsf{C}\mathbf{x} + \mathsf{D}^{-1}\mathbf{b}. \tag{11.33}$$

(2) Show that the above procedure implies the following matrix inversion formula.

$$\begin{pmatrix} \mathsf{A} & \mathsf{B} \\ \mathsf{C} & \mathsf{D} \end{pmatrix}^{-1} = \begin{pmatrix} (\mathsf{A} - \mathsf{B}\mathsf{D}^{-1}\mathsf{C})^{-1} & -(\mathsf{A} - \mathsf{B}\mathsf{D}^{-1}\mathsf{C})^{-1}\mathsf{B}\mathsf{D}^{-1} \\ -\mathsf{D}^{-1}\mathsf{C}(\mathsf{A} - \mathsf{B}\mathsf{D}^{-1}\mathsf{C})^{-1} & \mathsf{D}^{-1} + \mathsf{D}^{-1}\mathsf{C}(\mathsf{A} - \mathsf{B}\mathsf{D}^{-1}\mathsf{C})^{-1}\mathsf{B}\mathsf{D}^{-1} \end{pmatrix}. \tag{11.34}$$

References

1. K. Kanatani, Y. Sugaya, Implementation and evaluation of bundle adjustment for 3-D reconstruction, in *Proceedings of the 17th Symposium on Sensing via Imaging Information* (2011), pp. IS4-02-1–IS4-02-8
2. M.I.A. Lourakis, A.A. Argyros, SBA: a software package for generic sparse bundle adjustment. ACM Trans. Math. Softw. **36**(1), 2:1–30 (2009)
3. N. Snavely, S. Seitz, R. Szeliski, Photo tourism: exploring photo collections in 3D. ACM Trans. Graph. **25**(8), 835–846 (1995)
4. N. Snavely, S. Seitz, R. Szeliski, Modeling the world from internet photo collections. Int. J. Comput. Vis. **80**(22), 189–210 (2008)
5. B. Triggs, P.F. McLauchlan, R.I. Hartley, A. Fitzgibbon, Bundle adjustment—a modern synthesis, in *Vision Algorithms: Theory and Practice*, ed. by B. Triggs, A. Zisserman, R. Szeliski (Springer, Berlin, 2000), pp. 298–375

Self-calibration of Affine Cameras

12

Abstract

For doing the bundle adjustment of the preceding chapter, we need an initial reconstruction from which to start. To this end, this chapter presents a simple procedure for computing an approximate reconstruction. This is done by approximating the perspective camera projection model by a simplified affine camera model. We describe the procedure for affine reconstruction using affine cameras and discuss the metric condition for upgrading the result to a correct Euclidean reconstruction. Because the computation involves the singular value decomposition (SVD) of matrices, this procedure is widely known as the factorization method. We introduce symmetric affine cameras that best mimic perspective projection in the affine camera framework. They include, as special cases, paraperspective cameras, weak perspective cameras, and orthographic cameras. For each camera modeling, we describe the self-calibration procedure for reconstructing the 3D shape and the camera motion.

12.1 Affine Cameras

The camera imaging geometry of perspective projection is described by Eq. (11.1). Replacing it by

$$\begin{pmatrix} x_{\alpha\kappa}/f_0 \\ y_{\alpha\kappa}/f_0 \\ 1 \end{pmatrix} = \mathsf{P}_\kappa \begin{pmatrix} X_\alpha \\ Y_\alpha \\ Z_\alpha \\ 1 \end{pmatrix} \tag{12.1}$$

is called *affine camera* modeling. The difference is that the relation symbol $\simeq$ is replaced by the equality sign $=$. As a result, Eq. (12.1) no longer describes perspective projection, but it is known to be a good approximation if the scene is in the distance or the portion we want to reconstruct is imaged around the principal point within a

© Springer International Publishing AG 2016 163
K. Kanatani et al., *Guide to 3D Vision Computation*, Advances in Computer
Vision and Pattern Recognition, DOI 10.1007/978-3-319-48493-8_12

small region as compared with the focal length. For perspective projection, we obtain the nonlinear relationship of Eq. (11.5) by eliminating the unknown proportionality constant implied by the symbol $\simeq$. For affine camera modeling, in contrast, Eq. (11.5) is a *linear* equation, therefore subsequent analysis becomes very simple.

If all points are observed in all images, Eq. (12.1) is further simplified. First, we translate the image coordinate system for each camera so that the centroid of the observed positions $(x_{\alpha\kappa}, y_{\alpha\kappa})$ coincides with the image origin of that camera. Thus,

$$\sum_{\alpha=1}^{N} x_{\alpha\kappa} = 0, \qquad \sum_{\alpha=1}^{N} y_{\alpha\kappa} = 0, \qquad \kappa = 1, ..., M. \tag{12.2}$$

Next, we choose the world coordinate origin to be the centroid of the points $(X_\alpha, Y_\alpha, Z_\alpha)$. Then,

$$\sum_{\alpha=1}^{N} X_\alpha = 0, \qquad \sum_{\alpha=1}^{N} Y_\alpha = 0, \qquad \sum_{\alpha=1}^{N} Z_\alpha = 0. \tag{12.3}$$

Applying $(1/N)\sum_{\alpha=1}^{N}$ on both sides of Eq. (12.1), we obtain $(0, 0, 1)^\top = \mathsf{P}_\kappa(0, 0, 0, 1)^\top$. This means that Eq. (12.1) has the form

$$\begin{pmatrix} x_{\alpha\kappa}/f_0 \\ y_{\alpha\kappa}/f_0 \\ 1 \end{pmatrix} = \begin{pmatrix} & \boldsymbol{\Pi}_\kappa & & \begin{matrix} 0 \\ 0 \end{matrix} \\ 0 & 0 & 0 & 1 \end{pmatrix} \begin{pmatrix} X_\alpha \\ Y_\alpha \\ Z_\alpha \\ 1 \end{pmatrix}, \tag{12.4}$$

where $\boldsymbol{\Pi}_\kappa$ is some 2×3 matrix. This equation is further simplified to

$$\begin{pmatrix} x_{\alpha\kappa} \\ y_{\alpha\kappa} \end{pmatrix} = \boldsymbol{\Pi}_\kappa \begin{pmatrix} X_\alpha \\ Y_\alpha \\ Z_\alpha \end{pmatrix}. \tag{12.5}$$

Here, we omit the scaling constant f_0, inasmuch as its role is merely to make the components of the vector on the left-hand side of Eq. (12.4) have the same order of magnitude; in the form of Eq. (12.5), we can let $f_0 = 1$. We call the matrix $\boldsymbol{\Pi}_\kappa$ in Eq. (12.5) the *camera matrix* of the *affine camera* described by Eq. (12.5). By the *self-calibration* of affine cameras, we mean reconstructing all 3D positions $(X_\alpha, Y_\alpha, Z_\alpha)$, $\alpha = 1, ..., N$, and all camera matrices $\boldsymbol{\Pi}_\kappa$, $\kappa = 1, ..., M$, from observed points $(x_{\alpha\kappa}, y_{\alpha\kappa})$, $\alpha = 1, ..., N$.

12.2 Factorization and Affine Reconstruction

We arrange all the camera matrices $\boldsymbol{\Pi}_\kappa$ and all the 3D positions $(X_\alpha, Y_\alpha, Z_\alpha)$ in the matrix form:

$$\mathsf{M} = \begin{pmatrix} \boldsymbol{\Pi}_1 \\ \vdots \\ \boldsymbol{\Pi}_M \end{pmatrix}, \qquad \mathsf{S} = \begin{pmatrix} X_1 & \cdots & X_N \\ Y_1 & \cdots & Y_N \\ Z_1 & \cdots & Z_N \end{pmatrix}. \tag{12.6}$$

The $2M \times 3$ matrix M is called the *motion matrix*, and the $3 \times N$ matrix S the *shape matrix*. On the other hand, we arrange all the observed coordinates $(x_{\alpha\kappa}, y_{\alpha\kappa})$ in the matrix form:

$$\mathsf{W} = \begin{pmatrix} x_{11} & x_{21} & \cdots & x_{N1} \\ y_{11} & y_{21} & \cdots & y_{N1} \\ x_{12} & x_{22} & \cdots & x_{N2} \\ y_{12} & y_{22} & \cdots & y_{N2} \\ \vdots & \vdots & \ddots & \vdots \\ x_{1M} & x_{2M} & \cdots & x_{NM} \\ y_{1M} & y_{2M} & \cdots & y_{NM} \end{pmatrix}. \tag{12.7}$$

This $2M \times N$ matrix is called the *observation matrix*. From Eq. (12.6), the following equality holds ($\hookrightarrow$ Problem 12.1).

$$\mathsf{W} = \mathsf{MS}. \tag{12.8}$$

However, the actual images we observe are taken by perspective cameras, therefore the matrix W cannot be factored into matrices S and M in this form. Here, we introduce the following approximation.

Procedure 12.1 (Factorization)

1. Compute the SVD in the form
$$\mathsf{W} = \mathsf{U}_{2M\times L}\boldsymbol{\Sigma}_L\mathsf{V}_{N\times L}^\top, \qquad \boldsymbol{\Sigma}_L = \begin{pmatrix} \sigma_1 & \cdots & 0 \\ \vdots & \ddots & \vdots \\ 0 & \cdots & \sigma_L \end{pmatrix}, \tag{12.9}$$
where $L = \min(2M, N)$. Here, $\mathsf{U}_{2M\times L}$ and $\mathsf{V}_{N\times L}$ are, respectively, $2M \times L$ and $N \times L$ matrices consisting of orthonormal columns, and $\boldsymbol{\Sigma}_L$ is a diagonal matrix with singular values $\sigma_1 \geq \cdots \geq \sigma_L$ (≥ 0) as diagonal elements in that order.
2. Determine the motion matrix M and the shape matrix S by
$$\mathsf{M} = \mathsf{U}, \qquad \mathsf{S} = \boldsymbol{\Sigma}\mathsf{V}^\top, \qquad \boldsymbol{\Sigma} = \begin{pmatrix} \sigma_1 & 0 & 0 \\ 0 & \sigma_2 & 0 \\ 0 & 0 & \sigma_3 \end{pmatrix}, \tag{12.10}$$
where U is the $2M \times 3$ matrix consisting of the first three columns of $\mathsf{U}_{2M\times L}$, and V is the $N \times 3$ matrix consisting of the first three columns of $\mathsf{V}_{N\times L}$.

Comments The observation matrix W would be factored in the form of Eq. (12.8) if the images were taken by affine cameras of Eq. (12.5). This means that W would have at most rank 3; that is, the singular values in Eq. (12.9) are such that $\sigma_i = 0$, $i = 4, ..., L$. This does not hold for perspective cameras, but if the scene is at a sufficient distance or the object we want to reconstruct is imaged in a small region near the origin, the affine camera modeling approximately holds, therefore $\sigma_i \approx 0$, $i = 4, ..., L$. Hence, we obtain an approximate factorization $\mathsf{W} \approx \mathsf{MS}$, using

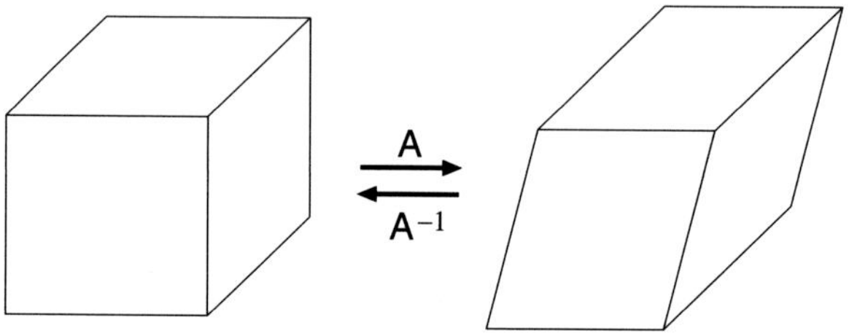

Fig. 12.1 Affine reconstruction. The reconstructed 3D shape and the true shape are related by an unknown affine transformation A. Lengths and angles may alter, but the ratio of lengths is preserved, and lines and planes are mapped to lines and planes

Eqs. (12.9) and (12.10). Because we decompose a matrix into two using SVD, this procedure is called the *factorization method*, or simply *factorization*. However, this decomposition is not unique. In fact, if we modify the resulting M and S to

$$\mathsf{M}' = \mathsf{M}\mathsf{A}, \qquad \mathsf{S}' = \mathsf{A}^{-1}\mathsf{S}, \tag{12.11}$$

using some 3×3 nonsingular matrix A, the equality $\mathsf{M}\mathsf{S} = \mathsf{M}'\mathsf{S}'$ holds. The second equation of Eq. (12.11) means multiplying each column of S, that is, $(X_\alpha, Y_\alpha, Z_\alpha)^\top$, by matrix A^{-1}, resulting in a linear mapping of the object shape. In other words, what is reconstructed by factorization is a 3D shape obtained by applying some unknown linear mapping to the true shape. Recall that the absolute 3D position is also indeterminate from images alone. Such a linear mapping with indeterminate position is called an *affine transformation*. Affine transformations preserve collinearity and coplanarity (Fig. 12.1); that is, collinear points are mapped to collinear points, and coplanar points are mapped to coplanar points. Hence, parallel lines are mapped to parallel lines, and parallel planes are mapped to parallel lines. The ratio of lengths is also preserved by affine transformations. However, scale and angle are not preserved. As a result, a cube, for example, is mapped to a parallelepiped. This type of reconstruction up to an unknown affine transformation is called *affine reconstruction*.

There exist two approaches for resolving this indeterminacy of affine transformations.

Use of knowledge about the scene If we know the lengths and angles (e.g., being orthogonal) of some parts, we multiply the shape matrix S by some matrix A^{-1} from the left so that the lengths and angles become correct.

Use of knowledge about the cameras If we know some properties about the cameras, we multiply the motion matrix M by some matrix A from the right such that the expected properties hold.

The resulting shape is called a *Euclidean reconstruction*, although this term is not strictly accurate because the absolute scale and size are indeterminate for reconstruction from images alone. However, this terminology is widely used for convenience. The process of obtaining a Euclidean reconstruction from an affine reconstruction is called *Euclidean upgrading*.

12.3 Metric Condition for Affine Cameras

We consider Euclidean upgrading using knowledge about the cameras. In view of Eq. (12.11), we compute the matrices M and S not in the form of Eq. (12.10) but in the form

$$\mathsf{M} = \mathsf{U}\mathsf{A}, \qquad \mathsf{S} = \mathsf{A}^{-1}\boldsymbol{\Sigma}\mathsf{V}^\top, \tag{12.12}$$

and derive the condition that the 3×3 matrix A should satisfy, considering the properties of the cameras. We define the 3D vectors $\mathbf{u}_{\kappa(1)}$ and $\mathbf{u}_{\kappa(2)}$ by

$$\mathsf{U} = \begin{pmatrix} \mathbf{u}_{1(1)}^\top \\ \mathbf{u}_{1(2)}^\top \\ \vdots \\ \mathbf{u}_{M(1)}^\top \\ \mathbf{u}_{M(2)}^\top \end{pmatrix}. \tag{12.13}$$

In other words, we let the $(2\kappa - 1)$th and the 2κth row of the $2M \times 3$ matrix U be the transposes $\mathbf{u}_{\kappa(1)}^\top$ and $\mathbf{u}_{\kappa(2)}^\top$ of $\mathbf{u}_{\kappa(1)}$ and $\mathbf{u}_{\kappa(2)}$, respectively. Then, the relation $\mathsf{M} = \mathsf{U}\mathsf{A}$ is rewritten in the form

$$\mathsf{M} = \begin{pmatrix} \begin{pmatrix} \mathbf{u}_{1(1)}^\top \\ \mathbf{u}_{1(2)}^\top \end{pmatrix} \mathsf{A} \\ \vdots \\ \begin{pmatrix} \mathbf{u}_{2M(1)}^\top \\ \mathbf{u}_{2M(2)}^\top \end{pmatrix} \mathsf{A} \end{pmatrix}. \tag{12.14}$$

Comparing this and the definition of the matrix M of Eq. (12.6), we find that the camera matrix $\boldsymbol{\Pi}_\kappa$ has the form

$$\boldsymbol{\Pi}_\kappa = \begin{pmatrix} \mathbf{u}_{\kappa(1)}^\top \\ \mathbf{u}_{\kappa(2)}^\top \end{pmatrix} \mathsf{A}. \tag{12.15}$$

Multiplying this with its transpose on both sides, we obtain

$$\boldsymbol{\Pi}_\kappa \boldsymbol{\Pi}_\kappa^\top = \begin{pmatrix} \mathbf{u}_{\kappa(1)}^\top \\ \mathbf{u}_{\kappa(2)}^\top \end{pmatrix} \mathsf{A}\mathsf{A}^\top \begin{pmatrix} \mathbf{u}_{\kappa(1)} & \mathbf{u}_{\kappa(2)} \end{pmatrix} = \begin{pmatrix} (\mathbf{u}_{\kappa(1)}, \mathsf{T}\mathbf{u}_{\kappa(1)}) & (\mathbf{u}_{\kappa(1)}, \mathsf{T}\mathbf{u}_{\kappa(2)}) \\ (\mathbf{u}_{\kappa(2)}, \mathsf{T}\mathbf{u}_{\kappa(1)}) & (\mathbf{u}_{\kappa(2)}, \mathsf{T}\mathbf{u}_{\kappa(2)}) \end{pmatrix}, \tag{12.16}$$

where we put

$$\mathsf{T} = \mathsf{A}\mathsf{A}^\top. \tag{12.17}$$

If we write the (i, j) element of $\boldsymbol{\Pi}_\kappa$ as $\Pi_{\kappa(ij)}$, we obtain from Eq. (12.16) the following three equalities.

$$(\mathbf{u}_{\kappa(1)}, \mathsf{T}\mathbf{u}_{\kappa(1)}) = \sum_{i=1}^{3} \Pi_{\kappa(1i)}^2, \quad (\mathbf{u}_{\kappa(1)}, \mathsf{T}\mathbf{u}_{\kappa(2)}) = \sum_{i=1}^{3} \Pi_{\kappa(1i)} \Pi_{\kappa(2i)},$$

$$(\mathbf{u}_{\kappa(2)}, \mathsf{T}\mathbf{u}_{\kappa(2)}) = \sum_{i=1}^{3} \Pi_{\kappa(2i)}^2. \tag{12.18}$$

These are called the *metric condition* of the κth affine camera, and the matrix T its *metric matrix*. We obtain, in total, $3M$ such equalities for $\kappa = 1, \ldots, M$. We cannot solve these for T, because the right-hand sides contain the unknown $\Pi_{\kappa(ij)}$. However, if we express the camera matrix $\boldsymbol{\Pi}_\kappa$ in terms of some unknown parameters (we do this in the subsequent sections), we can eliminate the parameters from the $3M$ equalities and obtain equations for T. Then we solve them for T and determine a matrix A such that Eq. (12.17) holds. A straightforward method for this is the use of Cholesky decomposition, but we can also determine A from the eigenvalues and eigenvectors of T ($\hookrightarrow$ Problem 12.2).

However, the matrix A and thus the metric matrix T have scale indeterminacy. In fact, if we multiply S by a constant c, the object shape expands, but if we divide M by that c at the same time, which means that the camera moves closer to the object, the product MS remains the same. After all, as pointed out in Chap. 5, we cannot determine the absolute size of the scene from images alone. Thus the metric matrix T needs to be determined only up to scale.

12.4 Description in the Camera Coordinate System

In order to do Euclidean upgrading using knowledge about the cameras, we first need a camera imaging model. For this, the use of the camera coordinate system, rather than the world coordinate system, is more convenient. This is because the "camera imaging model" is, by definition, the description of how the scene appears in the image of that camera.

Consider an XYZ coordinate system with origin O at the camera viewpoint and the Z-axis along the optical axis. Suppose point $\mathbf{r}$ in the scene described in this coordinate system is imaged at (x, y). Affine camera modeling means expressing the observed (x, y) as a linear expression in the 3D position $\mathbf{r}$. Therefore we write

$$\begin{pmatrix} x \\ y \end{pmatrix} = \mathsf{C}\mathbf{r} + \mathbf{d}. \tag{12.19}$$

In other words, affine camera imaging is specified by a 2×3 matrix C and a 2D vector $\mathbf{d}$.

We consider how this matrix C and the vector $\mathbf{d}$ are related to the camera matrix $\boldsymbol{\Pi}_\kappa$ described in the world coordinate system. As in the preceding chapter, we regard the κth camera as displaced from the world coordinate origin by $\mathbf{t}_\kappa$ (with respect to the world coordinate system) and rotated by R_κ (with respect to the world coordinate system). Because the camera is rotated by R_κ, the orientation of $\mathbf{r}_{\alpha\kappa}$ (with respect to the camera coordinate system) of the αth point $(X_\alpha, Y_\alpha, Z_\alpha)$ (with respect to the world coordinate system) is $\mathsf{R}_\kappa \mathbf{r}_{\alpha\kappa}$ in the world coordinate system. Hence, as described in Eq. (5.4), the following relation holds (Fig. 5.2).

$$\begin{pmatrix} X_\alpha \\ Y_\alpha \\ Z_\alpha \end{pmatrix} = \mathbf{t}_\kappa + \mathsf{R}_\kappa \mathbf{r}_{\alpha\kappa}. \tag{12.20}$$

From Eq. (12.5), this point is imaged at

$$\begin{pmatrix} x_{\alpha\kappa} \\ y_{\alpha\kappa} \end{pmatrix} = \boldsymbol{\Pi}_\kappa(\mathbf{t}_\kappa + \mathsf{R}_\kappa \mathbf{r}_{\alpha\kappa}). \tag{12.21}$$

On the other hand, we have

$$\begin{pmatrix} x_{\alpha\kappa} \\ y_{\alpha\kappa} \end{pmatrix} = \mathsf{C}_\kappa \mathbf{r}_{\alpha\kappa} + \mathbf{d}_\kappa, \tag{12.22}$$

in terms of the matrix C_κ and the vector $\mathbf{d}_\kappa$ of the κth affine camera. Therefore we obtain

$$\mathsf{C}_\kappa = \boldsymbol{\Pi}_\kappa \mathsf{R}_\kappa, \qquad \mathbf{d}_\kappa = \boldsymbol{\Pi}_\kappa \mathbf{t}_\kappa, \tag{12.23}$$

which means $\boldsymbol{\Pi}_\kappa = \mathsf{C}_\kappa \mathsf{R}^\top$. Equation (12.16) implies that the metric condition is specified by $\boldsymbol{\Pi}_\kappa \boldsymbol{\Pi}_\kappa^\top$. Thus R_κ is canceled, and we obtain

$$\boldsymbol{\Pi}_\kappa \boldsymbol{\Pi}_\kappa^\top = \mathsf{C}_\kappa \mathsf{C}_\kappa^\top. \tag{12.24}$$

In other words, for imposing the metric condition, we only need the matrix C_κ described in the coordinate system of each camera.

12.5 Symmetric Affine Camera

Let $\mathbf{g}$ be the centroid of the 3D positions $\mathbf{r}_\alpha$, $\alpha = 1, \ldots, N$, in the camera coordinate system:

$$\mathbf{g} = \frac{1}{N}\sum_{\alpha=1}^{N} \mathbf{r}_\alpha. \tag{12.25}$$

From our convention of Sect. 12.1, this is the position of the world coordinate origin viewed from the camera. The perspective projection of the camera can be approximated by an affine camera when all the points that we are viewing are concentrated in the neighborhood of $\mathbf{g}$. In order to describe the matrix C and the vector $\mathbf{d}$ of that affine camera in a general form, we consider the following minimum requirements that the affine camera we are considering should satisfy.

Requirement 1: The matrix C and the vector $\mathbf{d}$ are functions of the centroid $\mathbf{g} = (g_x, g_y, g_z)^\top$.

Requirement 2: If we view a planar surface that passes through the centroid $\mathbf{g}$ and is parallel to the XY plane of the camera coordinate system, we observe the same image as viewed by a perspective camera.

Requirement 3: The camera imaging geometry is symmetric around the Z-axis.

The first condition requires that the imaging geometry depend on the world coordinate origin (set to the centroid of the points) but not on the orientation of the world coordinate system. This is reasonable, because the world coordinate system is arbitrarily defined for the convenience of description. The second condition requires that a planar surface parallel to the image plane is imaged in the same way as a perspective camera. This is a minimum requirement that the camera "mimics" perspective projection. The third condition requires that if the scene rotates around the optical axis by an angle θ, the projected image should also rotate by the same angle θ. This is also a very natural condition, because all commercially available cameras are axially symmetric to a high degree. It can be shown from the theory of invariants that the only affine projection model that satisfies all these conditions is given by

$$\begin{pmatrix} x \\ y \end{pmatrix} = \frac{1}{\zeta}\left(\begin{pmatrix} X \\ Y \end{pmatrix} + \beta(g_z - Z)\begin{pmatrix} g_x \\ g_y \end{pmatrix}\right), \tag{12.26}$$

where $\mathbf{r} = (X, Y, Z)^\top$ is the 3D position (with respect to the camera coordinate system) of the point we are viewing and ζ and β are arbitrary functions of $g_x^2 + g_y^2$ and g_z. In the matrix form of Eq. (12.19), we obtain the following matrix C and the vector $\mathbf{d}$.

$$\mathsf{C} = \begin{pmatrix} 1/\zeta & 0 & -\beta g_x/\zeta \\ 0 & 1/\zeta & -\beta g_y/\zeta \end{pmatrix}, \qquad \mathbf{d} = \begin{pmatrix} \beta g_x g_z/\zeta \\ \beta g_y g_z/\zeta \end{pmatrix}. \tag{12.27}$$

We call this class of camera modeling the *symmetric affine camera*. It includes as special cases the following simplified models, which have frequently been used in the past.

Paraperspective projection: If we let ζ and β be

$$\zeta = \frac{g_z}{f}, \qquad \beta = \frac{1}{g_z}, \tag{12.28}$$

Eq. (12.26) becomes

$$\begin{pmatrix} x \\ y \end{pmatrix} = \frac{f}{g_z}\left(\begin{pmatrix} X \\ Y \end{pmatrix} + \frac{g_z - Z}{g_z}\begin{pmatrix} g_x \\ g_y \end{pmatrix}\right). \tag{12.29}$$

In the form of Eq. (12.19), the matrix C and the vector $\mathbf{d}$ are

$$\mathsf{C} = \begin{pmatrix} f/g_z & 0 & -fg_x/g_z^2 \\ 0 & f/g_z & -fg_y/g_z^2 \end{pmatrix}, \qquad \mathbf{d} = \begin{pmatrix} fg_x/g_z \\ fg_y/g_z \end{pmatrix}. \tag{12.30}$$

We can interpret this as Fig. 12.2a: Point (X, Y, Z) is first projected parallel to $\mathbf{g}$ onto the plane $Z = g_z$ and then perspectively projected onto the image plane $Z = f$.

Weak perspective projection: If we let ζ and β be

$$\zeta = \frac{g_z}{f}, \qquad \beta = 0, \tag{12.31}$$

Equation (12.26) becomes

$$\begin{pmatrix} x \\ y \end{pmatrix} = \frac{f}{g_z}\begin{pmatrix} X \\ Y \end{pmatrix}. \tag{12.32}$$

In the form of Eq. (12.19), the matrix C and the vector $\mathbf{d}$ are

$$\mathsf{C} = \begin{pmatrix} f/g_z & 0 & 0 \\ 0 & f/g_z & 0 \end{pmatrix}, \qquad \mathbf{d} = \begin{pmatrix} 0 \\ 0 \end{pmatrix}. \tag{12.33}$$

We can interpret this as Fig. 12.2b: Point (X, Y, Z) is first projected parallel to the Z-axis onto the plane $Z = g_z$ and then perspectively projected onto the image plane $Z = f$.

Fig. 12.2 Simplified affine camera models

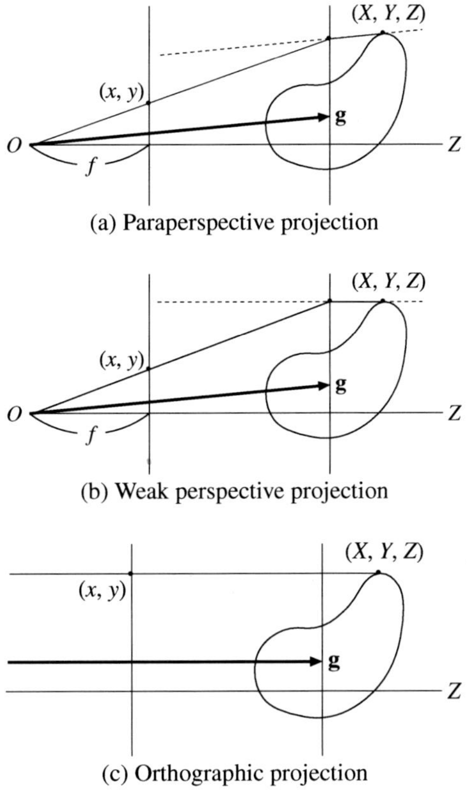

Orthographic projection: If we let ζ and β be

$$\zeta = 1, \qquad \beta = 0, \tag{12.34}$$

Equation (12.26) becomes

$$\begin{pmatrix} x \\ y \end{pmatrix} = \begin{pmatrix} X \\ Y \end{pmatrix}. \tag{12.35}$$

In the form of Eq. (12.19), the matrix C and the vector $\mathbf{d}$ are

$$\mathsf{C} = \begin{pmatrix} 1 & 0 & 0 \\ 0 & 1 & 0 \end{pmatrix}, \qquad \mathbf{d} = \begin{pmatrix} 0 \\ 0 \end{pmatrix}. \tag{12.36}$$

We can interpret this as Fig. 12.2c: Point (X, Y, Z) is projected parallel to the Z-axis onto a plane orthogonal to it. This can be seen as the limit of the focal length $f \to \infty$ of perspective projection.

12.6 Self-calibration of Symmetric Affine Cameras

If we assume the affine camera modeling of Eqs. (12.26) and (12.27), the following self-calibration procedure is obtained.

Procedure 12.2 **(Self-calibration of symmetric affine cameras)**

1. Compute the centroid $(t_{x\kappa}, t_{y\kappa})$ of the points $(x_{\alpha\kappa}, y_{\alpha\kappa})$ viewed in the κth image by

$$t_{x\kappa} = \frac{1}{N}\sum_{\alpha=1}^{N} x_{\alpha\kappa}, \qquad t_{y\kappa} = \frac{1}{N}\sum_{\alpha=1}^{N} y_{\alpha\kappa}. \tag{12.37}$$

2. Let

$$A_\kappa = t_{x\kappa}t_{y\kappa}, \qquad C_\kappa = t_{x\kappa}^2 - t_{y\kappa}^2. \tag{12.38}$$

3. Determine the matrix U by the SVD of Eq. (12.9), and obtain the 3D vectors $\mathbf{u}_{\kappa(1)}$ and $\mathbf{u}_{\kappa(2)}$, $\kappa = 1, \ldots, M$, by Eq. (12.13).
4. Define the $3 \times 3 \times 3 \times 3$ array $\mathscr{B} = (B_{ijkl})$ by

$$\begin{aligned}
B_{ijkl} = \sum_{\kappa=1}^{M}\Big[& A_\kappa^2\Big((\mathbf{u}_{\kappa(1)})_i(\mathbf{u}_{\kappa(1)})_j(\mathbf{u}_{\kappa(1)})_k(\mathbf{u}_{\kappa(1)})_l + (\mathbf{u}_{\kappa(2)})_i(\mathbf{u}_{\kappa(2)})_j(\mathbf{u}_{\kappa(2)})_k(\mathbf{u}_{\kappa(2)})_l \\
& - (\mathbf{u}_{\kappa(1)})_i(\mathbf{u}_{\kappa(1)})_j(\mathbf{u}_{\kappa(2)})_k(\mathbf{u}_{\kappa(2)})_l - (\mathbf{u}_{\kappa(2)})_i(\mathbf{u}_{\kappa(2)})_j(\mathbf{u}_{\kappa(1)})_k(\mathbf{u}_{\kappa(1)})_l\Big) \\
& + \frac{1}{4}C_\kappa^2\Big((\mathbf{u}_{\kappa(1)})_i(\mathbf{u}_{\kappa(2)})_j(\mathbf{u}_{\kappa(1)})_k(\mathbf{u}_{\kappa(2)})_l + (\mathbf{u}_{\kappa(2)})_i(\mathbf{u}_{\kappa(1)})_j(\mathbf{u}_{\kappa(1)})_k(\mathbf{u}_{\kappa(2)})_l \\
& + (\mathbf{u}_{\kappa(1)})_i(\mathbf{u}_{\kappa(2)})_j(\mathbf{u}_{\kappa(2)})_k(\mathbf{u}_{\kappa(1)})_l + (\mathbf{u}_{\kappa(2)})_i(\mathbf{u}_{\kappa(1)})_j(\mathbf{u}_{\kappa(2)})_k(\mathbf{u}_{\kappa(1)})_l\Big) \\
& - \frac{1}{2}A_\kappa C_\kappa\Big((\mathbf{u}_{\kappa(1)})_i(\mathbf{u}_{\kappa(1)})_j(\mathbf{u}_{\kappa(1)})_k(\mathbf{u}_{\kappa(2)})_l + (\mathbf{u}_{\kappa(1)})_i(\mathbf{u}_{\kappa(1)})_j(\mathbf{u}_{\kappa(2)})_k(\mathbf{u}_{\kappa(1)})_l \\
& + (\mathbf{u}_{\kappa(1)})_i(\mathbf{u}_{\kappa(2)})_j(\mathbf{u}_{\kappa(1)})_k(\mathbf{u}_{\kappa(1)})_l + (\mathbf{u}_{\kappa(2)})_i(\mathbf{u}_{\kappa(1)})_j(\mathbf{u}_{\kappa(1)})_k(\mathbf{u}_{\kappa(1)})_l \\
& - (\mathbf{u}_{\kappa(1)})_i(\mathbf{u}_{\kappa(2)})_j(\mathbf{u}_{\kappa(2)})_k(\mathbf{u}_{\kappa(2)})_l - (\mathbf{u}_{\kappa(2)})_i(\mathbf{u}_{\kappa(1)})_j(\mathbf{u}_{\kappa(2)})_k(\mathbf{u}_{\kappa(2)})_l \\
& - (\mathbf{u}_{\kappa(2)})_i(\mathbf{u}_{\kappa(2)})_j(\mathbf{u}_{\kappa(1)})_k(\mathbf{u}_{\kappa(2)})_l - (\mathbf{u}_{\kappa(2)})_i(\mathbf{u}_{\kappa(2)})_j(\mathbf{u}_{\kappa(2)})_k(\mathbf{u}_{\kappa(1)})_l\Big)\Big],
\end{aligned} \tag{12.39}$$

where $(\mathbf{u}_{\kappa(a)})_i$ is the ith component of $\mathbf{u}_{\kappa(a)}$, $a = 1, 2$, $i = 1, 2, 3$.

5. Define the following 6×6 matrix B.

$$\mathsf{B} = \begin{pmatrix}
B_{1111} & B_{1122} & B_{1133} & \sqrt{2}B_{1123} & \sqrt{2}B_{1131} & \sqrt{2}B_{1112} \\
B_{2211} & B_{2222} & B_{2233} & \sqrt{2}B_{2223} & \sqrt{2}B_{2231} & \sqrt{2}B_{2212} \\
B_{3311} & B_{3322} & B_{3333} & \sqrt{2}B_{3323} & \sqrt{2}B_{3331} & \sqrt{2}B_{3312} \\
\sqrt{2}B_{2311} & \sqrt{2}B_{2322} & \sqrt{2}B_{2333} & 2B_{2323} & 2B_{2331} & 2B_{2312} \\
\sqrt{2}B_{3111} & \sqrt{2}B_{3122} & \sqrt{2}B_{3133} & 2B_{3123} & 2B_{3131} & 2B_{3112} \\
\sqrt{2}B_{1211} & \sqrt{2}B_{1222} & \sqrt{2}B_{1233} & 2B_{1223} & 2B_{1231} & 2B_{1212}
\end{pmatrix}. \tag{12.40}$$

6. Compute the 6D unit eigenvector $\boldsymbol{\tau} = (\tau_i)$ of the matrix B for the smallest eigenvalue.

7. Determine the metric matrix T by

$$\mathsf{T} = \begin{pmatrix} \tau_1 & \tau_6/\sqrt{2} & \tau_5/\sqrt{2} \\ \tau_6/\sqrt{2} & \tau_2 & \tau_4/\sqrt{2} \\ \tau_5/\sqrt{2} & \tau_4/\sqrt{2} & \tau_3 \end{pmatrix}. \tag{12.41}$$

8. If $\det \mathsf{T} < 0$, change the sign of T.
9. Determine a matrix A such that $\mathsf{T} = \mathsf{A}\mathsf{A}^\top$ ($\hookrightarrow$ Problem 12.2).
10. Return the matrices M and S that satisfy Eq. (12.12) in the following form with corresponding double signs $\pm$.

$$\mathsf{M} = \pm\mathsf{U}\mathsf{A}, \qquad \mathsf{S} = \pm\mathsf{A}^{-1}\boldsymbol{\Sigma}\mathsf{V}^\top. \tag{12.42}$$

Comments The above procedure is based on the following observations. If we use the subscript κ to express quantities for the κth camera, Eq. (12.27) implies that Eq. (12.24) has the form

$$\begin{aligned}\boldsymbol{\Pi}_\kappa\boldsymbol{\Pi}_\kappa^\top &= \begin{pmatrix} 1/\zeta_\kappa & 0 & -\beta_\kappa g_{x\kappa}/\zeta_\kappa \\ 0 & 1/\zeta_\kappa & -\beta_\kappa g_{y\kappa}/\zeta_\kappa \end{pmatrix} \begin{pmatrix} 1/\zeta_\kappa & 0 \\ 0 & 1/\zeta_\kappa \\ -\beta g_{x\kappa}/\zeta_\kappa & -\beta_\kappa g_{y\kappa}/\zeta_\kappa \end{pmatrix} \\ &= \begin{pmatrix} 1/\zeta_\kappa^2 + \beta_\kappa^2 g_{x\kappa}^2/\zeta_\kappa^2 & \beta_\kappa^2 g_{x\kappa} g_{y\kappa}/\zeta_\kappa^2 \\ \beta_\kappa^2 g_{x\kappa} g_{y\kappa}/\zeta_\kappa^2 & 1/\zeta_\kappa^2 + \beta_\kappa^2 g_{y\kappa}^2/\zeta_\kappa^2 \end{pmatrix}.\end{aligned} \tag{12.43}$$

From this, we can write the metric condition of Eq. (12.18) in the form:

$$(\mathbf{u}_{\kappa(1)}, \mathsf{T}\mathbf{u}_{\kappa(1)}) = \frac{1}{\zeta_\kappa^2} + \beta_\kappa^2 t_{x\kappa}^2, \quad (\mathbf{u}_{\kappa(1)}, \mathsf{T}\mathbf{u}_{\kappa(2)}) = \beta_\kappa^2 t_{x\kappa} t_{y\kappa},$$

$$(\mathbf{u}_{\kappa(2)}, \mathsf{T}\mathbf{u}_{\kappa(2)}) = \frac{1}{\zeta_\kappa^2} + \beta_\kappa^2 t_{y\kappa}^2, \tag{12.44}$$

where we define

$$t_{x\kappa} = \frac{g_{x\kappa}}{\zeta_\kappa}, \qquad t_{y\kappa} = \frac{g_{y\kappa}}{\zeta_\kappa}. \tag{12.45}$$

If we let $X = g_x$, $Y = g_y$, and $Z = g_z$ in Eq. (12.26), we obtain $(x, y) = (g_x/\zeta, g_y/\zeta) = (t_x, t_y)$. Therefore $(t_{x\kappa}, t_{y\kappa})$ is the image position of the centroid $\mathbf{g}_\kappa$ of the N points in the scene viewed by the κth camera. Thus, $(t_{x\kappa}, t_{y\kappa})$ are given by Eq. (12.37).

From the second equation of Eq. (12.44), we can write

$$\beta_\kappa^2 = \frac{(\mathbf{u}_{\kappa(1)}, \mathsf{T}\mathbf{u}_{\kappa(2)})}{t_{x\kappa} t_{y\kappa}}. \tag{12.46}$$

Substituting this into the first and the third equations of Eq. (12.44), we obtain

$$(\mathbf{u}_{\kappa(1)}, \mathsf{T}\mathbf{u}_{\kappa(1)}) = \frac{1}{\zeta_\kappa^2} + \frac{t_{x\kappa}}{t_{y\kappa}}(\mathbf{u}_{\kappa(1)}, \mathsf{T}\mathbf{u}_{\kappa(2)}),$$

$$(\mathbf{u}_{\kappa(2)}, \mathsf{T}\mathbf{u}_{\kappa(2)}) = \frac{1}{\zeta_\kappa^2} + \frac{t_{y\kappa}}{t_{x\kappa}}(\mathbf{u}_{\kappa(1)}, \mathsf{T}\mathbf{u}_{\kappa(2)}). \tag{12.47}$$

Eliminating $1/\zeta_\kappa^2$ by subtraction on both sides, we obtain

$$A_\kappa(\mathbf{u}_{\kappa(1)}, \mathsf{T}\mathbf{u}_{\kappa(1)}) - C_\kappa(\mathbf{u}_{\kappa(1)}, \mathsf{T}\mathbf{u}_{\kappa(2)}) - A_\kappa(\mathbf{u}_{\kappa(2)}, \mathsf{T}\mathbf{u}_{\kappa(2)}) = 0, \tag{12.48}$$

where A_κ and C_κ are defined by Eq. (12.38). We determine the metric matrix T by least squares, minimizing the following function K.

$$K = \sum_{\kappa=1}^{M} \Big(A_\kappa(\mathbf{u}_{\kappa(1)}, \mathsf{T}\mathbf{u}_{\kappa(1)}) - C_\kappa(\mathbf{u}_{\kappa(1)}, \mathsf{T}\mathbf{u}_{\kappa(2)}) - A_\kappa(\mathbf{u}_{\kappa(2)}, \mathsf{T}\mathbf{u}_{\kappa(2)}) \Big)^2, \tag{12.49}$$

Expanding this and using the array B_{ijkl} defined by Eq. (12.39), we can write

$$K = \sum_{\kappa=1}^{M} \sum_{i,j,k,l=1}^{3} B_{ijkl} T_{ij} T_{kl}. \tag{12.50}$$

Because the metric matrix $\mathsf{T} = (T_{ij})$ has scale indeterminacy, we only need to determine it up to scale. Therefore we normalized it to $\|\mathsf{T}\|^2 = \sum_{i,j=1}^{3} T_{ij}^2 = 1$. If we define the 6D vector $\boldsymbol{\tau}$ by

$$\boldsymbol{\tau} = \begin{pmatrix} T_{11} & T_{22} & T_{33} & \sqrt{2}T_{23} & \sqrt{2}T_{31} & \sqrt{2}T_{12} \end{pmatrix}^\top, \tag{12.51}$$

the normalization $\|\mathsf{T}\|^2 = 1$ is equivalent to the normalization $\|\boldsymbol{\tau}\|^2 = 1$. If we define the 6×6 matrix B by Eq. (12.40), the function K of Eq. (12.49) is written as the following quadratic form in $\boldsymbol{\tau}$.

$$K = (\boldsymbol{\tau}, \mathsf{B}\boldsymbol{\tau}). \tag{12.52}$$

The unit vector $\boldsymbol{\tau}$ that minimizes this is given by the unit eigenvector of the matrix B for the smallest eigenvalue. Hence, the metric matrix $\mathsf{T} = (T_{ij})$ is given by Eq. (12.41). From the definition of Eq. (12.17), the matrix T is positive definite, but the eigenvector $\boldsymbol{\tau}$ has sign indeterminacy. Therefore we adjust its sign as described in Step 8. If we have determined the metric matrix T, the motion matrix M, and the shape matrix S are obtained by Procedure 12.2, we can also determine the matrix C_κ and the rotation R_κ of each camera ($\hookrightarrow$ Problem 12.3).

The reason that we obtain two solutions in Step 10 is due to the sign indeterminacy of the matrix A computed in Step 9. If we change the sign of the shape matrix S of Eq. (12.6), all the points change their signs, resulting in a mirror image shape. As shown in Chap. 5, the mirror image solution was also obtained by 3D reconstruction using perspective cameras. There, we removed the mirror image shape, using the condition that the object is in front of the cameras. For affine camera modeling, however, the depth is indeterminate, thus we cannot tell front from behind. The fact that we cannot remove the mirror image solution is one of the intrinsic properties of affine cameras.

Furthermore, the matrix A not only has sign indeterminacy but also rotation indeterminacy, for if we let $\mathsf{A}' = \mathsf{A}\mathsf{R}$ for any rotation matrix R, we have $\mathsf{A}\mathsf{A}^\top = \mathsf{A}'\mathsf{A}'^\top$. Using this A' instead of A, we obtain, as seen from Eq. (12.42), a shape rotated by $\mathsf{R}^\top$. This indeterminacy corresponds to the fact that the coordinates $(X_\alpha, Y_\alpha, Z_\alpha)$ are defined with respect to the world coordinate system, whose absolute orientation is indeterminate; we can define the world coordinate system in an arbitrary orientation.

12.7 Self-calibration of Simplified Affine Cameras

One problem of the above Procedure 12.2 is that the metric matrix T determined by Eq. (12.41) is not always positive definite. It is positive definite if and only if its eigenvalues are all positive, but $\det \mathsf{T} > 0$ is only a necessary condition for it. If T is not positive definite, no matrix A exists such that $\mathsf{T} = \mathsf{A}\mathsf{A}^\top$, so the computation of Step 9 fails. This typically occurs when:

- The perspective effect of the camera is too strong to be approximated by an affine camera.
- The camera motion is nearly collinear such that sufficient depth information is not obtained.
- Point correspondences among images are inaccurate, resulting in a large amount of error of the observation matrix W.

One way to cope with this is to use the simplified affine camera models described in Sect. 12.5. As we introduce more approximation, the solution becomes less accurate, but the chance of computational failure decreases as the conditions to be satisfied become weaker.

12.7.1 Paraperspective Projection Model

For the paraperspective projection model of Eqs. (12.29) and (12.30), we obtain the following procedure.

Procedure 12.3 **(Self-calibration of paraperspective cameras)**

1. Compute the centroid $(t_{x\kappa}, t_{y\kappa})$ of the points $(x_{\alpha\kappa}, y_{\alpha\kappa})$ viewed in the κth image by
$$t_{x\kappa} = \frac{1}{N}\sum_{\alpha=1}^{N} x_{\alpha\kappa}, \qquad t_{y\kappa} = \frac{1}{N}\sum_{\alpha=1}^{N} y_{\alpha\kappa}. \tag{12.53}$$
2. Arbitrarily give f_κ, $\kappa = 1, \ldots, M$, and let
$$\alpha_\kappa = \frac{1}{1 + t_{x\kappa}^2/f_\kappa^2}, \qquad \beta_\kappa = \frac{1}{1 + t_{y\kappa}^2/f_\kappa^2}, \qquad \gamma_\kappa = \frac{t_{x\kappa}t_{y\kappa}}{f_\kappa^2}. \tag{12.54}$$
3. Determine the matrix U by the SVD of Eq. (12.9), and obtain the 3D vectors $\mathbf{u}_{\kappa(1)}$ and $\mathbf{u}_{\kappa(2)}$, $\kappa = 1, \ldots, M$, by Eq. (12.13).
4. Define the $3 \times 3 \times 3 \times 3$ array $\mathcal{B} = (B_{ijkl})$ by
$$B_{ijkl} = \sum_{\kappa=1}^{M}\Big((\gamma_\kappa^2 + 1)\alpha_\kappa^2(\mathbf{u}_{\kappa(1)})_i(\mathbf{u}_{\kappa(1)})_j(\mathbf{u}_{\kappa(1)})_k(\mathbf{u}_{\kappa(1)})_l$$
$$+ (\gamma_\kappa^2 + 1)\beta_\kappa^2(\mathbf{u}_{\kappa(2)})_i(\mathbf{u}_{\kappa(2)})_j(\mathbf{u}_{\kappa(2)})_k(\mathbf{u}_{\kappa(2)})_l$$

$$
\begin{aligned}
&+ (\mathbf{u}_{\kappa(1)})_i(\mathbf{u}_{\kappa(2)})_j(\mathbf{u}_{\kappa(1)})_k(\mathbf{u}_{\kappa(2)})_l + (\mathbf{u}_{\kappa(1)})_i(\mathbf{u}_{\kappa(2)})_j(\mathbf{u}_{\kappa(2)})_k(\mathbf{u}_{\kappa(1)})_l \\
&+ (\mathbf{u}_{\kappa(2)})_i(\mathbf{u}_{\kappa(1)})_j(\mathbf{u}_{\kappa(1)})_k(\mathbf{u}_{\kappa(2)})_l + (\mathbf{u}_{\kappa(2)})_i(\mathbf{u}_{\kappa(1)})_j(\mathbf{u}_{\kappa(2)})_k(\mathbf{u}_{\kappa(1)})_l \\
&- \alpha_\kappa\gamma_\kappa(\mathbf{u}_{\kappa(1)})_i(\mathbf{u}_{\kappa(1)})_j(\mathbf{u}_{\kappa(1)})_k(\mathbf{u}_{\kappa(2)})_l \\
&- \alpha_\kappa\gamma_\kappa(\mathbf{u}_{\kappa(1)})_i(\mathbf{u}_{\kappa(1)})_j(\mathbf{u}_{\kappa(2)})_k(\mathbf{u}_{\kappa(1)})_l \\
&- \alpha_\kappa\gamma_\kappa(\mathbf{u}_{\kappa(1)})_i(\mathbf{u}_{\kappa(2)})_j(\mathbf{u}_{\kappa(1)})_k(\mathbf{u}_{\kappa(1)})_l \\
&- \alpha_\kappa\gamma_\kappa(\mathbf{u}_{\kappa(2)})_i(\mathbf{u}_{\kappa(1)})_j(\mathbf{u}_{\kappa(1)})_k(\mathbf{u}_{\kappa(1)})_l \\
&- \beta_\kappa\gamma_\kappa(\mathbf{u}_{\kappa(2)})_i(\mathbf{u}_{\kappa(2)})_j(\mathbf{u}_{\kappa(1)})_k(\mathbf{u}_{\kappa(2)})_l \\
&- \beta_\kappa\gamma_\kappa(\mathbf{u}_{\kappa(2)})_i(\mathbf{u}_{\kappa(2)})_j(\mathbf{u}_{\kappa(2)})_k(\mathbf{u}_{\kappa(1)})_l \\
&- \beta_\kappa\gamma_\kappa(\mathbf{u}_{\kappa(1)})_i(\mathbf{u}_{\kappa(2)})_j(\mathbf{u}_{\kappa(2)})_k(\mathbf{u}_{\kappa(2)})_l \\
&- \beta_\kappa\gamma_\kappa(\mathbf{u}_{\kappa(2)})_i(\mathbf{u}_{\kappa(1)})_j(\mathbf{u}_{\kappa(2)})_k(\mathbf{u}_{\kappa(2)})_l \\
&+ (\gamma_\kappa^2 - 1)\alpha_\kappa\beta_\kappa(\mathbf{u}_{\kappa(1)})_i(\mathbf{u}_{\kappa(1)})_j(\mathbf{u}_{\kappa(2)})_k(\mathbf{u}_{\kappa(2)})_l \\
&+ (\gamma_\kappa^2 - 1)\alpha_\kappa\beta_\kappa(\mathbf{u}_{\kappa(2)})_i(\mathbf{u}_{\kappa(2)})_j(\mathbf{u}_{\kappa(1)})_k(\mathbf{u}_{\kappa(1)})_l\Big),
\end{aligned}
\tag{12.55}
$$

where $(\mathbf{u}_{\kappa(a)})_i$ is the ith component of $\mathbf{u}_{\kappa(a)}$, $a = 1, 2$, $i = 1, 2, 3$.

5. Define the 6×6 matrix B by Eq. (12.40), and compute the 6D unit eigenvector $\boldsymbol{\tau} = (\tau_i)$ of B for the smallest eigenvalue.
6. Determine the metric matrix T by Eq. (12.41). If $\det \mathsf{T} < 0$, change the sign of T.
7. Determine a matrix A such that $\mathsf{T} = \mathsf{A}\mathsf{A}^\top$, and return the matrices M and S of Eq. (12.42).

Comments Paraperspective projection is modeled by Eq. (12.29), where unknowns are f and $\mathbf{g} = (g_x, g_y, g_z)^\top$. As shown by Eq. (12.24), however, the metric condition is described only by the matrix C. From Eq. (12.30), we see that C is specified by the ratio between f and $\mathbf{g}$. In other words, C is unchanged by multiplying f and $\mathbf{g}$ by the same constant, so the absolute value of f is indeterminate. Hence, we set f arbitrarily (the reconstructed shape depends on the assumed f, though, and large distortion results if f is very different from the focal length of the actual camera). If we use the subscript κ to express quantities for the κth camera, Eq. (12.30) implies that Eq. (12.24) has the form

$$
\begin{aligned}
\boldsymbol{\Pi}_\kappa\boldsymbol{\Pi}_\kappa^\top &= \begin{pmatrix} f_\kappa/g_{z\kappa} & 0 & -f_\kappa g_{x\kappa}/g_{z\kappa}^2 \\ 0 & f_\kappa/g_{z\kappa} & -f_\kappa g_{y\kappa}/g_{z\kappa}^2 \end{pmatrix} \begin{pmatrix} f_\kappa/g_{z\kappa} & 0 \\ 0 & f_\kappa/g_{z\kappa} \\ -f_\kappa g_{x\kappa}/g_{z\kappa}^2 & -f_\kappa g_{y\kappa}/g_{z\kappa}^2 \end{pmatrix} \\
&= \begin{pmatrix} f_\kappa^2/g_{z\kappa}^2 + f_\kappa^2 g_{x\kappa}^2/g_{z\kappa}^4 & f_\kappa^2 g_{x\kappa} g_{y\kappa}/g_{z\kappa}^4 \\ f_\kappa^2 g_{x\kappa} g_{y\kappa}/g_{z\kappa}^4 & f_\kappa^2/g_{z\kappa}^2 + f_\kappa^2 g_{y\kappa}^2/g_{z\kappa}^4 \end{pmatrix} \\
&= f_\kappa^2 \begin{pmatrix} (1 + t_{x\kappa}^2)/g_{z\kappa}^2 & t_{x\kappa}t_{y\kappa}/g_{z\kappa}^2 \\ t_{x\kappa}t_{y\kappa}/g_{z\kappa}^2 & (1 + t_{y\kappa}^2)/g_{z\kappa}^2 \end{pmatrix},
\end{aligned}
\tag{12.56}
$$

where we define

$$
t_{x\kappa} = f_\kappa \frac{g_{x\kappa}}{g_{z\kappa}}, \qquad t_{y\kappa} = f_\kappa \frac{g_{y\kappa}}{g_{z\kappa}}.
\tag{12.57}
$$

If we let $X = g_x$, $Y = g_y$, and $Z = g_z$ in Eq. (12.29), we obtain $(x, y) = (f_\kappa g_x/g_z, f_\kappa g_y/g_z)$. Namely, $(t_{x\kappa}, t_{y\kappa})$ is the image position of the centroid $\mathbf{g}_\kappa$ of the N points in the scene viewed by the κth camera. Hence, $(t_{x\kappa}, t_{y\kappa})$ is given by Eq. (12.53). From Eq. (12.56), the metric condition of Eq. (12.18) is written in the form

$$(\mathbf{u}_{\kappa(1)}, \mathsf{T}\mathbf{u}_{\kappa(1)}) = \frac{f_\kappa^2}{\alpha_\kappa t_{z\kappa}^2}, \qquad (\mathbf{u}_{\kappa(1)}, \mathsf{T}\mathbf{u}_{\kappa(2)}) = \frac{f_\kappa^2 \gamma_\kappa}{t_{z\kappa}^2}, \qquad (\mathbf{u}_{\kappa(2)}, \mathsf{T}\mathbf{u}_{\kappa(2)}) = \frac{f_\kappa^2}{\beta_\kappa t_{z\kappa}^2}, \tag{12.58}$$

where α_κ, β_κ, and γ_κ are defined by Eq. (12.54). Eliminating $t_{z\kappa}$ from Eq. (12.58), we obtain the following two equalities.

$$\alpha_\kappa(\mathbf{u}_{\kappa(1)}, \mathsf{T}\mathbf{u}_{\kappa(1)}) = \beta_\kappa(\mathbf{u}_{\kappa(2)}, \mathsf{T}\mathbf{u}_{\kappa(2)}),$$

$$\gamma_\kappa\Big(\alpha_\kappa(\mathbf{u}_{\kappa(1)}, \mathsf{T}\mathbf{u}_{\kappa(1)}) + \beta_\kappa(\mathbf{u}_{\kappa(2)}, \mathsf{T}\mathbf{u}_{\kappa(2)})\Big) = 2(\mathbf{u}_{\kappa(1)}, \mathsf{T}\mathbf{u}_{\kappa(2)}). \tag{12.59}$$

We determine the metric matrix T by least squares, minimizing the following function K.

$$K = \sum_{\kappa=1}^{M}\Big(\Big(\alpha_\kappa(\mathbf{u}_{\kappa(1)}, \mathsf{T}\mathbf{u}_{\kappa(1)}) - \beta_\kappa(\mathbf{u}_{\kappa(2)}, \mathsf{T}\mathbf{u}_{\kappa(2)})\Big)^2$$
$$+\Big(\gamma_\kappa\Big(\alpha_\kappa(\mathbf{u}_{\kappa(1)}, \mathsf{T}\mathbf{u}_{\kappa(1)}) + \beta_\kappa(\mathbf{u}_{\kappa(2)}, \mathsf{T}\mathbf{u}_{\kappa(2)})\Big) - 2(\mathbf{u}_{\kappa(1)}, \mathsf{T}\mathbf{u}_{\kappa(2)})\Big)^2\Big). \tag{12.60}$$

Expanding this and using the array B_{ijkl} defined by Eq. (12.55), we can write this in the form of Eq. (12.50). The subsequent steps are the same as in Procedure 12.2.

12.7.2 Weak Perspective Projection Model

For the weak perspective projection model of Eqs. (12.32) and (12.33), we obtain the following procedure.

Procedure 12.4 (Self-calibration of weak perspective cameras)

1. Determine the matrix U by the SVD of Eq. (12.9), and obtain the 3D vectors $\mathbf{u}_{\kappa(1)}$ and $\mathbf{u}_{\kappa(2)}$, $\kappa = 1, ..., M$, by Eq. (12.13).
2. Define the $3 \times 3 \times 3 \times 3$ array $\mathscr{B} = (B_{ijkl})$ by

$$B_{ijkl} = \sum_{\kappa=1}^{M}\Big((\mathbf{u}_{\kappa(1)})_i(\mathbf{u}_{\kappa(1)})_j(\mathbf{u}_{\kappa(1)})_k(\mathbf{u}_{\kappa(1)})_l - (\mathbf{u}_{\kappa(1)})_i(\mathbf{u}_{\kappa(1)})_j(\mathbf{u}_{\kappa(2)})_k(\mathbf{u}_{\kappa(2)})_l$$
$$-(\mathbf{u}_{\kappa(2)})_i(\mathbf{u}_{\kappa(2)})_j(\mathbf{u}_{\kappa(1)})_k(\mathbf{u}_{\kappa(1)})_l + (\mathbf{u}_{\kappa(2)})_i(\mathbf{u}_{\kappa(2)})_j(\mathbf{u}_{\kappa(2)})_k(\mathbf{u}_{\kappa(2)})_l$$
$$+\frac{1}{4}\Big((\mathbf{u}_{\kappa(1)})_i(\mathbf{u}_{\kappa(2)})_j(\mathbf{u}_{\kappa(1)})_k(\mathbf{u}_{\kappa(2)})_l + (\mathbf{u}_{\kappa(2)})_i(\mathbf{u}_{\kappa(1)})_j(\mathbf{u}_{\kappa(1)})_k(\mathbf{u}_{\kappa(2)})_l$$
$$+(\mathbf{u}_{\kappa(1)})_i(\mathbf{u}_{\kappa(2)})_j(\mathbf{u}_{\kappa(2)})_k(\mathbf{u}_{\kappa(1)})_l + (\mathbf{u}_{\kappa(2)})_i(\mathbf{u}_{\kappa(1)})_j(\mathbf{u}_{\kappa(2)})_k(\mathbf{u}_{\kappa(1)})_l\Big)\Big), \tag{12.61}$$

where $(\mathbf{u}_{\kappa(a)})_i$ is the ith component of $\mathbf{u}_{\kappa(a)}$, $a = 1, 2$, $i = 1, 2, 3$.

3. Define the 6×6 matrix B by Eq. (12.40), and compute the 6D unit eigenvector $\boldsymbol{\tau} = (\tau_i)$ of B for the smallest eigenvalue.
4. Determine the metric matrix T by Eq. (12.41). If $\det \mathsf{T} < 0$, change the sign of T.
5. Determine a matrix such that $\mathsf{T} = \mathsf{A}\mathsf{A}^\top$, and return the matrices M and S of Eq. (12.42).

Comments If we use the subscript κ to express quantities for the κth camera, Eq. (12.33) implies that Eq. (12.24) has the form

$$\boldsymbol{\Pi}_\kappa \boldsymbol{\Pi}_\kappa^\top = \begin{pmatrix} f_\kappa/g_{z\kappa} & 0 & 0 \\ 0 & f_\kappa/g_{z\kappa} & 0 \end{pmatrix} \begin{pmatrix} f_\kappa/g_{z\kappa} & 0 \\ 0 & f_\kappa/g_{z\kappa} \\ 0 & 0 \end{pmatrix} = \begin{pmatrix} f_\kappa^2/g_{z\kappa}^2 & 0 \\ 0 & f_\kappa^2/g_{z\kappa}^2 \end{pmatrix}. \tag{12.62}$$

The metric condition of Eq. (12.18) is written in the form

$$(\mathbf{u}_{\kappa(1)}, \mathsf{T}\mathbf{u}_{\kappa(1)}) = (\mathbf{u}_{\kappa(2)}, \mathsf{T}\mathbf{u}_{\kappa(2)}) = \frac{f_\kappa^2}{g_{z\kappa}^2}, \qquad (\mathbf{u}_{\kappa(1)}, \mathsf{T}\mathbf{u}_{\kappa(2)}) = 0. \tag{12.63}$$

We determine the metric matrix T by least squares, minimizing the following function K.

$$K = \sum_{\kappa=1}^{M} \Big(\big((\mathbf{u}_{\kappa(1)}, \mathsf{T}\mathbf{u}_{\kappa(1)}) - (\mathbf{u}_{\kappa(2)}, \mathsf{T}\mathbf{u}_{\kappa(2)}) \big)^2 + (\mathbf{u}_{\kappa(1)}, \mathsf{T}\mathbf{u}_{\kappa(2)})^2 \Big). \tag{12.64}$$

Expanding this and using the array B_{ijkl} defined by Eq. (12.61), we can write this in the form of Eq. (12.50). The subsequent steps are the same as Procedure 12.2.

12.7.3 Orthographic Projection Model

For the orthographic projection model of Eqs. (12.35) and (12.36), we obtain the following procedure.

Procedure 12.5 (Self-calibration of orthographic cameras)

1. Determine the matrix U by the SVD of Eq. (12.9), and obtain the 3D vectors $\mathbf{u}_{\kappa(1)}$ and $\mathbf{u}_{\kappa(2)}$, $\kappa = 1, ..., M$ from Eq. (12.13).
2. Define the $3 \times 3 \times 3 \times 3$ array $\mathscr{B} = (B_{ijkl})$ by

$$\begin{aligned} B_{ijkl} = \sum_{\kappa=1}^{M} \Big(& (\mathbf{u}_{\kappa(1)})_i (\mathbf{u}_{\kappa(1)})_j (\mathbf{u}_{\kappa(1)})_k (\mathbf{u}_{\kappa(1)})_l + (\mathbf{u}_{\kappa(2)})_i (\mathbf{u}_{\kappa(2)})_j (\mathbf{u}_{\kappa(2)})_k (\mathbf{u}_{\kappa(2)})_l \\ & + \frac{1}{4} \big((\mathbf{u}_{\kappa(1)})_i (\mathbf{u}_{\kappa(2)})_j + (\mathbf{u}_{\kappa(2)})_i (\mathbf{u}_{\kappa(1)})_j \big) \big((\mathbf{u}_{\kappa(1)})_k (\mathbf{u}_{\kappa(2)})_l \\ & + (\mathbf{u}_{\kappa(2)})_k (\mathbf{u}_{\kappa(1)})_l \big) \Big), \end{aligned} \tag{12.65}$$

where $(\mathbf{u}_{\kappa(a)})_i$ is the ith component of $\mathbf{u}_{\kappa(a)}$, $a = 1, 2$, $i = 1, 2, 3$.

3. Define the 6×6 matrix B by Eq. (12.40), and compute the 6D vector $\boldsymbol{\tau} = (\tau_i)$ by solving the following linear equation.

$$\mathsf{B}\boldsymbol{\tau} = \begin{pmatrix} 1 \\ 1 \\ 1 \\ 0 \\ 0 \\ 0 \end{pmatrix}. \tag{12.66}$$

4. Determine the metric matrix T by Eq. (12.41).
5. Determine a matrix A such that $\mathsf{T} = \mathsf{A}\mathsf{A}^\top$, and return the matrices M and S of Eq. (12.42).

Comments From Eq. (12.36), Eq. (12.24) for the κth camera has the form

$$\boldsymbol{\Pi}_\kappa \boldsymbol{\Pi}_\kappa^\top = \begin{pmatrix} 1 & 0 & 0 \\ 0 & 1 & 0 \end{pmatrix} \begin{pmatrix} 1 & 0 \\ 0 & 1 \\ 0 & 0 \end{pmatrix} = \begin{pmatrix} 1 & 0 \\ 0 & 1 \end{pmatrix}. \tag{12.67}$$

The metric condition of Eq. (12.18) is written in the form:

$$(\mathbf{u}_{\kappa(1)}, \mathsf{T}\mathbf{u}_{\kappa(1)}) = (\mathbf{u}_{\kappa(2)}, \mathsf{T}\mathbf{u}_{\kappa(2)}) = 1, \qquad (\mathbf{u}_{\kappa(1)}, \mathsf{T}\mathbf{u}_{\kappa(2)}) = 0. \tag{12.68}$$

We determine the metric matrix T by least squares, minimizing the following function K.

$$K = \sum_{\kappa=1}^{M} \Big(\big((\mathbf{u}_{\kappa(1)}, \mathsf{T}\mathbf{u}_{\kappa(1)}) - 1\big)^2 + \big((\mathbf{u}_{\kappa(2)}, \mathsf{T}\mathbf{u}_{\kappa(2)}) - 1\big)^2 + (\mathbf{u}_{\kappa(1)}, \mathsf{T}\mathbf{u}_{\kappa(2)})^2 \Big). \tag{12.69}$$

We expand this and differentiate it with respect to T_{ij}. Letting the result be 0, we obtain the linear equation

$$\sum_{i,j,k,l=1}^{3} B_{ijkl} T_{kl} = \delta_{ij}, \tag{12.70}$$

where B_{ijkl} is the array defined by Eq. (12.65), and δ_{ij} is the Kronecker delta, taking 1 for $i = j$ and 0 otherwise. Using the matrix B of Eq. (12.40) and the vector $\boldsymbol{\tau}$ of Eq. (12.50), we can write the above equation in the form of Eq. (12.66). Solving it for $\boldsymbol{\tau}$, we can obtain the metric matrix T in the form of Eq. (12.41).

12.8 Examples

Figure 12.3 shows five frames of simulated perspective images of a cylindrical object taken by a moving camera. The image size is assumed to be 600×600 pixels with focal length 600 pixels. Figure 12.4 shows the 3D shape reconstructed by assuming different affine camera modelings, where the size and the viewing direction are appropriately adjusted to make comparison easy. The upper row shows the shape viewed from the front and the lower row shows the base shape viewed along the cylinder axis; only the end points connected by line segments are shown. Column (a) shows the true shape, and columns (b)–(e) correspond to the symmetric affine camera, the paraperspective camera, the weak perspective camera, and the orthographic camera, respectively. For the paraperspective modeling, we set $f = 600$. We can see from these that although the reconstructed shape is not exact, a fairly good approximation is obtained using any affine camera modeling.

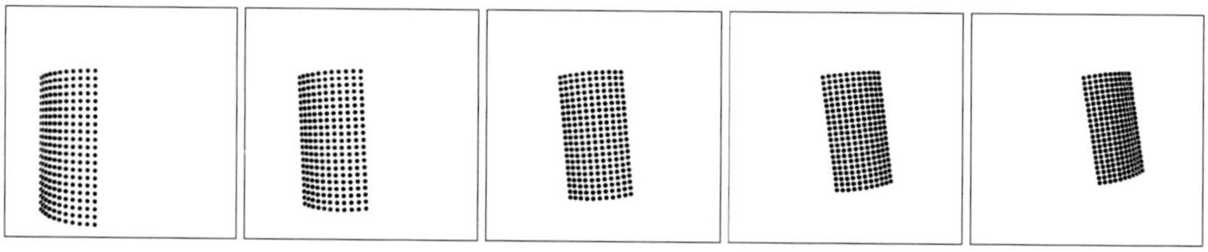

Fig. 12.3 Simulated images of a cylindrical object taken by a moving perspective camera

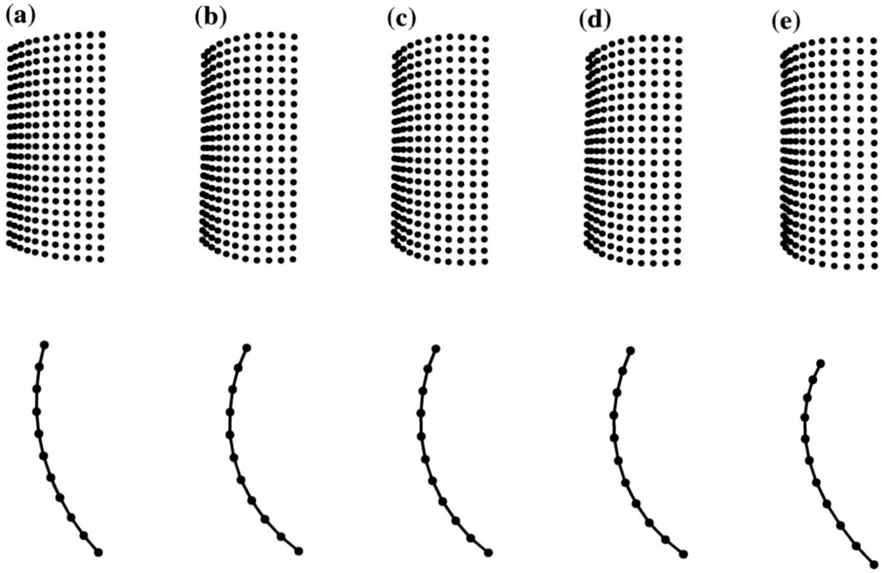

Fig. 12.4 The 3D shape reconstructed from the image sequence of Fig. 12.3. The upper row is for the front view, and the lower row is for the view along the cylinder axis. **a** The true shape, **b** Symmetric affine camera, **c** Paraperspective camera, **d** Weak perspective camera, **e** Orthographic camera

12.9 Supplemental Note

The factorization method for 3D reconstruction from multiple images using affine camera modeling was first presented by Tomasi and Kanade [5] for orthographic cameras. The technique was extended to weak perspective cameras and paraperspective cameras by Poelman and Kanade [3]. However, these camera models are mutually unrelated without any hierarchical relations. Namely, orthographic projection is not a special case of weak perspective or paraperspective projection, and weak perspective projection is not a special case of paraperspective projection. The symmetric affine camera model was proposed by the authors [2] as a model that includes all these models. They pointed out that orthographic, weak perspective, and paraperspective projections are special cases of the symmetric affine camera model and showed that self-calibration is possible using only the symmetric affine camera modeling.

For approximating perspective cameras by affine cameras, we assumed in this chapter that the 3D points we observe are concentrated in a neighborhood of the world coordinate origin, which we identified with the centroid of the 3D points. In our formulation, the vector $\mathbf{g}$ that specifies paraperspective projection is the world coordinate origin (= the centroid of the 3D points) viewed from the camera coordinate system. Basri [1] pointed out that the class of paraperspective projection coincides with the entire class of affine cameras if the vector $\mathbf{g}$ is arbitrarily defined and the 3D points undergo arbitrary rigid motions.

In this chapter, we assume that all 3D points are seen in all images. In real image processing, however, we usually determine point correspondences between images by detecting and tracing feature points over consecutive video frames. However, the tracking is disrupted when the points we are tracking are occluded by other objects or go out of the frame. Many studies have been done for estimating such missing points. The basic principle is to estimate the missing entries of the observation matrix W of Eq. (12.7) such that the decomposition of Eq. (12.8) holds with reasonable accuracy. Note that Eq. (12.8) is derived by choosing the image coordinate origin of each frame to be the centroid of the points seen in that image, but the centroid cannot be computed if missing points exist. In that case, we can extend the shape matrix S and the motion matrix M of Eq. (12.6) to $\tilde{\mathsf{S}}$ and $\tilde{\mathsf{M}}$ by adding a new column to S and a new row to $\tilde{\mathsf{M}}$ such that $\mathsf{W} = \tilde{\mathsf{M}}\tilde{\mathsf{S}}$ holds (hence W is ideally of rank 4). Many mathematical theories have been proposed for estimating the missing elements based on this formulation. This is a very difficult problem in a general setting, but the problem becomes easier if a sufficient number of points are visible in a sufficient number of images. For example, we first do tentative 3D reconstruction using visible points only and estimate the location of missing points such that they are compatible with the computed 3D shape and camera positions. For this 3D reconstruction, affine camera modeling is sufficient, and the metric condition is not necessary, because we observe the same images if we affinely transform the object shape and accordingly transform the cameras. It is shown by the authors [4] that this observation allows us to extend feature point tracking after it is interrupted. In their computation, weights for describing the uncertainty of the missing points are used.

Problems

12.1 (1) Because the αth column of Eq. (12.7) sequentially lists the $x-$ and $y-$coordinates of the αth point over the M images, it can be regarded as the *trajectory* of the αth point, which is identified with a point in a $2M$D space. Show that Eq. (12.8) implies that all the $2M$D points representing the trajectories of the M points are included in some 3D subspace.
(2) Show that if Eq. (12.8) holds, the orthonormal basis of that 3D subspace is given by the columns $\mathbf{u}_1$, $\mathbf{u}_2$, and $\mathbf{u}_3$ of the $2M \times 3$ matrix U obtained by the SVD of the matrix W in the form of $\mathsf{W} = \mathsf{U}\boldsymbol{\Sigma}\mathsf{V}^\top$.

12.2 Show how to compute a matrix B such that $\mathsf{A} = \mathsf{B}\mathsf{B}^\top$ for a given $n \times n$ symmetric semi-definite matrix A (the solution is not unique).

12.3 (1) Show how the parameters ζ_κ, β_κ, $g_{x\kappa}$, and $g_{y\kappa}$ are estimated from the metric matrix T computed by Procedure 12.2. Hint: Apply least squares to the metric condition of Eq. (12.44).
(2) Show how the rotation R_κ of the κth camera is estimated from the camera matrix $\boldsymbol{\Pi}_\kappa$ computed by Procedure 12.2 and from the parameters ζ_κ, β_κ, $g_{x\kappa}$, and $g_{y\kappa}$ obtained in the above (1).

References

1. R. Basri, Paraperspective ≡ affine. Int. J. Comput. Vision **19**(2), 169–179 (1996)
2. K. Kanatani, Y. Sugaya, H. Ackermann, Uncalibrated factorization using a variable symmetric affine camera. IEICE Trans. Inf. Syst. **E89-D**(10), 2653–2660 (2006)
3. C.J. Poelman, T. Kanade, A paraperspective factorization method for shape and motion recovery. IEEE Trans. Pattern Anal. Mach. Intell. **19**(3), 206–218 (1997)
4. Y. Sugaya, K. Kanatani, Extending interrupted feature point tracking for 3-D affine reconstruction. IEICE Trans. Inf. Syst. **E87-D**(4), 1031–1033 (2004)
5. C. Tomasi, T. Kanade, Shape and motion from image streams under orthography–A factorization method. Int. J. Comput. Vision **9**(2), 137–154 (1992)

Self-calibration of Perspective Cameras

13

Abstract

We extend the affine camera self-calibration technique of the preceding chapter to perspective projection. We show that the factorization of the preceding chapter can be applied to perspective cameras if we introduce new unknowns called projective depths. They are determined so that the observation matrix can be factorized, for which two approaches exist. One, called the primary method, iteratively determines the projective depths with the result that the observation matrix has column rank 4, and the other, called the dual method, iteratively determines them with the result that it has row rank 4. However, the reconstructed 3D shape is a projective transformation of the true shape, called projective reconstruction. The correct shape is obtained by applying an appropriate Euclidean upgrading. The entire self-calibration procedure requires a large number of iterations over a large number of unknowns, therefore the computational efficiency is the main concern. We discuss the complexity and efficiency of the involved computation.

13.1 Homogeneous Coordinates and Projective Reconstruction

In the preceding chapter, we considered the affine camera modeling of Eq. (12.4), whereas the perspective projection is modeled by Eq. (11.1). If we write the unknown constant implied by the relation $\simeq$ as $z_{\alpha\kappa}$, Eq. (11.1) is rewritten in the form

$$z_{\alpha\kappa}\begin{pmatrix} x_{\alpha\kappa}/f_0 \\ y_{\alpha\kappa}/f_0 \\ 1 \end{pmatrix} = \mathsf{P}_\kappa \mathsf{X}_\alpha. \tag{13.1}$$

The unknown constant $z_{\alpha\kappa}$ is called the *projective depth*. The vector X_α on the right-hand side is a 4D vector consisting of the 3D coordinates X_α, Y_α, and Z_α and a constant 1. However, if we introduce the projective depth $z_{\alpha\kappa}$ as an unknown, we

K. Kanatani et al., *Guide to 3D Vision Computation*, Advances in Computer Vision and Pattern Recognition, DOI 10.1007/978-3-319-48493-8_13

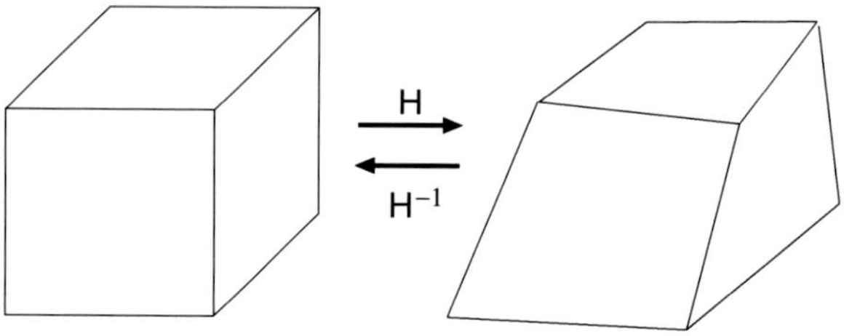

Fig. 13.1 Projective reconstruction. The reconstructed 3D shape is related to the true shape by an unknown homography H. The lengths, angles, and ratios may alter, but lines and planes are mapped to lines and planes

need not consider that the constraint's fourth component is 1, because multiplying X_α by a constant is equivalent to multiplying the unknowns $z_{\alpha\kappa}$ by that constant. In view of this, we specify the 3D position by the ratio $X_{\alpha(1)} : X_{\alpha(2)} : X_{\alpha(3)} : X_{\alpha(4)}$ of the components of X. This ratio is called the *homogeneous coordinate*. The actual 3D $(X_\alpha, Y_\alpha, Z_\alpha)$ is given by

$$X_\alpha = \frac{X_{\alpha(1)}}{X_{\alpha(4)}}, \qquad Y_\alpha = \frac{X_{\alpha(2)}}{X_{\alpha(4)}}, \qquad Z_\alpha = \frac{X_{\alpha(3)}}{X_{\alpha(4)}}. \tag{13.2}$$

If the fourth component $X_{\alpha(4)}$ of X_α is 0, we interpret X_α as representing a point infinitely far away, or a *point at infinity*, in the direction $(X_\alpha, Y_\alpha, Z_\alpha)$. By *self-calibration* of a perspective camera, we mean computing from points $(x_{\alpha\kappa}, y_{\alpha\kappa})$, $\alpha = 1, \ldots, N, \kappa = 1, \ldots, M$, observed in images, all the homogeneous coordinates X_α of the 3D positions and all the camera matrices P_κ, $\kappa = 1, ..., M$.

However, the matrix P_κ and the vector X_α that satisfy Eq. (13.1) are not unique, because if they are multiplied by an arbitrary 4×4 nonsingular matrix H in the form

$$\mathsf{P}'_\kappa = \mathsf{P}_\kappa \mathsf{H}, \qquad \mathsf{X}'_\alpha = \mathsf{H}^{-1}\mathsf{X}_\alpha. \tag{13.3}$$

the relation $\mathsf{P}'_\kappa \mathsf{X}'_\alpha = \mathsf{P}_\kappa \mathsf{X}_\alpha$ holds. Multiplying the homogeneous coordinate vector X_α by matrix H^{-1} means mapping the scene by a *homography*, or a *projective transformation*, which is an extension of the 2D homography of Eq. (7.1) to 3D. This means that the reconstructed 3D shape is related to the true shape by an unknown homography H. As in 2D, linearity and coplanarity are preserved by a homography (Fig. 13.1); that is, collinear points are mapped to collinear points, and coplanar points are mapped to coplanar points. However, lengths, angles, and ratios are not preserved such that a cube is mapped to a general hexahedron and a sphere is mapped to a general ellipsoid. Thus, the 3D shape reconstructed using Eq. (13.1) has indeterminacy of a homography and is called *projective reconstruction*. The operation of modifying it to the correct shape, called *Euclidean reconstruction*, is called *Euclidean upgrading*. Although the absolute lengths cannot be determined from images alone, the term "Euclidean" is commonly used for convenience, as in the case of affine reconstruction. Corresponding to the Euclidean upgrading of affine reconstruction, the following two approaches exist.

Use of knowledge about the scene If we know the lengths and angles (e.g., being orthogonal) of some parts, we multiply X_α by a homography matrix H^{-1} from the left such that the lengths and angles become correct.

Use of knowledge about the cameras If we know some properties about the cameras, we multiply the camera matrix P_κ by a homography matrix H from the right such that the expected properties hold.

In the following, we first show procedures for projective reconstruction and then describe procedures for Euclidean upgrading using knowledge about the cameras.

13.2 Projective Reconstruction by Factorization

We first summarize the principle of projective reconstruction by factorization and then describe the procedure of the primary and dual methods for computing it.

13.2.1 Principle of Factorization

We arrange the image coordinates $(x_{\alpha\kappa}, y_{\alpha\kappa})$ and the projective depth $z_{\alpha\kappa}$ of all points observed in all images in the following matrix form.

$$\mathsf{W} = \begin{pmatrix} z_{11}x_{11}/f_0 & z_{21}x_{21}/f_0 & \cdots & z_{N1}x_{N1}/f_0 \\ z_{11}y_{11}/f_0 & z_{21}y_{21}/f_0 & \cdots & z_{N1}y_{N1}/f_0 \\ z_{11} & z_{21} & \cdots & z_{N1} \\ \vdots & \vdots & \ddots & \vdots \\ z_{1M}x_{1M}/f_0 & z_{2M}x_{2M}/f_0 & \cdots & z_{NM}x_{NM}/f_0 \\ z_{1M}y_{1M}/f_0 & z_{2M}y_{2M}/f_0 & \cdots & z_{NM}y_{NM}/f_0 \\ z_{1M} & z_{2M} & \cdots & z_{NM} \end{pmatrix}. \tag{13.4}$$

Adopting the same terminology as in the case of affine cameras, we call this $3M \times N$ matrix the *observation matrix*. We also arrange the matrix P_κ of all the cameras and the homogeneous coordinate vector X_α of all the points as matrices in the form

$$\mathsf{M} = \begin{pmatrix} \mathsf{P}_1 \\ \vdots \\ \mathsf{P}_M \end{pmatrix}, \qquad \mathsf{S} = \begin{pmatrix} \mathsf{X}_1 & \cdots & \mathsf{X}_N \end{pmatrix}. \tag{13.5}$$

As in the case of affine reconstruction, we call the $3M \times 4$ matrix M and the $4 \times N$ matrix S the *motion matrix* and the *shape matrix*, respectively. From Eq. (13.1) and the above definition of the matrices S and M, we obtain

$$\mathsf{W} = \mathsf{MS}. \tag{13.6}$$

Hence, if the projective depths $z_{\alpha\kappa}$ are known, we can determine M and S by the singular value decomposition (SVD) of W, as we did in the preceding chapter, and obtain the camera matrices P_κ and the 3D positions X_α. Thus, the task is to find $z_{\alpha\kappa}$ such that the matrix M of Eq. (13.4) can be factored into the product of some $3M \times 4$ matrix M and some $4 \times N$ matrix S. The matrix W is factored in this way if and only if W has rank 4. The rank of a matrix is the number of its independent columns or independent rows. Thus we can take either of the following two approaches.

Primary method: We determine $z_{\alpha\kappa}$ so that the N columns of Eq. (13.4) span a 4D subspace; that is, they can be expressed as linear combinations of four basis column vectors.

Dual method: We determine $z_{\alpha\kappa}$ so that the $3M$ rows of Eq. (13.4) span a 4D subspace; that is, they can be expressed as linear combinations of four basis row vectors.

Once the projective depths $z_{\alpha\kappa}$ are given, we can determine the camera matrices P_κ and the 3D positions X_α, as in the case of affine cameras, by the following factorization of W.

Procedure 13.1 (Factorization)

1. Compute the SVD of the observation matrix W in the form

$$\mathsf{W} = \mathsf{U}_{3M\times L}\boldsymbol{\Sigma}_L\mathsf{V}_{N\times L}^\top, \qquad \boldsymbol{\Sigma}_L = \begin{pmatrix} \sigma_1 & \cdots & 0 \\ \vdots & \ddots & \vdots \\ 0 & \cdots & \sigma_L \end{pmatrix}, \tag{13.7}$$

where $L = \min(3M, N)$. Here, $\mathsf{U}_{3M\times L}$ and $\mathsf{V}_{N\times L}$ are, respectively, the $3M \times L$ and $N \times L$ matrices consisting of orthogonal columns, and $\boldsymbol{\Sigma}_L$ is a diagonal matrix with singular values $\sigma_1 \geq \cdots \geq \sigma_L$ (≥ 0) as diagonal elements in that order.

2. Let U be the $3M \times 4$ matrix consisting of the first four columns of $\mathsf{U}_{3M\times L}$, and V the $N \times 4$ matrix consisting of the first four columns of $\mathsf{V}_{N\times L}$. Let

$$\boldsymbol{\Sigma} = \begin{pmatrix} \sigma_1 & 0 & 0 & 0 \\ 0 & \sigma_2 & 0 & 0 \\ 0 & 0 & \sigma_3 & 0 \\ 0 & 0 & 0 & \sigma_4 \end{pmatrix}. \tag{13.8}$$

3. Determine the motion matrix M and the shape matrix S by

$$\mathsf{M} = \mathsf{U}, \qquad \mathsf{S} = \boldsymbol{\Sigma}\mathsf{V}^\top, \qquad \text{or} \tag{13.9}$$

$$\mathsf{M} = \mathsf{U}\boldsymbol{\Sigma}, \qquad \mathsf{S} = \mathsf{V}^\top. \tag{13.10}$$

4. Determine the camera matrices P_κ and the 3D positions X_α from Eq. (13.5).

Comments If the projective depths $z_{\alpha\kappa}$ are correctly computed, the observation matrix W should have rank 4, and the singular values of Eq. (13.7) should be $\sigma_i = 0$, $i = 5, 6, \ldots, L$. However, this does not hold if $z_{\alpha\kappa}$ are not exact. Hence, we use SVD to obtain an approximate decomposition $\mathsf{W} \approx \mathsf{MS}$, letting $\sigma_i = 0$, $i = 5, 6, \ldots, L$. For projective reconstruction, M and S have indeterminacy of a homography, thus we can use either Eq. (13.9) or (13.10). Alternatively, we can let $\mathsf{M} = \mathsf{U}\boldsymbol{\Sigma}^{1/2}$ and $\mathsf{S} = \boldsymbol{\Sigma}^{1/2}\mathsf{V}^\top$, where $\boldsymbol{\Sigma}^{1/2}$ is the diagonal matrix obtained by replacing in Eq. (13.8) σ_1, ..., σ_4 by $\sqrt{\sigma_1}$, ..., $\sqrt{\sigma_4}$.

13.2.2 Primary Method

If we define the 3D vector $\mathbf{x}_{\alpha\kappa}$ by

$$\mathbf{x}_{\alpha\kappa} = \begin{pmatrix} x_{\alpha\kappa}/f_0 \\ y_{\alpha\kappa}/f_0 \\ 1 \end{pmatrix}, \tag{13.11}$$

the observation matrix of Eq. (13.4) is written as

$$\mathsf{W} = \begin{pmatrix} z_{11}\mathbf{x}_{11} & \cdots & z_{N1}\mathbf{x}_{N1} \\ \vdots & \ddots & \vdots \\ z_{1M}\mathbf{x}_{1M} & \cdots & z_{NM}\mathbf{x}_{NM} \end{pmatrix}. \tag{13.12}$$

The primary method for computing the projective depths $z_{\alpha\kappa}$ is given as follows. We first describe a procedure faithful to the principle. Techniques for improving efficiency are discussed later.

Procedure 13.2 (Primary method)

1. Give an admissible reprojection error ε (in pixels), and initialize the projective depths to $z_{\alpha\kappa} = 1$, $\alpha = 1, \ldots, N$, $\kappa = 1, \ldots, M$.
2. Normalize each column of the observation matrix W of Eq. (13.12) to a unit vector, and compute its SVD in the form of Eq. (13.7). Let $\mathbf{u}_1$, ..., $\mathbf{u}_4$ be the first four columns of the matrix $\mathsf{U}_{3M\times L}$.
3. Do the following computation for $\alpha = 1, \ldots, N$.

   a. Define the $M \times M$ matrix $\mathsf{A}^{(\alpha)} = (A^{(\alpha)}_{\kappa\lambda})$ by

   $$A^{(\alpha)}_{\kappa\lambda} = \frac{\sum_{i=1}^4 (\mathbf{x}_{\alpha\kappa}, \mathbf{u}_{i\kappa})(\mathbf{x}_{\alpha\lambda}, \mathbf{u}_{i\lambda})}{\|\mathbf{x}_{\alpha\kappa}\|\|\mathbf{x}_{\alpha\lambda}\|}, \tag{13.13}$$

   where $\mathbf{u}_{i\kappa}$ is the 3D vector consisting of the $(3(\kappa-1)+1)$, the $(3(\kappa-1)+2)$, and the $(3(\kappa-1)+3)$ components of the $3M$D vector $\mathbf{u}_i$ (i.e., the κth 3D segment of $\mathbf{u}_i$), $i = 1, 2, 3, 4$.

   b. Compute the unit eigenvector $\boldsymbol{\xi}_\alpha = (\xi_{\alpha\kappa})$ of the matrix $\mathsf{A}^{(\alpha)}$ for the largest eigenvalue, where the sign is chosen so that

   $$\sum_{\kappa=1}^{M} \xi_{\alpha\kappa} \geq 0. \tag{13.14}$$

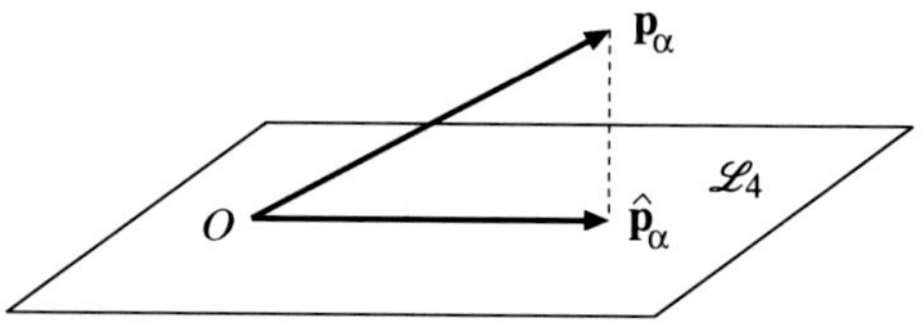

Fig. 13.2 Projection of vector $\mathbf{p}_\alpha$ onto the 4D space $\mathscr{L}_4$ spanned by $\mathbf{u}_1$, ..., $\mathbf{u}_4$

c. Update the projective depths $z_{\alpha\kappa}$ to

$$z_{\alpha\kappa} \leftarrow \frac{\xi_{\alpha\kappa}}{\|\mathbf{x}_{\alpha\kappa}\|}. \tag{13.15}$$

4. Determine the camera matrices P_κ and 3D positions X_α as in Steps 3 and 4 of Procedure 13.1.
5. Compute the reprojection error E by

$$E = f_0 \sqrt{\frac{1}{MN} \sum_{\alpha=1}^{N} \sum_{\kappa=1}^{M} \|\mathbf{x}_{\alpha\kappa} - \mathscr{Z}[\mathsf{P}_\kappa \mathsf{X}_\alpha]\|^2}, \tag{13.16}$$

where $\mathscr{Z}[\,\cdot\,]$ denotes normalization to make the third component 1 ($\mathscr{Z}[(a, b, c)^\top] = (a/c, b/c, 1)^\top$).
6. If $E < \varepsilon$, stop. Else, go back to Step 2.

Comments Initially letting $z_{\alpha\kappa} = 1$ is equivalent to assuming affine cameras. The homogeneous coordinate vectors X_α have scale indeterminacy, therefore the projective depths $z_{\alpha\kappa}$ do also. Hence, we do not lose generality if we normalize the columns of W of Eq. (13.12) to unit norm. Inasmuch as the space spanned by the N columns of W is equal to the space spanned by the N columns of the matrix $\mathsf{U}_{3M\times L}$ of Eq. (13.7), all the columns of W are included by the 4D space $\mathscr{L}_4$ spanned by $\mathbf{u}_1$, ..., $\mathbf{u}_4$ computed in Step 2, if the computed $z_{\alpha\kappa}$ are correct. Let $\mathbf{p}_\alpha$ be the αth column of W. As is well known, its projection onto the 4D space $\mathscr{L}_4$ has the form (Fig. 13.2):

$$\hat{\mathbf{p}}_\alpha = \sum_{i=1}^{4} (\mathbf{p}_\alpha, \mathbf{u}_i)\mathbf{u}_i. \tag{13.17}$$

because $\mathbf{u}_1, \ldots, \mathbf{u}_4$ are an orthonormal basis of $\mathscr{L}_4$, the projected length of $\mathbf{p}_\alpha$, which is normalized to $\|\mathbf{p}_\alpha\| = 1$, onto $\mathscr{L}_4$ is given by

$$\sqrt{\|\mathbf{p}_\alpha\|^2 - \|\hat{\mathbf{p}}_\alpha\|^2} = \sqrt{1 - \sum_{i=1}^{4} (\mathbf{p}_\alpha, \mathbf{u}_i)^2}. \tag{13.18}$$

This should be 0 if $z_{\alpha\kappa}$ are correct. Therefore we determine $z_{\alpha\kappa}$ by maximizing

$$J_\alpha = \sum_{i=1}^{4} (\mathbf{p}_\alpha, \mathbf{u}_i)^2 = \sum_{i=1}^{4} \Big(\sum_{\kappa=1}^{M} (z_{\alpha\kappa} \mathbf{x}_{\alpha\kappa}, \mathbf{u}_{i\kappa}) \Big)^2$$
$$= \sum_{\kappa,\lambda=1}^{M} \Big(\sum_{i=1}^{4} (\mathbf{x}_{\alpha\kappa}, \mathbf{u}_{i\kappa})(\mathbf{x}_{\alpha\lambda}, \mathbf{u}_{i\lambda}) \Big) z_{\alpha\kappa} z_{\alpha\lambda}, \tag{13.19}$$

subject to the normalization condition

$$\|\mathbf{p}_\alpha\|^2 = \sum_{\kappa=1}^{M} z_{\alpha\kappa}^2 \|\mathbf{x}_{\alpha\kappa}\|^2 = 1. \tag{13.20}$$

If we define new variables $\xi_{\alpha\kappa}$ by

$$\xi_{\alpha\kappa} = \|\mathbf{x}_{\alpha\kappa}\| z_{\alpha\kappa}, \tag{13.21}$$

and write $\boldsymbol{\xi}_\alpha$ for the MD vector consisting of $\xi_{\alpha 1}, \ldots, \xi_{\alpha M}$, Eq. (13.20) is equivalently written as $\|\boldsymbol{\xi}_\alpha\| = 1$. If we define the matrix $\mathsf{A}^{(\alpha)}$ by Eq. (13.13), Eq. (13.19) is rewritten as

$$J_\alpha = \sum_{\kappa,\lambda=1}^{M} A_{\kappa\lambda}^{(\alpha)} \xi_{\alpha\kappa} \xi_{\alpha\lambda} = (\boldsymbol{\xi}_\alpha, \mathsf{A}^{(\alpha)} \boldsymbol{\xi}_\alpha). \tag{13.22}$$

This is a quadratic form in $\boldsymbol{\xi}_\alpha$ and thus is maximized by the unit eigenvector $\boldsymbol{\xi}_\alpha$ of the matrix $\mathsf{A}^{(\alpha)}$ for the largest eigenvalue, but eigenvectors have sign indeterminacy. Considering that $z_{\alpha\kappa} = 1$ for affine cameras, we choose the sign so that Eq. (13.14) holds.

In each iteration, we update the basis $\mathbf{u}_1, \ldots, \mathbf{u}_4$ of $\mathscr{L}_4$ and recompute all the $\boldsymbol{\xi}_\alpha$. This is a variant of what is called the *EM algorithm*, which guarantees the reprojection error E to decrease monotonically. However, the convergence is very slow. There exist many techniques for improving efficiency. One is the improvement of the eigenvalue computation of the $M \times M$ matrix $\mathsf{A}^{(\alpha)}$ in Step 3(b). We repeat this computation N times for $\alpha = 1, \ldots, N$, therefore the burden of straightforward computation becomes very heavy when M and N are large. This can be alleviated by reducing the eigenvalue computation of an $M \times M$ matrix to the SVD of an $M \times 4$ matrix (↪ Problem 13.1). This greatly improves the efficiency.

Another consideration is the SVD of the $3M \times N$ matrix W in Step 2, which is done in each iteration step. The computational burden is heavy when M and N are large, but what we need to update are only the first four columns $\mathbf{u}_1, \ldots, \mathbf{u}_4$. These are eigenvectors of $\mathsf{W}\mathsf{W}^\top$, and all the eigenvalues for $\mathbf{u}_5, \ldots, \mathbf{u}_L$ are expected to be very small. We can improve the efficiency by taking advantage of this fact (↪ Problem 13.2). Note that in Step 4 we can update P_κ and X_α from $\mathbf{u}_1, \ldots, \mathbf{u}_4$ alone without factorizing W (↪ Exercise 13.3). How these techniques are effective depends on the number N of the points and the number M of the images. In general, more efficiency is achieved as N and M become large.

13.2.3 Dual Method

Let $\mathbf{q}_{\kappa(1)}$, $\mathbf{q}_{\kappa(2)}$, and $\mathbf{q}_{\kappa(3)}$ be the ND vectors obtained by vertically arranging the $3(\kappa-1)+1$, the $3(\kappa-1)+2$, and the $3(\kappa-1)+3$ rows of the observation matrix W of Eq. (13.4), respectively (i.e., the κth triplet of rows transposed to columns):

$$\mathbf{q}_{\kappa(1)} = \begin{pmatrix} z_{1\kappa}x_{1\kappa}/f_0 \\ \vdots \\ z_{N\kappa}x_{N\kappa}/f_0 \end{pmatrix}, \quad \mathbf{q}_{\kappa(2)} = \begin{pmatrix} z_{1\kappa}y_{1\kappa}/f_0 \\ \vdots \\ z_{N\kappa}y_{N\kappa}/f_0 \end{pmatrix}, \quad \mathbf{q}_{\kappa(3)} = \begin{pmatrix} z_{1\kappa} \\ \vdots \\ z_{N\kappa} \end{pmatrix}. \tag{13.23}$$

Then, W is written as

$$\mathsf{W} = \begin{pmatrix} \mathbf{q}_{1(1)}^\top \\ \mathbf{q}_{1(2)}^\top \\ \mathbf{q}_{1(3)}^\top \\ \vdots \\ \mathbf{q}_{N(1)}^\top \\ \mathbf{q}_{N(2)}^\top \\ \mathbf{q}_{N(3)}^\top \end{pmatrix}. \tag{13.24}$$

The dual method for computing the projective depths $z_{\alpha\kappa}$ is given as follows. We first describe a procedure faithful to the principle. Techniques for improving efficiency are discussed later.

Procedure 13.3 (Dual method)

1. Give an admissible reprojection error ε (in pixels), and initialize the projective depths to $z_{\alpha\kappa} = 1$, $\alpha = 1, \ldots, N$, $\kappa = 1, \ldots, M$.
2. For each κ, normalize the three vectors of Eq. (13.23) to

$$\sum_{i=1}^{3} \|\mathbf{q}_{\kappa(i)}\|^2 = 1, \tag{13.25}$$

   and compute the SVD of the observation matrix W of Eq. (13.24) in the form of Eq. (13.7). Let $\mathbf{v}_1$, ..., $\mathbf{v}_4$ be the first four columns of the matrix $\mathsf{V}_{N\times L}$.
3. Do the following computation for $\alpha = 1, \ldots, N$.

   a. Define the $N \times N$ matrix $\mathsf{B}^{(\kappa)} = (B_{\alpha\beta}^{(\kappa)})$ by

$$B_{\alpha\beta}^{(\kappa)} = \frac{(\mathbf{v}_\alpha, \mathbf{v}_\beta)(\mathbf{x}_{\alpha\kappa}, \mathbf{x}_{\beta\kappa})}{\|\mathbf{x}_{\alpha\kappa}\|\|\mathbf{x}_{\beta\kappa}\|}, \tag{13.26}$$

   where $\mathbf{v}_\alpha$ is the 4D vector consisting of the αth components of the vectors $\mathbf{v}_i$, $i = 1, 2, 3, 4$, and $\mathbf{x}_{\alpha\kappa}$ is the 3D vector of Eq. (13.11).

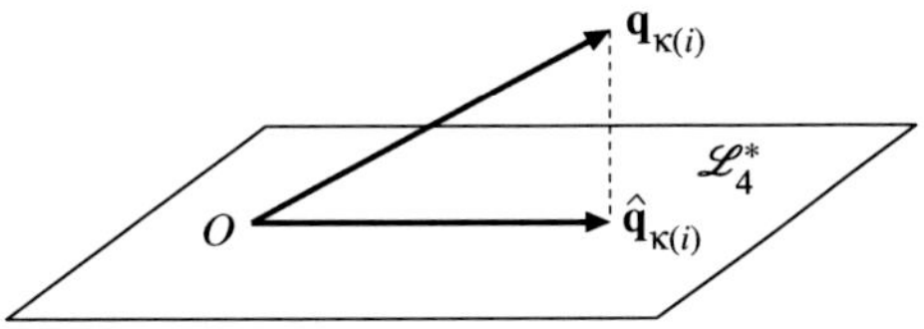

Fig. 13.3 Projection of vector $\mathbf{q}_{\kappa(i)}$ onto the 4D space $\mathscr{L}_4^*$ spanned by $\mathbf{v}_1, \ldots, \mathbf{v}_4$

b. Compute the unit eigenvector $\boldsymbol{\xi}_\kappa = (\xi_{\alpha\kappa})$ of the matrix $\mathsf{B}^{(\kappa)}$ for the largest eigenvalue, where the sign is chosen so that

$$\sum_{\kappa=1}^{M} \xi_{\alpha\kappa} \geq 0. \tag{13.27}$$

c. Update the projective depths $z_{\alpha\kappa}$ to

$$z_{\alpha\kappa} \leftarrow \frac{\xi_{\alpha\kappa}}{\|\mathbf{x}_{\alpha\kappa}\|}. \tag{13.28}$$

4. Determine the camera matrices P_κ and the 3D positions X_α as in Steps 3 and 4 of Procedure 13.1.
5. Compute the reprojection error E by

$$E = f_0 \sqrt{\frac{1}{MN} \sum_{\alpha=1}^{N} \sum_{\kappa=1}^{M} \|\mathbf{x}_{\alpha\kappa} - \mathscr{Z}[\mathsf{P}_\kappa \mathsf{X}_\alpha]\|^2}. \tag{13.29}$$

6. If $E < \varepsilon$, stop. Else, go back to Step 2.

Comments As in the case of the primary method, we initially let $z_{\alpha\kappa} = 1$ start from the affine camera modeling. Because the projective depths $z_{\alpha\kappa}$ have scale indeterminacy, we do not lose generality if we normalize each triplet of the rows of the observation matrix W of Eq. (13.24) as in Eq. (13.25) for each κ. Inasmuch as the space spanned by the $3M$ rows of W is equal to the space spanned by the $3M$ columns of the matrix $\mathsf{V}_{N\times L}$ of Eq. (13.7), all the rows of W, regarded as columns, are included in the 4D space $\mathscr{L}_4^*$ spanned by $\mathbf{v}_1, \ldots, \mathbf{v}_4$ computed in Step 2, if the computed $z_{\alpha\kappa}$ are correct. The projection of the vector $\mathbf{q}_{\kappa(i)}$, which is the row of W regarded as a column, onto the 3D space $\mathscr{L}_4^*$ has the following form (Fig. 13.3).

$$\hat{\mathbf{q}}_{\kappa(i)} = \sum_{k=1}^{4} (\mathbf{q}_{\kappa(i)}, \mathbf{v}_k)\mathbf{v}_k. \tag{13.30}$$

The sum of the square distances of $\mathbf{q}_{\kappa(i)}$ from $\hat{\mathbf{q}}_{\kappa(i)}$, $i = 1, 2, 3$, is

$$\sum_{i=1}^{3} \left(\|\mathbf{q}_{\kappa(i)}\|^2 - \|\hat{\mathbf{q}}_{\kappa(i)}\|^2 \right) = \sum_{i=1}^{3} \|\mathbf{q}_{\kappa(i)}\|^2 - \sum_{i=1}^{3} \sum_{k=1}^{4} (\mathbf{q}_{\kappa(i)}, \mathbf{v}_k)^2. \tag{13.31}$$

If $z_{\alpha\kappa}$ are correct, $\mathbf{q}_{\kappa(i)}$ are included in $\mathscr{L}_4^*$, and hence the above sum of squares should be 0. Therefore we minimize Eq. (13.31) subject to the normalization condition of Eq. (13.25). This is equivalent to maximizing

$$J_\kappa = \sum_{i=1}^{3}\sum_{k=1}^{4}(\mathbf{q}_{\kappa(i)}, \mathbf{v}_k)^2. \tag{13.32}$$

If we define $\xi_{\alpha\kappa}$ by Eq. (13.21) and write $\boldsymbol{\xi}_\kappa$ for the ND vector consisting of $\xi_{1\kappa}$, ..., $\xi_{N\kappa}$, we can see from Eqs. (13.23) and (13.21) that Eq. (13.25) is written as $\|\boldsymbol{\xi}_\kappa\| = 1$ (↪ Problem 13.4). If we define the matrix $\mathsf{B}^{(\kappa)}$ by Eq. (13.26), Eq. (13.32) is rewritten as follows (↪ Problem 13.5).

$$J_\kappa = (\boldsymbol{\xi}_\kappa, \mathsf{B}^{(\kappa)}\boldsymbol{\xi}_\kappa). \tag{13.33}$$

The unit vector $\boldsymbol{\xi}_\kappa$ that maximizes this is given by the unit eigenvector of the matrix $\mathsf{B}^{(\kappa)}$ for the largest eigenvalue, but eigenvalues have sign indeterminacy. Considering that $z_{\alpha\kappa} = 1$ for affine cameras we choose the sign so that Eq. (13.27) holds.

The above computation is basically the same as the primary method, except that the roles of the columns and the rows of the observation matrix W are interchanged. As with the primary method, the convergence is slow. We can introduce the same type of technique for improving efficiency that we used for the primary method. First, we can decrease the amount of computation by reducing the eigenvalue computation of the $N \times N$ matrix $\mathsf{B}^{(\kappa)}$ in Step 3(b) to the SVD (↪ Problem 13.6). We can also introduce to the SVD of the $3M \times N$ matrix W in Step 2 a technique similar to that for the primary method, noting that we need to update only the first four columns $\mathbf{v}_1$, ..., $\mathbf{v}_4$ of the matrix $\mathsf{V}_{N\times L}$ of Eq. (13.7). They are eigenvectors of $\mathsf{W}^\top\mathsf{W}$ and can be updated without factoring W (↪ Problem 13.7). Note that in Step 4 we can update P_κ and X_α from $\mathbf{v}_1$, ..., $\mathbf{v}_4$ alone without factorizing W (↪ Problem 13.8). The efficiency gain by these techniques depends on the number N of the points and the number M of the images; more efficiency is obtained as N and M become large.

13.3 Euclidean Upgrading

The projective reconstruction obtained above by the primary or dual method is a projective transformation of the true shape. We now describe the details of the Euclidean upgrading procedure that transforms it to a correct Euclidean reconstruction.

13.3.1 Principle of Euclidean Upgrading

As pointed out in Sect. 13.1, the 3D positions X_α and the camera matrices P_κ obtained by projective reconstruction are related to their true values by a homography H. Therefore their correct values (Euclidean reconstruction) are obtained by applying the inverse homography H^{-1} to them. We can find an H such the 3D positions X'_α of

Eq. (13.3) satisfy some known properties (knowledge about the scene) or the camera matrices P'_κ satisfy some known properties (knowledge about the cameras). Here, we adopt the latter approach.

The correct camera matrix has the form of Eq. (11.3). We first eliminate the translation $\mathbf{t}_\kappa$ and the rotation R_κ from it. The translation $\mathbf{t}_\kappa$ is eliminated by removing the fourth column of the 3×4 matrix P_κ. In Eq.

$$\mathsf{P}_\kappa \begin{pmatrix} 1 & 0 & 0 & 0 \\ 0 & 1 & 0 & 0 \\ 0 & 0 & 1 & 0 \\ 0 & 0 & 0 & 0 \end{pmatrix} = \mathsf{K}_\kappa \begin{pmatrix} \mathsf{R}_\kappa^\top & \mathbf{0} \end{pmatrix}. \tag{13.34}$$

The rotation R_κ is eliminated by multiplying this by its transpose; because $\mathsf{R}_\kappa^\top \mathsf{R}_\kappa = \mathsf{I}$, we obtain

$$\mathsf{P}_\kappa \begin{pmatrix} 1 & 0 & 0 & 0 \\ 0 & 1 & 0 & 0 \\ 0 & 0 & 1 & 0 \\ 0 & 0 & 0 & 0 \end{pmatrix} \mathsf{P}_\kappa^\top = \mathsf{K}_\kappa \mathsf{K}_\kappa^\top. \tag{13.35}$$

The camera matrix P_κ resulting from projective reconstruction may not satisfy this, therefore we compute a homography H such that the transformed matrix $\mathsf{P}'_\kappa = \mathsf{P}_\kappa \mathsf{H}$ of Eq. (13.3) satisfies this. Replacing P_κ of the above equation by P'_κ, and noting that the homography matrix H has scale indeterminacy, we obtain

$$\mathsf{P}_\kappa \boldsymbol{\Omega} \mathsf{P}_\kappa^\top \simeq \mathsf{K}_\kappa \mathsf{K}_\kappa^\top, \tag{13.36}$$

where we define the 4×4 matrix $\boldsymbol{\Omega}$ by

$$\boldsymbol{\Omega} \equiv \mathsf{H} \begin{pmatrix} 1 & 0 & 0 & 0 \\ 0 & 1 & 0 & 0 \\ 0 & 0 & 1 & 0 \\ 0 & 0 & 0 & 0 \end{pmatrix} \mathsf{H}^\top. \tag{13.37}$$

We first determine the matrix $\boldsymbol{\Omega}$ that satisfies Eq. (13.36) and then determine the matrix H that satisfies (13.37). From Eq. (13.36), we obtain in total M equations of this form for $\kappa = 1, \ldots, M$. However, we cannot solve them for $\boldsymbol{\Omega}$, because the right sides contain the intrinsic parameter matrices K_κ that are not yet known. Therefore we do the following iterations.

1. Using approximate values of the intrinsic parameter matrices K_κ for all κ, we compute $\boldsymbol{\Omega}$ from Eq. (13.36).
2. Using the computed $\boldsymbol{\Omega}$, we modify K_κ so that Eq. (13.36) is better satisfied.
3. Using the modified K_κ, we recompute $\boldsymbol{\Omega}$. We repeat this computation until Eq. (13.36) is well satisfied for all κ.

After $\boldsymbol{\Omega}$ is determined, we compute the matrix H that satisfies Eq. (13.37).

13.3.2 Computation of $\boldsymbol{\Omega}$

Assuming an approximate focal length f_κ and an approximate principal point $(u_{0\kappa}, v_{0\kappa})$ for each camera, we put the intrinsic parameter matrix of Eq. (11.3) in the form

$$\mathsf{K}_\kappa = \begin{pmatrix} f_\kappa & 0 & u_{0\kappa} \\ 0 & f_\kappa & v_{0\kappa} \\ 0 & 0 & f_0 \end{pmatrix}. \tag{13.38}$$

Using this and the camera matrix P_κ obtained by projective reconstruction, we compute the matrix $\boldsymbol{\Omega}$ that satisfies Eq. (13.36). Equation (13.36) is rewritten in the form:

$$\mathsf{K}_\kappa^{-1}\mathsf{P}_\kappa\boldsymbol{\Omega}^\top\mathsf{P}_\kappa^\top\mathsf{K}_\kappa^{-1\top} \simeq \mathsf{I}. \tag{13.39}$$

We obtain for each κ four equalities by requiring that the $(1, 1)$ and $(1, 2)$ elements on the left-hand side are the same and that nondiagonal elements are 0. These equalities are all linear in $\boldsymbol{\Omega} = (\Omega_{ij})$. Thus we obtain in total $4M$ linear equations in $\boldsymbol{\Omega}$. We solve them by the following least squares.

Procedure 13.4 (Computation of $\boldsymbol{\Omega}$)

1. Compute the following 3×4 matrix Q_κ, $\kappa = 1, \ldots, M$.

$$\mathsf{Q}_\kappa = \mathsf{K}_\kappa^{-1}\mathsf{P}_\kappa. \tag{13.40}$$

2. Define the $4 \times 4 \times 4 \times 4$ array $\mathscr{A} = (A_{ijkl})$ by

$$\begin{aligned} A_{ijkl} = \sum_{\kappa=1}^{M}\Big(& Q_{\kappa(1i)}Q_{\kappa(1j)}Q_{\kappa(1k)}Q_{\kappa(1l)} - Q_{\kappa(1i)}Q_{\kappa(1j)}Q_{\kappa(2k)}Q_{\kappa(2l)} \\ & - Q_{\kappa(2i)}Q_{\kappa(2j)}Q_{\kappa(1k)}Q_{\kappa(1l)} + Q_{\kappa(2i)}Q_{\kappa(2j)}Q_{\kappa(2k)}Q_{\kappa(2l)} \\ & + \frac{1}{4}(Q_{\kappa(1i)}Q_{\kappa(2j)}Q_{\kappa(1k)}Q_{\kappa(2l)} + Q_{\kappa(2i)}Q_{\kappa(1j)}Q_{\kappa(1k)}Q_{\kappa(2l)} \\ & + Q_{\kappa(1i)}Q_{\kappa(2j)}Q_{\kappa(2k)}Q_{\kappa(1l)} + Q_{\kappa(2i)}Q_{\kappa(1j)}Q_{\kappa(2k)}Q_{\kappa(1l)}) \\ & + \frac{1}{4}(Q_{\kappa(2i)}Q_{\kappa(3j)}Q_{\kappa(2k)}Q_{\kappa(3l)} + Q_{\kappa(3i)}Q_{\kappa(2j)}Q_{\kappa(2k)}Q_{\kappa(3l)} \\ & + Q_{\kappa(2i)}Q_{\kappa(3j)}Q_{\kappa(3k)}Q_{\kappa(2l)} + Q_{\kappa(3i)}Q_{\kappa(2j)}Q_{\kappa(3k)}Q_{\kappa(2l)}) \\ & + \frac{1}{4}(Q_{\kappa(3i)}Q_{\kappa(1j)}Q_{\kappa(3k)}Q_{\kappa(1l)} + Q_{\kappa(1i)}Q_{\kappa(3j)}Q_{\kappa(3k)}Q_{\kappa(1l)} \\ & + Q_{\kappa(3i)}Q_{\kappa(1j)}Q_{\kappa(1k)}Q_{\kappa(3l)} + Q_{\kappa(1i)}Q_{\kappa(3j)}Q_{\kappa(1k)}Q_{\kappa(3l)})\Big), \end{aligned} \tag{13.41}$$

where $Q_{\kappa(ij)}$ is the (i, j) element of Q_κ.

3. Define the following 10×10 matrix A.

$$
\mathsf{A} = \begin{pmatrix}
A_{1111} & A_{1122} & A_{1133} & A_{1144} & \sqrt{2}A_{1112} & \sqrt{2}A_{1113} & \sqrt{2}A_{1114} & \sqrt{2}A_{1123} & \sqrt{2}A_{1124} & \sqrt{2}A_{1134} \\
A_{2211} & A_{2222} & A_{2233} & A_{2244} & \sqrt{2}A_{2212} & \sqrt{2}A_{2213} & \sqrt{2}A_{2214} & \sqrt{2}A_{2223} & \sqrt{2}A_{2224} & \sqrt{2}A_{2234} \\
A_{3311} & A_{3322} & A_{3333} & A_{3344} & \sqrt{2}A_{3312} & \sqrt{2}A_{3313} & \sqrt{2}A_{3314} & \sqrt{2}A_{3323} & \sqrt{2}A_{3324} & \sqrt{2}A_{3334} \\
A_{4411} & A_{4422} & A_{4433} & A_{4444} & \sqrt{2}A_{4412} & \sqrt{2}A_{4413} & \sqrt{2}A_{4414} & \sqrt{2}A_{4423} & \sqrt{2}A_{4424} & \sqrt{2}A_{4434} \\
\sqrt{2}A_{1211} & \sqrt{2}A_{1222} & \sqrt{2}A_{1233} & \sqrt{2}A_{1244} & 2A_{1212} & 2A_{1213} & 2A_{1214} & 2A_{1223} & 2A_{1224} & 2A_{1234} \\
\sqrt{2}A_{1311} & \sqrt{2}A_{1322} & \sqrt{2}A_{1333} & \sqrt{2}A_{1344} & 2A_{1312} & 2A_{1313} & 2A_{1314} & 2A_{1323} & 2A_{1324} & 2A_{1334} \\
\sqrt{2}A_{1411} & \sqrt{2}A_{1422} & \sqrt{2}A_{1433} & \sqrt{2}A_{1444} & 2A_{1412} & 2A_{1413} & 2A_{1414} & 2A_{1423} & 2A_{1424} & 2A_{1434} \\
\sqrt{2}A_{2311} & \sqrt{2}A_{2322} & \sqrt{2}A_{2333} & \sqrt{2}A_{2344} & 2A_{2312} & 2A_{2313} & 2A_{2314} & 2A_{2323} & 2A_{2324} & 2A_{2334} \\
\sqrt{2}A_{2411} & \sqrt{2}A_{2422} & \sqrt{2}A_{2433} & \sqrt{2}A_{2444} & 2A_{2412} & 2A_{2413} & 2A_{2414} & 2A_{2423} & 2A_{2424} & 2A_{2434} \\
\sqrt{2}A_{3411} & \sqrt{2}A_{3422} & \sqrt{2}A_{3433} & \sqrt{2}A_{3444} & 2A_{3412} & 2A_{3413} & 2A_{3414} & 2A_{3423} & 2A_{3424} & 2A_{3434}
\end{pmatrix}. \tag{13.42}
$$

4. Compute the 10D unit eigenvector $\boldsymbol{\omega} = (\omega_i)$ of the matrix A for the smallest eigenvalue.
5. Determine the matrix $\boldsymbol{\Omega}$ as follows.

$$
\boldsymbol{\Omega} = \begin{pmatrix}
\omega_1 & \omega_5/\sqrt{2} & \omega_6/\sqrt{2} & \omega_7/\sqrt{2} \\
\omega_5/\sqrt{2} & \omega_2 & \omega_8/\sqrt{2} & \omega_9/\sqrt{2} \\
\omega_6/\sqrt{2} & \omega_8/\sqrt{2} & \omega_3 & \omega_{10}/\sqrt{2} \\
\omega_7/\sqrt{2} & \omega_9/\sqrt{2} & \omega_{10}/\sqrt{2} & \omega_4
\end{pmatrix}. \tag{13.43}
$$

6. Compute the unit eigenvectors $\mathbf{w}_1$, $\mathbf{w}_2$, $\mathbf{w}_3$, and $\mathbf{w}_4$ of $\boldsymbol{\Omega}$ for eigenvalues $\sigma_1 \geq \sigma_2 \geq \sigma_3 \geq \sigma_4$.
7. Redefine $\boldsymbol{\Omega}$ in the form

$$
\boldsymbol{\Omega} = \begin{cases} \sigma_1\mathbf{w}_1\mathbf{w}_1^\top + \sigma_2\mathbf{w}_2\mathbf{w}_2^\top + \sigma_3\mathbf{w}_3\mathbf{w}_3^\top & \sigma_3 > 0 \\ -\sigma_4\mathbf{w}_4\mathbf{w}_4^\top - \sigma_3\mathbf{w}_3\mathbf{w}_3^\top - \sigma_2\mathbf{w}_2\mathbf{w}_2^\top & \sigma_2 < 0 \end{cases}. \tag{13.44}
$$

Comments If we define the matrix Q_κ by Eq. (13.40), Eq. (13.39) is written as

$$
\mathsf{Q}_\kappa \boldsymbol{\Omega} \mathsf{Q}_\kappa^\top \simeq \mathsf{I}. \tag{13.45}
$$

Equating the (1,1) and (1,2) elements on the left-hand side and letting the nondiagonal elements be 0, we obtain

$$
\sum_{i,j=1}^{4} Q_{\kappa(1i)} Q_{\kappa(1j)} \Omega_{ij} - \sum_{i,j=1}^{4} Q_{\kappa(2i)} Q_{\kappa(2j)} \Omega_{ij} = 0,
$$

$$\sum_{i,j=1}^{4} Q_{\kappa(1i)} Q_{\kappa(2j)} \Omega_{ij} = 0, \quad \sum_{i,j=1}^{4} Q_{\kappa(2i)} Q_{\kappa(3j)} \Omega_{ij} = 0, \quad \sum_{i,j=1}^{4} Q_{\kappa(3i)} Q_{\kappa(1j)} \Omega_{ij} = 0. \tag{13.46}$$

There may not exist a $\boldsymbol{\Omega}$ that exactly satisfies all these, therefore we minimize the sum of the squares of the left-hand sides for $\kappa = 1, \ldots, M$. In terms of the array A_{ijkl} defined by (13.41), the sum of squares is written as

$$K = \sum_{i,j,k,l=1}^{4} A_{ijkl} \Omega_{ij} \Omega_{kl}. \tag{13.47}$$

If we define the 10×10 symmetric matrix A by Eq. (13.42) and the 10D vector $\boldsymbol{\omega} = (\omega_i)$ by Eq. (13.43), Eq. (13.47) is expressed as a quadratic form in $\boldsymbol{\omega}$ in the form

$$K = (\boldsymbol{\omega}, \mathsf{A}\boldsymbol{\omega}). \tag{13.48}$$

As seen from Eq. (13.45), the matrix $\boldsymbol{\Omega}$ is determined only up to scale, therefore we normalize it to $\|\boldsymbol{\Omega}\|^2 = \sum_{i,j=1}^{4} \Omega_{ij}^2 = 1$. From Eq. (13.43), this is equivalent to $\|\boldsymbol{\omega}\|^2 = 1$. The unit vector $\boldsymbol{\omega}$ that minimizes Eq. (13.48) is given by the unit eigenvector of the matrix A for the smallest eigenvalue. Hence, we obtain $\boldsymbol{\Omega}$ by arranging the components of $\boldsymbol{\omega}$ in the form of Eq. (13.43). However, the definition of Eq. (13.37) implies that $\boldsymbol{\Omega}$ is a symmetric semi-definite matrix of rank 3. Letting $\sigma_1 \geq \cdots \geq \sigma_4$ be the eigenvalues of $\boldsymbol{\Omega}$, and $\boldsymbol{\omega}_1, \ldots, \boldsymbol{\omega}_4$ the corresponding unit eigenvectors, we can write $\boldsymbol{\Omega}$ in the form

$$\boldsymbol{\Omega} = \sigma_1 \mathbf{w}_1 \mathbf{w}_1^\top + \sigma_2 \mathbf{w}_2 \mathbf{w}_2^\top + \sigma_3 \mathbf{w}_3 \mathbf{w}_3^\top + \sigma_4 \mathbf{w}_4 \mathbf{w}_4^\top. \tag{13.49}$$

Then we remove the eigenvector for the smallest eigenvalue and redefine $\boldsymbol{\Omega}$ in terms of the eigenvectors for the largest three eigenvalues. However, the eigenvector $\boldsymbol{\omega}$ has an indeterminate sign, and hence the matrix $\boldsymbol{\Omega}$ of Eq. (13.43) also has sign indeterminacy. Therefore we choose the sign of $\boldsymbol{\Omega}$ so that its eigenvalues are positive and force its rank to be 3 in the form of Eq. (13.44).

13.3.3 Modification of K_κ

Because the assumed value of K_κ may not be exact, Eq. (13.36) may not be strictly satisfied for the computed $\boldsymbol{\Omega}$. Therefore we modify K_κ by a small amount such that Eq. (13.36) is better satisfied. To this end, we multiply the current value K_κ by a correction term $\delta\mathsf{K}_\kappa$, which is close to the identity matrix, in the form of $\mathsf{K}_\kappa \delta\mathsf{K}_\kappa$ and determine $\delta\mathsf{K}_\kappa$ such that

$$\mathsf{P}_\kappa \boldsymbol{\Omega} \mathsf{P}_\kappa \simeq (\mathsf{K}_\kappa \delta\mathsf{K}_\kappa)(\mathsf{K}_\kappa \delta\mathsf{K}_\kappa)^\top \tag{13.50}$$

is better satisfied. This is rewritten as

$$\mathsf{K}_\kappa^{-1} \mathsf{P}_\kappa \boldsymbol{\Omega} \mathsf{P}_\kappa \mathsf{K}_\kappa^{\top -1} \simeq \delta\mathsf{K}_\kappa \delta\mathsf{K}_\kappa. \tag{13.51}$$

Inasmuch as the left-hand side equals $\mathsf{Q}_\kappa \boldsymbol{\Omega} \mathsf{Q}_\kappa^\top$, we determine $\delta \mathsf{K}_\kappa$ such that $\mathsf{Q}_\kappa \boldsymbol{\Omega} \mathsf{Q}_\kappa^\top$ is close to a scalar multiple of $\delta \mathsf{K}_\kappa \delta \mathsf{K}_\kappa$. The procedure is as follows.

Procedure 13.5 (Modification of K_κ)

1. Let the elements of the computed $\mathsf{Q}_\kappa \boldsymbol{\Omega} \mathsf{Q}_\kappa^\top$ be
$$\mathsf{Q}_\kappa \boldsymbol{\Omega} \mathsf{Q}_\kappa^\top = \begin{pmatrix} c_{\kappa(11)} & c_{\kappa(12)} & c_{\kappa(13)} \\ c_{\kappa(21)} & c_{\kappa(22)} & c_{\kappa(23)} \\ c_{\kappa(31)} & c_{\kappa(32)} & c_{\kappa(33)} \end{pmatrix}. \tag{13.52}$$
2. Compute the following F_κ.
$$F_\kappa = \frac{c_{\kappa(11)} + c_{\kappa(22)}}{c_{\kappa(33)}} - \left(\frac{c_{\kappa(13)}}{c_{\kappa(33)}}\right)^2 - \left(\frac{c_{\kappa(23)}}{c_{\kappa(33)}}\right)^2. \tag{13.53}$$
3. If $c_{\kappa(33)} \le 0$ or $F_\kappa \le 0$, stop without modifying K_κ.
4. Else, compute the correction terms of the principal point $(u_{0\kappa}, v_{0\kappa})$ and the focal length f_κ in the form
$$\delta u_{0\kappa} = \frac{c_{\kappa(13)}}{c_{\kappa(33)}}, \qquad \delta v_{0\kappa} = \frac{c_{\kappa(23)}}{c_{\kappa(33)}},$$
$$\delta f_\kappa = \sqrt{\frac{1}{2}\left(\frac{c_{\kappa(11)} + c_{\kappa(22)}}{c_{\kappa(33)}} - \delta u_{0\kappa}^2 - \delta v_{0\kappa}^2\right)}. \tag{13.54}$$
5. Compute $\delta \mathsf{K}_\kappa$ in the form
$$\delta \mathsf{K}_\kappa = \begin{pmatrix} \delta f_\kappa & 0 & \delta u_{0\kappa} \\ 0 & \delta f_\kappa & \delta v_{0\kappa} \\ 0 & 0 & 1 \end{pmatrix}. \tag{13.55}$$
6. Modify K_κ in the form
$$\mathsf{K}_\kappa \leftarrow \mathsf{K}_\kappa \delta \mathsf{K}_\kappa, \qquad \mathsf{K}_\kappa \leftarrow \sqrt{c_{\kappa(33)}}\mathsf{K}_\kappa. \tag{13.56}$$

Comments Considering Eq. (13.38), we put the correction term of K_κ in the form of Eq. (13.55), where $\delta u_{0\kappa}$ and $\delta v_{0\kappa}$ are close to 0 and δf_κ is close to 1. Then $\delta \mathsf{K}_\kappa \delta \mathsf{K}_\kappa^\top$ has the form
$$\delta \mathsf{K}_\kappa \delta \mathsf{K}_\kappa^\top = \begin{pmatrix} \delta f_\kappa^2 + \delta u_{0\kappa}^2 & \delta u_{0\kappa} \delta v_{0\kappa} & \delta u_{0\kappa} \\ \delta u_{0\kappa} \delta v_{0\kappa} & \delta f_\kappa^2 + \delta v_{0\kappa}^2 & \delta v_{0\kappa} \\ \delta u_{0\kappa} & \delta v_{0\kappa} & 1 \end{pmatrix}. \tag{13.57}$$
In order to make this close to a scalar multiple of Eq. (13.52), we determine $\delta u_{0\kappa}$ and $\delta v_{0\kappa}$ by the first and second equations of Eq. (13.54). For δf_κ, we take the average of the (1, 1) and (2, 2) elements and determine it by the third equation of Eq. (13.54). However, we do not modify K_κ if the (3, 3) element of Eq. (13.52) is not positive or if the inside of the square root of Eq. (13.54) is not positive. Note that due to scale indeterminacy only ratios between elements of Eq. (13.52) have meaning. However, it is not computationally desirable if the elements become too large or too small even though the ratios are the same. Therefore we divide Q_κ by $\sqrt{c_{\kappa(33)}}$ so that Eq. (13.52) becomes closer to the identity matrix. Eq. (13.40) implies that this is equivalent to multiplying K_κ by $\sqrt{c_{\kappa(33)}}$, which we do at the end of the computation of Eq. (13.56).

13.3.4 Computation of H

We need to determine the 4×4 homography matrix H from the computed $\boldsymbol{\Omega}$ such that Eq. (13.37) holds. If we let $\mathbf{h}_1$, $\mathbf{h}_2$, $\mathbf{h}_3$, $\mathbf{h}_4$ be the columns of the matrix H, we can write Eq. (13.37) in the form

$$\boldsymbol{\Omega} = \mathbf{h}_1\mathbf{h}_1^\top + \mathbf{h}_2\mathbf{h}_2^\top + \mathbf{h}_3\mathbf{h}_3^\top. \tag{13.58}$$

This implies that we cannot determine the fourth column $\mathbf{h}_4$. This is because we eliminated the translation $\mathbf{t}_\kappa$ by Eq. (13.34). This corresponds to the fact that the absolute position of the world coordinate system is indeterminate. Therefore we can arbitrarily define $\mathbf{h}_4$ such that H is nonsingular. If we note that $\boldsymbol{\Omega}$ can be expressed in the form of Eq. (13.49) in terms of its eigenvalues and eigenvectors, the simplest way is to choose $\boldsymbol{\Omega}$ such that σ_1, σ_2, and σ_3 are positive and regard $\sqrt{\sigma_1}\mathbf{w}_1$, $\sqrt{\sigma_2}\mathbf{w}_2$, and $\sqrt{\sigma_3}\mathbf{w}_3$ as $\mathbf{h}_1$, $\mathbf{h}_2$, and $\mathbf{h}_3$, respectively. Then, we let $\mathbf{w}_4$ be $\mathbf{h}_4$. To be specific, we let

$$\mathsf{H} = \begin{cases} \begin{pmatrix} \sqrt{\sigma_1}\mathbf{w}_1 & \sqrt{\sigma_2}\mathbf{w}_2 & \sqrt{\sigma_3}\mathbf{w}_3 & \mathbf{w}_4 \end{pmatrix} & \sigma_3 > 0 \\ \begin{pmatrix} \sqrt{-\sigma_4}\mathbf{w}_4 & \sqrt{-\sigma_3}\mathbf{w}_3 & \sqrt{-\sigma_2}\mathbf{w}_2 & \mathbf{w}_1 \end{pmatrix} & \sigma_2 < 0 \end{cases}, \tag{13.59}$$

in correspondence with Eq. (13.44).

13.3.5 Procedure for Euclidean Upgrading

Euclidean upgrading can be done by combining the above procedures. Many possibilities and variants are conceivable. The following is a typical example.

Procedure 13.6 (Euclidean upgrading)

1. Let $\hat{J}_{\text{med}} = \infty$ (a sufficiently large number), and give an initial estimate of the internal parameter matrix K_κ in the form of Eq. (13.38), $\kappa = 1, \ldots, M$.
2. Determine $\boldsymbol{\Omega}$ by Procedure 13.4.
3. Using the values obtained in Procedure 13.4, determine H by Eq. (13.59).
4. Modify K_κ by Procedure 13.5 for each κ.
5. If K_κ is modified by Procedure 13.5, compute the following J_κ for each κ, using the values obtained in the procedure.

$$J_\kappa = \left(\frac{c_{\kappa(11)}}{c_{\kappa(33)}} - 1\right)^2 + \left(\frac{c_{\kappa(22)}}{c_{\kappa(33)}} - 1\right)^2 + 2\frac{c_{\kappa(12)}^2 + c_{\kappa(23)}^2 + c_{\kappa(31)}^2}{c_{\kappa(33)}^2}. \tag{13.60}$$

   If K_κ is not modified, let $J_\kappa = \infty$ (a sufficiently large number).
6. Compute the following median.

$$J_{\text{med}} = \text{med}_{\kappa=1}^{M} J_\kappa. \tag{13.61}$$

7. If $J_{\text{med}} \approx 0$, return H and K_κ, and stop.
8. Else, if $J_{\text{med}} \geq \hat{J}_{\text{med}}$, return H and K_κ, and Step
9. Else, let $\hat{J}_{\text{med}} \leftarrow J_{\text{med}}$, and go back to Step 2.

Comments The value F_κ of Eq. (13.53) measures how close Eq. (13.52) is to a scalar multiple to the identity matrix. Ideally, this procedure should be iterated until $F_\kappa \approx 0$ for all κ. However, this is often not achieved for real data computation. There are many causes for this. For one thing, the image positions of the points may not be correct due to errors in image matching. For another, the camera matrices P_κ computed by the projective reconstruction of Sect. 13.2, whether using the primary or the dual method, may not necessarily be sufficiently accurate. The use of the median in Step 6 is for removing "bad" images for which the point positions or the camera matrices P_κ are not sufficiently correct. Step 7 requires that F_κ be smaller than the specified threshold for a majority of the images. However, even that threshold may not be reached in the presence of large errors. Therefore we stop the iterations when the median no longer decreases.

One technique for improving the convergence is introduction of a weight W_κ to each term in $\sum_{\kappa=1}^{M}(\cdots)$ of Eq. (13.41) for computing A_{ijkl}. We require the weight W_κ to be small for "bad" images that may involve errors and large for other "good" images. One such choice is

$$W_\kappa = e^{-J_\kappa/J_{\rm med}}, \tag{13.62}$$

using the values J_κ and $J_{\rm med}$ in the preceding steps.

13.4 3D Reconstruction Computation

Once the homography matrix H for Euclidean upgrading and the internal parameter matrices K_κ have been obtained, we can compute the 3D position $(X_\alpha, Y_\alpha, Z_\alpha)$ of each point and the translation $\mathbf{t}_\kappa$ and the rotation R_κ of each camera by transforming the homogeneous coordinate vector X_α and the camera matrix P_κ obtained by projective reconstruction in the following way.

Procedure 13.7 (3D reconstruction)

1. Compute the projective transformation of each X_α in the form
$$\mathsf{X}_\alpha \leftarrow \mathsf{H}^{-1}\mathsf{X}_\alpha. \tag{13.63}$$
2. Let $X_{\alpha(1)}$, $X_{\alpha(2)}$, $X_{\alpha(3)}$, and $X_{\alpha(4)}$ be the components of the resulting X_α, and compute the 3D coordinates $(X_\alpha, Y_\alpha, Z_\alpha)$ by Eq. (13.2).
3. Compute the following projective transformation of each P_κ in the form
$$\mathsf{P}_\kappa \leftarrow \mathsf{P}_\kappa\mathsf{H}. \tag{13.64}$$
4. Compute the following matrix A_κ and the vector $\mathbf{b}_\kappa$.
$$\mathsf{K}_\kappa^{-1}\mathsf{P}_\kappa = \begin{pmatrix} \mathsf{A}_\kappa & \mathbf{b}_\kappa \end{pmatrix}. \tag{13.65}$$

5. Let the scaling constant s be

$$s = \sqrt[3]{\det \mathsf{A}_\kappa}, \tag{13.66}$$

and normalize A_κ and $\mathbf{b}_\kappa$ as follows.

$$\mathsf{A}_\kappa \leftarrow \frac{\mathsf{A}_\kappa}{s}, \qquad \mathbf{b}_\kappa \leftarrow \frac{\mathbf{b}_\kappa}{s}. \tag{13.67}$$

6. Compute the following SVD (U_A and V_A are orthogonal matrices, and $\boldsymbol{\Sigma}_A$ is a diagonal matrix consisting of singular values).

$$\mathsf{A}_\kappa = \mathsf{U}_A \boldsymbol{\Sigma}_A \mathsf{V}_A^\top. \tag{13.68}$$

7. Let the rotation R_κ be

$$\mathsf{R}_\kappa = \mathsf{V}_A \mathsf{U}_A^\top. \tag{13.69}$$

8. Let the translation $\mathbf{t}_\kappa$ be

$$\mathbf{t}_\kappa = -\mathsf{R}_\kappa \mathbf{b}. \tag{13.70}$$

9. Compute the 3D position of the point $(X_\alpha, Y_\alpha, Z_\alpha)$ relative to the camera coordinate system of the κth camera by

$$\begin{pmatrix} X_{\alpha\kappa} \\ Y_{\alpha\kappa} \\ Z_{\alpha\kappa} \end{pmatrix} = \mathsf{R}_\kappa^\top \left(\begin{pmatrix} X_\alpha \\ Y_\alpha \\ Z_\alpha \end{pmatrix} - \mathbf{t}_\kappa \right). \tag{13.71}$$

10. If

$$\sum_{\alpha=1}^{N} \mathrm{sgn}(Z_{\alpha 1}) > 0 \tag{13.72}$$

is not satisfied, change the signs of all $\mathbf{t}_\kappa$ and all $(X_\alpha, Y_\alpha, Z_\alpha)$, where $\mathrm{sgn}(x)$ is the signature function, returning 1, 0, and -1 for $x > 0$, $x = 0$, and $x < 0$, respectively.

Comments As pointed out earlier, the 3D position X_α and the camera matrix P_κ computed by projective reconstruction are corrected to their true values in the form $\mathsf{H}^{-1}\mathsf{X}_\alpha$ and $\mathsf{P}_\kappa\mathsf{H}^{-1}$, respectively. On the other hand, Eq. (11.3) implies that the correct camera matrix P_κ has the form

$$\mathsf{P}_\kappa = \mathsf{K}_\kappa \begin{pmatrix} \mathsf{R}_\kappa^\top & -\mathsf{R}_\kappa^\top \mathbf{t}_\kappa \end{pmatrix}. \tag{13.73}$$

However, projective reconstruction produces the camera matrix P_κ only up to scale. Therefore if the projective reconstruction and the Euclidean upgrading are correctly computed, $\mathsf{K}_\kappa^{-1}\mathsf{P}_\kappa$ should be a scalar multiple of $\begin{pmatrix} \mathsf{R}_\kappa^\top & -\mathsf{R}_\kappa^\top \mathbf{t}_\kappa \end{pmatrix}$. This means that if we normalize the matrix A_κ in Eq. (13.65) to $\det \mathsf{A}_\kappa = 1$, it should equal $\mathsf{R}_\kappa^\top$. However, the computation of projective reconstruction and Euclidean upgrading may not necessarily be exact, and hence the obtained A_κ may not be a rotation matrix. If it is a rotation matrix, its SVD in the form of Eq. (13.68) should result in $\boldsymbol{\Sigma}_A = \mathsf{I}$ (the identity matrix). This is not guaranteed, therefore we replace $\boldsymbol{\Sigma}_A$ by I to obtain $\mathsf{R}_\kappa^\top$. Namely, we obtain R_κ in the form of Eq. (13.69) after transposition. Then, we compute $\mathbf{b}_\kappa$ in the form of Eq. (13.70).

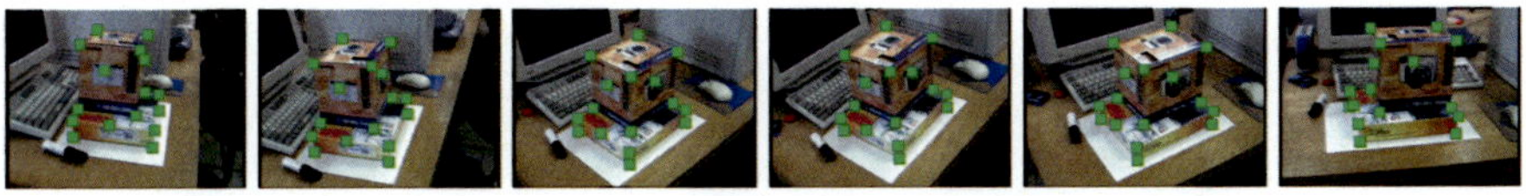

Fig. 13.4 Six frames of a real video sequence, in which 16 points are tracked

However, we face the mirror image solution problem just as in the case of two-view reconstruction (Chap. 5) and planar surface reconstruction (Chap. 8): a mirror image shape can be computed behind the camera. This is because the mathematical expression of perspective projection in the form of Eq. (5.1) does not consider the fact that the point (X, Y, Z) we are viewing is in front of the camera. To judge if the solution is a mirror image, we compute the 3D position $(X_\alpha, Y_\alpha, Z_\alpha)$ relative to the κth camera by Eq. (13.71) ($\hookrightarrow$ Fig. 5.2, and Eqs. (5.4) and (12.20)) and check if the sign of $Z_{\alpha 1}$ is positive for all the points. We do not write the left-hand side as $\sum_{\alpha=1}^{N} Z_{\alpha 1}$ because, as in Eq. (5.24), we consider the possibility that points for $Z_\alpha \approx \infty$ may be computed to be $Z_\alpha \approx -\infty$ in the presence of noise. Theoretically, we should consider all the cameras and write Eq. (13.72) as $\sum_{\kappa=1}^{M} \sum_{\alpha=1}^{N} \text{sgn}(Z_{\alpha\kappa})$, but in practice it is sufficient to consider only the first camera.

13.5 Examples

Figure 13.4 shows six decimated frames, a real video frame of 200 frames, in which 16 feature points are tracked. These points are manually specified in the first frame and tracked over the subsequent frames.

Figure 13.5 shows the 3D shape obtained by computing the 3D positions of the feature points and mapping the texture of the original images. On the left is shown an oblique view, and on the right a top view. Figure 13.6 shows the shape reconstructed by assuming affine camera modeling (we used the weak perspective model); it is placed such that it is viewed from the same directions. The top face and bottom faces of the true shape of this object (a box) are rectangles. If a perspective camera is approximated by an affine camera, the perspective effect on depth is not considered. As a result, parts farther away from the camera are reconstructed to be smaller, and hence angles that should be rectangular are reconstructed to be nonrectangular. In contrast, the self-calibration method of this chapter is based on perspective camera modeling, therefore it reconstructs angles that should be rectangular to be rectangular, producing a correct 3D shape.

In Procedures 13.2 and 13.3 for projective reconstruction, the reprojection error computed by Eqs. (13.16) and (13.29) evaluates the magnitude of error per tracked feature point. In this example, it converged to around two pixels, meaning that the feature point tracking over the input video sequence has an error of around two pixels. For this noise level, we can obtain a 3D shape of this quality without applying the bundle adjustment of Chap. 11.

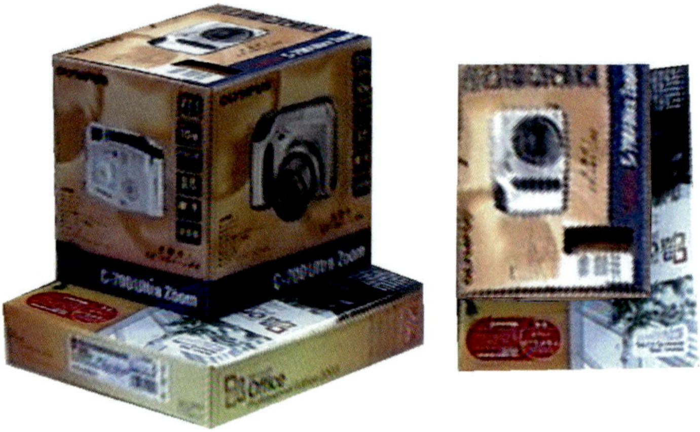

Fig. 13.5 3D shape reconstructed from the video sequence of Fig. 13.4 using the method of this chapter

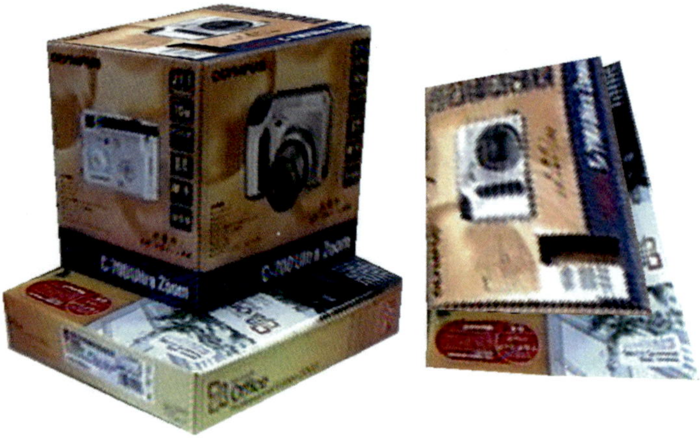

Fig. 13.6 3D shape reconstructed by self-calibration of affine cameras from the video sequence of Fig. 13.4

13.6 Supplemental Notes

The self-calibration technique for reconstructing the 3D shape from feature point correspondences over multiple images taken by uncalibrated cameras made considerable progress in the 1990s, and sophisticated mathematical theories based on projective geometry were established. This is one of the greatest achievements in the history of computer vision study. Initially, a wide variety of ideas was proposed for the computational procedure by different researchers, but in the end they were refined in the form of the two stages of projective reconstruction and Euclidean upgrading, although many variations exist for both of these stages. The two-stage approach of self-calibration has the advantage that the camera parameters and the 3D shape are computed from observed data from scratch, using algebraic procedures. However, the computation is based only on principles of projective geometry and no statistical properties of noise in the data are considered. As a result, the accuracy is not very high. For real problems, we need to apply the bundle adjustment described in Chap. 11 afterward; self-calibration techniques are suited for initializing bundle adjustment.

The primary method of projective reconstruction described in the chapter is based on Mahamud and Hebert [1]. The dual method is based on Heyden et al. [2]. Whichever method is used, projective reconstruction computation requires much more computation than Euclidean upgrading. In fact, almost all the self-calibration computation time is spent on the iterations for projective reconstruction. Thus computational complexity is a major concern.

Consider the primary method of Procedure 13.2. First, we need to compute the basis vectors $\mathbf{u}_1, \ldots, \mathbf{u}_4$ of the 4D space spanned by the columns of the matrix W of Eq. (13.12). They are the eigenvectors of $\mathsf{W}\mathsf{W}^\top$. If we compute the SVD of W in the form of Eq. (13.7), the columns of the matrix $\mathsf{U}_{3M\times L}$ are the eigenvectors of $\mathsf{W}\mathsf{W}^\top$. Here, SVD is far more efficient than eigenvalue computation. This is shown by the following consideration. Computation involving matrices and vectors essentially consists of evaluation of sums of products. The number of additions/subtractions is (the number of terms)-1, but if we ignore the -1, the number of multiplications is equal to the number of additions/subtractions for evaluating sums of products. Hence, we can focus only on the number of multiplications for complexity evaluation. Note that $(3M)^2N$ multiplications are necessary for computing $\mathsf{W}\mathsf{W}^\top$. The complexity of eigenvalue computation slightly differs from algorithm to algorithm, but the required number of multiplications is approximately the cube of the dimension of the matrix. Therefore the complexity of the eigenvalue computation of $\mathsf{W}\mathsf{W}^\top$ is around $(3M)^2N + (3M)^3 = (3M)^2(3M + N)$.

On the other hand, the complexity of the SVD is about (the number of rows)^2 times (the number of columns) or (the number of columns)^2 times (the number of rows). The SVD of a transpose is the transpose of the SVD, thus the complexity for the matrix W of Eq. (13.12) is about $3MN^2$ if $3M > N$ and about $(3M)^2N$ if $3M \leq N$. In either case, this is far more efficient than eigenvalue computation. For the same reason, the eigenvectors of the matrix $\mathsf{A}^{(\alpha)}$ in Step 3(b) of Procedure 13.2 can be computed more efficiently using the SVD of the $M \times 4$ matrix $\mathsf{C}^{(\alpha)}$ as shown in Problem 13.1 than by direct computation.

The technique of updating the basis of the subspace $\mathscr{L}_4$ shown in Problem 13.2 is based on the well-known *power method* for computing the eigenvector for the largest eigenvalue of a symmetric matrix. If we let $\lambda_1 \geq \cdots \geq \lambda_n$ be the eigenvalues of a symmetric matrix A, and $\mathbf{u}_1, \ldots, \mathbf{u}_n$ the corresponding unit eigenvectors, we can write

$$\mathsf{A} = \lambda_1\mathbf{u}_1\mathbf{u}_1^\top + \cdots + \lambda_n\mathbf{u}_n\mathbf{u}_n^\top. \tag{13.74}$$

Because $\mathbf{u}_1, \ldots, \mathbf{u}_n$ are orthonormal vectors, the Nth power of A has the form

$$\mathsf{A}^N = \lambda_1^N\mathbf{u}_1\mathbf{u}_1^\top + \cdots + \lambda_n^N\mathbf{u}_n\mathbf{u}_n^\top. \tag{13.75}$$

Suppose we increase N. Because $\lambda_1 \geq \cdots \geq \lambda_n$, the power λ_1^N rapidly grows in magnitude relative to other terms, and λ_n^N most rapidly decreases. Thus $\mathsf{A}^N\mathbf{a}$ for any vector $\mathbf{a}$ can be well approximated by $\lambda_1(\mathbf{a}, \mathbf{u}_1)\mathbf{u}_1$ when N is large. This is the principle of the power method. In Problem 13.2, we update the matrix $\mathsf{U}_{3M\times L}$ of the SVD of Eq. (13.7), using this principle. The columns of $\mathsf{U}_{3M\times L}$ are the eigenvectors of $\mathsf{WW}^\top$, therefore we multiply $\mathbf{u}_i$ by $\mathsf{WW}^\top$ and approximate the eigenvectors for the largest eigenvalues after their orthogonalization. In this computation, directly computing $\mathsf{WW}^\top$ requires $(3M)^2N$ multiplications. For $\mathsf{WW}^\top\mathbf{u}_i$ for $i = 1, 2, 3, 4$, we need $4(3M)^2$ multiplications. In total, we need $(3M)^2(N + 4)$ multiplications, whereas the computation of Eq. (13.82) requires $4(3MN + 3MN) = 24MN$ multiplications. Therefore using Eq. (13.82) is more efficient when $M \geq 2N/3$. Here we did not consider that $\mathsf{WW}^\top$ is symmetric and thus only its upper (or lower) triangular part needs to be determined, but this consideration little affects the above complexity analysis.

The preceding complexity analysis also holds for the dual method. The space spanned by the rows of the matrix W in Eq. (13.24) is the space spanned by the columns of $\mathsf{W}^\top$, and its basis vectors $\mathbf{v}_1, \ldots, \mathbf{v}_4$ are eigenvectors of $\mathsf{W}^\top\mathsf{W}$. Directly computing them needs $3MN^2$ multiplications for the matrix product and around N^3 multiplications for eigenvalue computation, requiring $N^2(3M + N)$ multiplications in total. On the other hand the SVD of W is the same as in the case of the primary method, therefore using SVD is far more favorable. The eigenvector computation of the $N \times N$ matrix $\mathsf{B}^{(\kappa)}$ in Step 3(b) of Procedure 13.3 is also made efficient by reducing it to the SVD of the $N \times 12$ matrix $\mathsf{C}^{(\kappa)}$ shown in Problem 13.6. Also, the update of the basis $\mathbf{v}_1, \ldots, \mathbf{v}_4$ of the subspace $\mathscr{L}_4^*$ can be made efficient by using the power method principle, as shown in Problem 13.7, for which the use of Eq. (13.88) is more favorable than direct computation when $N > 6M$.

For Euclidean upgrading, many different methods have been proposed in the past. Today, the one widely regarded as the most effective is the use of Eq. (13.36) proposed by Triggs [3]. From a projective geometrical interpretation, the matrix $\boldsymbol{\Omega}$ in Eq. (13.36) is called the *dual absolute quadric (DAQ)*. The principle is parallel to the Euclidean upgrading of affine reconstruction for affine cameras; Eq. (13.36) plays the role of the metric condition of Eq. (12.18) for affine cameras, the matrix $\boldsymbol{\Omega}$ of Eq. (13.37) corresponding to the matrix T of Eq. (12.17). The main problem is how to determine $\boldsymbol{\Omega}$ from the constraint of Eq. (13.36). Initially, attempts were made to use the same strategy as computing T from the metric condition of Eq. (12.18) for affine cameras: the camera intrinsic parameters are eliminated from Eq. (12.18) using particular properties of K_κ, and $\boldsymbol{\Omega}$ is computed using least squares. As the particular properties of K_κ to be used, one of the following conditions or some of their combinations were adopted.

- Camera parameters remain the same over different frames.
- The (1, 2) and (2, 1) elements of K_κ are 0. This is equivalent to assuming that the pixels are rectangularly arranged with no directional distortions.

- The (1, 1) and (2, 1) elements are equal. This is equivalent to assuming that the pixel array has the same spacing in the two orthogonal directions, the aspect ratio being 1.

Eliminating the camera intrinsic parameters from Eq. (13.36) in this way, we can determine $\boldsymbol{\Omega}$ by solving the resulting equations by least squares. Then, each K_α can be obtained by doing the Choleski decomposition in the form of the right-hand side of Eq. (13.36) (↪ Problem 13.10).

In this chapter, on the other hand, we first compute $\boldsymbol{\Omega}$ by assuming approximate values for K_κ and iteratively modify each K_κ such that Eq. (13.36) is well satisfied for the computed $\boldsymbol{\Omega}$. This is the strategy first proposed by Seo and Heyden [4] and slightly modified by the authors [5]. The use of (13.38) is equivalent to assuming that no directional distortions exist with aspect ratio 1. In this method, K_κ is also obtained at the time of the convergence of the iterations, thus no Choleski decomposition is necessary.

The Choleski decomposition is a variant of the *LU decomposition* for solving linear equations, which decomposes a given square matrix A in the form

$$\mathsf{A} = \mathsf{LU}, \tag{13.76}$$

where L and U are lower and upper triangular matrices, respectively. The matrix A is often symmetric and positive definite in many practical problems of physics and engineering. In such a case, LU decomposition reduces to Choleski decomposition. For this reason, many software tools define Choleski decomposition to be the process of computing for a given positive definite symmetric matrix A a "lower" triangular matrix L such that

$$\mathsf{A} = \mathsf{LL}^\top. \tag{13.77}$$

For computing the upper triangular matrix K in the form of Eq. (13.91), we let $\mathsf{K} = \mathsf{L}^\top$ after computing L in the above form.

One thing to note about Euclidean upgrading is the fact that the shape obtained by projective reconstruction does not necessarily have a resemblance to the true shape. In some cases, the computed shape is completely different from the true one. In the case of affine reconstruction assuming an affine camera, a cube is merely distorted to a parallelogram, but a homography mapping of a cube may not necessarily be a closed box. Consider, for example, an ellipsoid. If we apply a homography such that a point in it is mapped to a point at infinity, the resulting shape is a hyperboloid of two sheets, consisting of two convex surfaces facing each other (Fig. 13.7).

If the noise in the data is not very large, the iterations of Procedure 13.6 for Euclidean upgrading converge after a few iterations, returning a more or less correct shape. In the presence of large noise, however, Eq. (13.44) sometimes results in $\sigma_2 > 0 > \sigma_3$ with two positive and two negative eigenvalues. This means that we cannot obtain a positive semi-definite $\boldsymbol{\Omega}$ of rank 3; Eq. (13.44) is based on the assumption that at least three eigenvalues of $\boldsymbol{\Omega}$ are all positive or all negative. We can reduce this possibility by computing an affine reconstruction first and applying to the projective reconstruction a homography that approximately maps it to that affine reconstruction; such a homography can be obtained by least squares.

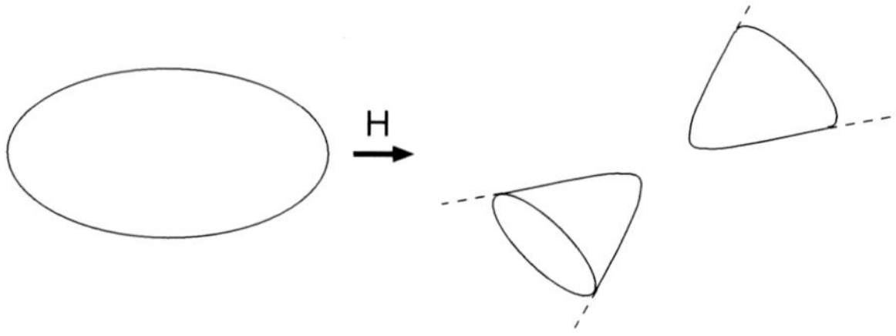

Fig. 13.7 By a homography H of the 3D space, a closed surface can be mapped to two open surfaces that contain points at infinity

The core part of the procedure of Problem 13.11 for decomposing the camera matrix P_κ into K_κ, $\mathbf{t}_\kappa$, and R_κ is to decompose a given matrix Q into an upper triangular matrix and a rotation matrix in the form $\mathsf{Q} = \mathsf{K}_\kappa \mathsf{R}_\kappa^\top$. This can also be done using what is known as *QR decomposition*: decomposing a given matrix A into the product of an orthogonal matrix Q and an upper triangular matrix R in the form $\mathsf{A} = \mathsf{QR}$ (confusion may arise, because today an upper triangular matrix is usually denoted by U, and a rotation matrix by R, but the notation QR is used here for historical reasons). For Problem 13.11, we can consider the inverse of $\mathsf{Q} = \mathsf{K}_\kappa \mathsf{R}_\kappa^\top$ in the form $\mathsf{Q}^{-1} = \mathsf{R}\mathsf{K}_\kappa^{-1\top}$, which can be computed by the QR decomposition. As shown in Problem 13.11, we can reduce the QR decomposition to the Choleski decomposition, but we can also reduce it to Schmidt orthogonalization (↪ Problem 13.2). It is described as follows. Given nD vectors $\mathbf{u}_1, \ldots, \mathbf{u}_n$, the Schmidt orthogonalization for obtaining orthogonal unit vectors $\mathbf{e}_1, \ldots, \mathbf{e}_n$ from them is as follows (note that only n vectors at most can be orthogonalized in nD). First, normalize $\mathbf{u}_1$ to a unit vector in the form $\mathbf{e}_1 = c_{10}\mathbf{u}_1$, where c_{10} is a normalization constant. Next, choose constants c_{20} and c_{21} such that $\mathbf{e}_2 = c_{20}\mathbf{u}_2 - c_{21}\mathbf{e}_1$ is a unit vector orthogonal to $\mathbf{e}_1$. Then, choose constants c_{30}, c_{31}, and c_{32} such that $\mathbf{e}_3 = c_{30}\mathbf{u}_3 - c_{31}\mathbf{e}_1 - c_{32}\mathbf{e}_2$ is a unit vector orthogonal to both $\mathbf{e}_1$ and $\mathbf{e}_2$. If we continue this, we obtain

$$\mathbf{u}_1 = \frac{1}{c_{10}}\mathbf{e}_1, \quad \mathbf{u}_2 = \frac{1}{c_{20}}(c_{21}\mathbf{e}_1 + \mathbf{e}_2), \quad \mathbf{u}_3 = \frac{1}{c_{30}}(c_{31}\mathbf{e}_1 + c_{32}\mathbf{e}_2 + \mathbf{e}_3), . \quad (13.78)$$

These are rearranged in the matrix form:

$$(\mathbf{u}_1\ \mathbf{u}_2\ \mathbf{u}_3\ \cdots) = (\mathbf{e}_1\ \mathbf{e}_2\ \mathbf{e}_3\ \cdots)\begin{pmatrix} 1/c_{10} & c_{21}/c_{20} & c_{31}/c_{30} & \cdots \\ 0 & 1/c_{20} & c_{32}/c_{30} & \cdots \\ 0 & 0 & 1/c_{30} & \cdots \\ \vdots & \vdots & \vdots & \ddots \end{pmatrix}. \quad (13.79)$$

This indicates that a given matrix on the left-hand side is decomposed into the product of an orthogonal matrix and an upper triangular matrix. See [6,7] for details of SVD, LU decomposition, Choleski decomposition, and QR decomposition.

The self-calibration procedure of this chapter assumes, as the affine reconstruction procedure of the preceding section, that all 3D points are observed in all images. In real applications, however, some points may be out of the visible frame or occluded by other objects. The treatment of missing points has been studied in the past just as in the affine reconstruction case. Some researchers studied, from a purely mathematical viewpoint, methods for estimating missing elements of the observation matrix W of

Eq. (13.4) from its rank requirement. Others first reconstructed a tentative shape only from visible points and used it to estimate missing points iteratively, introducing weights for describing the uncertainty of the current estimate.

Some researchers studied the applicability limit of self-calibration techniques from a theoretical point of view. The Euclidean upgrading procedure of this chapter is based on the dual absolute quadric constraint of Eq. (13.36), and it is known that there exist special camera motions for which the matrix $\boldsymbol{\Omega}$ cannot be determined. Such motions are called *critical motions*. The fixating camera configuration for the two-view reconstruction of Chap. 5 is a typical instance of critical motion. For the self-calibration of this chapter, critical motions include pure translation of the camera with its intrinsic parameters fixed and rotation of the camera around a fixed axis in such a way that the camera is always fixating on the axis. For such mathematical analysis based on projective geometry, see Kahl et al. [8] and Sturm [9].

Problems

13.1 Show that Step 3(b) of Procedure 13.2 is computed by the following SVD.

1. Compute the following $M \times 4$ matrix $\mathsf{C}_\alpha = (C^{(\alpha)}_{\kappa(i)})$, where its (κ, i) element $C^{(\alpha)}_{\kappa(i)}$ is given by
$$C^{(\alpha)}_{\kappa(i)} = \frac{(\mathbf{x}_{\alpha\kappa}, \mathbf{u}_{i\kappa})}{\|\mathbf{x}_{\alpha\kappa}\|}. \tag{13.80}$$
2. Compute the following SVD of $\mathsf{C}^{(\alpha)}$.
$$\mathsf{C}^{(\alpha)} = \mathsf{U}^{(\alpha)}\boldsymbol{\Sigma}^{(\alpha)}\mathsf{V}^{(\alpha)\top}. \tag{13.81}$$
3. Let $\boldsymbol{\xi}_\alpha$ be the first column of the $M \times 4$ matrix $\mathsf{U}^{(\alpha)}$, and adjust the sign so that Eq. (13.14) holds.

13.2 (1) Show that the ith column of the matrix $\mathsf{U}_{3M\times L}$ in the SVD of Eq. (13.7) is the eigenvector of $\mathsf{W}\mathsf{W}^\top$ for eigenvalue σ_i^2.
(2) Show that the iterative update of $\mathbf{u}_1$, ..., $\mathbf{u}_4$ in Step 2 of Procedure 13.2 can be done after the second iteration by the following two steps without doing SVD. We first let
$$\mathbf{v}_i = \mathsf{W}^\top\mathbf{u}_i, \qquad \mathbf{u}_i \leftarrow \mathsf{W}\mathbf{v}_i, \qquad i = 1, \ldots, 4, \tag{13.82}$$
and then compute the Schmidt orthogonalization of $\mathbf{u}_1$, ..., $\mathbf{u}_4$ in the form
$$\begin{aligned}
\mathbf{u}_1 &\leftarrow \mathscr{N}[\mathbf{u}_1],\\
\mathbf{u}_2 &\leftarrow \mathscr{N}[\mathbf{u}_2 - (\mathbf{u}_1, \mathbf{u}_2)\mathbf{u}_1],\\
\mathbf{u}_3 &\leftarrow \mathscr{N}[\mathbf{u}_3 - (\mathbf{u}_1, \mathbf{u}_3)\mathbf{u}_1 - (\mathbf{u}_2, \mathbf{u}_3)\mathbf{u}_2],\\
\mathbf{u}_4 &\leftarrow \mathscr{N}[\mathbf{u}_4 - (\mathbf{u}_1, \mathbf{u}_4)\mathbf{u}_1 - (\mathbf{u}_2, \mathbf{u}_4)\mathbf{u}_2 - (\mathbf{u}_3, \mathbf{u}_4)\mathbf{u}_3],
\end{aligned} \tag{13.83}$$
where $\mathscr{N}[\,\cdot\,]$ designate normalization to unit norm.
(3) Show that the above Schmidt orthogonalization of Eq. (13.83) indeed produces orthonormal vectors.

13.3 Show that the camera matrices P_κ and 3D positions X_α are obtained from the first four columns $\mathbf{u}_1$, ..., $\mathbf{u}_4$ of $\mathsf{U}_{3M\times L}$ in the SVD of Eq. (13.7) by the following computation.

1. Determine P_κ in the form of Eq. (13.5) by regarding the $3M \times 4$ matrix consisting of $\mathbf{u}_1$, ..., $\mathbf{u}_4$ as the matrix M.
2. Let $\mathbf{p}_\alpha$ be the αth column of the observation matrix W of Eq. (13.4), and determine the 3D position $\mathsf{X}_\alpha = (X_{\alpha(k)})$, $k = 1, 2, 3, 4$, by

$$X_{\alpha(k)} = (\mathbf{p}_\alpha, \mathbf{u}_k). \tag{13.84}$$

13.4 Show that the normalization condition of Eq. (13.25) is equivalent to $\|\boldsymbol{\xi}_\kappa\| = 1$.

13.5 Show that if the matrix $\mathsf{B}^{(\kappa)}$ is defined by Eq. (13.26), we can rewrite Eq. (13.32) in the form of Eq. (13.33).

13.6 Show that Step 3(b) of Procedure 13.3 can be computed by the following SVD.

1. Define the $N \times 4$ matrices $\mathsf{C}^{(\kappa 1)} = (C^{(\kappa 1)}_{\alpha i})$, $\mathsf{C}^{(\kappa 2)} = (C^{(\kappa 3)}_{\alpha i})$, and $\mathsf{C}^{(\kappa 3)} = (C^{(\kappa 3)}_{\alpha i})$ by

$$C^{(\kappa 1)}_{\alpha i} = \frac{x_{\alpha\kappa} v_{i\alpha}}{f_0\|\mathbf{x}_{\alpha\kappa}\|}, \quad C^{(\kappa 2)}_{\alpha i} = \frac{y_{\alpha\kappa} v_{i\alpha}}{f_0\|\mathbf{x}_{\alpha\kappa}\|}, \quad C^{(\kappa 3)}_{\alpha i} = \frac{v_{i\alpha}}{\|\mathbf{x}_{\alpha\kappa}\|}. \tag{13.85}$$

2. Define the following $N \times 12$ matrix $\mathsf{C}^{(\kappa)}$.

$$\mathsf{C}^{(\kappa)} = \begin{pmatrix} \mathsf{C}^{(\kappa 1)} & \mathsf{C}^{(\kappa 2)} & \mathsf{C}^{(\kappa 3)} \end{pmatrix}. \tag{13.86}$$

3. Compute the following SVD of $\mathsf{C}^{(\kappa)}$.

$$\mathsf{C}^{(\kappa)} = \mathsf{U}^{(\kappa)}\boldsymbol{\Sigma}^{(\kappa)}\mathsf{V}^{(\kappa)\top}. \tag{13.87}$$

4. Let $\boldsymbol{\xi}_\kappa$ be the first column of the $N \times 12$ matrix $\mathsf{U}^{(\kappa)}$, and adjust the sign so that Eq. (13.27) holds.

13.7 (1) Show that the ith column of the matrix $\mathsf{V}_{N\times L}$ in the SVD of Eq. (13.7) is the eigenvector of $\mathsf{W}^\top\mathsf{W}$ for eigenvalue σ_i^2.
(2) Show that the iterative update of $\mathbf{v}_1$, ..., $\mathbf{v}_4$ in Step 2 of Procedure 13.3 can be done after the second iteration by the following steps without using SVD. We first let

$$\mathbf{u}_i = \mathsf{W}\mathbf{v}_i, \qquad \mathbf{v}_i \leftarrow \mathsf{W}^\top\mathbf{u}_i, \qquad i = 1, \ldots, 4, \tag{13.88}$$

and then compute the Schmidt orthogonalization of $\mathbf{v}_1$, ..., $\mathbf{v}_4$ in the form

$$\begin{aligned}
&\mathbf{v}_1 \leftarrow \mathcal{N}[\mathbf{v}_1], \\
&\mathbf{v}_2 \leftarrow \mathcal{N}[\mathbf{v}_2 - (\mathbf{v}_1, \mathbf{v}_2)\mathbf{v}_1], \\
&\mathbf{v}_3 \leftarrow \mathcal{N}[\mathbf{v}_3 - (\mathbf{v}_1, \mathbf{v}_3)\mathbf{v}_1 - (\mathbf{v}_2, \mathbf{v}_3)\mathbf{v}_2], \\
&\mathbf{v}_4 \leftarrow \mathcal{N}[\mathbf{v}_4 - (\mathbf{v}_1, \mathbf{v}_4)\mathbf{v}_1 - (\mathbf{v}_2, \mathbf{v}_4)\mathbf{v}_2 - (\mathbf{v}_3, \mathbf{v}_4)\mathbf{v}_3],
\end{aligned} \tag{13.89}$$

where $\mathcal{N}[\,\cdot\,]$ designates normalization to the unit norm.

13.8 Show that the camera matrices P_κ and the 3D positions X_α are obtained from the first four columns $\mathbf{v}_1, \ldots, \mathbf{v}_4$ of $\mathsf{V}_{N\times L}$ in the SVD of Eq. (13.7) by the following computation.

1. Let X_α be the αth column in the form of Eq. (13.5) by regarding the transpose $\mathsf{V}^\top$ of the $N \times 4$ matrix V consisting of $\mathbf{v}_1, \ldots, \mathbf{v}_4$ as the matrix S.
2. Let $\mathbf{q}_{\kappa(i)}^\top$ be the $(3(\kappa - 1) + i)$th row of the observation matrix W of Eq. (13.4), and determine the (i, j) element of the camera matrix $\mathsf{P}_\kappa = (P_{\kappa(ij)})$ by

$$P_{\kappa(ij)} = (\mathbf{q}_{\kappa(i)}, \mathbf{v}_j). \tag{13.90}$$

13.9 The set of nonsingular upper triangular matrices is known to have a group structure. Show this in the 3×3 case. Namely, show that
(1) If K and K' are upper triangular matrices, so is their product $\mathsf{K}'' = \mathsf{K}\mathsf{K}'$.
(2) If K is a nonsingular upper triangular matrix, so is its inverse K^{-1}.

13.10 It is known that for an arbitrary positive definite symmetric matrix A there exists a nonsingular upper triangle K such that

$$\mathsf{A} = \mathsf{K}^\top \mathsf{K}. \tag{13.91}$$

This is known as the *Choleski decomposition*. Show that in the 3×3 case this is computed as follows.

$$\begin{pmatrix} a_{11} & a_{12} & a_{13} \\ a_{12} & a_{22} & a_{23} \\ a_{13} & a_{23} & a_{33} \end{pmatrix} = \begin{pmatrix} x_{11} & 0 & 0 \\ x_{12} & x_{22} & 0 \\ x_{13} & x_{23} & x_{33} \end{pmatrix} \begin{pmatrix} x_{11} & x_{12} & x_{13} \\ 0 & x_{22} & x_{23} \\ 0 & 0 & x_{33} \end{pmatrix}, \tag{13.92}$$

$$x_{11} = \sqrt{a_{11}}, 12 = \frac{a_{12}}{x_{11}}, 13 = \frac{a_{13}}{x_{11}},$$

$$x_{22} = \sqrt{a_{22} - x_{12}^2}, \qquad x_{23} = \frac{x_{23} - x_{12}x_{13}}{x_{22}}, \qquad x_{33} = \sqrt{a_{33} - x_{13}^2 - x_{23}^2}. \tag{13.93}$$

13.11 Show that for a given 3×4 camera matrix P_κ, there exist an upper triangular matrix K_κ, a translation vector $\mathbf{t}_\kappa$, and a rotation matrix R_κ such that Eq. (13.73) holds and that they are computed by the following procedure.

1. Let $\mathsf{P}_\kappa = \begin{pmatrix} \mathsf{Q} & \mathbf{q} \end{pmatrix}$, P_κ being the first 3×3 block and $\mathbf{q}$ the fourth column.
2. If $\det \mathsf{Q} < 0$, change the sign of Q and $\mathbf{q}$.
3. Determine the translation $\mathbf{t}_\kappa$ by

$$\mathbf{t}_\kappa = -\mathsf{Q}^{-1}\mathbf{q}. \tag{13.94}$$

4. Compute the Choleski decomposition of $(\mathsf{Q}\mathsf{Q}^\top)^{-1}$ and express it in terms of an upper triangle C in the form

$$(\mathsf{Q}\mathsf{Q}^\top)^{-1} = \mathsf{C}^\top \mathsf{C}. \tag{13.95}$$

5. Determine K_κ by

$$\mathsf{K}_\kappa = \mathsf{C}^{-1}. \tag{13.96}$$

6. Determine the rotation R_κ by

$$\mathsf{R}_\kappa = \mathsf{Q}^\top \mathsf{C}^\top \tag{13.97}$$

References

1. S. Mahamud, M. Hebert, Iterative projective reconstruction from multiple views, in *Proceedings of the IEEE Conference on Computer Vision and Pattern Recognition*, Hilton Head Island, SC, U.S., Vol. 2, pp. 430–437 (2000)
2. A. Heyden, R. Berthilsson, G. Sparr, An iterative factorization method for projective structure and motion from image sequences. Image Vis. Comput. **17**(13), 981–991 (1999)
3. B. Triggs, Autocalibration and the absolute quadric, in *Proceedings of the IEEE Conference on Computer Vision and Pattern Recognition*, San Juan, Puerto Rico, pp. 609–614 (1997)
4. Y. Seo, A. Heyden, Auto-calibration by linear iteration using the DAC equation. Image Vis. Comput. **22**(11), 919–926 (2004)
5. K. Kanatani, Latest progress of 3-D reconstruction from moving camera images, in *Robotics Research Trends*, ed. by X.P. Guo (Nova Science Publishers, Hauppauge, N.Y, U.S., 2008), pp. 33–75
6. G.H. Golub, C.F. Van Loan, *Matrix Computation*, 4th edn. (Johns Hopkins University Press, Baltimore, MD, U.S., 2012)
7. W.H. Press, S.A. Teukolsky, W.T. Vetterling, B.P. Flannery, *Numerical Recipes: The Art of Scientific Computing*, 3rd edn. (Cambridge University Press, Cambridge, U.K., 2007)
8. F. Kahl, B. Triggs, K. Åström, Critical motions for auto-calibration when some intrinsic parameters can vary. J. Math. Imaging Vis. **13**(2), 131–146 (2000)
9. P. Sturm, Critical motion sequences for the self-calibration of cameras and stereo systems with variable focal length. Image Vis. Comput. **20**(5/6), 415–426 (2002)

Part III
Mathematical Foundation of Geometric Estimation

Accuracy of Geometric Estimation 14

Abstract

We focus here on algebraic methods for ellipse fitting, fundamental matrix computation, and homography computation described in Chaps. 2, 3, and 6 in a more generalized mathematical framework. We do a detailed error analysis in general terms and derive explicit expressions for the covariance and bias of the solution. The hyper-renormalization procedure is derived in this mathematical framework.

14.1 Constraint of the Problem

The basic equation for ellipse fitting, fundamental matrix computation, and homography computation has the form

$$(\boldsymbol{\xi}^{(k)}, \boldsymbol{\theta}) = 0, \qquad k = 1, \ldots, L, \tag{14.1}$$

which should be satisfied if there is no noise, that is, if all the data are exact. For ellipse fitting, Eq. (14.1) is the ellipse equation of Eq. (2.1) with $L = 1$, and $\boldsymbol{\xi}$ and $\boldsymbol{\theta}$ are the vectors given by Eq. (2.3). For fundamental matrix computation, Eq. (14.1) is the epipolar equation of Eq. (3.1) with $L = 1$, and $\boldsymbol{\xi}$ and $\boldsymbol{\theta}$ are given by Eq. (3.3). For homography computation, Eq. (14.1) describes the homography of Eq. (6.1, or equivalently Eq. (6.2) or (6.3)) with $L = 3$, and $\boldsymbol{\xi}^{(k)}$ and $\boldsymbol{\theta}$ given by Eq. (6.5). In order to fix the scale indeterminacy of Eq. (14.1), we normalize $\boldsymbol{\theta}$ to the unit norm.

Generally, the equation that should hold in the absence of noise is called the *constraint* of the problem. Equation (14.1) generalizes the constraint for ellipse fitting, fundamental matrix computation, and homography computation. The task is: given noisy observations $\boldsymbol{\xi}_\alpha^{(k)}$, ..., $\boldsymbol{\xi}_N^{(k)}$ of the vectors $\boldsymbol{\xi}^{(k)}$, $k = 1, 2, 3$, we infer the unit vector $\boldsymbol{\theta}$ for which the constraint of Eq. (14.1) would be satisfied if the observations were noiseless.

K. Kanatani et al., *Guide to 3D Vision Computation*, Advances in Computer Vision and Pattern Recognition, DOI 10.1007/978-3-319-48493-8_14

14.2 Noise and Covariance Matrices

This type of inference relies on our knowledge about the properties of the noise in the observations $\boldsymbol{\xi}_\alpha^{(k)}$. Regarding $\boldsymbol{\xi}_\alpha^{(k)}$ as random variables, we generalize our assumptions for ellipse fitting, fundamental matrix computation, and homography computation as follows. We assume that $\boldsymbol{\xi}_\alpha^{(k)}$ is a (generally nonlinear) function of the coordinates (x_α, y_α) of the αth point or the coordinates (x_α, y_α) and (x'_α, y'_α) of the αth point pair. We assume that each coordinate is perturbed from its true value by independent Gaussian noise of mean 0 and standard deviation σ (pixels). Then we can expand $\boldsymbol{\xi}_\alpha^{(k)}$ in σ in the form (see Eqs. (2.11), (3.7), and (6.9))

$$\boldsymbol{\xi}_\alpha^{(k)} = \bar{\boldsymbol{\xi}}_\alpha^{(k)} + \Delta_1 \boldsymbol{\xi}_\alpha^{(k)} + \Delta_2 \boldsymbol{\xi}_\alpha^{(k)} + \cdots, \tag{14.2}$$

where the bar denotes the "true" value (i.e., the value that we would observe in the absence of noise) and Δ_i denotes the ith-order term in σ (see Eqs. (2.12), (3.8), and (6.10)).

We define the *covariance matrices* of the observations by

$$V^{(kl)}[\boldsymbol{\xi}_\alpha] = E[\Delta_1 \boldsymbol{\xi}_\alpha^{(k)} \Delta_1 \boldsymbol{\xi}_\alpha^{(l)\top}], \tag{14.3}$$

where $E[\,\cdot\,]$ denotes the expectation over the probability distribution of the observations $\boldsymbol{\xi}_\alpha^{(k)}$ (see Eqs. (2.13), (3.9), and (6.13)). Because $\Delta_1 \boldsymbol{\xi}_\alpha^{(k)}$ is linear in σ, we can write Eq. (14.2) in the form

$$V^{(kl)}[\boldsymbol{\xi}_\alpha] = \sigma^2 V_0^{(kl)}[\boldsymbol{\xi}_\alpha], \tag{14.4}$$

where $V_0^{(kl)}[\boldsymbol{\xi}_\alpha]$ does not depend on σ; we call $V_0^{(kl)}[\boldsymbol{\xi}_\alpha]$ the *normalized covariance matrices* and σ the *noise level*. We assume that σ is unknown but $V_0^{(kl)}[\boldsymbol{\xi}_\alpha]$ is known (see Eqs. (2.15), (3.12), and (6.15)).

14.3 Error Analysis

The basic algebraic approach for ellipse fitting, fundamental matrix computation, and homography computation is to solve an algebraic equation of $\boldsymbol{\theta}$ in the form

$$\mathsf{M}\boldsymbol{\theta} = \lambda \mathsf{N}\boldsymbol{\theta}. \tag{14.5}$$

The matrix M has the form

$$\mathsf{M} = \frac{1}{N} \sum_{\alpha=1}^{N} \sum_{k,l=1}^{L} W_\alpha^{(kl)} \boldsymbol{\xi}_\alpha^{(k)} \boldsymbol{\xi}_\alpha^{(k)\top}, \quad W_\alpha^{(kl)} = \Big((\boldsymbol{\theta}, V_0^{(kl)}[\boldsymbol{\xi}_\alpha]\boldsymbol{\theta}) \Big)_r^-, \tag{14.6}$$

where $\Big((\boldsymbol{\theta}, V_0^{(kl)}[\boldsymbol{\xi}_\alpha]\boldsymbol{\theta}) \Big)_r^-$ denotes the (k, l) element of the pseudoinverse with truncated rank r of the matrix whose (k, l) element is $(\boldsymbol{\theta}, V_0^{(kl)}[\boldsymbol{\xi}_\alpha]\boldsymbol{\theta})$. The rank r is the number of independent vectors among $\boldsymbol{\xi}_\alpha^{(1)}, \ldots, \boldsymbol{\xi}_\alpha^{(L)}$ in Eq. (14.1); $r = 1$ for ellipse fitting and fundamental matrix computation, and $r = 2$ for homography computation.

The expression in Eq. (14.6) generalizes the matrix M for ellipse fitting, fundamental matrix computation, and homography computation (see Eqs. (2.32), (6.16), (6.22), and (6.25)).

When the observation $\boldsymbol{\xi}_\alpha^{(k)}$ is perturbed from its true value $\bar{\boldsymbol{\xi}}_\alpha^{(k)}$ in the form of Eq. (14.2), the solution $\boldsymbol{\theta}$ of Eq. (14.5) is also perturbed from its true value $\bar{\boldsymbol{\theta}}$ in the form

$$\boldsymbol{\theta} = \bar{\boldsymbol{\theta}} + \Delta_1\boldsymbol{\theta} + \Delta_2\boldsymbol{\theta} + \cdots . \tag{14.7}$$

The scalar λ and the matrix M on the right-hand side of Eq. (14.5) depend on the particular fitting method, but here we do error analysis by regarding them as yet to be determined. Substituting Eqs. (14.2) and (14.7) into Eq. (14.6), we can expand M and $W_\alpha^{(kl)}$ in the form

$$\mathsf{M} = \bar{\mathsf{M}} + \Delta_1\mathsf{M} + \Delta_2\mathsf{M} + \cdots , \tag{14.8}$$

$$W_\alpha^{(kl)} = \bar{W}_\alpha^{(kl)} + \Delta_1 W_\alpha^{(kl)} + \Delta_2 W_\alpha^{(kl)} + \cdots , \tag{14.9}$$

where $\Delta_1\mathsf{M}$ and $\Delta_2\mathsf{M}$ are given by

$$\begin{aligned}\Delta_1\mathsf{M} = &\frac{1}{N}\sum_{\alpha=1}^{N}\sum_{k,l=1}^{L}\bar{W}_\alpha^{(kl)}\Big(\Delta_1\boldsymbol{\xi}_\alpha^{(k)}\bar{\boldsymbol{\xi}}_\alpha^{(l)\top} + \bar{\boldsymbol{\xi}}_\alpha^{(k)}\Delta_1\boldsymbol{\xi}_\alpha^{(l)\top}\Big)\\ &+\frac{1}{N}\sum_{\alpha=1}^{N}\sum_{k,l=1}^{L}\Delta_1 \bar{W}_\alpha^{(kl)}\bar{\boldsymbol{\xi}}_\alpha^{(k)}\bar{\boldsymbol{\xi}}_\alpha^{(l)\top},\end{aligned} \tag{14.10}$$

$$\begin{aligned}\Delta_2\mathsf{M} = &\frac{1}{N}\sum_{\alpha=1}^{N}\sum_{k,l=1}^{L}\bar{W}_\alpha^{(kl)}\Big(\Delta_1\boldsymbol{\xi}_\alpha^{(k)}\Delta_1\boldsymbol{\xi}_\alpha^{(l)\top} + \Delta_2\boldsymbol{\xi}_\alpha^{(k)}\bar{\boldsymbol{\xi}}_\alpha^{(l)\top} + \bar{\boldsymbol{\xi}}_\alpha^{(k)}\Delta_2\boldsymbol{\xi}_\alpha^{(l)\top}\Big)\\ &+\frac{1}{N}\sum_{\alpha=1}^{N}\sum_{k,l=1}^{L}\Delta_1 W_\alpha^{(kl)}(\Delta_1\boldsymbol{\xi}_\alpha^{(k)}\bar{\boldsymbol{\xi}}_\alpha^{(l)\top} + \bar{\boldsymbol{\xi}}_\alpha^{(k)}\Delta_1\boldsymbol{\xi}_\alpha^{(l)\top})\\ &+\frac{1}{N}\sum_{\alpha=1}^{N}\sum_{k,l=1}^{L}\Delta_2 W_\alpha^{(kl)}\bar{\boldsymbol{\xi}}_\alpha^{(k)}\bar{\boldsymbol{\xi}}_\alpha^{(l)\top},\end{aligned} \tag{14.11}$$

and $\Delta_1 W_\alpha^{(kl)}$ and $\Delta_2 W_\alpha^{(kl)}$ are given by

$$\Delta_1 W_\alpha^{(kl)} = -2\sum_{m,n=1}^{L}\bar{W}_\alpha^{(km)}\bar{W}_\alpha^{(ln)}(\Delta_1\boldsymbol{\theta}, V_0^{(mn)}[\boldsymbol{\xi}_\alpha]\bar{\boldsymbol{\theta}}), \tag{14.12}$$

$$\begin{aligned}\Delta_2 W_\alpha^{(kl)} = &\sum_{m,n=1}^{L}\Delta_1 W_\alpha^{(km)}\Delta_1 W_\alpha^{(ln)}(\bar{\boldsymbol{\theta}}, V_0^{(mn)}[\boldsymbol{\xi}_\alpha]\bar{\boldsymbol{\theta}})\\ &-\sum_{m,n=1}^{L}\bar{W}_\alpha^{(km)}\bar{W}_\alpha^{(ln)}\Big((\Delta_1\boldsymbol{\theta}, V_0^{(mn)}[\boldsymbol{\xi}_\alpha]\Delta_1\boldsymbol{\theta}) + 2(\Delta_2\boldsymbol{\theta}, V_0^{(mn)}[\boldsymbol{\xi}_\alpha]\bar{\boldsymbol{\theta}})\Big).\end{aligned} \tag{14.13}$$

See Proposition 14.1 below for the derivation. Similarly expanding λ and N, we write Eq. (14.5) in the form

$$(\bar{\mathsf{M}} + \Delta_1\mathsf{M} + \Delta_2\mathsf{M} + \cdots)(\bar{\boldsymbol{\theta}} + \Delta_1\boldsymbol{\theta} + \Delta_2\boldsymbol{\theta} + \cdots)$$
$$= (\bar{\lambda} + \Delta_1\lambda + \Delta_2\lambda + \cdots)(\bar{\mathsf{N}} + \Delta_1\mathsf{N} + \Delta_2\mathsf{N} + \cdots)(\bar{\boldsymbol{\theta}} + \Delta_1\boldsymbol{\theta} + \Delta_2\boldsymbol{\theta} + \cdots). \tag{14.14}$$

The true solution $\bar{\boldsymbol{\theta}}$ satisfies the constraint of Eq. (14.1) for the true observations $\bar{\boldsymbol{\xi}}_\alpha^{(k)}$. Thus we see from Eq. (14.6) that $\bar{\mathsf{M}}\bar{\boldsymbol{\theta}} = \mathbf{0}$. Equating the noiseless terms on both sides of Eq. (14.14), we have $\bar{\mathsf{M}}\bar{\boldsymbol{\theta}} = \bar{\lambda}\bar{\mathsf{N}}\bar{\boldsymbol{\theta}}$ and hence $\bar{\lambda} = 0$, assuming that $\bar{\mathsf{N}}\bar{\boldsymbol{\theta}} \neq \mathbf{0}$. Equating the first-order terms on both sides of Eq. (14.14), we obtain

$$\bar{\mathsf{M}}\Delta_1\boldsymbol{\theta} + \Delta_1\mathsf{M}\bar{\boldsymbol{\theta}} = \Delta_1\lambda\bar{\mathsf{N}}\bar{\boldsymbol{\theta}}. \tag{14.15}$$

Computing the inner product with $\bar{\boldsymbol{\theta}}$ on both sides, we obtain

$$(\bar{\boldsymbol{\theta}}, \bar{\mathsf{M}}\Delta_1\boldsymbol{\theta}) + (\bar{\boldsymbol{\theta}}, \Delta_1\mathsf{M}\bar{\boldsymbol{\theta}}) = \Delta_1\lambda(\bar{\boldsymbol{\theta}}, \bar{\mathsf{N}}\bar{\boldsymbol{\theta}}). \tag{14.16}$$

Note that $(\bar{\boldsymbol{\theta}}, \bar{\mathsf{M}}\Delta_1\boldsymbol{\theta}) = (\bar{\mathsf{M}}\bar{\boldsymbol{\theta}}, \Delta_1\boldsymbol{\theta}) = 0$ and that Eq. (14.10) implies $(\bar{\boldsymbol{\theta}}, \Delta_1\mathsf{M}\bar{\boldsymbol{\theta}}) = 0$. We are assuming that $(\bar{\boldsymbol{\theta}}, \bar{\mathsf{N}}\bar{\boldsymbol{\theta}}) \neq 0$, therefore we see that $\Delta_1\lambda = 0$. Noting this and equating the second-order terms on both sides of Eq. (14.14), we obtain

$$\bar{\mathsf{M}}\Delta_2\boldsymbol{\theta} + \Delta_1\mathsf{M}\Delta_1\boldsymbol{\theta} + \Delta_2\mathsf{M}\bar{\boldsymbol{\theta}} = \Delta_2\lambda\bar{\mathsf{N}}\bar{\boldsymbol{\theta}}. \tag{14.17}$$

14.4 Covariance and Bias

We want to solve Eq. (14.15) (with $\Delta_1\lambda = 0$) for $\Delta_1\boldsymbol{\theta}$. However, because $\bar{\mathsf{M}}\bar{\boldsymbol{\theta}} = \mathbf{0}$, the matrix $\bar{\mathsf{M}}$ has rank $n-1$ (n is the dimension of $\boldsymbol{\theta}$), $\bar{\boldsymbol{\theta}}$ being its null vector. Therefore its inverse does not exist. Instead, we consider its pseudoinverse $\bar{\mathsf{M}}^-$ (also rank $n-1$). Note that the product $\bar{\mathsf{M}}^-\bar{\mathsf{M}}$ (↪ Problem 14.4) equals the projection matrix $\mathsf{P}_{\bar{\boldsymbol{\theta}}}$ in the direction of $\bar{\boldsymbol{\theta}}$ (↪ Problem 3.3). Thus multiplying Eq. (14.15) by the pseudoinverse $\bar{\mathsf{M}}^-$ on both sides from the left, we can express $\Delta_1\boldsymbol{\theta}$ in the form

$$\Delta_1\boldsymbol{\theta} = -\bar{\mathsf{M}}^-\Delta_1\mathsf{M}\bar{\boldsymbol{\theta}}, \tag{14.18}$$

where we have noted that because $\boldsymbol{\theta}$ is normalized to the unit norm, $\Delta_1\boldsymbol{\theta}$ is orthogonal to $\bar{\boldsymbol{\theta}}$ and therefore $\mathsf{P}_{\bar{\boldsymbol{\theta}}}\Delta_1\boldsymbol{\theta} = \Delta_1\boldsymbol{\theta}$. From Eq. (14.18), we can obtain the covariance matrix $V[\boldsymbol{\theta}]$ of the solution $\boldsymbol{\theta}$ to a first approximation in the form

$$V[\boldsymbol{\theta}] = E[\Delta_1\boldsymbol{\theta}\Delta_1\boldsymbol{\theta}^\top] = \frac{\sigma^2}{N}\bar{\mathsf{M}}^-. \tag{14.19}$$

See Proposition 14.2 below for the derivation. It turns out that Eq. (14.19) coincides with the theoretical accuracy limit called the *KCR (Kanatani-Cramer-Rao) lower bound* discussed in Chap. 16. Thus we can conclude that *the solution* $\boldsymbol{\theta}$ *of iterative reweight, renormalization, and hyper-renormalization has the same covariance matrix, which agrees with the theoretical accuracy limit to a first approximation.* This implies that we cannot substantially improve the covariance any further. However,

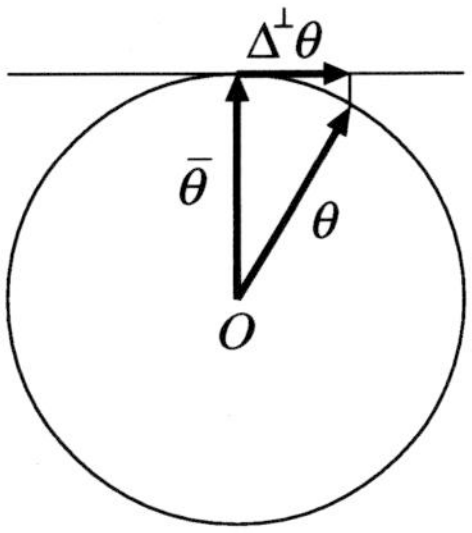

Fig. 14.1 The true value $\bar{\boldsymbol{\theta}}$, the computed value $\boldsymbol{\theta}$, and its orthogonal component $\Delta^{\perp}\boldsymbol{\theta}$ of $\bar{\boldsymbol{\theta}}$

the total error is the sum of the covariance terms and the bias terms, and we may be able to reduce the bias by a clever choice of the matrix N.

Evidently, the first-order term $\Delta_1\boldsymbol{\theta}$ in Eq. (14.18) is unbiased, because the expectation of odd-order error terms is zero: $E[\Delta_1\boldsymbol{\theta}] = \mathbf{0}$. Thus we focus on the second-order bias $E[\Delta_2^{\top}\boldsymbol{\theta}]$. Substituting Eq. (14.18) into Eq. (14.17), we obtain

$$\Delta_2\lambda\bar{\mathsf{N}}\bar{\boldsymbol{\theta}} = \bar{\mathsf{M}}\Delta_2\boldsymbol{\theta} - \Delta_1\mathsf{M}\bar{\mathsf{M}}^{-}\Delta_1\mathsf{M}\bar{\boldsymbol{\theta}} + \Delta_2\mathsf{M}\bar{\boldsymbol{\theta}} = \bar{\mathsf{M}}\Delta_2\boldsymbol{\theta} + \mathsf{T}\bar{\boldsymbol{\theta}}, \tag{14.20}$$

where we define the matrix T to be

$$\mathsf{T} \equiv \Delta_2\mathsf{M} - \Delta_1\mathsf{M}\bar{\mathsf{M}}^{-}\Delta_1\mathsf{M}. \tag{14.21}$$

Because $\boldsymbol{\theta}$ is a unit vector, it has no error in the direction of itself; we are interested in the error orthogonal to it. Therefore we define the second-order error of $\boldsymbol{\theta}$ to be its orthogonal component (Fig. 14.1):

$$\Delta_2^{\perp}\boldsymbol{\theta} \equiv \mathsf{P}_{\bar{\boldsymbol{\theta}}}\Delta_2\boldsymbol{\theta} = \bar{\mathsf{M}}^{-}\bar{\mathsf{M}}\Delta_2\boldsymbol{\theta}. \tag{14.22}$$

Note that the first-order error $\Delta_1\boldsymbol{\theta}$ in Eq. (14.18) is itself orthogonal to $\bar{\boldsymbol{\theta}}$. Multiplying Eq. (14.20) by $\bar{\mathsf{M}}^{-}$ on both, we obtain $\Delta_2^{\perp}\boldsymbol{\theta}$ in the form:

$$\Delta_2^{\perp}\boldsymbol{\theta} = \bar{\mathsf{M}}^{-}(\Delta_2\lambda\bar{\mathsf{N}} - \mathsf{T})\bar{\boldsymbol{\theta}}. \tag{14.23}$$

Computing the inner product of Eq. (14.20) and $\bar{\boldsymbol{\theta}}$ on both sides and noting that $(\bar{\boldsymbol{\theta}}, \bar{\mathsf{M}}\Delta_2\boldsymbol{\theta}) = 0$, we obtain $\Delta_2\lambda$ in the form

$$\Delta_2\lambda = \frac{(\bar{\boldsymbol{\theta}}, \mathsf{T}\bar{\boldsymbol{\theta}})}{(\bar{\boldsymbol{\theta}}, \bar{\mathsf{N}}\bar{\boldsymbol{\theta}})}. \tag{14.24}$$

Hence Eq. (14.23) is rewritten as follows.

$$\Delta_2^{\perp}\boldsymbol{\theta} = \bar{\mathsf{M}}^{-}\Big(\frac{(\bar{\boldsymbol{\theta}}, \mathsf{T}\bar{\boldsymbol{\theta}})}{(\bar{\boldsymbol{\theta}}, \bar{\mathsf{N}}\bar{\boldsymbol{\theta}})}\bar{\mathsf{N}}\bar{\boldsymbol{\theta}} - \mathsf{T}\bar{\boldsymbol{\theta}}\Big). \tag{14.25}$$

Thus, the second-order bias is

$$E[\Delta_2^{\perp}\boldsymbol{\theta}] = \bar{\mathsf{M}}^{-}\Big(\frac{(\bar{\boldsymbol{\theta}}, E[\mathsf{T}\bar{\boldsymbol{\theta}}])}{(\bar{\boldsymbol{\theta}}, \bar{\mathsf{N}}\bar{\boldsymbol{\theta}})}\bar{\mathsf{N}}\bar{\boldsymbol{\theta}} - E[\mathsf{T}\bar{\boldsymbol{\theta}}]\Big). \tag{14.26}$$

14.5 Bias Elimination and Hyper-Renormalization

From Eq. (14.26), we find a crucial fact: *if we can choose such an* N that its noiseless value $\bar{\mathsf{N}}$ satisfied $E[\mathsf{T}\bar{\boldsymbol{\theta}}] = c\bar{\mathsf{N}}\bar{\boldsymbol{\theta}}$ for some constant c, we would have

$$E[\Delta_2^{\perp}\boldsymbol{\theta}] = \bar{\mathsf{M}}^{-}\Big(\frac{(\bar{\boldsymbol{\theta}}, c\bar{\mathsf{N}}\boldsymbol{\theta})}{(\bar{\boldsymbol{\theta}}, \bar{\mathsf{N}}\bar{\boldsymbol{\theta}})}\bar{\mathsf{N}}\bar{\boldsymbol{\theta}} - c\bar{\mathsf{N}}\bar{\boldsymbol{\theta}}\Big) = \mathbf{0}. \tag{14.27}$$

Then, the bias would be $O(\sigma^4)$, because the expectation of odd-order noise terms is zero. In order to choose such an N, we need to evaluate the expectation $E[\mathsf{T}\bar{\boldsymbol{\theta}}]$. We evaluate the expectation of $\mathsf{T}\bar{\boldsymbol{\theta}} = \Delta_2\mathsf{M}\bar{\boldsymbol{\theta}} - \Delta_1\mathsf{M}\bar{\mathsf{M}}^{-}\Delta_1\mathsf{M}\bar{\boldsymbol{\theta}}$ term by term, using the identity

$$E[\Delta_1\boldsymbol{\xi}_\alpha^{(k)}\Delta_1\boldsymbol{\xi}_\beta^{(l)\top}] = \delta_{\alpha\beta}V_0^{(kl)}[\boldsymbol{\xi}_\alpha], \tag{14.28}$$

which results from our assumption of independent noise for different points and the definition of the covariance matrices $V_0^{(kl)}[\boldsymbol{\xi}_\alpha]$, where $\delta_{\alpha\beta}$ is the Kronecker delta. We also use the definition of the vectors $\mathbf{e}_\alpha^{(k)}$ via

$$E[\Delta_2\boldsymbol{\xi}_\alpha^{(k)}] = \sigma^2\mathbf{e}_\alpha^{(k)}. \tag{14.29}$$

In the end, we find that $E[\mathsf{T}\bar{\boldsymbol{\theta}}] = \sigma^2\bar{\mathsf{N}}\bar{\boldsymbol{\theta}}$ holds if we define $\bar{\mathsf{N}}$ in the form (see Lemmas 14.1, 14.2, and Proposition 14.3 below for the proof):

$$\begin{aligned}\bar{\mathsf{N}} = &\frac{1}{N}\sum_{\alpha=1}^{N}\sum_{k,l=1}^{L}\bar{W}_\alpha^{(kl)}\Big(V_0^{(kl)}[\boldsymbol{\xi}_\alpha] + 2\mathscr{S}[\bar{\boldsymbol{\xi}}_\alpha^{(k)}\mathbf{e}_\alpha^{(l)\top}]\Big)\\ &-\frac{1}{N^2}\sum_{\alpha=1}^{N}\sum_{k,l,m,n=1}^{L}\bar{W}_\alpha^{(kl)}\bar{W}_\alpha^{(mn)}\Big((\bar{\boldsymbol{\xi}}_\alpha^{(k)}, \bar{\mathsf{M}}^{-}\bar{\boldsymbol{\xi}}_\alpha^{(m)})V_0^{(ln)}[\boldsymbol{\xi}_\alpha]\\ &+2\mathscr{S}[V_0^{(km)}[\boldsymbol{\xi}_\alpha]\bar{\mathsf{M}}^{-}\bar{\boldsymbol{\xi}}_\alpha^{(l)}\bar{\boldsymbol{\xi}}_\alpha^{(n)\top}]\Big),\end{aligned} \tag{14.30}$$

where $\mathscr{S}[\,\cdot\,]$ denotes symmetrization ($\mathscr{S}[\mathsf{A}] = (\mathsf{A} + \mathsf{A}^\top)/2$).

Because the matrix $\bar{\mathsf{N}}$ in Eq. (14.30) is defined in terms of true values, we replace them by observations in actual computation. In the presence of noise, however, the matrix M defined by observations $\boldsymbol{\xi}_\alpha^{(k)}$ no longer has rank $n-1$; it is generally positive definite with positive eigenvalues. Therefore we replace $\bar{\mathsf{M}}^{-}$ by the pseudoinverse M_{n-1}^{-} of truncated rank $n-1$ (see Eq. (2.30)). Using the resulting matrix N, we obtain the following hyper-renormalization scheme (see Procedures 2.4 and 6.3).

1. Initialize $\boldsymbol{\theta}$ and compute the matrices M and N.
2. Solve the generalized eigenvalue problem $\mathsf{M}\boldsymbol{\theta} = \lambda\mathsf{N}\boldsymbol{\theta}$, and compute the unit generalized eigenvector $\boldsymbol{\theta}$ for the generalized eigenvalue λ of the smallest absolute value.
3. Using the computed $\boldsymbol{\theta}$, recompute the matrices M and N and go back to Step 2. Repeat this until $\boldsymbol{\theta}$ converges.

The matrix N is not positive definite in general. In order to use a standard numerical tool, we modify $\mathsf{M}\boldsymbol{\theta} = \lambda\mathsf{N}\boldsymbol{\theta}$ into $\mathsf{N}\boldsymbol{\theta} = (1/\lambda)\mathsf{M}\boldsymbol{\theta}$ (see Eq. (2.24)). The bias of the resulting solution $\boldsymbol{\theta}$ is $O(\sigma^4)$; replacing $\bar{\boldsymbol{\xi}}_\alpha^{(k)}$ and $\bar{\mathsf{M}}$ by $\boldsymbol{\xi}_\alpha^{(k)}$ and M introduces only errors of $O(\sigma^4)$, inasmuch as the third-order noise terms have expectation 0.

14.6 Derivations

For deriving various identities, we first note important identities involving pseudoinverses. Let $V_\alpha^{(kl)} = (\boldsymbol{\theta}, V_0^{(kl)}\boldsymbol{\theta})$, and V_α the matrix whose (k, l) element is $V_\alpha^{(kl)}$. Let W_α be the matrix whose (k, l) element is $W_\alpha^{(kl)}$. From Eq. (14.6), we see that $\mathsf{W}_\alpha = (\mathsf{V}_\alpha)_r^-$. Hence, the identities

$$\mathsf{V}_\alpha\mathsf{W}_\alpha\mathsf{V}_\alpha = \mathsf{V}_\alpha, \qquad \mathsf{W}_\alpha\mathsf{V}_\alpha\mathsf{W}_\alpha = \mathsf{W}_\alpha. \tag{14.31}$$

hold (↪ Problem 14.3). In elements, these are

$$\sum_{m,n=1}^{L} V_\alpha^{(km)}W_\alpha^{(mn)}V_\alpha^{(nl)} = V_\alpha^{(kl)}, \quad \sum_{m,n=1}^{L} W_\alpha^{(km)}V_\alpha^{(mn)}W_\alpha^{(nl)} = W_\alpha^{(kl)}. \tag{14.32}$$

These identities hold for the true values $\bar{\mathsf{W}}_\alpha = \left(\bar{W}_\alpha^{(kl)}\right)$ and $\bar{\mathsf{V}}_\alpha = \left(\bar{V}_\alpha^{(kl)}\right)$, too.

Proposition 14.1 (Errors of weights) *The first and second error terms of $W_\alpha^{(kl)}$ are given by Eqs. (14.12) and (14.13).*

Proof Let $\bar{\mathsf{W}}_\alpha$ and $\bar{\mathsf{V}}_\alpha$ be the true values of W_α and V_α, respectively. In the presence of noise, we expand the first equation of Eq. (14.31) in the form

$$\begin{aligned}&(\bar{\mathsf{V}}_\alpha + \Delta_1\mathsf{V}_\alpha + \Delta_2\mathsf{V}_\alpha + \cdots)(\bar{\mathsf{W}}_\alpha + \Delta_1\mathsf{W}_\alpha + \Delta_2\mathsf{W}_\alpha + \cdots)(\bar{\mathsf{V}}_\alpha + \Delta_1\mathsf{V}_\alpha \\ &+\Delta_2\mathsf{V}_\alpha + \cdots) = (\bar{\mathsf{V}}_\alpha + \Delta_1\mathsf{V}_\alpha + \Delta_2\mathsf{V}_\alpha + \cdots).\end{aligned} \tag{14.33}$$

We now show that Eqs. (14.12) and (14.13) are obtained by equating the terms of the same order on both sides.

Equating the first-order terms on both sides, we obtain

$$\Delta_1\mathsf{V}_\alpha\bar{\mathsf{W}}_\alpha\bar{\mathsf{V}}_\alpha + \bar{\mathsf{V}}_\alpha\Delta_1\mathsf{W}_\alpha\bar{\mathsf{V}}_\alpha + \bar{\mathsf{V}}_\alpha\bar{\mathsf{W}}_\alpha\Delta_1\mathsf{V}_\alpha = \Delta_1\mathsf{V}_\alpha. \tag{14.34}$$

Multiplying this by $\bar{\mathsf{W}}_\alpha$ on both sides from the left and right, we have

$$\bar{\mathsf{W}}_\alpha\Delta_1\mathsf{V}_\alpha\bar{\mathsf{W}}_\alpha\bar{\mathsf{V}}_\alpha\bar{\mathsf{W}}_\alpha + \bar{\mathsf{W}}_\alpha\bar{\mathsf{V}}_\alpha\Delta_1\mathsf{W}_\alpha\bar{\mathsf{V}}_\alpha\bar{\mathsf{W}}_\alpha + \bar{\mathsf{W}}_\alpha\bar{\mathsf{V}}_\alpha\bar{\mathsf{W}}_\alpha\Delta_1\mathsf{V}_\alpha\bar{\mathsf{W}}_\alpha = \bar{\mathsf{W}}_\alpha\Delta_1\mathsf{V}_\alpha\bar{\mathsf{W}}_\alpha. \tag{14.35}$$

The product $\bar{\mathsf{V}}_\alpha\bar{\mathsf{W}}_\alpha = \bar{\mathsf{W}}_\alpha\bar{\mathsf{V}}_\alpha$ operates as the projection onto the orthogonal complement of the (common) null space of $\bar{\mathsf{V}}_\alpha$ and $\bar{\mathsf{W}}_\alpha$ (↪ Problem 14.4), and the variations $\Delta_1\mathsf{V}_\alpha$ and $\Delta_1\mathsf{W}_\alpha$ take place within that domain. Using the identities of Eq. (14.31), we write Eq. (14.35) in the form

$$\bar{\mathsf{W}}_\alpha\Delta_1\mathsf{V}_\alpha\bar{\mathsf{W}}_\alpha + \Delta_1\mathsf{W}_\alpha + \bar{\mathsf{W}}_\alpha\Delta_1\mathsf{V}_\alpha\bar{\mathsf{W}}_\alpha = \bar{\mathsf{W}}_\alpha\Delta_1\mathsf{V}_\alpha\bar{\mathsf{W}}_\alpha, \tag{14.36}$$

from which we obtain $\Delta_1 \mathsf{W}_\alpha$ in the form

$$\Delta_1 \mathsf{W}_\alpha = -\bar{\mathsf{W}}_\alpha \Delta_1 \mathsf{V}_\alpha \bar{\mathsf{W}}_\alpha. \tag{14.37}$$

Its (k, l) element is written as

$$\Delta_1 W_\alpha^{(kl)} = -\sum_{m,n=1}^{L} \bar{W}_\alpha^{(km)} \bar{W}_\alpha^{ln} \Delta_1 V_\alpha^{(mn)} = -2 \sum_{m,n=1}^{L} \bar{W}_\alpha^{(km)} \bar{W}_\alpha^{(ln)} (\Delta_1 \boldsymbol{\theta}, V_0^{(mn)}[\boldsymbol{\xi}_\alpha]\bar{\boldsymbol{\theta}}). \tag{14.38}$$

Thus we obtain Eq. (14.12).

Equating the second-order terms from both sides of Eq. (14.33), we have

$$\Delta_2 \mathsf{V}_\alpha \bar{\mathsf{W}}_\alpha \bar{\mathsf{V}}_\alpha + \bar{\mathsf{V}}_\alpha \Delta_2 \mathsf{W}_\alpha \bar{\mathsf{V}}_\alpha + \bar{\mathsf{V}}_\alpha \bar{\mathsf{W}}_\alpha \Delta_2 \mathsf{V}_\alpha + \bar{\mathsf{V}}_\alpha \Delta_1 \mathsf{W}_\alpha \Delta_1 \mathsf{V}_\alpha + \Delta_1 \mathsf{V}_\alpha \bar{\mathsf{W}}_\alpha \Delta_1 \mathsf{V}_\alpha \\ + \Delta_1 \mathsf{V}_\alpha \Delta_1 \mathsf{W}_\alpha \bar{\mathsf{V}}_\alpha = \Delta_2 \mathsf{V}_\alpha. \tag{14.39}$$

Multiplying this by $\bar{\mathsf{W}}_\alpha$ on both sides from the left and right, we obtain

$$\bar{\mathsf{W}}_\alpha \Delta_2 \mathsf{V}_\alpha \bar{\mathsf{W}}_\alpha \bar{\mathsf{V}}_\alpha \bar{\mathsf{W}}_\alpha + \bar{\mathsf{W}}_\alpha \bar{\mathsf{V}}_\alpha \Delta_2 \mathsf{W}_\alpha \bar{\mathsf{V}}_\alpha \bar{\mathsf{W}}_\alpha + \bar{\mathsf{W}}_\alpha \bar{\mathsf{V}}_\alpha \bar{\mathsf{W}}_\alpha \Delta_2 \mathsf{V}_\alpha \bar{\mathsf{W}}_\alpha \\ + \bar{\mathsf{W}}_\alpha \bar{\mathsf{V}}_\alpha \Delta_1 \mathsf{W}_\alpha \Delta_1 \mathsf{V}_\alpha \bar{\mathsf{W}}_\alpha + \bar{\mathsf{W}}_\alpha \Delta_1 \mathsf{V}_\alpha \bar{\mathsf{W}}_\alpha \Delta_1 \mathsf{V}_\alpha \bar{\mathsf{W}}_\alpha + \bar{\mathsf{W}}_\alpha \Delta_1 \mathsf{V}_\alpha \Delta_1 \mathsf{W}_\alpha \bar{\mathsf{V}}_\alpha \bar{\mathsf{W}}_\alpha \\ = \bar{\mathsf{W}}_\alpha \Delta_2 \mathsf{V}_\alpha \bar{\mathsf{W}}_\alpha, \tag{14.40}$$

which is rewritten as

$$\bar{\mathsf{W}}_\alpha \Delta_2 \mathsf{V}_\alpha \bar{\mathsf{W}}_\alpha + \Delta_2 \mathsf{W}_\alpha + \bar{\mathsf{W}}_\alpha \Delta_2 \mathsf{V}_\alpha \bar{\mathsf{W}}_\alpha + \Delta_1 \mathsf{W}_\alpha \Delta_1 \mathsf{V}_\alpha \bar{\mathsf{W}}_\alpha \\ + \bar{\mathsf{W}}_\alpha \Delta_1 \mathsf{V}_\alpha \bar{\mathsf{W}}_\alpha \Delta_1 \mathsf{V}_\alpha \bar{\mathsf{W}}_\alpha + \bar{\mathsf{W}}_\alpha \Delta_1 \mathsf{V}_\alpha \Delta_1 \mathsf{W}_\alpha = \bar{\mathsf{W}}_\alpha \Delta_2 \mathsf{V}_\alpha \bar{\mathsf{W}}_\alpha. \tag{14.41}$$

This is further rewritten as follows.

$$\bar{\mathsf{W}}_\alpha \Delta_2 \mathsf{V}_\alpha \bar{\mathsf{W}}_\alpha + \Delta_2 \mathsf{W}_\alpha + \bar{\mathsf{W}}_\alpha \Delta_2 \mathsf{V}_\alpha \bar{\mathsf{W}}_\alpha + \Delta_1 \mathsf{W}_\alpha (\bar{\mathsf{V}}_\alpha \bar{\mathsf{W}}_\alpha) \Delta_1 \mathsf{V}_\alpha \bar{\mathsf{W}}_\alpha \\ + \bar{\mathsf{W}}_\alpha \Delta_1 \mathsf{V}_\alpha \bar{\mathsf{W}}_\alpha (\bar{\mathsf{V}}_\alpha \bar{\mathsf{W}}_\alpha) \Delta_1 \mathsf{V}_\alpha \bar{\mathsf{W}}_\alpha + \bar{\mathsf{W}}_\alpha \Delta_1 \mathsf{V}_\alpha (\bar{\mathsf{W}}_\alpha \bar{\mathsf{V}}_\alpha) \Delta_1 \mathsf{W}_\alpha \\ = \bar{\mathsf{W}}_\alpha \Delta_2 \mathsf{V}_\alpha \bar{\mathsf{W}}_\alpha. \tag{14.42}$$

Substituting Eq. (14.37), we obtain

$$\bar{\mathsf{W}}_\alpha \Delta_2 \mathsf{V}_\alpha \bar{\mathsf{W}}_\alpha + \Delta_2 \mathsf{W}_\alpha + \bar{\mathsf{W}}_\alpha \Delta_2 \mathsf{V}_\alpha \bar{\mathsf{W}}_\alpha - \Delta_1 \mathsf{W}_\alpha \bar{\mathsf{V}}_\alpha \Delta_1 \mathsf{W}_\alpha + \Delta_1 \mathsf{W}_\alpha \bar{\mathsf{V}}_\alpha \Delta_1 \mathsf{W}_\alpha \\ - \Delta_1 \mathsf{W}_\alpha \bar{\mathsf{V}}_\alpha \Delta_1 \mathsf{W}_\alpha = \bar{\mathsf{W}}_\alpha \Delta_2 \mathsf{V}_\alpha \bar{\mathsf{W}}_\alpha. \tag{14.43}$$

Hence, $\Delta_2 \mathsf{W}_\alpha$ is written as

$$\Delta_2 \mathsf{W}_\alpha = \Delta_1 \mathsf{W}_\alpha \bar{\mathsf{V}}_\alpha \Delta_1 \mathsf{W}_\alpha - \bar{\mathsf{W}}_\alpha \Delta_2 \mathsf{V}_\alpha \bar{\mathsf{W}}_\alpha, \tag{14.44}$$

whose (k, l) element is

$$\begin{aligned} \Delta_2 W_\alpha^{(kl)} &= \sum_{m,n=1}^{L} \Delta_1 W_\alpha^{(km)} \bar{V}_\alpha^{(mn)} \Delta_1 W_\alpha^{(nl)} - \sum_{m,n=1}^{L} \bar{W}_\alpha^{(km)} \Delta_2 V_\alpha^{(mn)} \bar{W}_\alpha^{(nl)} \\ &= \sum_{m,n=1}^{L} \Delta_1 W_\alpha^{(km)} \Delta_1 W_\alpha^{(ln)} (\bar{\boldsymbol{\theta}}, V_0^{(mn)}[\boldsymbol{\xi}_\alpha]\bar{\boldsymbol{\theta}}) \\ &\quad - \sum_{m,n=1}^{L} \bar{W}_\alpha^{(km)} \bar{W}_\alpha^{(ln)} \Big((\Delta_1 \boldsymbol{\theta}, V_0^{(mn)}[\boldsymbol{\xi}_\alpha] \Delta_1 \boldsymbol{\theta}) + 2(\Delta_2 \boldsymbol{\theta}, V_0^{(mn)}[\boldsymbol{\xi}_\alpha]\bar{\boldsymbol{\theta}}) \Big). \end{aligned} \tag{14.45}$$

Thus, we obtain Eq. (14.13). □

Proposition 14.2 (Covariance of the solution) *The covariance matrix of $\boldsymbol{\theta}$ is given by Eq. (14.19).*

Proof Substituting Eq. (14.10) into Eq. (14.18) and noting that $\boldsymbol{\xi}_\alpha^{(k)\top}\boldsymbol{\theta} = 0$, we can write $\Delta_1\boldsymbol{\theta}$ as follows.

$$\Delta_1\boldsymbol{\theta} = -\bar{\mathsf{M}}^-\Delta_1\mathsf{M}\bar{\boldsymbol{\theta}} = -\bar{\mathsf{M}}^-\Big(\frac{1}{N}\sum_{\alpha=1}^{N}\sum_{k,l=1}^{L}\bar{W}_\alpha^{(kl)}(\Delta_1\boldsymbol{\xi}_\alpha^{(l)},\bar{\boldsymbol{\theta}})\bar{\boldsymbol{\xi}}_\alpha^{(k)}\Big). \tag{14.46}$$

We now evaluate $V[\boldsymbol{\theta}] = E[\Delta_1\boldsymbol{\theta}\Delta_1\boldsymbol{\theta}^\top]$, the covariance matrix of $\boldsymbol{\theta}$ to a first approximation, by eliminating $\Delta_1\boldsymbol{\xi}_\alpha^{(k)}$ in the expectation, using Eqs. (14.31) and (14.32). From Eq. (14.28), we obtain

$$\begin{aligned}
V[\boldsymbol{\theta}] &= E[\bar{\mathsf{M}}^-\Big(\frac{1}{N}\sum_{\alpha=1}^{N}\sum_{k,l=1}^{L}\bar{W}_\alpha^{(kl)}(\Delta_1\boldsymbol{\xi}_\alpha^{(l)},\bar{\boldsymbol{\theta}})\bar{\boldsymbol{\xi}}_\alpha^{(k)}\Big) \\
&\quad \frac{1}{N}\sum_{\beta=1}^{N}\sum_{m,n=1}^{L}\bar{W}_\beta^{(mn)}(\Delta_1\boldsymbol{\xi}_\beta^{(n)},\bar{\boldsymbol{\theta}})\bar{\boldsymbol{\xi}}_\beta^{(m)\top}\Big)\bar{\mathsf{M}}^-] \\
&= E[\bar{\mathsf{M}}^-\Big(\frac{1}{N^2}\sum_{\alpha,\beta=1}^{N}\sum_{k,l,m,n=1}^{L}\bar{W}_\alpha^{(kl)}\bar{W}_\beta^{(mn)}(\bar{\boldsymbol{\theta}},\Delta_1\boldsymbol{\xi}_\alpha^{(l)})(\Delta_1\boldsymbol{\xi}_\beta^{(n)},\bar{\boldsymbol{\theta}})\bar{\boldsymbol{\xi}}_\alpha^{(k)}\bar{\boldsymbol{\xi}}_\beta^{(m)\top}\Big)\bar{\mathsf{M}}^-] \\
&= \bar{\mathsf{M}}^-\Big(\frac{1}{N^2}\sum_{\alpha,\beta=1}^{N}\sum_{k,l,m,n=1}^{L}\bar{W}_\alpha^{(kl)}\bar{W}_\beta^{(mn)}(\bar{\boldsymbol{\theta}},E[\Delta_1\boldsymbol{\xi}_\alpha^{(l)}\Delta_1\boldsymbol{\xi}_\beta^{\top(n)}]\bar{\boldsymbol{\theta}})\bar{\boldsymbol{\xi}}_\alpha^{(k)}\bar{\boldsymbol{\xi}}_\beta^{(m)\top}\Big)\bar{\mathsf{M}}^- \\
&= \bar{\mathsf{M}}^-\Big(\frac{1}{N^2}\sum_{\alpha,\beta=1}^{N}\sum_{k,l,m,n=1}^{L}\bar{W}_\alpha^{(kl)}\bar{W}_\beta^{(mn)}(\bar{\boldsymbol{\theta}},\sigma^2\delta_{\alpha\beta}V_0^{(ln)}[\boldsymbol{\xi}_\alpha]\bar{\boldsymbol{\theta}})\bar{\boldsymbol{\xi}}_\alpha^{(k)}\bar{\boldsymbol{\xi}}_\beta^{(m)\top}\Big)\bar{\mathsf{M}}^- \\
&= \bar{\mathsf{M}}^-\Big(\frac{\sigma^2}{N^2}\sum_{\alpha=1}^{N}\sum_{k,m=1}^{L}\Big(\sum_{l,n=1}^{L}\bar{W}_\alpha^{(kl)}(\bar{\boldsymbol{\theta}},V_0^{(ln)}[\boldsymbol{\xi}_\alpha]\bar{\boldsymbol{\theta}})\bar{W}_\alpha^{(mn)}\Big)\bar{\boldsymbol{\xi}}_\alpha^{(k)}\bar{\boldsymbol{\xi}}_\alpha^{(m)\top}\Big)\bar{\mathsf{M}}^- \\
&= \bar{\mathsf{M}}^-\Big(\frac{\sigma^2}{N^2}\sum_{\alpha=1}^{N}\sum_{k,m=1}^{L}\Big(\sum_{l,n=1}^{L}\bar{W}_\alpha^{(kl)}\bar{V}_\alpha^{(ln)}\bar{W}_\alpha^{(mn)}\Big)\bar{\boldsymbol{\xi}}_\alpha^{(k)}\bar{\boldsymbol{\xi}}_\alpha^{(m)\top}\Big)\bar{\mathsf{M}}^- \\
&= \frac{\sigma^2}{N}\bar{\mathsf{M}}^-\Big(\frac{1}{N}\sum_{\alpha=1}^{N}\sum_{k,m=1}^{L}\bar{W}_\alpha^{(km)}\bar{\boldsymbol{\xi}}_\alpha^{(k)}\bar{\boldsymbol{\xi}}_\alpha^{(m)\top}\Big)\bar{\mathsf{M}}^- = \frac{\sigma^2}{N}\bar{\mathsf{M}}^-\bar{\mathsf{M}}\bar{\mathsf{M}}^- = \frac{\sigma^2}{N}\bar{\mathsf{M}}^-.
\end{aligned} \tag{14.47}$$

□

In order to derive Eq. (14.30), we evaluate the expression of Eq. (14.21) term by term.

Lemma 14.1 (Expectation of $\Delta_2 \mathsf{M}\bar{\boldsymbol{\theta}}$)

$$E[\Delta_2 \mathsf{M}\bar{\boldsymbol{\theta}}] = \frac{\sigma^2}{N}\sum_{\alpha=1}^{N}\sum_{k,l=1}^{L} \bar{W}_\alpha^{(kl)}\Big(V_0^{(kl)}[\boldsymbol{\xi}_\alpha]\bar{\boldsymbol{\theta}} + (\mathbf{e}_\alpha^{(l)}, \bar{\boldsymbol{\theta}})\bar{\boldsymbol{\xi}}_\alpha^{(k)}\Big) + \frac{2\sigma^2}{N^2}\sum_{\alpha=1}^{N}\sum_{k,m,n,p=1}^{L} \bar{W}_\alpha^{(km)}\bar{W}_\alpha^{(np)}(\bar{\boldsymbol{\xi}}_\alpha^{(p)}, \bar{\mathsf{M}}^- V_0^{(mn)}[\boldsymbol{\xi}_\alpha]\bar{\boldsymbol{\theta}})\bar{\boldsymbol{\xi}}_\alpha^{(k)}. \tag{14.48}$$

Proof From $(\bar{\boldsymbol{\xi}}_\alpha, \bar{\boldsymbol{\theta}}) = 0$ and Eq. (14.11), we obtain

$$\begin{aligned}
\Delta_2 \mathsf{M}\bar{\boldsymbol{\theta}} &= \frac{1}{N}\sum_{\alpha=1}^{N}\sum_{k,l=1}^{L} \bar{W}_\alpha^{(kl)}\Big(\Delta_1\boldsymbol{\xi}_\alpha^{(k)}\Delta_1\boldsymbol{\xi}_\alpha^{(l)\top} + \Delta_2\boldsymbol{\xi}_\alpha^{(k)}\bar{\boldsymbol{\xi}}_\alpha^{(l)\top} + \bar{\boldsymbol{\xi}}_\alpha^{(k)}\Delta_2\boldsymbol{\xi}_\alpha^{(l)\top}\Big)\bar{\boldsymbol{\theta}} \\
&+ \frac{1}{N}\sum_{\alpha=1}^{N}\sum_{k,l=1}^{L} \Delta_1 W_\alpha^{(kl)}(\Delta_1\boldsymbol{\xi}_\alpha^{(k)}\bar{\boldsymbol{\xi}}_\alpha^{(l)\top} + \bar{\boldsymbol{\xi}}_\alpha^{(k)}\Delta_1\boldsymbol{\xi}_\alpha^{(l)\top})\bar{\boldsymbol{\theta}} \\
&+ \frac{1}{N}\sum_{\alpha=1}^{N}\sum_{k,l=1}^{L} \Delta_2 W_\alpha^{(kl)}\bar{\boldsymbol{\xi}}_\alpha^{(k)}\bar{\boldsymbol{\xi}}_\alpha^{(l)\top}\bar{\boldsymbol{\theta}} \\
&= \frac{1}{N}\sum_{\alpha=1}^{N}\sum_{k,l=1}^{L} \bar{W}_\alpha^{(kl)}((\Delta_1\boldsymbol{\xi}_\alpha^{(l)}, \bar{\boldsymbol{\theta}})\Delta_1\boldsymbol{\xi}_\alpha^{(k)} + (\Delta_2\boldsymbol{\xi}_\alpha^{(l)}, \bar{\boldsymbol{\theta}})\bar{\boldsymbol{\xi}}_\alpha^{(k)}) \\
&+ \frac{1}{N}\sum_{\alpha=1}^{N}\sum_{k,l=1}^{L} \Delta_1 W_\alpha^{(kl)}(\Delta_1\boldsymbol{\xi}_\alpha^{(l)}, \bar{\boldsymbol{\theta}})\bar{\boldsymbol{\xi}}_\alpha^{(k)}.
\end{aligned} \tag{14.49}$$

Therefore

$$\begin{aligned}
E[\Delta_2 \mathsf{M}\bar{\boldsymbol{\theta}}] &= \frac{1}{N}\sum_{\alpha=1}^{N}\sum_{k,l=1}^{L} \bar{W}_\alpha^{(kl)}\Big(E[\Delta_1\boldsymbol{\xi}_\alpha^{(k)}\Delta_1\boldsymbol{\xi}_\alpha^{(l)\top}]\bar{\boldsymbol{\theta}} + (E[\Delta_2\boldsymbol{\xi}_\alpha^{(l)}], \bar{\boldsymbol{\theta}})\bar{\boldsymbol{\xi}}_\alpha^{(k)}\Big) \\
&+ \frac{1}{N}\sum_{\alpha=1}^{N}\sum_{k,l=1}^{L} (E[\Delta_1 W_\alpha^{(kl)}\Delta_1\boldsymbol{\xi}_\alpha^{(l)}], \bar{\boldsymbol{\theta}})\bar{\boldsymbol{\xi}}_\alpha^{(k)} \\
&= \frac{\sigma^2}{N}\sum_{\alpha=1}^{N}\sum_{k,l=1}^{L} \bar{W}_\alpha^{(kl)}\Big(V_0^{(kl)}[\boldsymbol{\xi}_\alpha]\bar{\boldsymbol{\theta}} + (\mathbf{e}_\alpha^{(l)}, \bar{\boldsymbol{\theta}})\bar{\boldsymbol{\xi}}_\alpha^{(k)}\Big) \\
&+ \frac{1}{N}\sum_{\alpha=1}^{N}\sum_{k,l=1}^{L} (E[\Delta_1 W_\alpha^{(kl)}\Delta_1\boldsymbol{\xi}_\alpha^{(l)}], \bar{\boldsymbol{\theta}})\bar{\boldsymbol{\xi}}_\alpha^{(k)}.
\end{aligned} \tag{14.50}$$

Consider the expectation of $\Delta_1 W_\alpha \Delta_1 \boldsymbol{\xi}_\alpha$. From Eqs. (14.10), (14.12), and (14.18), we see that

$$\Delta_1 W_\alpha^{(kl)} = -2\sum_{m,n=1}^{L} \bar{W}_\alpha^{(km)}\bar{W}_\alpha^{(ln)}(\Delta_1\boldsymbol{\theta}, V_0^{(mn)}[\boldsymbol{\xi}_\alpha]\bar{\boldsymbol{\theta}})$$

$$
\begin{aligned}
&= 2\sum_{m,n=1}^{L} \bar{W}_\alpha^{(km)}\bar{W}_\alpha^{(ln)}(\bar{\mathsf{M}}^-\Delta_1\mathsf{M}\bar{\boldsymbol{\theta}}, V_0^{(mn)}[\boldsymbol{\xi}_\alpha]\bar{\boldsymbol{\theta}}) \\
&= 2\sum_{m,n=1}^{L} \bar{W}_\alpha^{(km)}\bar{W}_\alpha^{(ln)}(\bar{\mathsf{M}}^-\Big(\frac{1}{N}\sum_{\beta=1}^{N}\sum_{p,q=1}^{L} \bar{W}_\beta^{(pq)}\Big(\Delta_1\boldsymbol{\xi}_\beta^{(p)}\bar{\boldsymbol{\xi}}_\beta^{(q)\top} + \bar{\boldsymbol{\xi}}_\beta^{(p)}\Delta_1\boldsymbol{\xi}_\beta^{(q)\top}\Big) \\
&\quad + \frac{1}{N}\sum_{\beta=1}^{N}\sum_{p,q=1}^{L} \Delta_1\bar{W}_\beta^{(pq)}\bar{\boldsymbol{\xi}}_\beta^{(p)}\bar{\boldsymbol{\xi}}_\beta^{(q)\top}\Big)\bar{\boldsymbol{\theta}}, V_0^{(mn)}[\boldsymbol{\xi}_\alpha]\bar{\boldsymbol{\theta}}) \\
&= 2\sum_{m,n=1}^{L} \bar{W}_\alpha^{(km)}\bar{W}_\alpha^{(ln)}(\bar{\mathsf{M}}^-\Big(\frac{1}{N}\sum_{\beta=1}^{N}\sum_{p,q=1}^{L} \bar{W}_\beta^{(pq)}\bar{\boldsymbol{\xi}}_\beta^{(p)}(\Delta_1\boldsymbol{\xi}_\beta^{(q)}, \bar{\boldsymbol{\theta}})\Big), V_0^{(mn)}[\boldsymbol{\xi}_\alpha]\bar{\boldsymbol{\theta}}) \\
&= \frac{2}{N}\sum_{\beta=1}^{N}\sum_{m,n,p,q=1}^{L} \bar{W}_\alpha^{(km)}\bar{W}_\alpha^{(ln)}\bar{W}_\beta^{(pq)}(\Delta_1\boldsymbol{\xi}_\beta^{(q)}, \bar{\boldsymbol{\theta}})(\bar{\mathsf{M}}^-\bar{\boldsymbol{\xi}}_\beta^{(p)}, V_0^{(mn)}[\boldsymbol{\xi}_\alpha]\bar{\boldsymbol{\theta}}) \\
&= \frac{2}{N}\sum_{\beta=1}^{N}\sum_{m,n,p,q=1}^{L} \bar{W}_\alpha^{(km)}\bar{W}_\alpha^{(ln)}\bar{W}_\beta^{(pq)}(\bar{\boldsymbol{\xi}}_\beta^{(p)}, \bar{\mathsf{M}}^- V_0^{(mn)}[\boldsymbol{\xi}_\alpha]\bar{\boldsymbol{\theta}})(\Delta_1\boldsymbol{\xi}_\beta^{(q)}, \bar{\boldsymbol{\theta}}).
\end{aligned}
\tag{14.51}
$$

Therefore

$$
\begin{aligned}
&E[\Delta_1 W_\alpha^{(kl)}\Delta_1\boldsymbol{\xi}_\alpha^{(l)}] \\
&= E[\frac{2}{N}\sum_{\beta=1}^{N}\sum_{m,n,p,q=1}^{L} \bar{W}_\alpha^{(km)}\bar{W}_\alpha^{(ln)}\bar{W}_\beta^{(pq)}(\bar{\boldsymbol{\xi}}_\beta^{(p)}, \bar{\mathsf{M}}^- V_0^{(mn)}[\boldsymbol{\xi}_\alpha]\bar{\boldsymbol{\theta}})(\Delta_1\boldsymbol{\xi}_\beta^{(q)}, \bar{\boldsymbol{\theta}})\Delta_1\boldsymbol{\xi}_\alpha^{(l)}] \\
&= \frac{2}{N}\sum_{\beta=1}^{N}\sum_{m,n,p,q=1}^{L} \bar{W}_\alpha^{(km)}\bar{W}_\alpha^{(ln)}\bar{W}_\beta^{(pq)}(\bar{\boldsymbol{\xi}}_\beta^{(p)}, \bar{\mathsf{M}}^- V_0^{(mn)}[\boldsymbol{\xi}_\alpha]\bar{\boldsymbol{\theta}})E[\Delta_1\boldsymbol{\xi}_\alpha^{(l)}\Delta_1\boldsymbol{\xi}_\beta^{(q)\top}]\bar{\boldsymbol{\theta}} \\
&= \frac{2}{N}\sum_{\beta=1}^{N}\sum_{m,n,p,q=1}^{L} \bar{W}_\alpha^{(km)}\bar{W}_\alpha^{(ln)}\bar{W}_\beta^{(pq)}(\bar{\boldsymbol{\xi}}_\beta^{(p)}, \bar{\mathsf{M}}^- V_0^{(mn)}[\boldsymbol{\xi}_\alpha]\bar{\boldsymbol{\theta}})\sigma^2\delta_{\alpha\beta}V_0^{(lq)}[\boldsymbol{\xi}_\alpha]\bar{\boldsymbol{\theta}} \\
&= \frac{2\sigma^2}{N}\sum_{m,n,p,q=1}^{L} \bar{W}_\alpha^{(km)}\bar{W}_\alpha^{(ln)}\bar{W}_\alpha^{(pq)}(\bar{\boldsymbol{\xi}}_\alpha^{(p)}, \bar{\mathsf{M}}^- V_0^{(mn)}[\boldsymbol{\xi}_\alpha]\bar{\boldsymbol{\theta}})V_0^{(lq)}[\boldsymbol{\xi}_\alpha]\bar{\boldsymbol{\theta}}.
\end{aligned}
\tag{14.52}
$$

It follows that

$$
\begin{aligned}
&\frac{1}{N}\sum_{\alpha=1}^{N}\sum_{k,l=1}^{L}(E[\Delta_1 W_\alpha^{(kl)}\Delta_1\boldsymbol{\xi}_\alpha^{(l)}], \bar{\boldsymbol{\theta}})\bar{\boldsymbol{\xi}}_\alpha^{(k)} \\
&= \frac{1}{N}\sum_{\alpha=1}^{N}\sum_{k,l=1}^{L}(\Big(\frac{2\sigma^2}{N}\sum_{m,n,p,q=1}^{L} \bar{W}_\alpha^{(km)}\bar{W}_\alpha^{(ln)}\bar{W}_\alpha^{(pq)} \\
&\quad (\bar{\boldsymbol{\xi}}_\alpha^{(p)}, \bar{\mathsf{M}}^- V_0^{(mn)}[\boldsymbol{\xi}_\alpha]\bar{\boldsymbol{\theta}})V_0^{(lq)}[\boldsymbol{\xi}_\alpha]\bar{\boldsymbol{\theta}}\Big), \bar{\boldsymbol{\theta}})\bar{\boldsymbol{\xi}}_\alpha^{(k)} \\
&= \frac{2\sigma^2}{N^2}\sum_{\alpha=1}^{N}\sum_{k,l,m,n,p,q=1}^{L} \bar{W}_\alpha^{(km)}\bar{W}_\alpha^{(ln)}\bar{W}_\alpha^{(pq)}(\bar{\boldsymbol{\xi}}_\alpha^{(p)}, \bar{\mathsf{M}}^- V_0^{(mn)}[\boldsymbol{\xi}_\alpha]\bar{\boldsymbol{\theta}})(\bar{\boldsymbol{\theta}}, V_0^{(lq)}[\boldsymbol{\xi}_\alpha]\bar{\boldsymbol{\theta}})\bar{\boldsymbol{\xi}}_\alpha^{(k)}
\end{aligned}
$$

$$
=\frac{2\sigma^2}{N^2}\sum_{\alpha=1}^{N}\sum_{k,m,n,p=1}^{L}\bar{W}_\alpha^{(km)}(\bar{\xi}_\alpha^{(p)},\bar{\mathsf{M}}^-V_0^{(mn)}[\xi_\alpha]\bar{\theta})
$$

$$
\Big(\sum_{l,q=1}^{L}\bar{W}_\alpha^{(nl)}(\bar{\theta},V_0^{(lq)}[\xi_\alpha]\bar{\theta})\bar{W}_\alpha^{(qp)}\Big)\bar{\xi}_\alpha^{(k)}
$$

$$
=\frac{2\sigma^2}{N^2}\sum_{\alpha=1}^{N}\sum_{k,m,n,p=1}^{L}\bar{W}_\alpha^{(km)}(\bar{\xi}_\alpha^{(p)},\bar{\mathsf{M}}^-V_0^{(mn)}[\xi_\alpha]\bar{\theta})\bar{W}_\alpha^{(np)}\bar{\xi}_\alpha^{(k)}
$$

$$
=\frac{2\sigma^2}{N^2}\sum_{\alpha=1}^{N}\sum_{k,m,n,p=1}^{L}\bar{W}_\alpha^{(km)}\bar{W}_\alpha^{(np)}(\bar{\xi}_\alpha^{(p)},\bar{\mathsf{M}}^-V_0^{(mn)}[\xi_\alpha]\bar{\theta})\bar{\xi}_\alpha^{(k)}. \tag{14.53}
$$

Substituting Eq. (14.53) into Eq. (14.50), we obtain Eq. (14.48). □

Lemma 14.2 (Expectation of $\Delta_1\mathsf{M}\bar{\mathsf{M}}^-\Delta_1\mathsf{M}\bar{\theta}$)

$$
E[\Delta_1\mathsf{M}\bar{\mathsf{M}}^-\Delta_1\mathsf{M}\bar{\theta}]=\frac{\sigma^2}{N^2}\sum_{\alpha=1}^{N}\sum_{k,l,m,n=1}^{L}\bar{W}_\alpha^{(kl)}\bar{W}_\alpha^{(mn)}(\bar{\xi}_\alpha^{(m)},\bar{\mathsf{M}}^-\bar{\xi}_\alpha^{(k)})V_0^{(ln)}[\xi_\alpha]\bar{\theta}
$$

$$
+\frac{3\sigma^2}{N^2}\sum_{\alpha=1}^{N}\sum_{k,l,m,n=1}^{L}\bar{W}_\alpha^{(kl)}\bar{W}_\alpha^{(mn)}(\bar{\xi}_\alpha^{(k)},\bar{\mathsf{M}}^-V_0^{(lm)}[\xi_\alpha]\bar{\theta})\bar{\xi}_\alpha^{(n)}. \tag{14.54}
$$

Proof From Eq. (14.10), we can write

$$
\Delta_1\mathsf{M}\bar{\theta}=\frac{1}{N}\sum_{\alpha=1}^{N}\sum_{k,l=1}^{L}\bar{W}_\alpha^{(kl)}\Big(\Delta_1\xi_\alpha^{(k)}\bar{\xi}_\alpha^{(l)\top}+\bar{\xi}_\alpha^{(k)}\Delta_1\xi_\alpha^{(l)\top}\Big)\bar{\theta}
$$

$$
+\frac{1}{N}\sum_{\alpha=1}^{N}\sum_{k,l=1}^{L}\Delta_1\bar{W}_\alpha^{(kl)}\bar{\xi}_\alpha^{(k)}\bar{\xi}_\alpha^{(l)\top}\bar{\theta}=\frac{1}{N}\sum_{\alpha=1}^{N}\sum_{k,l=1}^{L}\bar{W}_\alpha^{(kl)}(\Delta_1\xi_\alpha^{(l)},\bar{\theta})\bar{\xi}_\alpha^{(k)}. \tag{14.55}
$$

We can also write

$$
\Delta_1\mathsf{M}\bar{\mathsf{M}}^-\Delta_1\mathsf{M}\bar{\theta}=\Delta_1\mathsf{M}\bar{\mathsf{M}}^-\Big(\frac{1}{N}\sum_{\alpha=1}^{N}\sum_{k,l=1}^{L}\bar{W}_\alpha^{(kl)}(\Delta_1\xi_\alpha^{(l)},\bar{\theta})\bar{\xi}_\alpha^{(k)}\Big)
$$

$$
=\Big(\frac{1}{N}\sum_{\beta=1}^{N}\sum_{p,q=1}^{L}\bar{W}_\beta^{(pq)}\Big(\Delta_1\xi_\beta^{(p)}\bar{\xi}_\beta^{(q)\top}+\bar{\xi}_\beta^{(p)}\Delta_1\xi_\beta^{(q)\top}\Big)
$$

$$
+\frac{1}{N}\sum_{\beta=1}^{N}\sum_{p,q=1}^{L}\Delta_1\bar{W}_\beta^{(pq)}\bar{\xi}_\beta^{(p)}\bar{\xi}_\beta^{(q)\top}\Big)\bar{\mathsf{M}}^-\Big(\frac{1}{N}\sum_{\alpha=1}^{N}\sum_{k,l=1}^{L}\bar{W}_\alpha^{(kl)}(\Delta_1\xi_\alpha^{(l)},\bar{\theta})\bar{\xi}_\alpha^{(k)}\Big)
$$

$$
=\frac{1}{N}\sum_{\beta=1}^{N}\sum_{p,q=1}^{L}\bar{W}_\beta^{(pq)}\Big(\Delta_1\xi_\beta^{(p)}\bar{\xi}_\beta^{(q)\top}+\bar{\xi}_\beta^{(p)}\Delta_1\xi_\beta^{(q)\top}\Big)\bar{\mathsf{M}}^-
$$

$$
\Big(\frac{1}{N}\sum_{\alpha=1}^{N}\sum_{k,l=1}^{L}\bar{W}_\alpha^{(kl)}(\Delta_1\boldsymbol{\xi}_\alpha^{(l)},\bar{\boldsymbol{\theta}})\bar{\boldsymbol{\xi}}_\alpha^{(k)}\Big)
+\frac{1}{N}\sum_{\beta=1}^{N}\sum_{p,q=1}^{L}\Delta_1\bar{W}_\beta^{(pq)}\bar{\boldsymbol{\xi}}_\beta^{(p)}\bar{\boldsymbol{\xi}}_\beta^{(q)\top}\bar{\mathsf{M}}^-\Big(\frac{1}{N}\sum_{\alpha=1}^{N}\sum_{k,l=1}^{L}\bar{W}_\alpha^{(kl)}(\Delta_1\boldsymbol{\xi}_\alpha^{(l)},\bar{\boldsymbol{\theta}})\bar{\boldsymbol{\xi}}_\alpha^{(k)}\Big)
$$

$$
=\frac{1}{N^2}\sum_{\alpha,\beta=1}^{N}\sum_{k,l,p,q=1}^{L}\bar{W}_\alpha^{(kl)}\bar{W}_\beta^{(pq)}(\Delta_1\boldsymbol{\xi}_\alpha^{(l)},\bar{\boldsymbol{\theta}})\Big(\Delta_1\boldsymbol{\xi}_\beta^{(p)}\bar{\boldsymbol{\xi}}_\beta^{(q)\top}+\bar{\boldsymbol{\xi}}_\beta^{(p)}\Delta_1\boldsymbol{\xi}_\beta^{(q)\top}\Big)\bar{\mathsf{M}}^-\bar{\boldsymbol{\xi}}_\alpha^{(k)}
+\frac{1}{N^2}\sum_{\alpha,\beta=1}^{N}\sum_{k,l,p,q=1}^{L}\bar{W}_\alpha^{(kl)}\Delta_1\bar{W}_\beta^{(pq)}(\Delta_1\boldsymbol{\xi}_\alpha^{(l)},\bar{\boldsymbol{\theta}})\bar{\boldsymbol{\xi}}_\beta^{(p)}\bar{\boldsymbol{\xi}}_\beta^{(q)\top}\bar{\mathsf{M}}^-\bar{\boldsymbol{\xi}}_\alpha^{(k)}
$$

$$
=\frac{1}{N^2}\sum_{\alpha,\beta=1}^{N}\sum_{k,l,p,q=1}^{L}\bar{W}_\alpha^{(kl)}\bar{W}_\beta^{(pq)}(\Delta_1\boldsymbol{\xi}_\alpha^{(l)},\bar{\boldsymbol{\theta}})(\bar{\boldsymbol{\xi}}_\beta^{(q)},\bar{\mathsf{M}}^-\bar{\boldsymbol{\xi}}_\alpha^{(k)})\Delta_1\boldsymbol{\xi}_\beta^{(p)}\ (\equiv\mathbf{t}_1)
+\frac{1}{N^2}\sum_{\alpha,\beta=1}^{N}\sum_{k,l,p,q=1}^{L}\bar{W}_\alpha^{(kl)}\bar{W}_\beta^{(pq)}(\Delta_1\boldsymbol{\xi}_\alpha^{(l)},\bar{\boldsymbol{\theta}})(\Delta_1\boldsymbol{\xi}_\beta^{(q)},\bar{\mathsf{M}}^-\bar{\boldsymbol{\xi}}_\alpha^{(k)})\bar{\boldsymbol{\xi}}_\beta^{(p)}\ (\equiv\mathbf{t}_2)
+\frac{1}{N^2}\sum_{\alpha,\beta=1}^{N}\sum_{k,l,p,q=1}^{L}\bar{W}_\alpha^{(kl)}\Delta_1\bar{W}_\beta^{(pq)}(\Delta_1\boldsymbol{\xi}_\alpha^{(l)},\bar{\boldsymbol{\theta}})(\bar{\boldsymbol{\xi}}_\beta^{(q)},\bar{\mathsf{M}}^-\bar{\boldsymbol{\xi}}_\alpha^{(k)})\bar{\boldsymbol{\xi}}_\beta^{(p)}\ (\equiv\mathbf{t}_3).
\tag{14.56}
$$

Consider the three terms $\mathbf{t}_1$, $\mathbf{t}_2$, and $\mathbf{t}_3$ separately. The expectation of $\mathbf{t}_1$ is

$$
\begin{aligned}
E[\mathbf{t}_1] &= \frac{1}{N^2}\sum_{\alpha,\beta=1}^{N}\sum_{k,l,p,q=1}^{L}\bar{W}_\alpha^{(kl)}\bar{W}_\beta^{(pq)}(\bar{\boldsymbol{\xi}}_\beta^{(q)},\bar{\mathsf{M}}^-\bar{\boldsymbol{\xi}}_\alpha^{(k)})E[\Delta_1\boldsymbol{\xi}_\beta^{(p)}\Delta_1\boldsymbol{\xi}_\alpha^{(l)\top}]\bar{\boldsymbol{\theta}} \\
&= \frac{1}{N^2}\sum_{\alpha,\beta=1}^{N}\sum_{k,l,p,q=1}^{L}\bar{W}_\alpha^{(kl)}\bar{W}_\beta^{(pq)}(\bar{\boldsymbol{\xi}}_\beta^{(q)},\bar{\mathsf{M}}^-\bar{\boldsymbol{\xi}}_\alpha^{(k)})\sigma^2\delta_{\alpha\beta}V_0^{(pl)}[\boldsymbol{\xi}_\alpha]\bar{\boldsymbol{\theta}} \\
&= \frac{\sigma^2}{N^2}\sum_{\alpha=1}^{N}\sum_{k,l,p,q=1}^{L}\bar{W}_\alpha^{(kl)}\bar{W}_\alpha^{(pq)}(\bar{\boldsymbol{\xi}}_\alpha^{(q)},\bar{\mathsf{M}}^-\bar{\boldsymbol{\xi}}_\alpha^{(k)})V_0^{(pl)}[\boldsymbol{\xi}_\alpha]\bar{\boldsymbol{\theta}}.
\end{aligned}
\tag{14.57}
$$

The expectation of $\mathbf{t}_2$ is

$$
\begin{aligned}
E[\mathbf{t}_2] &= \frac{1}{N^2}\sum_{\alpha,\beta=1}^{N}\sum_{k,l,p,q=1}^{L}\bar{W}_\alpha^{(kl)}\bar{W}_\beta^{(pq)}(\bar{\boldsymbol{\theta}},E[\Delta_1\boldsymbol{\xi}_\alpha^{(l)}\Delta_1\boldsymbol{\xi}_\beta^{(q)\top}]\bar{\mathsf{M}}^-\bar{\boldsymbol{\xi}}_\alpha^{(k)})\bar{\boldsymbol{\xi}}_\beta^{(p)} \\
&= \frac{1}{N^2}\sum_{\alpha,\beta=1}^{N}\sum_{k,l,p,q=1}^{L}\bar{W}_\alpha^{(kl)}\bar{W}_\beta^{(pq)}(\bar{\boldsymbol{\theta}},\sigma^2\delta_{\alpha\beta}V_0^{(lq)}[\boldsymbol{\xi}_\alpha]\bar{\mathsf{M}}^-\bar{\boldsymbol{\xi}}_\alpha^{(k)})\bar{\boldsymbol{\xi}}_\beta^{(p)} \\
&= \frac{\sigma^2}{N^2}\sum_{\alpha=1}^{N}\sum_{k,l,p,q=1}^{L}\bar{W}_\alpha^{(kl)}\bar{W}_\alpha^{(pq)}(\bar{\boldsymbol{\xi}}_\alpha^{(k)},\bar{\mathsf{M}}^-V_0^{(lq)}[\boldsymbol{\xi}_\alpha]\bar{\boldsymbol{\theta}})\bar{\boldsymbol{\xi}}_\alpha^{(p)}.
\end{aligned}
\tag{14.58}
$$

Finally, we consider the expectation of $\mathbf{t}_3$. From Eq. (14.51), we can write

$$\Delta_1 W_\beta^{(pq)} = \frac{2}{N}\sum_{\gamma=1}^{N}\sum_{m,n,r,s=1}^{L} \bar{W}_\beta^{(pm)}\bar{W}_\beta^{(qn)}\bar{W}_\gamma^{(rs)}(\bar{\boldsymbol{\xi}}_\gamma^{(r)}, \bar{\mathsf{M}}^- V_0^{(mn)}[\boldsymbol{\xi}_\beta]\bar{\boldsymbol{\theta}})(\Delta_1\boldsymbol{\xi}_\gamma^{(s)}, \bar{\boldsymbol{\theta}}). \tag{14.59}$$

Therefore

$$\begin{aligned}
&E[\Delta_1 W_\beta^{(pq)}\Delta_1\boldsymbol{\xi}_\alpha^{(l)}] \\
&= E[\frac{2}{N}\sum_{\gamma=1}^{N}\sum_{m,n,r,s=1}^{L} \bar{W}_\beta^{(pm)}\bar{W}_\beta^{(qn)}\bar{W}_\gamma^{(rs)}(\bar{\boldsymbol{\xi}}_\gamma^{(r)}, \bar{\mathsf{M}}^- V_0^{(mn)}[\boldsymbol{\xi}_\beta]\bar{\boldsymbol{\theta}})(\Delta_1\boldsymbol{\xi}_\gamma^{(s)}, \bar{\boldsymbol{\theta}})\Delta_1\boldsymbol{\xi}_\alpha^{(l)}] \\
&= \frac{2}{N}\sum_{\gamma=1}^{N}\sum_{m,n,r,s=1}^{L} \bar{W}_\beta^{(pm)}\bar{W}_\beta^{(qn)}\bar{W}_\gamma^{(rs)}(\bar{\boldsymbol{\xi}}_\gamma^{(r)}, \bar{\mathsf{M}}^- V_0^{(mn)}[\boldsymbol{\xi}_\beta]\bar{\boldsymbol{\theta}})E[\Delta_1\boldsymbol{\xi}_\alpha^{(l)}\Delta_1\boldsymbol{\xi}_\gamma^{(s)\top}]\bar{\boldsymbol{\theta}} \\
&= \frac{2}{N}\sum_{\gamma=1}^{N}\sum_{m,n,r,s=1}^{L} \bar{W}_\beta^{(pm)}\bar{W}_\beta^{(qn)}\bar{W}_\gamma^{(rs)}(\bar{\boldsymbol{\xi}}_\gamma^{(r)}, \bar{\mathsf{M}}^- V_0^{(mn)}[\boldsymbol{\xi}_\beta]\bar{\boldsymbol{\theta}})\sigma^2\delta_{\alpha\gamma}V_0^{(ls)}[\boldsymbol{\xi}_\alpha]\bar{\boldsymbol{\theta}} \\
&= \frac{2\sigma^2}{N}\sum_{m,n,r,s=1}^{L} \bar{W}_\beta^{(pm)}\bar{W}_\beta^{(qn)}\bar{W}_\alpha^{(rs)}(\bar{\boldsymbol{\xi}}_\alpha^{(r)}, \bar{\mathsf{M}}^- V_0^{(mn)}[\boldsymbol{\xi}_\beta]\bar{\boldsymbol{\theta}})V_0^{(ls)}[\boldsymbol{\xi}_\alpha]\bar{\boldsymbol{\theta}}.
\end{aligned} \tag{14.60}$$

It follows that

$$\begin{aligned}
&(E[\Delta_1 \bar{W}_\beta^{(pq)}\Delta_1\boldsymbol{\xi}_\alpha^{(l)}], \bar{\boldsymbol{\theta}})(\bar{\boldsymbol{\xi}}_\beta^{(q)}, \bar{\mathsf{M}}^-\bar{\boldsymbol{\xi}}_\alpha^{(k)}) \\
&=(\frac{2\sigma^2}{N}\sum_{m,n,r,s=1}^{L} \bar{W}_\beta^{(pm)}\bar{W}_\beta^{(qn)}\bar{W}_\alpha^{(rs)}(\bar{\boldsymbol{\xi}}_\alpha^{(r)}, \bar{\mathsf{M}}^- V_0^{(mn)}[\boldsymbol{\xi}_\beta]\bar{\boldsymbol{\theta}})V_0^{(ls)}[\boldsymbol{\xi}_\alpha]\bar{\boldsymbol{\theta}}, \bar{\boldsymbol{\theta}}) \\
&(\bar{\boldsymbol{\xi}}_\beta^{(q)}, \bar{\mathsf{M}}^-\bar{\boldsymbol{\xi}}_\alpha^{(k)}) \\
&=\frac{2\sigma^2}{N}\sum_{m,n,r,s=1}^{L} \bar{W}_\beta^{(pm)}\bar{W}_\beta^{(qn)}\bar{W}_\alpha^{(rs)}(\bar{\boldsymbol{\xi}}_\alpha^{(r)}, \bar{\mathsf{M}}^- V_0^{(mn)}[\boldsymbol{\xi}_\beta]\bar{\boldsymbol{\theta}})(\bar{\boldsymbol{\theta}}, V_0^{(ls)}[\boldsymbol{\xi}_\alpha]\bar{\boldsymbol{\theta}}) \\
&(\bar{\boldsymbol{\xi}}_\beta^{(q)}, \bar{\mathsf{M}}^-\bar{\boldsymbol{\xi}}_\alpha^{(k)}).
\end{aligned} \tag{14.61}$$

Thus, the expectation of $\mathbf{t}_3$ is

$$\begin{aligned}
E[\mathbf{t}_3] &= \frac{1}{N^2}\sum_{\alpha,\beta=1}^{N}\sum_{k,l,p,q=1}^{L} \bar{W}_\alpha^{(kl)}(E[\Delta_1\bar{W}_\beta^{(pq)}\Delta_1\boldsymbol{\xi}_\alpha^{(l)}], \bar{\boldsymbol{\theta}})(\bar{\boldsymbol{\xi}}_\beta^{(q)}, \bar{\mathsf{M}}^-\bar{\boldsymbol{\xi}}_\alpha^{(k)})\bar{\boldsymbol{\xi}}_\beta^{(p)} \\
&= \frac{1}{N^2}\sum_{\alpha,\beta=1}^{N}\sum_{k,l,p,q=1}^{L} \bar{W}_\alpha^{(kl)}\Big(\frac{2\sigma^2}{N}\sum_{m,n,r,s=1}^{L} \bar{W}_\beta^{(pm)}\bar{W}_\beta^{(qn)}\bar{W}_\alpha^{(rs)}(\bar{\boldsymbol{\xi}}_\alpha^{(r)}, \bar{\mathsf{M}}^- V_0^{(mn)}[\boldsymbol{\xi}_\beta]\bar{\boldsymbol{\theta}}) \\
&\quad (\bar{\boldsymbol{\theta}}, V_0^{(ls)}[\boldsymbol{\xi}_\alpha]\bar{\boldsymbol{\theta}})(\bar{\boldsymbol{\xi}}_\beta^{(q)}, \bar{\mathsf{M}}^-\bar{\boldsymbol{\xi}}_\alpha^{(k)})\Big)\bar{\boldsymbol{\xi}}_\beta^{(p)} \\
&= \frac{2\sigma^2}{N^3}\sum_{\alpha,\beta=1}^{N}\sum_{k,l,m,n,p,q,r,s=1}^{L} \bar{W}_\alpha^{(kl)}\bar{W}_\beta^{(pm)}\bar{W}_\beta^{(qn)}\bar{W}_\alpha^{(rs)}\bar{V}_\alpha^{(ls)}(\bar{\boldsymbol{\xi}}_\alpha^{(r)}, \bar{\mathsf{M}}^- V_0^{(mn)}[\boldsymbol{\xi}_\beta]\bar{\boldsymbol{\theta}})
\end{aligned}$$

$$
\begin{aligned}
&(\bar{\xi}_\beta^{(q)}, \bar{\mathsf{M}}^-\bar{\xi}_\alpha^{(k)})\bar{\xi}_\beta^{(p)} \\
&= \frac{2\sigma^2}{N^3}\sum_{\alpha,\beta=1}^{N}\sum_{k,m,n,p,q,r=1}^{L} \bar{W}_\beta^{(pm)}\bar{W}_\beta^{(qn)}(\bar{\xi}_\alpha^{(r)}, \bar{\mathsf{M}}^- V_0^{(mn)}[\xi_\beta]\bar{\theta})\bar{W}_\alpha^{(kr)} \\
&(\bar{\xi}_\beta^{(q)}, \bar{\mathsf{M}}^-\bar{\xi}_\alpha^{(k)})\bar{\xi}_\beta^{(p)} \\
&= \frac{2\sigma^2}{N^2}\sum_{\beta=1}^{N}\sum_{m,n,p,q=1}^{L} \bar{W}_\beta^{(pm)}\bar{W}_\beta^{(qn)}(\bar{\xi}_\beta^{(q)}, \bar{\mathsf{M}}^-\Big(\frac{1}{N}\sum_{\alpha=1}^{N}\sum_{k,r=1}^{L}\bar{W}_\alpha^{(kr)}\bar{\xi}_\alpha^{(k)}\bar{\xi}_\alpha^{(r)\top}\Big) \\
&\bar{\mathsf{M}}^- V_0^{(mn)}[\xi_\beta]\bar{\theta})\bar{\xi}_\beta^{(p)} \\
&= \frac{2\sigma^2}{N^2}\sum_{\beta=1}^{N}\sum_{m,n,p,q=1}^{L} \bar{W}_\beta^{(pm)}\bar{W}_\beta^{(qn)}(\bar{\xi}_\beta^{(q)}, \bar{\mathsf{M}}^-\bar{\mathsf{M}}\bar{\mathsf{M}}^- V_0^{(mn)}[\xi_\beta]\bar{\theta})\bar{\xi}_\beta^{(p)} \\
&= \frac{2\sigma^2}{N^2}\sum_{\beta=1}^{N}\sum_{m,n,p,q=1}^{L} \bar{W}_\beta^{(pm)}\bar{W}_\beta^{(qn)}(\bar{\xi}_\beta^{(q)}, \bar{\mathsf{M}}^- V_0^{(mn)}[\xi_\beta]\bar{\theta})\bar{\xi}_\beta^{(p)}, \qquad (14.62)
\end{aligned}
$$

where we have used Eq. (14.31). Adding the expectations of $\mathbf{t}_1$, $\mathbf{t}_2$, and $\mathbf{t}_3$ together, we obtain the expectation of $\Delta_1\mathsf{M}\bar{\mathsf{M}}^-\Delta_1\mathsf{M}\bar{\theta}$ in the form of Eq. (14.54). □

Proposition 14.3 (Optimal $\bar{\mathsf{N}}$) *The identity $E[\mathsf{T}\bar{\theta}] = \sigma^2\bar{\mathsf{N}}\bar{\theta}$ holds if $\bar{\mathsf{N}}$ is defined by Eq. (14.30).*

Proof Combining the above expectations $E[\Delta_2\mathsf{M}\bar{\theta}]$ and $E[\Delta_1\mathsf{M}\bar{\mathsf{M}}^-\Delta_1\mathsf{M}\bar{\theta}]$, we obtain the expectation of $\mathsf{T}\bar{\theta}$ in the form

$$
\begin{aligned}
&E[\mathsf{T}\bar{\theta}] = E[\Delta_2\mathsf{M}\bar{\theta}] - E[\Delta_1\mathsf{M}\bar{\mathsf{M}}^-\Delta_1\mathsf{M}\bar{\theta}] \\
&= \frac{\sigma^2}{N}\sum_{\alpha=1}^{N}\sum_{k,l=1}^{L}\bar{W}_\alpha^{(kl)}\Big(V_0^{(kl)}[\xi_\alpha]\bar{\theta} + (\mathbf{e}_\alpha^{(l)}, \bar{\theta})\bar{\xi}_\alpha^{(k)}\Big) \\
&\quad + \frac{2\sigma^2}{N^2}\sum_{\alpha=1}^{N}\sum_{k,m,n,p=1}^{L}\bar{W}_\alpha^{(km)}\bar{W}_\alpha^{(np)}(\bar{\xi}_\alpha^{(p)}, \bar{\mathsf{M}}^- V_0^{(mn)}[\xi_\alpha]\bar{\theta})\bar{\xi}_\alpha^{(k)} \\
&\quad - \frac{\sigma^2}{N^2}\sum_{\alpha=1}^{N}\sum_{k,l,m,n=1}^{L}\bar{W}_\alpha^{(kl)}\bar{W}_\alpha^{(mn)}(\bar{\xi}_\alpha^{(m)}, \bar{\mathsf{M}}^-\bar{\xi}_\alpha^{(k)})V_0^{(ln)}[\xi_\alpha]\bar{\theta} \\
&\quad - \frac{3\sigma^2}{N^2}\sum_{\alpha=1}^{N}\sum_{k,l,m,n=1}^{L}\bar{W}_\alpha^{(kl)}\bar{W}_\alpha^{(mn)}(\bar{\xi}_\alpha^{(k)}, \bar{\mathsf{M}}^- V_0^{(lm)}[\xi_\alpha]\bar{\theta})\bar{\xi}_\alpha^{(n)} \\
&= \frac{\sigma^2}{N}\sum_{\alpha=1}^{N}\sum_{k,l=1}^{L}\bar{W}_\alpha^{(kl)}\Big(V_0^{(kl)}[\xi_\alpha]\bar{\theta} + (\mathbf{e}_\alpha^{(l)}, \bar{\theta})\bar{\xi}_\alpha^{(k)}\Big) \\
&\quad - \frac{\sigma^2}{N^2}\sum_{\alpha=1}^{N}\sum_{k,l,m,n=1}^{L}\bar{W}_\alpha^{(kl)}\bar{W}_\alpha^{(mn)}(\bar{\xi}_\alpha^{(m)}, \bar{\mathsf{M}}^-\bar{\xi}_\alpha^{(k)})V_0^{(ln)}[\xi_\alpha]\bar{\theta}
\end{aligned}
$$

$$
\begin{aligned}
& -\frac{\sigma^2}{N^2}\sum_{\alpha=1}^{N}\sum_{k,l,m,n=1}^{L} \bar{W}_\alpha^{(kl)}\bar{W}_\alpha^{(mn)}(\bar{\xi}_\alpha^{(k)}, \bar{\mathsf{M}}^- V_0^{(lm)}[\xi_\alpha]\bar{\theta})\bar{\xi}_\alpha^{(n)} \\
&= \frac{\sigma^2}{N}\sum_{\alpha=1}^{N}\sum_{k,l=1}^{L} \bar{W}_\alpha^{(kl)}\Big(V_0^{(kl)}[\xi_\alpha]\bar{\theta} + \bar{\xi}_\alpha^{(k)}\mathbf{e}_\alpha^{(l)\top}\bar{\theta}\Big) \\
& -\frac{\sigma^2}{N^2}\sum_{\alpha=1}^{N}\sum_{k,l,m,n=1}^{L} \bar{W}_\alpha^{(kl)}\bar{W}_\alpha^{(mn)}(\bar{\xi}_\alpha^{(m)}, \bar{\mathsf{M}}^- \bar{\xi}_\alpha^{(k)}) V_0^{(ln)}[\xi_\alpha]\bar{\theta} \\
& -\frac{\sigma^2}{N^2}\sum_{\alpha=1}^{N}\sum_{k,l,m,n=1}^{L} \bar{W}_\alpha^{(kl)}\bar{W}_\alpha^{(mn)}\bar{\xi}_\alpha^{(n)}\bar{\xi}_\alpha^{(k)\top}\bar{\mathsf{M}}^- V_0^{(lm)}[\xi_\alpha]\bar{\theta} \\
&= \frac{\sigma^2}{N}\sum_{\alpha=1}^{N}\sum_{k,l=1}^{L} \bar{W}_\alpha^{(kl)}\Big(V_0^{(kl)}[\xi_\alpha]\bar{\theta} + (\bar{\xi}_\alpha^{(k)}\mathbf{e}_\alpha^{(l)\top} + \mathbf{e}_\alpha^{(l)}\bar{\xi}_\alpha^{(k)\top})\bar{\theta}\Big) \\
& -\frac{\sigma^2}{N^2}\sum_{\alpha=1}^{N}\sum_{k,l,m,n=1}^{L} \bar{W}_\alpha^{(kl)}\bar{W}_\alpha^{(mn)}(\bar{\xi}_\alpha^{(m)}, \bar{\mathsf{M}}^- \bar{\xi}_\alpha^{(k)}) V_0^{(ln)}[\xi_\alpha]\bar{\theta} \\
& -\frac{\sigma^2}{N^2}\sum_{\alpha=1}^{N}\sum_{k,l,m,n=1}^{L} \bar{W}_\alpha^{(kl)}\bar{W}_\alpha^{(mn)}(\bar{\xi}_\alpha^{(n)}\bar{\xi}_\alpha^{(k)\top}\bar{\mathsf{M}}^- V_0^{(lm)}[\xi_\alpha] + V_0^{(lm)}[\xi_\alpha]\bar{\mathsf{M}}^-\bar{\xi}_\alpha^{(k)}\bar{\xi}_\alpha^{(n)\top})\bar{\theta}.
\end{aligned}
\tag{14.63}
$$

Thus if we define the matrix $\bar{\mathsf{N}}$ by Eq. (14.30), the above equation is written as $E[\mathsf{T}\bar{\theta}] = \sigma^2\bar{\mathsf{N}}\bar{\theta}$. □

14.7 Supplemental Note

The covariance and bias analysis of ellipse fitting and fundamental matrix computation was first presented by Kanatani [1]. Based on this analysis, the hyper-renormalization procedure was obtained by Kanatani et al. [2]. The result was generalized by Kanatani et al. [3] to multiple constraint cases in the form of Eq. (14.1), which included homography computation. As pointed out in the Supplemental Note of Chap. 6, however, the vectors $\mathbf{e}_\alpha^{(k)}$ defined by Eq. (14.29) vanish due to Eqs. (6.12) for homography computation. This is the case for many estimation problems from multiple images, including fundamental matrix computation, where noise in each image is assumed to be independent of the noise in other images.

Problems

14.1 Let

$$
\mathsf{A} = \mathsf{U}\begin{pmatrix} \sigma_1 & & \\ & \ddots & \\ & & \sigma_r \end{pmatrix}\mathsf{V}^\top \tag{14.64}
$$

be the singular value decomposition of an $m \times n$ matrix A, where the second matrix on the right-hand side is an $r \times r$ diagonal matrix, $r \leq \min(m, n)$, with singular values $\sigma_1 \geq \cdots \geq \sigma_r$ (> 0) as diagonal elements in that order (all nondiagonal elements are 0). The matrices U and V are, respectively, $m \times r$ and $n \times r$ matrices consisting of orthonormal columns. Show that Eq. (14.64) can be written in the form

$$\mathsf{A} = \sigma_1 \mathbf{u}_1 \mathbf{v}_1^\top + \sigma_2 \mathbf{u}_2 \mathbf{v}_2^\top + \cdots + \sigma_r \mathbf{u}_r \mathbf{v}_r^\top, \tag{14.65}$$

where $\mathbf{u}_i$ and $\mathbf{v}_i$ are the ith columns of U and V, respectively.

14.2 The pseudoinverse of the matrix A of Eq. (14.64) is defined to be

$$\mathsf{A}^- = \mathsf{V} \begin{pmatrix} 1/\sigma_1 & & \\ & \ddots & \\ & & 1/\sigma_r \end{pmatrix} \mathsf{U}^\top. \tag{14.66}$$

Show that this can be written in the form

$$\mathsf{A}^- = \frac{1}{\sigma_1} \mathbf{v}_1 \mathbf{u}_1^\top + \frac{1}{\sigma_2} \mathbf{v}_2 \mathbf{u}_2^\top + \cdots + \frac{1}{\sigma_r} \mathbf{v}_r \mathbf{u}_r^\top. \tag{14.67}$$

14.3 Show that the following identities hold for the pseudoinverse.

$$\mathsf{A}\mathsf{A}^-\mathsf{A} = \mathsf{A}, \qquad \mathsf{A}^-\mathsf{A}\mathsf{A}^- = \mathsf{A}^-. \tag{14.68}$$

14.4 For the matrix A of Eq. (14.64), show that the product $\mathsf{A}\mathsf{A}^-$ is the projection matrix onto the space spanned by $\mathbf{u}_1$, ..., $\mathbf{u}_r$ and that $\mathsf{A}^-\mathsf{A}$ is the projection matrix onto the space spanned by $\mathbf{v}_1$, ..., $\mathbf{v}_r$.

References

1. K. Kanatani, Statistical optimization for geometric fitting: Theoretical accuracy bound and high order error analysis. Int. J. Comput. Vis. **80**(2), 167–188 (2008)
2. K. Kanatani, A. Al-Sharadqah, N. Chernov, and Y. Sugaya, Renormalization returns: Hyper-renormalization and its applications, in*Proceedings 12th European Conference on Computer Vision*, Firenze, Italy, October 2012 (2012)
3. K. Kanatani, A. Al-Sharadqah, N. Chernov, Y. Sugaya, Hyper-renormalization: Non-minimization approach for geometric estimation. IPSJ Trans. Comput. Vis. Appl. **6**, 143–159 (2014)

Maximum Likelihood of Geometric Estimation

15

Abstract

We discuss here maximum likelihood (ML) estimation and Sampson error minimization in the general mathematical framework of the preceding chapter. We first derive the Sampson error as a first approximation to the Mahalanobis distance (a generalization of the geometric distance or the reprojection error) of ML. Then we do high-order error analysis to derive explicit expressions for the covariance and bias of the solution. The hyperaccurate correction procedure is derived in this framework.

15.1 Maximum Likelihood

We continue statistical analysis in the mathematical framework defined in the preceding chapter. Let $\mathbf{x}_\alpha$ be the αth raw observation. For ellipse fitting, it is a vector consisting of the coordinates x_α and y_α of the αth point. For fundamental matrix computation and homography computation, it consists of coordinates x_α, y_α, $x_{\alpha'}$, and y'_α of the αth pair of points. Let $\bar{\mathbf{x}}_\alpha$ be the true value of $\mathbf{x}_\alpha$, that is, the value we would observe in the absence of noise. As in the preceding chapter, we assume that the geometric constraint is given in the form

$$(\boldsymbol{\xi}^{(k)}(\bar{\mathbf{x}}_\alpha), \boldsymbol{\theta}) = 0, \qquad k = 1, \ldots, L, \quad \alpha = 1, \ldots, N, \tag{15.1}$$

where $\boldsymbol{\xi}^{(k)}(\bar{\mathbf{x}}_\alpha)$ is some nonlinear vector function of $\bar{\mathbf{x}}_\alpha$, and $\boldsymbol{\theta}$ is the parameter vector we want to estimate. Equation (15.1) generalizes the ellipse equation ($L = 1$), the epipolar equation ($L = 1$), and the homography equation ($L = 3$).

Let $V[\mathbf{x}_\alpha]$ be the covariance matrix of $\mathbf{x}_\alpha$ that describes the uncertainty of observing $\mathbf{x}_\alpha$. We assume that the statistical properties of noise are independent for different α. If σ is the average magnitude of noise, the covariance matrix can be written, as shown in the preceding chapter, in the form $V[\mathbf{x}_\alpha] = \sigma^2 V_0[\mathbf{x}_\alpha]$; we call σ the *noise*

K. Kanatani et al., *Guide to 3D Vision Computation*, Advances in Computer Vision and Pattern Recognition, DOI 10.1007/978-3-319-48493-8_15

level, and $V_0[\mathbf{x}_\alpha]$ the *normalized covariance matrix*. As in the preceding chapter, we assume that noise level σ is unknown but the normalized covariance matrix $V_0[\mathbf{x}_\alpha]$ is known. The *Mahalanobis distance* between N observations $\mathbf{x}_\alpha$, $\alpha = 1, \ldots, N$ and their true values is defined by

$$S = \frac{1}{N}\sum_{\alpha=1}^{N}(\mathbf{x}_\alpha - \bar{\mathbf{x}}_\alpha, V_0[\mathbf{x}_\alpha]^{-1}(\mathbf{x}_\alpha - \bar{\mathbf{x}}_\alpha)). \tag{15.2}$$

Strictly speaking, it should be called the "square Mahalanobis distance," because it has the dimension of square length, but the term "Mahalanobis distance" is widely used for simplicity. In the following, we assume that the noise distribution is Gaussian (the discussions in the preceding chapter did not specifically use this assumption). Then, estimating the value of $\boldsymbol{\theta}$ that minimizes Eq. (15.2) subject to Eq. (15.1) is called *maximum likelihood* (*ML*) estimation. In fact, the probability density, or the *likelihood*, of the total noise is (constant) $\times$ $\mathrm{e}^{-\sigma^2 NS/2}$, therefore maximizing this is equivalent to minimizing the Mahalanobis distance S ($\hookrightarrow$ Problem 15.2). If the noise in $\mathbf{x}_\alpha$ is identical and isotropic for all α, in particular, Eq. (15.2) reduces to the *geometric distance* (although it has the dimension of square length, this term is widely used for simplicity) or *reprojection error* $(1/N)\sum_{\alpha=1}^{N}\|\mathbf{x}_\alpha - \bar{\mathbf{x}}_\alpha\|^2$ (see Eqs. (2.34), (3.15), and (6.28)).

15.2 Sampson Error

The difficulty of minimizing Eq. (15.2) is that Eq. (15.2) itself does not contain the parameter $\boldsymbol{\theta}$ with respect to which Eq. (15.2) is to be minimized but instead it is included in the "constraint" of Eq. (15.1). This problem could be handled by introducing auxiliary variables (we discuss this later), but all the difficulty is resolved if we do maximum likelihood *with respect to* $\boldsymbol{\xi}^{(k)}(\mathbf{x}_\alpha)$ *rather than* $\mathbf{x}_\alpha$. Let us write $\boldsymbol{\xi}_\alpha^{(k)} = \boldsymbol{\xi}^{(k)}(\mathbf{x}_\alpha)$. The covariance matrix for $\boldsymbol{\xi}_\alpha^{(k)}$ and $\boldsymbol{\xi}_\alpha^{(k)}$ is written, up to $O(\sigma^4)$ terms, in the form:

$$V^{(kl)}[\boldsymbol{\xi}_\alpha] = \Big(\frac{\partial \boldsymbol{\xi}^{(k)}(\mathbf{x})}{\partial \mathbf{x}}\Big|_{\mathbf{x}=\bar{\mathbf{x}}_\alpha}\Big)V[\boldsymbol{\xi}_\alpha]\Big(\frac{\partial \boldsymbol{\xi}^{(l)}(\mathbf{x})}{\partial \mathbf{x}}\Big|_{\mathbf{x}=\bar{\mathbf{x}}_\alpha}\Big)^\top, \tag{15.3}$$

where $\partial\boldsymbol{\xi}^{(k)}(\mathbf{x})/\partial\mathbf{x}$ is the Jacobi matrix of $\boldsymbol{\xi}^{(k)}(\mathbf{x})$ with respect to $\mathbf{x}$. We can factor out σ^2 from Eq. (15.3) and write $V^{(kl)}[\boldsymbol{\xi}_\alpha] = \sigma^2 V_0^{(kl)}[\boldsymbol{\xi}_\alpha]$; as before, we call $V_0^{(kl)}[\boldsymbol{\xi}_\alpha]$ the normalized covariance matrices.

Let us write $\bar{\boldsymbol{\xi}}_\alpha^{(k)} = \boldsymbol{\xi}^{(k)}(\bar{\mathbf{x}}_\alpha)$ for short. If we assume that noise in $\boldsymbol{\xi}_\alpha^{(k)}$ is Gaussian with mean $\mathbf{0}$ and covariance matrices $V^{(kl)}[\boldsymbol{\xi}_\alpha]$, maximum likelihood estimation of $\boldsymbol{\theta}$ is to minimize the Mahalanobis distance

$$J = \frac{1}{N}\sum_{\alpha=1}^{N}\left(\begin{pmatrix}\boldsymbol{\xi}_\alpha^{(1)} - \bar{\boldsymbol{\xi}}_\alpha^{(1)} \\ \vdots \\ \boldsymbol{\xi}_\alpha^{(L)} - \bar{\boldsymbol{\xi}}_\alpha^{(L)}\end{pmatrix}, \begin{pmatrix} V_0^{(11)}[\boldsymbol{\xi}_\alpha] & \cdots & V_0^{(1L)}[\boldsymbol{\xi}_\alpha] \\ \vdots & \ddots & \vdots \\ V_0^{(L1)}[\boldsymbol{\xi}_\alpha] & \cdots & V_0^{(LL)}[\boldsymbol{\xi}_\alpha]\end{pmatrix}^{-} \begin{pmatrix}\boldsymbol{\xi}_\alpha^{(1)} - \bar{\boldsymbol{\xi}}_\alpha^{(1)} \\ \vdots \\ \boldsymbol{\xi}_\alpha^{(L)} - \bar{\boldsymbol{\xi}}_\alpha^{(L)}\end{pmatrix}\right), \tag{15.4}$$

subject to the constraint

$$(\bar{\boldsymbol{\xi}}_\alpha^{(k)}, \boldsymbol{\theta}) = 0, \quad k = 1, \ldots, L, \quad \alpha = 1, \ldots, N. \tag{15.5}$$

In Eq. (15.4), we use the pseudoinverse for the covariance matrix, allowing the possibility that it is not positive definite. This occurs, for example, when $\boldsymbol{\xi}_\alpha^{(1)}, \ldots, \boldsymbol{\xi}_\alpha^{(L)}$ are not linearly independent, which is the case for homography computation (L = 3). Now that the constraint of Eq. (15.5) is *linear* in $\bar{\boldsymbol{\xi}}_\alpha^{(k)}$, we can eliminate $\bar{\boldsymbol{\xi}}_\alpha^{(k)}$ from Eq. (15.4), using Lagrange multipliers, to obtain

$$J = \frac{1}{N} \sum_{\alpha=1}^{N} \sum_{k,l=1}^{L} W_\alpha^{(kl)} (\boldsymbol{\xi}_\alpha^{(k)}, \boldsymbol{\theta})(\boldsymbol{\xi}_\alpha^{(l)}, \boldsymbol{\theta}) \tag{15.6}$$

where $W_\alpha^{(kl)}$ is defined as in Eq. (14.6) (see Proposition 15.1 below for the derivation). We call Eq. (15.6) the *Sampson error* (see Eqs. (2.36), (3.23), and (6.31)). Because this is a function of $\boldsymbol{\theta}$, we can minimize this by searching the $\boldsymbol{\theta}$-space.

The Sampson error J of Eq. (15.6) should be identical to the Mahalanobis distance S of Eq. (15.2) if noise in $\boldsymbol{\xi}_\alpha^{(k)}$ is Gaussian. However, $\boldsymbol{\xi}_\alpha^{(k)}$ is a nonlinear mapping of $\mathbf{x}_\alpha$, therefore the assumption that $\boldsymbol{\xi}_\alpha^{(k)}$ are Gaussian variables of mean $\mathbf{0}$ and covariance matrices $V^{(kl)}[\boldsymbol{\xi}_\alpha]$ $(= \sigma^2 V_0^{(kl)}[\boldsymbol{\xi}_\alpha])$ does not strictly hold, even if the noise in $\mathbf{x}_\alpha$ is zero-mean Gaussian. Still, if the noise in $\mathbf{x}_\alpha$ is small, the noise distribution of $\boldsymbol{\xi}_\alpha^{(k)}$ is expected to have an approximately "Gaussian-like" form, concentrating around $\mathbf{0}$ with covariance matrices $V^{(kl)}[\boldsymbol{\xi}_\alpha]$ $(= \sigma^2 V_0^{(kl)}[\boldsymbol{\xi}_\alpha])$. It has been experimentally confirmed that is indeed the case; the Sampson error minimization solution and the Mahalanobis/geometric distance minimization solution are almost identical. In fact, we can minimize the geometric distance by repeating the Sampson error minimization, as shown in Chaps. 2, 3, and 6, and the resulting solution is the same over several significant digits. Thus we can in practice identify the Sampson error with the Mahalanobis/geometric distance. Moreover, the Gaussian noise assumption for $\boldsymbol{\xi}_\alpha^{(k)}$ makes mathematical error analysis very easy, enabling us to evaluate higher-order covariance and bias and derive a theoretical accuracy limit. In the following, we adopt this Gaussian noise assumption for $\boldsymbol{\xi}_\alpha^{(k)}$.

15.3 Error Analysis

We now analyze the error of the solution $\boldsymbol{\theta}$ that minimizes the Sampson error J of Eq. (15.6). The derivative of Eq. (15.6) with respect to $\boldsymbol{\theta}$ has the form ($\hookrightarrow$ Problems 2.6(1) and 6.5(2))

$$\nabla_{\boldsymbol{\theta}} J = 2(\mathsf{M} - \mathsf{L})\boldsymbol{\theta}, \tag{15.7}$$

where

$$\mathsf{M} = \frac{1}{N}\sum_{\alpha=1}^{N}\sum_{k,l=1}^{L} W_\alpha^{(kl)} \boldsymbol{\xi}_\alpha^{(k)} \boldsymbol{\xi}_\alpha^{(l)\top}, \tag{15.8}$$

$$\mathsf{L} = \frac{1}{N}\sum_{\alpha=1}^{N}\sum_{k,l,m,n=1}^{L} W_\alpha^{(km)} W_\alpha^{(ln)} (\boldsymbol{\xi}_\alpha^{(m)}, \boldsymbol{\theta})(\boldsymbol{\xi}_\alpha^{(n)}, \boldsymbol{\theta}) V_0^{(kl)}[\boldsymbol{\xi}_\alpha]. \tag{15.9}$$

As in the preceding chapter, we substitute Eq. (14.2) into the expressions of M and $W_\alpha^{(kl)}$ and expand them in the form of Eqs. (14.8) and (14.9), where $\Delta_i\mathsf{M}$ and $\Delta_i W_\alpha^{(kl)}$, $i = 1, 2$, are given by Eqs. (14.10), (14.11), (14.12), and (14.13). Similarly, we expand the matrix L in Eq. (15.9) in the form

$$\mathsf{L} = \bar{\mathsf{L}} + \Delta_1\mathsf{L} + \Delta_2\mathsf{L} + \cdots \tag{15.10}$$

Because $(\bar{\boldsymbol{\xi}}_\alpha^{(k)}, \bar{\boldsymbol{\theta}}) = 0$ in the absence of noise, we can see from Eq. (15.9) that $\bar{\mathsf{L}} = \Delta_1\mathsf{L} = \mathsf{O}$. The second-order term $\Delta_2\mathsf{L}$ is given by

$$\Delta_2\mathsf{L} = \frac{1}{N}\sum_{\alpha=1}^{N}\sum_{k,l,m,n=1}^{L} \bar{W}_\alpha^{(km)} \bar{W}_\alpha^{(ln)} \Big((\bar{\boldsymbol{\xi}}_\alpha^{(m)}, \Delta_1\boldsymbol{\theta})(\bar{\boldsymbol{\xi}}_\alpha^{(n)}, \Delta_1\boldsymbol{\theta}) + (\bar{\boldsymbol{\xi}}_\alpha^{(m)}, \Delta_1\boldsymbol{\theta})(\Delta_1\boldsymbol{\xi}_\alpha^{(n)}, \bar{\boldsymbol{\theta}})$$
$$+ (\Delta_1\boldsymbol{\xi}_\alpha^{(m)}, \bar{\boldsymbol{\theta}})(\bar{\boldsymbol{\xi}}_\alpha^{(n)}, \Delta_1\boldsymbol{\theta}) + (\Delta_1\boldsymbol{\xi}_\alpha^{(m)}, \bar{\boldsymbol{\theta}})(\Delta_1\boldsymbol{\xi}_\alpha^{(n)}, \bar{\boldsymbol{\theta}})\Big) V_0^{(kl)}[\boldsymbol{\xi}_\alpha]. \tag{15.11}$$

At the value $\boldsymbol{\theta}$ that minimizes the Sampson error J, the derivative $\nabla_{\boldsymbol{\theta}} J$ in Eq. (15.7) vanishes, thus

$$\mathsf{M}\boldsymbol{\theta} = \mathsf{L}\boldsymbol{\theta}. \tag{15.12}$$

Substituting Eqs. (14.8), (14.7), and (15.10), we extend this equality in the form

$$(\bar{\mathsf{M}} + \Delta_1\mathsf{M} + \Delta_2\mathsf{M} + \cdots)(\bar{\boldsymbol{\theta}} + \Delta_1\boldsymbol{\theta} + \Delta_2\boldsymbol{\theta} + \cdots) = (\Delta_2\mathsf{L} + \cdots)(\bar{\boldsymbol{\theta}} + \Delta_1\boldsymbol{\theta} + \Delta_2\boldsymbol{\theta} + \cdots). \tag{15.13}$$

Equating terms of the same order on both sides, we obtain

$$\bar{\mathsf{M}}\Delta_1\boldsymbol{\theta} + \Delta_1\mathsf{M}\bar{\boldsymbol{\theta}} = \mathbf{0}, \tag{15.14}$$

$$\bar{\mathsf{M}}\Delta_2\boldsymbol{\theta} + \Delta_1\mathsf{M}\Delta_1\boldsymbol{\theta} + \Delta_2\mathsf{M}\bar{\boldsymbol{\theta}} = \Delta_2\mathsf{L}\bar{\boldsymbol{\theta}}. \tag{15.15}$$

Recall that $\bar{\mathsf{M}}\bar{\boldsymbol{\theta}} = \mathbf{0}$ in the absence of noise. Therefore $\bar{\mathsf{M}}$ has rank $n - 1$, $\bar{\boldsymbol{\theta}}$ being its null vector (n is the dimension of $\boldsymbol{\theta}$). Multiply Eq. (15.14) by the pseudoinverse $\bar{\mathsf{M}}^-$ on both sides from the left. Noting that $\bar{\mathsf{M}}^-\bar{\mathsf{M}} = \mathsf{P}_{\bar{\boldsymbol{\theta}}}$ (the projection matrix along $\bar{\boldsymbol{\theta}}$ ↪ Problem 3.3) and that $\Delta_1\boldsymbol{\theta}$ is orthogonal to $\bar{\boldsymbol{\theta}}$, we obtain

$$\Delta_1\boldsymbol{\theta} = -\bar{\mathsf{M}}^-\Delta_1\mathsf{M}\bar{\boldsymbol{\theta}}, \tag{15.16}$$

which is the same as Eq. (14.18). This means that the *algebraic solution and the Sampson error minimization solution agree to a first approximation*. Hence, the covariance matrix $V[\boldsymbol{\theta}]$ of $\boldsymbol{\theta}$ is given by Eq. (14.19), achieving the KCR (Kanatani-Cramer-Rao)lower bound up to $O(\sigma^4)$. Therefore we focus on the second-order terms. Multiplying Eq. (15.15) by $\bar{\mathsf{M}}^-$ on both sides from the left, we obtain

$$\Delta_2^\perp\boldsymbol{\theta} = -\bar{\mathsf{M}}^-\Delta_1\mathsf{M}\Delta_1\boldsymbol{\theta} - \bar{\mathsf{M}}^-\Delta_2\mathsf{M}\bar{\boldsymbol{\theta}} + \bar{\mathsf{M}}^-\Delta_2\mathsf{L}\bar{\boldsymbol{\theta}}, \tag{15.17}$$

where we define $\Delta_2^\perp\boldsymbol{\theta}$ by Eq. (14.22).

15.4 Bias Analysis and Hyper-Accurate Correction

Because $E[\Delta_1\boldsymbol{\theta}] = \mathbf{0}$, consider the second-order bias $E[\Delta_2^{\top}\boldsymbol{\theta}]$. From Eq. (15.17), we obtain

$$E[\Delta_2^{\perp}\boldsymbol{\theta}] = -E[\bar{\mathsf{M}}^{-}\Delta_1\mathsf{M}\Delta_1\boldsymbol{\theta}] - E[\bar{\mathsf{M}}^{-}\Delta_2\mathsf{M}\bar{\boldsymbol{\theta}}] + E[\bar{\mathsf{M}}^{-}\Delta_2\mathsf{L}\bar{\boldsymbol{\theta}}]. \qquad (15.18)$$

We evaluated each term separately, using our noise assumption of Eq. (14.28) and the definition of the vectors $\mathbf{e}_\alpha^{(k)}$ in Eq. (14.29). In the end, we obtained the following expression (see Lemmas 15.1, 15.2, and 15.3 below for the derivation).

$$\begin{aligned} E[\Delta_2^{\perp}\boldsymbol{\theta}] &= -\frac{\sigma^2}{N}\bar{\mathsf{M}}^{-}\sum_{\alpha=1}^{N}\sum_{k,l=1}^{L}\bar{W}_\alpha^{(kl)}(\mathbf{e}_\alpha^{(k)},\bar{\boldsymbol{\theta}})\bar{\boldsymbol{\xi}}_\alpha^{(l)} \\ &+\frac{\sigma^2}{N^2}\bar{\mathsf{M}}^{-}\sum_{\alpha=1}^{N}\sum_{k,l=1}^{L}\bar{W}_\alpha^{(km)}\bar{W}_\alpha^{(ln)}(\bar{\boldsymbol{\xi}}_\alpha^{(l)},\bar{\mathsf{M}}^{-}V_0^{(mn)}[\boldsymbol{\xi}_\alpha]\bar{\boldsymbol{\theta}})\bar{\boldsymbol{\xi}}_\alpha^{(k)}. \end{aligned} \qquad (15.19)$$

We can improve the accuracy of the computed $\boldsymbol{\theta}$ by subtracting this expression, which we call *hyper-accurate correction*. Inasmuch as Eq. (15.19) is defined in terms of true values, we replace them by observations for actual computation. The matrix M defined by observed $\boldsymbol{\xi}_\alpha^{(k)}$ is generally of full rank, therefore we replace $\bar{\mathsf{M}}^{-}$ by the pseudoinverse M_{n-1}^{-} of truncated rank $n-1$. Thus, we obtain the following expression for hyper-accurate correction (see Eqs. (2.52) and (6.50c)).

$$\begin{aligned} \Delta_c\boldsymbol{\theta} &= -\frac{\sigma^2}{N}\mathsf{M}_{n-1}^{-}\sum_{\alpha=1}^{N}\sum_{k,l=1}^{L}W_\alpha^{(kl)}(\mathbf{e}_\alpha^{(k)},\boldsymbol{\theta})\boldsymbol{\xi}_\alpha^{(l)} \\ &+\frac{\sigma^2}{N^2}\mathsf{M}_{n-1}^{-}\sum_{\alpha=1}^{N}\sum_{k,l=1}^{L}W_\alpha^{(km)}W_\alpha^{(ln)}(\boldsymbol{\xi}_\alpha^{(l)},\mathsf{M}_{n-1}^{-}V_0^{(mn)}[\boldsymbol{\xi}_\alpha]\boldsymbol{\theta})\boldsymbol{\xi}_\alpha^{(k)}. \end{aligned} \qquad (15.20)$$

Using this, we correct $\boldsymbol{\theta}$ to

$$\boldsymbol{\theta} \leftarrow \mathscr{N}[\boldsymbol{\theta} - \Delta_c\boldsymbol{\theta}], \qquad (15.21)$$

where $\mathscr{N}[\,\cdot\,]$ denotes normalization to the unit norm. This operation is necessary because $\boldsymbol{\theta}$ is defined to be a unit vector. For evaluating Eq. (15.20), we also need to estimate σ^2. This is done using the well-known fact in statistics: if $S_{\min}$ is the minimum of Eq. (15.2), then $NS_{\min}/\sigma^2$ is a χ^2 variable of $N-(n-1)$ degrees of freedom (the vector $\boldsymbol{\theta}$ has n components but is normalized to the unit norm), and the expectation of a χ^2 variable equals its degree of freedom. Therefore if we identify $E[S_{min}]$ with the minimized value J of the Sampson error J in Eq. (15.6), we can estimate σ^2 in the following form (see Eqs. (2.51) and (6.46)).

$$\hat{\sigma}^2 = \frac{J}{1-(n-1)/N}. \qquad (15.22)$$

By this hyper-accurate correction, the bias is removed up to $O(\sigma^4)$; replacing true values by observations in Eq. (15.20) and estimating σ^2 by Eq. (15.22) introduces only errors of $O(\sigma^4)$.

15.5 Derivations

We first derive Eq. (15.6).

Proposition 15.1 (Sampson error) *Minimizing Eq. (15.4) subject to Eq. (15.5) reduces to minimization of the Sampson error of Eq. (15.6).*

Proof In order to minimize the function J of Eq. (15.4) with respect to $\bar{\boldsymbol{\xi}}_\alpha^{(k)}$ subject to Eq. (15.5), we introduce Lagrange multipliers $\lambda_\alpha^{(k)}$ and differentiate $J - \sum_{\alpha=1}^N \sum_{k=1}^L \lambda_\alpha^{(k)} (\bar{\boldsymbol{\xi}}_\alpha^{(k)}, \boldsymbol{\theta})$ with respect to $\bar{\boldsymbol{\xi}}_\alpha$. Letting the result be $\mathbf{0}$, we obtain

$$-\frac{2}{N}\begin{pmatrix} V_0^{(11)}[\boldsymbol{\xi}_\alpha] & \cdots & V_0^{(1L)}[\boldsymbol{\xi}_\alpha] \\ \vdots & \ddots & \vdots \\ V_0^{(L1)}[\boldsymbol{\xi}_\alpha] & \cdots & V_0^{(LL)}[\boldsymbol{\xi}_\alpha] \end{pmatrix}^- \begin{pmatrix} \boldsymbol{\xi}_\alpha^{(1)} - \bar{\boldsymbol{\xi}}_\alpha^{(1)} \\ \vdots \\ \boldsymbol{\xi}_\alpha^{(L)} - \bar{\boldsymbol{\xi}}_\alpha^{(L)} \end{pmatrix} - \begin{pmatrix} \lambda_\alpha^{(1)}\boldsymbol{\theta} \\ \vdots \\ \lambda_\alpha^{(L)}\boldsymbol{\theta} \end{pmatrix} = \mathbf{0}. \tag{15.23}$$

Note that

$$\begin{pmatrix} V_0^{(11)}[\boldsymbol{\xi}_\alpha] & \cdots & V_0^{(1L)}[\boldsymbol{\xi}_\alpha] \\ \vdots & \ddots & \vdots \\ V_0^{(L1)}[\boldsymbol{\xi}_\alpha] & \cdots & V_0^{(LL)}[\boldsymbol{\xi}_\alpha] \end{pmatrix} \begin{pmatrix} V_0^{(11)}[\boldsymbol{\xi}_\alpha] & \cdots & V_0^{(1L)}[\boldsymbol{\xi}_\alpha] \\ \vdots & \ddots & \vdots \\ V_0^{(L1)}[\boldsymbol{\xi}_\alpha] & \cdots & V_0^{(LL)}[\boldsymbol{\xi}_\alpha] \end{pmatrix}^- = \mathscr{P}_\alpha, \tag{15.24}$$

where $\mathscr{P}_\alpha$ is a projection matrix onto the domain of the noise in $\boldsymbol{\xi}_\alpha^{(1)}, \ldots, \boldsymbol{\xi}_\alpha^{(L)}$ ($\hookrightarrow$ Problem 14.4). Multiplying Eq. (15.23) from the left by the first matrix on the left-hand side of Eq. (15.24) and noting that $\mathscr{P}_\alpha$ acts on $\boldsymbol{\xi}_\alpha^{(1)} - \bar{\boldsymbol{\xi}}_\alpha^{(1)}, \ldots, \boldsymbol{\xi}_\alpha^{(L)} - \bar{\boldsymbol{\xi}}_\alpha^{(L)}$ as the identity (because they are all on the noise domain), we obtain

$$-\frac{2}{N}\begin{pmatrix} \boldsymbol{\xi}_\alpha^{(1)} - \bar{\boldsymbol{\xi}}_\alpha^{(1)} \\ \vdots \\ \boldsymbol{\xi}_\alpha^{(L)} - \bar{\boldsymbol{\xi}}_\alpha^{(L)} \end{pmatrix} - \begin{pmatrix} V_0^{(11)}[\boldsymbol{\xi}_\alpha] & \cdots & V_0^{(1L)}[\boldsymbol{\xi}_\alpha] \\ \vdots & \ddots & \vdots \\ V_0^{(L1)}[\boldsymbol{\xi}_\alpha] & \cdots & V_0^{(LL)}[\boldsymbol{\xi}_\alpha] \end{pmatrix} \begin{pmatrix} \lambda_\alpha^{(1)}\boldsymbol{\theta} \\ \vdots \\ \lambda_\alpha^{(L)}\boldsymbol{\theta} \end{pmatrix} = \mathbf{0}, \tag{15.25}$$

from which we get

$$\bar{\boldsymbol{\xi}}_\alpha^{(k)} = \boldsymbol{\xi}_\alpha^{(k)} + \frac{N}{2}\sum_{l=1}^{L} \lambda_\alpha^{(l)} V_0^{(kl)}[\boldsymbol{\xi}_\alpha]\boldsymbol{\theta}. \tag{15.26}$$

Substituting this into Eq. (15.5), we have

$$(\boldsymbol{\xi}_\alpha^{(k)}, \boldsymbol{\theta}) = -\frac{N}{2}\sum_{l=1}^{L} \lambda_\alpha^{(l)} (\boldsymbol{\theta}, V_0^{(kl)}[\boldsymbol{\xi}_\alpha]\boldsymbol{\theta}). \tag{15.27}$$

This is a set of linear equations in $\lambda_\alpha^{(1)}, \ldots \lambda_\alpha^{(L)}$, but the coefficient matrix consisting of $(\boldsymbol{\theta}, V_0^{(kl)}[\boldsymbol{\xi}_\alpha]\boldsymbol{\theta})$ ($\equiv V_\alpha^{(kl)}$) may be singular. Therefore we use the pseudoinverse of truncated rank r (= the dimension of the noise domain) to solve this in the form

$$\lambda_\alpha^{(l)} = -\frac{2}{N}\sum_{k=1}^{L} W_\alpha^{(kl)} (\boldsymbol{\xi}_\alpha^k, \boldsymbol{\theta}). \tag{15.28}$$

Using Eq. (15.25), we can rewrite Eq. (15.4) in the form

$$
\begin{aligned}
J &= \frac{N}{4}\sum_{\alpha=1}^{N}\left(\begin{pmatrix} V_0^{(11)}[\xi_\alpha] & \cdots & V_0^{(1L)}[\xi_\alpha] \\ \vdots & \ddots & \vdots \\ V_0^{(L1)}[\xi_\alpha] & \cdots & V_0^{(LL)}[\xi_\alpha] \end{pmatrix}\begin{pmatrix} \lambda_\alpha^{(1)}\theta \\ \vdots \\ \lambda_\alpha^{(L)}\theta \end{pmatrix}, \begin{pmatrix} V_0^{(11)}[\xi_\alpha] & \cdots & V_0^{(1L)}[\xi_\alpha] \\ \vdots & \ddots & \vdots \\ V_0^{(L1)}[\xi_\alpha] & \cdots & V_0^{(LL)}[\xi_\alpha] \end{pmatrix}^{-} \right. \\
&\quad \left. \begin{pmatrix} V_0^{(11)}[\xi_\alpha] & \cdots & V_0^{(1L)}[\xi_\alpha] \\ \vdots & \ddots & \vdots \\ V_0^{(L1)}[\xi_\alpha] & \cdots & V_0^{(LL)}[\xi_\alpha] \end{pmatrix}\begin{pmatrix} \lambda_\alpha^{(1)}\theta \\ \vdots \\ \lambda_\alpha^{(L)}\theta \end{pmatrix}\right) \\
&= \frac{N}{4}\sum_{\alpha=1}^{N}\left(\begin{pmatrix} \lambda_\alpha^{(1)}\theta \\ \vdots \\ \lambda_\alpha^{(L)}\theta \end{pmatrix}, \begin{pmatrix} V_0^{(11)}[\xi_\alpha] & \cdots & V_0^{(1L)}[\xi_\alpha] \\ \vdots & \ddots & \vdots \\ V_0^{(L1)}[\xi_\alpha] & \cdots & V_0^{(LL)}[\xi_\alpha] \end{pmatrix}\begin{pmatrix} V_0^{(11)}[\xi_\alpha] & \cdots & V_0^{(1L)}[\xi_\alpha] \\ \vdots & \ddots & \vdots \\ V_0^{(L1)}[\xi_\alpha] & \cdots & V_0^{(LL)}[\xi_\alpha] \end{pmatrix}^{-} \right. \\
&\quad \left. \begin{pmatrix} V_0^{(11)}[\xi_\alpha] & \cdots & V_0^{(1L)}[\xi_\alpha] \\ \vdots & \ddots & \vdots \\ V_0^{(L1)}[\xi_\alpha] & \cdots & V_0^{(LL)}[\xi_\alpha] \end{pmatrix}\begin{pmatrix} \lambda_\alpha^{(1)}\theta \\ \vdots \\ \lambda_\alpha^{(L)}\theta \end{pmatrix}\right) \\
&= \frac{N}{4}\sum_{\alpha=1}^{N}\left(\begin{pmatrix} \lambda_\alpha^{(1)}\theta \\ \vdots \\ \lambda_\alpha^{(L)}\theta \end{pmatrix}, \begin{pmatrix} V_0^{(11)}[\xi_\alpha] & \cdots & V_0^{(1L)}[\xi_\alpha] \\ \vdots & \ddots & \vdots \\ V_0^{(L1)}[\xi_\alpha] & \cdots & V_0^{(LL)}[\xi_\alpha] \end{pmatrix}\begin{pmatrix} \lambda_\alpha^{(1)}\theta \\ \vdots \\ \lambda_\alpha^{(L)}\theta \end{pmatrix}\right) \\
&= \frac{N}{4}\sum_{\alpha=1}^{N}\sum_{k,l=1}^{L}\lambda_\alpha^{(k)}\lambda_\alpha^{(l)}(\theta, V_0^{(kl)}[\xi_\alpha]\theta) = \frac{N}{4}\sum_{\alpha=1}^{N}\sum_{k,l=1}^{L}V_\alpha^{(kl)}\lambda_\alpha^{(k)}\lambda_\alpha^{(l)}
\end{aligned} \tag{15.29}
$$

where we have used the identities in Eq. (14.68) for the pseudoinverse. Substituting Eq. (15.28), we can write Eq. (15.29) in the form

$$
\begin{aligned}
J &= \frac{1}{N}\sum_{\alpha=1}^{N}\sum_{k,l,m,n=1}^{L}W_\alpha^{(km)}W_\alpha^{(ln)}V_\alpha^{(kl)}(\xi_\alpha^m, \theta)(\xi_\alpha^n, \theta) \\
&= \frac{1}{N}\sum_{\alpha=1}^{N}\sum_{k,l=1}^{L}W_\alpha^{(kl)}(\xi_\alpha^l, \theta)(\xi_\alpha^l, \theta),
\end{aligned} \tag{15.30}
$$

where we have used Eq. (14.32). □

We now derive Eq. (15.19) by evaluating the right-hand side of Eq. (15.18) term by term. To do this, we split the expression of $\Delta_1\mathsf{M}$ and $\Delta_2\mathsf{M}$ in Eqs. (14.10) and (14.11) into the form

$$
\Delta_1\mathsf{M} = \Delta_1^0\mathsf{M} + \Delta_1^*\mathsf{M}, \tag{15.31}
$$

$$
\Delta_2\mathsf{M} = \Delta_2^0\mathsf{M} + \Delta_2^*\mathsf{M} + \Delta_2^\dagger\mathsf{M}, \tag{15.32}
$$

where

$$\Delta_1^0\mathsf{M} \equiv \frac{1}{N}\sum_{\alpha=1}^{N}\sum_{k,l=1}^{L} \bar{W}_\alpha^{(kl)}(\Delta_1\boldsymbol{\xi}_\alpha^{(k)}\bar{\boldsymbol{\xi}}_\alpha^{(l)\top} + \bar{\boldsymbol{\xi}}_\alpha^{(k)}\Delta_1\boldsymbol{\xi}_\alpha^{(l)\top}), \tag{15.33}$$

$$\Delta_1^*\mathsf{M} \equiv \frac{1}{N}\sum_{\alpha=1}^{N}\sum_{k,l=1}^{L} \Delta_1 W_\alpha^{(kl)}\bar{\boldsymbol{\xi}}_\alpha^{(k)}\bar{\boldsymbol{\xi}}_\alpha^{(l)\top}, \tag{15.34}$$

$$\Delta_2^0\mathsf{M} \equiv \frac{1}{N}\sum_{\alpha=1}^{N}\sum_{k,l=1}^{L} \bar{W}_\alpha^{(kl)}(\Delta_1\boldsymbol{\xi}_\alpha^{(k)}\Delta_1\boldsymbol{\xi}_\alpha^{(l)\top} + \Delta_2\boldsymbol{\xi}_\alpha^{(k)}\bar{\boldsymbol{\xi}}_\alpha^{(l)\top} + \bar{\boldsymbol{\xi}}_\alpha^{(k)}\Delta_2\boldsymbol{\xi}_\alpha^{(l)\top}), \tag{15.35}$$

$$\Delta_2^*\mathsf{M} \equiv \frac{1}{N}\sum_{\alpha=1}^{N}\sum_{k,l=1}^{L} \Delta_1 W_\alpha^{(kl)}(\Delta_1\boldsymbol{\xi}_\alpha^{(k)}\bar{\boldsymbol{\xi}}_\alpha^{(l)\top} + \bar{\boldsymbol{\xi}}_\alpha^{(k)}\Delta_1\boldsymbol{\xi}_\alpha^{(l)\top}), \tag{15.36}$$

$$\Delta_2^\dagger\mathsf{M} \equiv \frac{1}{N}\sum_{\alpha=1}^{N}\sum_{k,l=1}^{L} \Delta_2 W_\alpha^{(kl)}\bar{\boldsymbol{\xi}}_\alpha^{(k)}\bar{\boldsymbol{\xi}}_\alpha^{(l)\top}. \tag{15.37}$$

Equation (15.19) is obtained by combining the Lemmas 15.1, 15.2, and 15.3:

Lemma 15.1 (Expectation of $-\bar{\mathsf{M}}^-\Delta_1\mathsf{M}\Delta_1\boldsymbol{\theta}$)

$$E[-\bar{\mathsf{M}}^-\Delta_1\mathsf{M}\Delta_1\boldsymbol{\theta}] = \frac{\sigma^2}{N^2}\bar{\mathsf{M}}^-\sum_{\alpha=1}^{N}\sum_{k,l,m,n=1}^{L} \bar{W}_\alpha^{(kl)}\bar{W}_\alpha^{(mn)}(\bar{\boldsymbol{\xi}}_\alpha^{(l)}, \bar{\mathsf{M}}^-\bar{\boldsymbol{\xi}}_\alpha^{(n)})V_0^{(km)}[\boldsymbol{\xi}_\alpha]\bar{\boldsymbol{\theta}}$$
$$+\frac{3\sigma^2}{N^2}\bar{\mathsf{M}}^-\sum_{\alpha=1}^{N}\sum_{k,l=1}^{L} \bar{W}_\alpha^{(km)}\bar{W}_\alpha^{(ln)}(\bar{\boldsymbol{\xi}}_\alpha^{(l)}, \bar{\mathsf{M}}^-V_0^{(mn)}[\boldsymbol{\xi}_\alpha]\bar{\boldsymbol{\theta}})\bar{\boldsymbol{\xi}}_\alpha^{(k)}. \tag{15.38}$$

Proof From Eq. (15.31), we can write

$$-E[\bar{\mathsf{M}}^-\Delta_1\mathsf{M}\Delta_1\boldsymbol{\theta}] = E[\bar{\mathsf{M}}^-\Delta_1^0\mathsf{M}\bar{\mathsf{M}}^-\Delta_1^0\mathsf{M}\bar{\boldsymbol{\theta}}] + E[\bar{\mathsf{M}}^-\Delta_1^*\mathsf{M}\bar{\mathsf{M}}^-\Delta_1^0\mathsf{M}\bar{\boldsymbol{\theta}}]. \tag{15.39}$$

The first term on the right-hand side is written as follows.

$$E[\bar{\mathsf{M}}^-\Delta_1^0\mathsf{M}\bar{\mathsf{M}}^-\Delta_1^0\mathsf{M}\bar{\boldsymbol{\theta}}]$$
$$= E\Big[\frac{1}{N^2}\bar{\mathsf{M}}^-\sum_{\alpha,\beta=1}^{N}\sum_{k,l,m,n=1}^{L} \bar{W}_\alpha^{(kl)}\bar{W}_\beta^{(mn)}\Big(\Delta_1\boldsymbol{\xi}_\alpha^{(k)}\bar{\boldsymbol{\xi}}_\alpha^{(l)\top}$$
$$+\bar{\boldsymbol{\xi}}_\alpha^{(k)}\Delta_1\boldsymbol{\xi}_\alpha^{(l)\top}\Big)\bar{\mathsf{M}}^-(\Delta_1\boldsymbol{\xi}_\beta^{(m)}, \bar{\boldsymbol{\theta}})\bar{\boldsymbol{\xi}}_\beta^{(n)}\Big]$$
$$= \frac{\sigma^2}{N^2}\bar{\mathsf{M}}^-\sum_{\alpha=1}^{N}\sum_{k,l,m,n=1}^{L} \bar{W}_\alpha^{(kl)}\bar{W}_\alpha^{(mn)}(\bar{\boldsymbol{\xi}}_\alpha^{(l)}, \bar{\mathsf{M}}^-\bar{\boldsymbol{\xi}}_\alpha^{(n)})V_0^{(km)}[\boldsymbol{\xi}_\alpha]\bar{\boldsymbol{\theta}}$$
$$+\frac{\sigma^2}{N^2}\bar{\mathsf{M}}^-\sum_{\alpha=1}^{N}\sum_{k,l,m,n=1}^{L} \bar{W}_\alpha^{(kl)}\bar{W}_\alpha^{(mn)}(\bar{\boldsymbol{\theta}}, V_0^{(ml)}[\boldsymbol{\xi}_\alpha]\bar{\mathsf{M}}^-\bar{\boldsymbol{\xi}}_\alpha^{(n)})\bar{\boldsymbol{\xi}}_\alpha^{(k)}. \tag{15.40}$$

From Eqs. (15.35) and (14.13), we obtain

$$\Delta_1^* \mathsf{M} = \frac{2}{N}\sum_{\alpha=1}^{N}\sum_{k,l,m,n=1}^{L} \bar{W}_\alpha^{(km)}\bar{W}_\alpha^{(ln)}(\bar{\mathsf{M}}^- \Delta_1^0 \mathsf{M}\bar{\boldsymbol{\theta}}, V_0^{(mn)}[\boldsymbol{\xi}_\alpha]\bar{\boldsymbol{\theta}})\bar{\boldsymbol{\xi}}_\alpha^{(k)}\bar{\boldsymbol{\xi}}_\alpha^{(l)\top}. \quad (15.41)$$

Hence the second term on the right-hand side of Eq. (15.39) is

$$\begin{aligned}
&E[\bar{\mathsf{M}}^- \Delta_1^* \mathsf{M}\bar{\mathsf{M}}^- \Delta_1^0 \mathsf{M}\bar{\boldsymbol{\theta}}] \\
&= E\Big[\frac{2}{N}\bar{\mathsf{M}}^- \sum_{\alpha=1}^{N}\sum_{k,l=1}^{L} \bar{W}_\alpha^{(km)}\bar{W}_\alpha^{(ln)}(\bar{\mathsf{M}}^- \Delta_1^0 \mathsf{M}\bar{\boldsymbol{\theta}}, V_0^{(mn)}[\boldsymbol{\xi}_\alpha]\bar{\boldsymbol{\theta}})(\bar{\boldsymbol{\xi}}_\alpha^{(l)}, \bar{\mathsf{M}}^- \Delta_1^0 \mathsf{M}\bar{\boldsymbol{\theta}})\bar{\boldsymbol{\xi}}_\alpha^{(k)}\Big] \\
&= \frac{2}{N}\bar{\mathsf{M}}^- \sum_{\alpha=1}^{N}\sum_{k,l=1}^{L} \bar{W}_\alpha^{(km)}\bar{W}_\alpha^{(ln)}(\bar{\boldsymbol{\xi}}_\alpha^{(l)}, E[\Delta_1\boldsymbol{\theta}\Delta_1\boldsymbol{\theta}^\top]V_0^{(mn)}[\boldsymbol{\xi}_\alpha]\bar{\boldsymbol{\theta}})\bar{\boldsymbol{\xi}}_\alpha^{(k)} \\
&= \frac{2\sigma^2}{N^2}\bar{\mathsf{M}}^- \sum_{\alpha=1}^{N}\sum_{k,l=1}^{L} \bar{W}_\alpha^{(km)}\bar{W}_\alpha^{(ln)}(\bar{\boldsymbol{\xi}}_\alpha^{(l)}, \bar{\mathsf{M}}^- V_0^{(mn)}[\boldsymbol{\xi}_\alpha]\bar{\boldsymbol{\theta}})\bar{\boldsymbol{\xi}}_\alpha^{(k)},
\end{aligned} \quad (15.42)$$

where we have used with our noise assumption in Eq. (14.28). Adding Eqs. (15.35) and (15.42), we obtain Eq. (15.38). □

Lemma 15.2 (Expectation of $-\bar{\mathsf{M}}^- \Delta_2 \mathsf{M}\bar{\boldsymbol{\theta}}$)

$$\begin{aligned}
E[-\bar{\mathsf{M}}^- \Delta_2 \mathsf{M}\bar{\boldsymbol{\theta}}] = &-\frac{\sigma^2}{N}\bar{\mathsf{M}}^- \sum_{\alpha=1}^{N}\sum_{k,l=1}^{L} \bar{W}_\alpha^{(kl)}\Big(V_0^{(kl)}[\boldsymbol{\xi}_\alpha]\bar{\boldsymbol{\theta}} + (\mathbf{e}_\alpha^{(k)}, \bar{\boldsymbol{\theta}})\bar{\boldsymbol{\xi}}_\alpha^{(l)}\Big) \\
&-\frac{2\sigma^2}{N^2}\bar{\mathsf{M}}^- \sum_{\alpha=1}^{N}\sum_{k,l,m,n=1}^{L} \bar{W}_\alpha^{(kl)}\bar{W}_\alpha^{(mn)}(\bar{\boldsymbol{\xi}}_\alpha^{(k)}, \bar{\mathsf{M}}^- V_0^{(lm)}[\boldsymbol{\xi}_\alpha]\bar{\boldsymbol{\theta}})\bar{\boldsymbol{\xi}}_\alpha^{(n)}.
\end{aligned} \quad (15.43)$$

Proof Using Eq. (15.32) and noting that $\Delta_2^\dagger \mathsf{M}\bar{\boldsymbol{\theta}} = \mathbf{0}$, we can write

$$-E[\bar{\mathsf{M}}^- \Delta_2 \mathsf{M}\bar{\boldsymbol{\theta}}] = -E[\bar{\mathsf{M}}^- \Delta_2^0 \mathsf{M}\bar{\boldsymbol{\theta}}] - E[\bar{\mathsf{M}}^- \Delta_2^* \mathsf{M}\bar{\boldsymbol{\theta}}]. \quad (15.44)$$

The first term on the right-hand side is written as

$$\begin{aligned}
-E[\bar{\mathsf{M}}^- \Delta_2^0 \mathsf{M}\bar{\boldsymbol{\theta}}] &= -\bar{\mathsf{M}}^- \frac{1}{N}\sum_{\alpha=1}^{N}\sum_{k,l=1}^{L} \bar{W}_\alpha^{(kl)}\Big(E[\Delta_1\boldsymbol{\xi}_\alpha^{(k)}\Delta_1\boldsymbol{\xi}_\alpha^{(l)\top}] + \bar{\boldsymbol{\xi}}_\alpha^{(k)}E[\Delta_2\boldsymbol{\xi}_\alpha^{(l)\top}]\Big)\bar{\boldsymbol{\theta}} \\
&= -\frac{\sigma^2}{N}\bar{\mathsf{M}}^- \sum_{\alpha=1}^{N}\sum_{k,l=1}^{L} \bar{W}_\alpha^{(kl)}\Big(V_0^{(kl)}[\boldsymbol{\xi}_\alpha]\bar{\boldsymbol{\theta}} + (\mathbf{e}_\alpha^{(k)}, \bar{\boldsymbol{\theta}})\bar{\boldsymbol{\xi}}_\alpha^{(l)}\Big),
\end{aligned} \quad (15.45)$$

where we used the definition of the vectors $\mathbf{e}_\alpha^{(k)}$ in Eq. (14.29). The second term on the right-hand side of Eq. (15.44) is written as

$$
\begin{aligned}
-E[\bar{\mathsf{M}}^- \Delta_2^* \mathsf{M}\bar{\boldsymbol{\theta}}] &= -E[\bar{\mathsf{M}}^- \frac{1}{N}\sum_{\alpha=1}^{N}\sum_{k,l=1}^{L} \Delta_1 W_\alpha^{(kl)} (\Delta_1 \boldsymbol{\xi}_\alpha^{(l)}, \bar{\boldsymbol{\theta}}) \bar{\boldsymbol{\xi}}_\alpha^{(k)}] \\
&= -2\bar{\mathsf{M}}^- \frac{1}{N}\sum_{\alpha=1}^{N}\sum_{k,l,m,n=1}^{L} \bar{W}_\alpha^{(km)} \bar{W}_\alpha^{(ln)} (\bar{\boldsymbol{\theta}}, E[\Delta_1 \boldsymbol{\xi}_\alpha^{(l)} (\Delta_1^0 \mathsf{M}\bar{\boldsymbol{\theta}})^\top] \bar{\mathsf{M}}^- V_0^{(mn)}[\boldsymbol{\xi}_\alpha] \bar{\boldsymbol{\theta}}) \bar{\boldsymbol{\xi}}_\alpha^{(k)}.
\end{aligned}
\tag{15.46}
$$

The expression $E[\Delta_1 \boldsymbol{\xi}_\alpha^{(l)} (\Delta_1^0 \mathsf{M}\bar{\boldsymbol{\theta}})^\top]$ in the above equation is evaluated as follows.

$$
\begin{aligned}
E[\Delta_1 \boldsymbol{\xi}_\alpha^{(l)} (\Delta_1^0 \mathsf{M}\bar{\boldsymbol{\theta}})^\top] &= E[\Delta_1 \boldsymbol{\xi}_\alpha^{(l)} \Big(\frac{1}{N}\sum_{\beta=1}^{N}\sum_{p,q=1}^{L} \bar{W}_\beta^{(pq)} \Big(\Delta_1 \boldsymbol{\xi}_\beta^{(p)} \bar{\boldsymbol{\xi}}_\beta^{(q)\top} \\
&\quad + \bar{\boldsymbol{\xi}}_\beta^{(p)} \Delta_1 \boldsymbol{\xi}_\beta^{(q)\top}\Big)\bar{\boldsymbol{\theta}}\Big)^\top] \\
&= \frac{1}{N}\sum_{\beta=1}^{N}\sum_{p,q=1}^{L} \bar{W}_\beta^{(pq)} E[\Delta_1 \boldsymbol{\xi}_\alpha^{(l)} \Delta_1 \boldsymbol{\xi}_\beta^{(q)\top}] \bar{\boldsymbol{\theta}} \bar{\boldsymbol{\xi}}_\beta^{(p)\top} = \frac{\sigma^2}{N}\sum_{p,q=1}^{L} \bar{W}_\alpha^{(pq)} V_0^{(lq)}[\boldsymbol{\xi}_\alpha] \bar{\boldsymbol{\theta}} \bar{\boldsymbol{\xi}}_\alpha^{(p)\top}.
\end{aligned}
\tag{15.47}
$$

Thus Eq. (15.46) has the form

$$
\begin{aligned}
&-E[\bar{\mathsf{M}}^- \Delta_2^* \mathsf{M}\bar{\boldsymbol{\theta}}] \\
&= -\frac{2\sigma^2}{N^2}\bar{\mathsf{M}}^- \sum_{\alpha=1}^{N}\sum_{k,l,m,n,p,q=1}^{L} \bar{W}_\alpha^{(km)} \bar{W}_\alpha^{(ln)} \bar{W}_\alpha^{(pq)} (\bar{\boldsymbol{\theta}}, V_0^{(lq)}[\boldsymbol{\xi}_\alpha] \bar{\boldsymbol{\theta}}) \\
&\quad (\bar{\boldsymbol{\xi}}_\alpha^{(p)}, \bar{\mathsf{M}}^- V_0^{(mn)}[\boldsymbol{\xi}_\alpha] \bar{\boldsymbol{\theta}}) \bar{\boldsymbol{\xi}}_\alpha^{(k)} \\
&= -\frac{2\sigma^2}{N^2}\bar{\mathsf{M}}^- \sum_{\alpha=1}^{N}\sum_{k,l,m,n,p,q=1}^{L} \bar{W}_\alpha^{(km)} \bar{W}_\alpha^{(ln)} \bar{W}_\alpha^{(pq)} \bar{V}_\alpha^{(lq)} \bar{\boldsymbol{\theta}}) (\bar{\boldsymbol{\xi}}_\alpha^{(p)}, \bar{\mathsf{M}}^- V_0^{(mn)}[\boldsymbol{\xi}_\alpha] \bar{\boldsymbol{\theta}}) \bar{\boldsymbol{\xi}}_\alpha^{(k)} \\
&= -\frac{2\sigma^2}{N^2}\bar{\mathsf{M}}^- \sum_{\alpha=1}^{N}\sum_{k,l,m,n=1}^{L} \bar{W}_\alpha^{(kl)} \bar{W}_\alpha^{(mn)} (\bar{\boldsymbol{\xi}}_\alpha^{(k)}, \bar{\mathsf{M}}^- V_0^{(lm)}[\boldsymbol{\xi}_\alpha] \bar{\boldsymbol{\theta}}) \bar{\boldsymbol{\xi}}_\alpha^{(n)},
\end{aligned}
\tag{15.48}
$$

where we have put $\bar{V}_\alpha = (\bar{\boldsymbol{\theta}}, V_0^{(kl)}[\boldsymbol{\xi}_\alpha]\bar{\boldsymbol{\theta}})$ and Eq. (14.32) for $\bar{W}_\alpha^{(kl)}$ and $\bar{V}_\alpha^{(kl)}$. Thus Eq. (15.44) can be written in the form of Eq. (15.43). □

Lemma 15.3 (Expectation of $\bar{\mathsf{M}}^- \Delta_2 \mathsf{L}\bar{\boldsymbol{\theta}}$)

$$
\begin{aligned}
E[\bar{\mathsf{M}}^- \Delta_2 \mathsf{L}\bar{\boldsymbol{\theta}}] &= -\frac{\sigma^2}{N^2}\bar{\mathsf{M}}^- \sum_{\alpha=1}^{N}\sum_{k,l,m,n=1}^{L} \bar{W}_\alpha^{(km)} \bar{W}_\alpha^{(ln)} (\bar{\boldsymbol{\xi}}_\alpha^{(m)}, \bar{\mathsf{M}}^- \bar{\boldsymbol{\xi}}_\alpha^{(n)}) V_0^{(kl)}[\boldsymbol{\xi}_\alpha] \bar{\boldsymbol{\theta}} \\
&\quad + \frac{\sigma^2}{N}\bar{\mathsf{M}}^- \sum_{\alpha=1}^{N}\sum_{k,l=1}^{L} \bar{W}^{(kl)} V_0^{(kl)}[\boldsymbol{\xi}_\alpha] \bar{\boldsymbol{\theta}}.
\end{aligned}
\tag{15.49}
$$

Proof Substituting Eq. (15.11), we can write

$$
\begin{aligned}
E[\bar{\mathsf{M}}^- \Delta_2 \mathsf{L}\bar{\boldsymbol{\theta}}] &= E\Big[\frac{1}{N}\bar{\mathsf{M}}^- \sum_{\alpha=1}^{N} \sum_{k,l,m,n=1}^{L} \bar{W}_\alpha^{(km)} \bar{W}_\alpha^{(ln)} (\bar{\boldsymbol{\xi}}_\alpha^{(m)}, \Delta_1\boldsymbol{\theta})(\bar{\boldsymbol{\xi}}_\alpha^{(n)}, \Delta_1\boldsymbol{\theta}) V_0^{(kl)}[\boldsymbol{\xi}_\alpha]\bar{\boldsymbol{\theta}}\Big] \\
&+ E\Big[\frac{1}{N}\bar{\mathsf{M}}^- \sum_{\alpha=1}^{N} \sum_{k,l,m,n=1}^{L} \bar{W}_\alpha^{(km)} \bar{W}_\alpha^{(ln)} (\bar{\boldsymbol{\xi}}_\alpha^{(m)}, \Delta_1\boldsymbol{\theta})(\Delta_1\boldsymbol{\xi}_\alpha^{(n)}, \bar{\boldsymbol{\theta}}) V_0^{(kl)}[\boldsymbol{\xi}_\alpha]\bar{\boldsymbol{\theta}}\Big] \\
&+ E\Big[\frac{1}{N}\bar{\mathsf{M}}^- \sum_{\alpha=1}^{N} \sum_{k,l,m,n=1}^{L} \bar{W}_\alpha^{(km)} \bar{W}_\alpha^{(ln)} (\Delta_1\boldsymbol{\xi}_\alpha^{(m)}, \bar{\boldsymbol{\theta}})(\bar{\boldsymbol{\xi}}_\alpha^{(n)}, \Delta_1\boldsymbol{\theta}) V_0^{(kl)}[\boldsymbol{\xi}_\alpha]\bar{\boldsymbol{\theta}}\Big] \\
&+ E\Big[\frac{1}{N}\bar{\mathsf{M}}^- \sum_{\alpha=1}^{N} \sum_{k,l,m,n=1}^{L} \bar{W}_\alpha^{(km)} \bar{W}_\alpha^{(ln)} (\Delta_1\boldsymbol{\xi}_\alpha^{(m)}, \bar{\boldsymbol{\theta}})(\Delta_1\boldsymbol{\xi}_\alpha^{(n)}, \bar{\boldsymbol{\theta}}) V_0^{(kl)}[\boldsymbol{\xi}_\alpha]\bar{\boldsymbol{\theta}}\Big] \\
&= \frac{\sigma^2}{N^2}\bar{\mathsf{M}}^- \sum_{\alpha=1}^{N} \sum_{k,l,m,n=1}^{L} \bar{W}_\alpha^{(km)} \bar{W}_\alpha^{(ln)} (\bar{\boldsymbol{\xi}}_\alpha^{(m)}, \bar{\mathsf{M}}^- \bar{\boldsymbol{\xi}}_\alpha^{(n)}) V_0^{(kl)}[\boldsymbol{\xi}_\alpha]\bar{\boldsymbol{\theta}} \\
&+ \frac{1}{N}\bar{\mathsf{M}}^- \sum_{\alpha=1}^{N} \sum_{k,l,m,n=1}^{L} \bar{W}_\alpha^{(km)} \bar{W}_\alpha^{(ln)} (\bar{\boldsymbol{\xi}}_\alpha^{(m)}, E[\Delta_1\boldsymbol{\theta}\Delta_1\boldsymbol{\xi}_\alpha^{(n)\top}]\bar{\boldsymbol{\theta}}) V_0^{(kl)}[\boldsymbol{\xi}_\alpha]\bar{\boldsymbol{\theta}} \\
&+ \frac{1}{N}\bar{\mathsf{M}}^- \sum_{\alpha=1}^{N} \sum_{k,l,m,n=1}^{L} \bar{W}_\alpha^{(km)} \bar{W}_\alpha^{(ln)} (\bar{\boldsymbol{\theta}}, E[\Delta_1\boldsymbol{\xi}_\alpha^{(m)}\Delta_1\boldsymbol{\theta}^\top]\bar{\boldsymbol{\xi}}_\alpha^{(n)}) V_0^{(kl)}[\boldsymbol{\xi}_\alpha]\bar{\boldsymbol{\theta}} \\
&+ \frac{\sigma^2}{N}\bar{\mathsf{M}}^- \sum_{\alpha=1}^{N} \sum_{k,l=1}^{L} \bar{W}^{(kl)} V_0^{(kl)}[\boldsymbol{\xi}_\alpha]\bar{\boldsymbol{\theta}},
\end{aligned}
\tag{15.50}
$$

where we have used Eqs. (14.19) and (14.32). The expression $E[\Delta_1\boldsymbol{\theta}\Delta_1\boldsymbol{\xi}_\alpha^{(n)\top}]$ in the above equation can be evaluated as follows.

$$
E[\Delta_1\boldsymbol{\theta}\Delta_1\boldsymbol{\xi}_\alpha^{(n)\top}] = -E[\bar{\mathsf{M}}^- \Delta_1^0 \mathsf{M}\bar{\boldsymbol{\theta}}\Delta_1\boldsymbol{\xi}_\alpha^{(n)\top}] = -\frac{\sigma^2}{N}\bar{\mathsf{M}}^- \sum_{p,q=1}^{L} \bar{W}_\alpha^{(pq)} \bar{\boldsymbol{\xi}}_\alpha^{(p)} \bar{\boldsymbol{\theta}}^\top V_0^{(qn)}[\boldsymbol{\xi}_\alpha].
\tag{15.51}
$$

Therefore Eq. (15.50) has the form of Eq. (15.49). □

Substituting Eqs. (15.38), (15.43), and (15.49) into Eq. (15.18), we obtain Eq. (15.19).

15.6 Supplemental Note

The hyper-accurate correction technique for single-constraint problems was first presented by Kanatani [1], and detailed covariance and bias analysis was also given by Kanatani [2]. Later, the theory was generalized by Kanatani and Sugaya [3] to multiple-constraint problems in the form of Eq. (15.1). In the early analysis, the terms that contain the vectors $\mathbf{e}_\alpha^{(k)}$ (see Eq. (14.29)) were omitted, but it has been experimentally confirmed that the vectors $\mathbf{e}_\alpha^{(k)}$ play no significant role in the final results. As pointed out in the Supplemental Note of the preceding chapter, the vectors $\mathbf{e}_\alpha^{(k)}$ vanish for the fundamental matrix and homography computations.

Problems

15.1 Let A be a nonsingular matrix. Consider its infinitesimal variation $\mathsf{A} \to \mathsf{A} + \delta\mathsf{A}$. Show that the resulting first variations (i.e., variations omitting higher-order terms in $\delta\mathsf{A}$) of the inverse A^{-1} and the determinant $|\mathsf{A}|$ are given by

$$\delta\mathsf{A}^{-1} = -\mathsf{A}^{-1}\delta\mathsf{A}\mathsf{A}^{-1}, \qquad \delta|\mathsf{A}| = |\mathsf{A}|\mathrm{tr}[\mathsf{A}^{-1}\delta\mathsf{A}]. \tag{15.52}$$

15.2 Let $\mathbf{x}_1, \ldots, \mathbf{x}_N$ be independent samples from a Gaussian distribution of mean $\mathbf{m}$ and covariance matrix $\boldsymbol{\Sigma}$. Show that maximum likelihood estimates of $\mathbf{m}$ and $\boldsymbol{\Sigma}$, respectively, are given by

$$\hat{\mathbf{m}} = \frac{1}{N}\sum_{\alpha=1}^{N}\mathbf{x}_\alpha, \qquad \hat{\boldsymbol{\Sigma}} = \frac{1}{N}\sum_{\alpha=1}^{N}(\mathbf{x}_\alpha - \hat{\mathbf{m}})(\mathbf{x}_\alpha - \hat{\mathbf{m}})^\top, \tag{15.53}$$

that is, by the sample mean and by the sample covariance matrix.

References

1. K. Kanatani, Ellipse fitting with hyperaccuracy. IEICE Trans. Inf. Syst. **E89-D**(10), 2653–2660 (2006)
2. K. Kanatani, Statistical optimization for geometric fitting: theoretical accuracy bound and high order error analysis. Int. J. Comput. Vis. **80**(2), 167–188 (2008)
3. K. Kanatani, Y. Sugaya, Hyperaccurate correction of maximum likelihood for geometric estimation. IPSJ Trans. Comput. Vis. Appl. **5**, 19–29 (2013)

Theoretical Accuracy Limit

16

Abstract

We derive here a theoretical accuracy limit of the geometric estimation problem in the general mathematical framework described in Chaps. 14 and 15. It is given in the form of a bound, called the KCR (Kanatani-Cramer-Rao) lower bound, on the covariance matrix of the solution $\boldsymbol{\theta}$. The resulting form indicates that all iterative algebraic and geometric methods achieve this bound up to higher-order terms in σ, meaning that these are all optimal with respect to covariance. As in Chaps. 14 and 15, we treat $\boldsymbol{\theta}$ and $\boldsymbol{\xi}_\alpha$ as nD vectors for generality.

16.1 Kanatani-Cramer-Rao (KCR) Lower Bound

We continue to work in the mathematical framework of Chaps. 14 and 15. Namely, we want to estimate from noisy observations $\boldsymbol{\xi}_\alpha^{(k)}$ of the true values $\bar{\boldsymbol{\xi}}_\alpha^{(k)}$ the parameter $\boldsymbol{\theta}$ that should satisfy $(\bar{\boldsymbol{\xi}}_\alpha^{(k)}, \boldsymbol{\theta}) = 0$, $k = 1, \ldots, L$, $\alpha = 1, \ldots, N$, in the absence of noise. Evidently, whatever method is used, the estimate $\hat{\boldsymbol{\theta}}$ we obtain is a "function" of the observations $\boldsymbol{\xi}_\alpha^{(k)}$ in the form $\hat{\boldsymbol{\theta}} = \hat{\boldsymbol{\theta}}(\{\boldsymbol{\xi}_\alpha^{(k)}\})$ (we always understand that the indices run over $k = 1, \ldots, L$ and $\alpha = 1, \ldots, N$). Such a function is called an *estimator* of $\boldsymbol{\theta}$. It is an *unbiased estimator* if

$$E[\hat{\boldsymbol{\theta}}] = \boldsymbol{\theta}, \tag{16.1}$$

where $E[\,\cdot\,]$ is the expectation over the noise distribution. The important fact to note is that Eq. (16.1) should be satisfied *identically* in $\bar{\boldsymbol{\xi}}_\alpha^{(k)}$ and $\boldsymbol{\theta}$, because unbiasedness is a property of the "estimation method," not the data.

Adopting the Gaussian noise assumption for $\boldsymbol{\xi}_\alpha^{(k)}$ (although this is not strictly true, as discussed in Chap. 15), we regard $\boldsymbol{\xi}_\alpha^{(k)}$ as Gaussian variables of mean $\mathbf{0}$ and

K. Kanatani et al., *Guide to 3D Vision Computation*, Advances in Computer Vision and Pattern Recognition, DOI 10.1007/978-3-319-48493-8_16

covariance matrices $V^{(kl)}[\boldsymbol{\xi}_\alpha] = (\sigma^2 V_0^{(kl}[\boldsymbol{\xi}_\alpha])$, independent for different/ α. Let the computed estimate be $\hat{\boldsymbol{\theta}} = \boldsymbol{\theta} + \Delta\boldsymbol{\theta}$, and define its covariance matrix by

$$V[\boldsymbol{\theta}] = \mathsf{P}_{\boldsymbol{\theta}} E[\Delta\boldsymbol{\theta}\Delta\boldsymbol{\theta}^\top]\mathsf{P}_{\boldsymbol{\theta}}, \tag{16.2}$$

where $\mathsf{P}_{\boldsymbol{\theta}} \equiv \mathsf{I} - \boldsymbol{\theta}\boldsymbol{\theta}^\top$ is the projection matrix onto the space perpendicular to $\boldsymbol{\theta}$ ($\hookrightarrow$ Problem 3.3(1)). We introduce this projection because $\boldsymbol{\theta}$ is normalized to the unit norm. To be specific, the nD vector $\boldsymbol{\theta}$ is constrained to be on the $(n-1)$D unit sphere S^{n-1} centered at the origin, and we are interested in the error behavior of $\boldsymbol{\theta}$ in the tangent space $T_{\boldsymbol{\theta}}(S^{n-1})$ to the unit sphere S^{n-1} at $\boldsymbol{\theta}$ (see Fig. 14.1). We now derive the following result.

Theorem 16.1 (KCR lower bound) *If $\hat{\boldsymbol{\theta}}$ is an unbiased estimator of $\boldsymbol{\theta}$, the inequality*

$$V[\boldsymbol{\theta}] \succ \frac{\sigma^2}{N}\Big(\frac{1}{N}\sum_{\alpha=1}^{N}\sum_{k,l=1}^{L} \bar{W}_\alpha^{(kl)}\bar{\boldsymbol{\xi}}_\alpha^{(k)}\bar{\boldsymbol{\xi}}_\alpha^{(l)}\Big)^-, \qquad \bar{W}_\alpha^{(kl)} = \Big((\boldsymbol{\theta}, V_0^{(kl)}[\boldsymbol{\xi}_\alpha]\boldsymbol{\theta})\Big)^-, \tag{16.3}$$

holds for the covariance matrix $V[\boldsymbol{\theta}]$ of $\boldsymbol{\theta}$ irrespective of the estimation method, where the inequality $\mathsf{A} \succ \mathsf{B}$ means that $\mathsf{A} - \mathsf{B}$ is positive semi-definite.

Comments The definition of $\bar{W}_\alpha^{(kl)}$ has the same meaning as in Eq. (14.6), but here all the variables are true values therefore no rank truncation is necessary; the matrix on the right-hand side has rank r (=the number of linearly independent equations among the L constraints). The right-hand side of the inequality in Eq. (16.3) has rank $n-1$ (n is the dimension of $\boldsymbol{\theta}$) and is called the *KCR* (*Kanatani-Cramer-Rao*) *lower bound*. We give the proof of Theorem 16.1 in subsequent sections.

16.2 Structure of Constraints

As mentioned in Chaps. 14 and 15, we are assuming that the L vectors $\bar{\boldsymbol{\xi}}_\alpha^{(1)}, \ldots, \bar{\boldsymbol{\xi}}_\alpha^{(L)}$ are not necessarily linearly independent; that is, there may exist a nonzero vector $(c_1, \ldots, c_L)^\top$ such that

$$\sum_{k=1}^{L} c_k \bar{\boldsymbol{\xi}}_\alpha^{(k)} = \mathbf{0}. \tag{16.4}$$

The set $\mathcal{N}_\alpha \subset \mathbb{R}^L$ of such vectors $(c_1, \ldots, c_L)^\top$ is called the *null space* (or *kernel*) of the subspace $\mathcal{L}_\alpha$ spanned by $\boldsymbol{\xi}_\alpha^{(1)}, \ldots, \bar{\boldsymbol{\xi}}_\alpha^{(L)}$. Inasmuch as we are assuming that only r of $\bar{\boldsymbol{\xi}}_\alpha^{(k)}$ are linearly independent, the subspace $\mathcal{L}_\alpha$ has dimension r, and its null space $\mathcal{N}_\alpha$ has dimension $L - r$.

Let $\mathcal{N}_\alpha^\top \subset \mathbb{R}^L$ be the orthogonal complement of $\mathcal{N}_\alpha$, that is, the set of all LD vectors $(d_1, \ldots, d_L)^\top$ that is orthogonal to all $(c_1, \ldots, c_L)^\top \in \mathcal{N}_\alpha$. Because $\mathcal{N}_\alpha$

has dimension $L - r$, its orthogonal complement $\mathscr{N}_\alpha^\top$ has dimension r. Consider a vector

$$\begin{pmatrix} (\bar{\boldsymbol{\xi}}_\alpha^{(1)}, \mathbf{v}) \\ \vdots \\ (\bar{\boldsymbol{\xi}}_\alpha^{(L)}, \mathbf{v}) \end{pmatrix} \tag{16.5}$$

for some vector $\mathbf{v}$. From Eq. (16.4), we see that this vector is orthogonal to all $(c_1, \ldots, c_L)^\top \in \mathscr{N}_\alpha$. Hence this vector belongs to $\mathscr{N}_\alpha^\top$ for any $\mathbf{v}$.

We assume that to a first approximation the observations $\boldsymbol{\xi}_\alpha^{(k)} = \bar{\boldsymbol{\xi}}_\alpha^{(k)} + \Delta\bar{\boldsymbol{\xi}}_\alpha^{(k)}$ are also in $\mathscr{L}_\alpha$; that is, Eq. (16.4) also holds for $\boldsymbol{\xi}_\alpha^{(k)}$. Hence, $\sum_{k=1}^{L} c_k \Delta\boldsymbol{\xi}_\alpha^{(k)} = \mathbf{0}$ for $(c_1, \ldots, c_L)^\top \in \mathscr{N}_\alpha$. It follows that for the covariance matrix $V^{(kl)}[\boldsymbol{\xi}_\alpha] = E[\Delta\boldsymbol{\xi}_\alpha^{(k)} \Delta\boldsymbol{\xi}_\alpha^{(l)\top}]$ $(= \sigma^2 V_0^{(kl)}[\boldsymbol{\xi}_\alpha])$, which is defined in terms of the first-order perturbations $\Delta\boldsymbol{\xi}_\alpha^{(k)}$, we have

$$\sum_{k=1}^{L} c_k V_0^{(kl)}[\boldsymbol{\xi}_\alpha] = \mathbf{0}, \qquad (c_1, \ldots, c_L)^\top \in \mathscr{N}_\alpha. \tag{16.6}$$

Define the symmetric matrix

$$\bar{\mathsf{V}}_\alpha = \Big((\boldsymbol{\theta}, V_0^{(kl)}[\boldsymbol{\xi}_\alpha]\boldsymbol{\theta}) \Big), \tag{16.7}$$

whose $(kl,)$ element is $(\boldsymbol{\theta}, V_0^{(kl)}[\boldsymbol{\xi}_\alpha]\boldsymbol{\theta})$. It follows from Eq. (16.6) that

$$\bar{\mathsf{V}}_\alpha \begin{pmatrix} c_1 \\ \vdots \\ c_L \end{pmatrix} = \mathbf{0}, \qquad (c_1, \ldots, c_L)^\top \in \mathscr{N}_\alpha; \tag{16.8}$$

that is, all vectors of the null space $\mathscr{N}_\alpha$ of dimension $L - r$ are null vectors of $\bar{\mathsf{V}}_\alpha$, meaning that $\bar{\mathsf{V}}_\alpha$ has rank r and its domain is $\mathscr{N}_\alpha^\top$. Hence, its pseudoinverse

$$\bar{\mathsf{W}}_\alpha = \Big((\boldsymbol{\theta}, V_0^{(kl)}[\boldsymbol{\xi}_\alpha]\boldsymbol{\theta}) \Big)^- \tag{16.9}$$

also has rank r, and its domain is $\mathscr{N}_\alpha^\top$.

Lemma 16.1 (Structure of constraints) *For any* $\mathbf{v}$, *the following identity holds.*

$$\bar{\mathsf{V}}_\alpha \bar{\mathsf{W}}_\alpha \begin{pmatrix} (\bar{\boldsymbol{\xi}}_\alpha^{(1)}, \mathbf{v}) \\ \vdots \\ (\bar{\boldsymbol{\xi}}_\alpha^{(L)}, \mathbf{v}) \end{pmatrix} = \bar{\mathsf{W}}_\alpha \bar{\mathsf{V}}_\alpha \begin{pmatrix} (\bar{\boldsymbol{\xi}}_\alpha^{(1)}, \mathbf{v}) \\ \vdots \\ (\bar{\boldsymbol{\xi}}_\alpha^{(L)}, \mathbf{v}) \end{pmatrix} = \begin{pmatrix} (\bar{\boldsymbol{\xi}}_\alpha^{(1)}, \mathbf{v}) \\ \vdots \\ (\bar{\boldsymbol{\xi}}_\alpha^{(L)}, \mathbf{v}) \end{pmatrix}. \tag{16.10}$$

Proof Because the $L \times L$ matrix $\bar{\mathsf{V}}_\alpha$ in Eq. (16.7) and its pseudoinverse $\bar{\mathsf{W}}_\alpha$ in Eq. (16.9) both have the r-dimensional subspace $\mathscr{N}^\top \subset \mathbb{R}^L$ as their domain, their product $\bar{\mathsf{W}}_\alpha \bar{\mathsf{V}}_\alpha = \bar{\mathsf{V}}_\alpha \bar{\mathsf{W}}_\alpha$ is the orthogonal projection of $\mathbb{R}^L$ onto $\mathscr{N}_\alpha^\top$ ($\hookrightarrow$ Problem 14.4). Hence it acts as the identity within $\mathscr{N}_\alpha^\top$. The vector of Eq. (16.5) is in $\mathscr{N}_\alpha^\top$ for any $\mathbf{v}$, therefore Eq. (16.10) holds. □

16.3 Derivation of the KCR Lower Bound

From our noise assumption, the probability density of $\boldsymbol{\xi}_\alpha^{(k)}$, $k = 1, \ldots, L$, is

$$p(\boldsymbol{\xi}_\alpha^{(1)}, \ldots, \boldsymbol{\xi}_\alpha^{(L)}) = C_\alpha \exp\Big(-\frac{1}{2}\big(\begin{pmatrix} \boldsymbol{\xi}_\alpha^{(1)} - \bar{\boldsymbol{\xi}}_\alpha^{(1)} \\ \vdots \\ \boldsymbol{\xi}_\alpha^{(L)} - \bar{\boldsymbol{\xi}}_\alpha^{(L)} \end{pmatrix}, \begin{pmatrix} V^{(11)}[\boldsymbol{\xi}_\alpha] & \cdots & V^{(1L)}[\boldsymbol{\xi}_\alpha] \\ \vdots & \ddots & \vdots \\ V^{(L1)}[\boldsymbol{\xi}_\alpha] & \cdots & V^{(LL)}[\boldsymbol{\xi}_\alpha] \end{pmatrix}^{-} \begin{pmatrix} \boldsymbol{\xi}_\alpha^{(1)} - \bar{\boldsymbol{\xi}}_\alpha^{(1)} \\ \vdots \\ \boldsymbol{\xi}_\alpha^{(L)} - \bar{\boldsymbol{\xi}}_\alpha^{(L)} \end{pmatrix}\big)\Big), \tag{16.11}$$

where C_α is a normalization constant. Here we use the pseudoinverse because we are assuming that $\boldsymbol{\xi}_\alpha^{(1)}, \ldots, \boldsymbol{\xi}_\alpha^{(L)}$ are not necessarily linearly independent, as mentioned in the preceding section. From our assumption of unbiasedness, the equality

$$E[\hat{\boldsymbol{\theta}} - \boldsymbol{\theta}] = \mathbf{0} \tag{16.12}$$

is an *identity* in $\bar{\boldsymbol{\xi}}_\alpha^{(k)}$ and $\boldsymbol{\theta}$. Hence, Eq. (16.12) is invariant to infinitesimal variations in $\bar{\boldsymbol{\xi}}_\alpha^{(k)}$ and $\boldsymbol{\theta}$. In other words,

$$\begin{aligned} \delta \int (\hat{\boldsymbol{\theta}} - \boldsymbol{\theta}) p_1 \cdots p_N \mathrm{d}\boldsymbol{\xi} &= -\int (\delta\boldsymbol{\theta}) p_1 \cdots p_N \mathrm{d}\boldsymbol{\xi} + \sum_{\alpha=1}^{N} \int (\hat{\boldsymbol{\theta}} - \boldsymbol{\theta}) p_1 \cdots \delta p_\alpha \cdots p_N \mathrm{d}\boldsymbol{\xi} \\ &= -\delta\boldsymbol{\theta} + \int (\hat{\boldsymbol{\theta}} - \boldsymbol{\theta}) \sum_{\alpha=1}^{N} (p_1 \cdots \delta p_\alpha \cdots p_N) \mathrm{d}\boldsymbol{\xi} = 0, \end{aligned} \tag{16.13}$$

where p_α is an abbreviation of $p(\boldsymbol{\xi}_\alpha^{(1)}, \ldots, \boldsymbol{\xi}_\alpha^{(L)})$ and we use the shorthand $\int \mathrm{d}\boldsymbol{\xi} = \int \cdots \int \prod_{\alpha=1}^{N} \prod_{k=1}^{L} \mathrm{d}\boldsymbol{\xi}_\alpha^{(k)}$. Note that we are considering variations in $\bar{\boldsymbol{\xi}}_\alpha^{(k)}$ (not $\boldsymbol{\xi}_\alpha^{(k)}$) and $\boldsymbol{\theta}$. Because the estimator $\hat{\boldsymbol{\theta}}$ is a function of the observations $\boldsymbol{\xi}_\alpha^{(k)}$, it is not affected by such variations. Note that the variations $\delta\boldsymbol{\theta}$ are independent of $\boldsymbol{\xi}_\alpha^{(k)}$, thus it can be moved outside the integral $\int \mathrm{d}\boldsymbol{\xi}$. Also note that $\int p_1 \cdots p_N \mathrm{d}\boldsymbol{\xi} = 1$. Using the logarithmic differentiation formula, we can write the infinitesimal variation of Eq. (16.11) with respect to $\bar{\boldsymbol{\xi}}_\alpha^{(k)}$ in the form

$$\delta p_\alpha = p_\alpha \delta \log p_\alpha = p_\alpha \sum_{k=1}^{L} (\nabla_{\bar{\boldsymbol{\xi}}_\alpha^{(k)}} \log p_\alpha, \delta\bar{\boldsymbol{\xi}}_\alpha^{(k)}) = \sum_{k=1}^{L} (\mathbf{l}_\alpha^{(k)}, \delta\bar{\boldsymbol{\xi}}_\alpha^{(k)}) p_\alpha, \tag{16.14}$$

where we define the *score functions* $\mathbf{l}_\alpha^{(k)}$ (↪ Problem 16.1) by

$$\mathbf{l}_\alpha^{(k)} \equiv \nabla_{\bar{\boldsymbol{\xi}}_\alpha^{(k)}} \log p_\alpha. \tag{16.15}$$

Substituting Eq. (16.14) into Eq. (16.13), we obtain

$$\delta\boldsymbol{\theta} = \int (\hat{\boldsymbol{\theta}} - \boldsymbol{\theta}) \sum_{\alpha=1}^{N} \sum_{k=1}^{L} (\mathbf{l}_\alpha^{(k)}, \delta\bar{\boldsymbol{\xi}}_\alpha^{(k)}) p_1 \cdots p_N \mathrm{d}\boldsymbol{\xi} = E[(\hat{\boldsymbol{\theta}} - \boldsymbol{\theta}) \sum_{\alpha=1}^{N} \sum_{k=1}^{L} (\mathbf{l}_\alpha^{(k)}, \delta\bar{\boldsymbol{\xi}}_\alpha^{(k)})]. \tag{16.16}$$

Due to the constraint $(\bar{\boldsymbol{\xi}}_\alpha^{(k)}, \boldsymbol{\theta}) = 0$, the infinitesimal variations $\delta\bar{\boldsymbol{\xi}}_\alpha^{(k)}$ and $\delta\boldsymbol{\theta}$ are constrained to be

$$(\bar{\boldsymbol{\theta}}, \delta\bar{\boldsymbol{\xi}}_\alpha^{(k)}) + (\bar{\boldsymbol{\xi}}_\alpha^{(k)}, \delta\boldsymbol{\theta}) = 0. \tag{16.17}$$

Consider the following particular variations $\delta\bar{\boldsymbol{\xi}}_\alpha^{(k)}$.

$$\delta\bar{\boldsymbol{\xi}}_\alpha^{(k)} = -\sum_{l,m=1}^{L} \bar{W}_\alpha^{(lm)}(\bar{\boldsymbol{\xi}}_\alpha^{(l)}, \delta\boldsymbol{\theta}) V_0^{(mk)}[\boldsymbol{\xi}_\alpha]\bar{\boldsymbol{\theta}}. \tag{16.18}$$

This variation $\delta\bar{\boldsymbol{\xi}}_\alpha^{(k)}$ satisfies Eq. (16.17) for any variation $\delta\boldsymbol{\theta}$ of $\boldsymbol{\theta}$. In fact, we see that

$$\begin{aligned}(\bar{\boldsymbol{\theta}}, \delta\bar{\boldsymbol{\xi}}_\alpha^{(k)}) &= -\sum_{l,m=1}^{L} \bar{W}_\alpha^{(lm)}(\bar{\boldsymbol{\xi}}_\alpha^{(l)}, \delta\boldsymbol{\theta})(\bar{\boldsymbol{\theta}}, V_0^{(mk)}[\boldsymbol{\xi}_\alpha]\bar{\boldsymbol{\theta}}) \\ &= -\sum_{l,m=1}^{L} \bar{W}_\alpha^{(lm)} \bar{V}_\alpha^{(mk)}(\bar{\boldsymbol{\xi}}_\alpha^{(l)}, \delta\boldsymbol{\theta}) = -(\bar{\boldsymbol{\xi}}_\alpha^{(k)}, \delta\boldsymbol{\theta}),\end{aligned} \tag{16.19}$$

where we note from Eq. (16.10) that the matrix $\bar{\mathsf{V}}_\alpha\bar{\mathsf{W}}_\alpha$, whose (k, l) element is $\sum_{m=1}^{L} \bar{V}_\alpha^{(km)}\bar{W}_\alpha^{(ml)}$, acts as the identity on the vector $((\bar{\boldsymbol{\xi}}_\alpha^{(1)}, \delta\boldsymbol{\theta}), \ldots, (\bar{\boldsymbol{\xi}}_\alpha^{(L)}, \delta\boldsymbol{\theta}))^\top$. Substituting Eq. (16.18) into Eq. (16.16), we obtain

$$\begin{aligned}\delta\boldsymbol{\theta} &= -E[(\hat{\boldsymbol{\theta}} - \boldsymbol{\theta})\sum_{\alpha=1}^{N}\sum_{k,l,m=1}^{L} \bar{W}_\alpha^{(lm)}(\bar{\boldsymbol{\xi}}_\alpha^{(l)}, \delta\boldsymbol{\theta})(\mathbf{l}_\alpha^{(k)}, V_0^{(mk)}[\boldsymbol{\xi}_\alpha]\bar{\boldsymbol{\theta}})] \\ &= -E[(\hat{\boldsymbol{\theta}} - \boldsymbol{\theta})\sum_{\alpha=1}^{N}\sum_{k,l,m=1}^{L} \bar{W}_\alpha^{(lm)}(\mathbf{l}_\alpha^{(k)}, V_0^{(mk)}[\boldsymbol{\xi}_\alpha]\bar{\boldsymbol{\theta}})\bar{\boldsymbol{\xi}}_\alpha^{(l)\top}]\delta\boldsymbol{\theta} \\ &= -E[(\hat{\boldsymbol{\theta}} - \boldsymbol{\theta})\sum_{\alpha=1}^{N}\mathbf{m}_\alpha^\top]\delta\boldsymbol{\theta},\end{aligned} \tag{16.20}$$

where we define

$$\mathbf{m}_\alpha = \sum_{k,l,m=1}^{L} \bar{W}_\alpha^{(lm)}(\mathbf{l}_\alpha^{(k)}, V_0^{(mk)}[\boldsymbol{\xi}_\alpha]\bar{\boldsymbol{\theta}})\bar{\boldsymbol{\xi}}_\alpha^{(l)}. \tag{16.21}$$

Proposition 16.1 (Bound on the covariance matrix $V[\boldsymbol{\theta}]$)

$$V[\boldsymbol{\theta}] \succ \mathsf{M}^-, \tag{16.22}$$

where

$$\mathsf{M} = E\Big[\Big(\sum_{\alpha=1}^{N}\mathbf{m}_\alpha\Big)\Big(\sum_{\beta=1}^{N}\mathbf{m}_\beta\Big)^\top\Big]. \tag{16.23}$$

Proof Equation (16.20) should hold for an arbitrary infinitesimal variation $\delta\boldsymbol{\theta}$ of $\boldsymbol{\theta}$. However, we must not forget that $\boldsymbol{\theta}$ is normalized to the unit norm. Thus $\delta\boldsymbol{\theta}$ is constrained to be orthogonal to $\boldsymbol{\theta}$. Therefore we can write $\delta\boldsymbol{\theta} = \mathsf{P}_{\boldsymbol{\theta}}\delta\mathbf{u}$, where $\delta\mathbf{u}$ is an arbitrary infinitesimal variation and $\mathsf{P}_{\boldsymbol{\theta}}$ ($\equiv \mathsf{I} - \boldsymbol{\theta}\boldsymbol{\theta}^\top$) is the orthogonal projection matrix onto the tangent space $T_{\boldsymbol{\theta}}(S^{n-1})$ to the unit sphere S^{n-1} at $\boldsymbol{\theta}$. We obtain from Eq. (16.20)

$$\mathsf{P}_{\boldsymbol{\theta}} E[(\hat{\boldsymbol{\theta}} - \boldsymbol{\theta})\sum_{\alpha=1}^{N} \mathbf{m}_\alpha^\top]\mathsf{P}_{\boldsymbol{\theta}}\delta\mathbf{u} = -\mathsf{P}_{\boldsymbol{\theta}}\delta\mathbf{u}. \tag{16.24}$$

Note that the constraint $(\boldsymbol{\xi}_\alpha^{(k)}, \boldsymbol{\theta}) = 0$ implies $\mathsf{P}_{\boldsymbol{\theta}}\boldsymbol{\xi}_\alpha^{(k)} = \boldsymbol{\xi}_\alpha^{(k)}$, therefore from Eq. (16.21) we see that $\mathsf{P}_{\boldsymbol{\theta}}\mathbf{m}_\alpha = \mathbf{m}_\alpha$. Because Eq. (16.24) should hold for an arbitrary (unconstrained) $\delta\mathbf{u}$, we conclude that

$$\mathsf{P}_{\boldsymbol{\theta}} E[(\hat{\boldsymbol{\theta}} - \boldsymbol{\theta})\sum_{\alpha=1}^{N} \mathbf{m}_\alpha^\top] = -\mathsf{P}_{\boldsymbol{\theta}}, \tag{16.25}$$

from which we obtain the identity

$$\begin{aligned}
&E\left[\begin{pmatrix} \mathsf{P}_{\boldsymbol{\theta}}(\hat{\boldsymbol{\theta}} - \boldsymbol{\theta}) \\ \sum_{\alpha=1}^{N} \mathbf{m}_\alpha \end{pmatrix}\begin{pmatrix} \mathsf{P}_{\boldsymbol{\theta}}(\hat{\boldsymbol{\theta}} - \boldsymbol{\theta}) \\ \sum_{\beta=1}^{N} \mathbf{m}_\beta \end{pmatrix}^\top\right] \\
&= \begin{pmatrix} \mathsf{P}_{\boldsymbol{\theta}} E[(\hat{\boldsymbol{\theta}} - \boldsymbol{\theta})(\hat{\boldsymbol{\theta}} - \boldsymbol{\theta})^\top]\mathsf{P}_{\boldsymbol{\theta}} & \mathsf{P}_{\boldsymbol{\theta}} E[(\hat{\boldsymbol{\theta}} - \boldsymbol{\theta})\sum_{\alpha=1}^{N} \mathbf{m}_\alpha^\top] \\ (\mathsf{P}_{\boldsymbol{\theta}} E[(\hat{\boldsymbol{\theta}} - \boldsymbol{\theta})\sum_{\alpha=1}^{N} \mathbf{m}_\alpha^\top])^\top & E\left[(\sum_{\alpha=1}^{N} \mathbf{m}_\alpha)(\sum_{\beta=1}^{N} \mathbf{m}_\beta)^\top\right] \end{pmatrix} \\
&= \begin{pmatrix} V[\boldsymbol{\theta}] & -\mathsf{P}_{\boldsymbol{\theta}} \\ -\mathsf{P}_{\boldsymbol{\theta}} & \mathsf{M} \end{pmatrix}.
\end{aligned} \tag{16.26}$$

The inside of the expectation $E[\,\cdot\,]$ is positive semi-definite, therefore the right side is also. A positive semi-definite matrix sandwiched by any matrix and its transpose is also positive semi-definite, so the following is positive semi-definite.

$$\begin{aligned}
&\begin{pmatrix} \mathsf{P}_{\boldsymbol{\theta}} & \mathsf{M}^- \\ & \mathsf{M}^- \end{pmatrix}\begin{pmatrix} V[\boldsymbol{\theta}] & -\mathsf{P}_{\boldsymbol{\theta}} \\ -\mathsf{P}_{\boldsymbol{\theta}} & \mathsf{M} \end{pmatrix}\begin{pmatrix} \mathsf{P}_{\boldsymbol{\theta}} & \\ \mathsf{M}^- & \mathsf{M}^- \end{pmatrix} \\
&= \begin{pmatrix} \mathsf{P}_{\boldsymbol{\theta}} V[\boldsymbol{\theta}]\mathsf{P}_{\boldsymbol{\theta}} - \mathsf{P}_{\boldsymbol{\theta}}^2\mathsf{M}^- + \mathsf{M}^-\mathsf{P}_{\boldsymbol{\theta}}^2 & -\mathsf{P}_{\boldsymbol{\theta}}^2\mathsf{M}^- + \mathsf{M}^-\mathsf{M}\mathsf{M}^- \\ -\mathsf{M}^-\mathsf{P}_{\boldsymbol{\theta}}^2 + \mathsf{M}^-\mathsf{M}\mathsf{M}^- & \mathsf{M}^-\mathsf{M}\mathsf{M}^- \end{pmatrix} = \begin{pmatrix} V[\boldsymbol{\theta}] - \mathsf{M}^- & \\ & \mathsf{M}^- \end{pmatrix}.
\end{aligned} \tag{16.27}$$

Here, we have noted that $\mathsf{P}_{\boldsymbol{\theta}}^2 = \mathsf{P}_{\boldsymbol{\theta}}$ ($\hookrightarrow$ Problem 3.3(b)) and $\mathsf{P}_{\boldsymbol{\theta}}\mathbf{m}_\alpha = \mathbf{m}_\alpha$, thus $\mathsf{P}_{\boldsymbol{\theta}}\mathsf{M} = \mathsf{M}\mathsf{P}_{\boldsymbol{\theta}} = \mathsf{M}$ from Eq. (16.23). We have also used the identity $\mathsf{M}^-\mathsf{M}\mathsf{M}^- = \mathsf{M}^-$ ($\hookrightarrow$ Problem 14.3) and noted that the domain of M whose domain is $T_{\boldsymbol{\theta}}(S^{n-1})$ and thus $\mathsf{M}\mathsf{M}^- = \mathsf{M}^-\mathsf{M} = \mathsf{P}_{\boldsymbol{\theta}}$ ($\hookrightarrow$ Problem 14.4). The fact that Eq. (16.27) is positive semi-definite implies the inequality of Eq. (16.22).

16.4 Expression of the KCR Lower Bound

What remains is to show that the pseudoinverse M^- of the matrix M in Eq. (16.23) has its expression in Eq. (16.3). To show this, we use the following facts.

Lemma 16.2 (Expectation of score) *The score functions* $\mathbf{l}_\alpha^{(k)}$ *have expectation* $\mathbf{0}$*:*

$$E[\mathbf{l}_\alpha^{(k)}] = \mathbf{0}. \tag{16.28}$$

Proof Because

$$\int \cdots \int p_\alpha \mathrm{d}\boldsymbol{\xi}_\alpha \mathrm{d}\boldsymbol{\xi}_\alpha^{(1)} \cdots \mathrm{d}\boldsymbol{\xi}_\alpha \mathrm{d}\boldsymbol{\xi}_\alpha^{(L)} = 1 \tag{16.29}$$

is an identity in $\bar{\boldsymbol{\xi}}_\alpha^{(k)}$, the gradient of the left-hand side with respect to $\bar{\boldsymbol{\xi}}_\alpha^{(k)}$ (not $\boldsymbol{\xi}_\alpha^{(k)}$) is identically $\mathbf{0}$. Using the logarithmic differentiation formula (see Eq. (16.14)), we have

$$\begin{aligned}
\nabla_{\bar{\boldsymbol{\theta}}_\alpha^{(k)}} \int \cdots p_\alpha \mathrm{d}\boldsymbol{\xi}_\alpha \mathrm{d}\boldsymbol{\xi}_\alpha^{(1)} \cdots \mathrm{d}\boldsymbol{\xi}_\alpha \mathrm{d}\boldsymbol{\xi}_\alpha^{(L)} &= \int \cdots \int \nabla_{\bar{\boldsymbol{\theta}}_\alpha^{(k)}} p_\alpha \mathrm{d}\boldsymbol{\xi}_\alpha \mathrm{d}\boldsymbol{\xi}_\alpha^{(1)} \cdots \mathrm{d}\boldsymbol{\xi}_\alpha \mathrm{d}\boldsymbol{\xi}_\alpha^{(L)} \\
= \int \cdots p_\alpha \nabla_{\bar{\boldsymbol{\theta}}_\alpha^{(k)}} \log p_\alpha \mathrm{d}\boldsymbol{\xi}_\alpha \mathrm{d}\boldsymbol{\xi}_\alpha^{(1)} \cdots \mathrm{d}\boldsymbol{\xi}_\alpha \mathrm{d}\boldsymbol{\xi}_\alpha^{(L)} \\
= \int \cdots \int p_\alpha \mathbf{l}_\alpha^{(k)} \mathrm{d}\boldsymbol{\xi}_\alpha \mathrm{d}\boldsymbol{\xi}_\alpha^{(1)} \cdots \mathrm{d}\boldsymbol{\xi}_\alpha \mathrm{d}\boldsymbol{\xi}_\alpha^{(L)} &= E[\mathbf{l}_\alpha^{(k)}].
\end{aligned} \tag{16.30}$$

Therefore $\mathbf{l}_\alpha^{(k)}$ has expectation $\mathbf{0}$. □

The matrices $E[\mathbf{l}_\alpha^{(k)} \mathbf{l}_\alpha^{(l)\top}]$, $k, l = 1, \ldots, L$, are called the *Fisher information matrices* of $\bar{\boldsymbol{\xi}}_\alpha^{(1)}, \ldots, \bar{\boldsymbol{\xi}}_\alpha^{(L)}$ (↪ Problem 16.1). They are related to the covariance matrices $V_0^{(kl)}[\boldsymbol{\xi}_\alpha]$ as follows.

Lemma 16.3 (Fisher information and covariance) *The Fisher information matrices* $E[\mathbf{l}_\alpha^{(k)} \mathbf{l}_\alpha^{(l)\top}]$ *satisfy the identity:*

$$\sum_{m,n=1}^{L} V_0^{(km)}[\boldsymbol{\xi}_\alpha] E[\mathbf{l}_\alpha^{(m)} \mathbf{l}_\alpha^{(n)\top}] V_0^{(nl)}[\boldsymbol{\xi}_\alpha] = \frac{1}{\sigma^2} V_0^{(kl)}[\boldsymbol{\xi}_\alpha]. \tag{16.31}$$

Proof From Eq. (16.11), the score functions $\mathbf{l}_\alpha^{(k)}$ in Eq. (16.15) satisfy

$$\begin{pmatrix} \mathbf{l}_\alpha^{(1)} \\ \vdots \\ \mathbf{l}_\alpha^{(L)} \end{pmatrix} = \begin{pmatrix} \nabla_{\bar{\boldsymbol{\xi}}_\alpha^{(1)}} \log p_\alpha \\ \vdots \\ \nabla_{\bar{\boldsymbol{\xi}}_\alpha^{(L)}} \log p_\alpha \end{pmatrix} = - \begin{pmatrix} V^{(11)}[\boldsymbol{\xi}_\alpha] & \cdots & V^{(1L)}[\boldsymbol{\xi}_\alpha] \\ \vdots & \ddots & \vdots \\ V^{(L1)}[\boldsymbol{\xi}_\alpha] & \cdots & V^{(LL)}[\boldsymbol{\xi}_\alpha] \end{pmatrix}^{-} \begin{pmatrix} \Delta\boldsymbol{\xi}_\alpha^{(1)} \\ \vdots \\ \Delta\boldsymbol{\xi}_\alpha^{(L)} \end{pmatrix}. \tag{16.32}$$

Therefore

$$
\begin{pmatrix} E[\mathbf{l}_\alpha^{(1)}\mathbf{l}_\alpha^{(1)\top}] & \cdots & E[\mathbf{l}_\alpha^{(1)}\mathbf{l}_\alpha^{(L)\top}] \\ \vdots & \ddots & \vdots \\ E[\mathbf{l}_\alpha^{(L)}\mathbf{l}_\alpha^{(1)\top}] & \cdots & E[\mathbf{l}_\alpha^{(L)}\mathbf{l}_\alpha^{(L)\top}] \end{pmatrix} = E[\begin{pmatrix} \mathbf{l}_\alpha^{(1)} \\ \vdots \\ \mathbf{l}_\alpha^{(L)} \end{pmatrix}\begin{pmatrix} \mathbf{l}_\alpha^{(1)} \\ \vdots \\ \mathbf{l}_\alpha^{(L)} \end{pmatrix}^\top]
$$

$$
= E[\begin{pmatrix} V^{(11)}[\boldsymbol{\xi}_\alpha] & \cdots & V^{(1L)}[\boldsymbol{\xi}_\alpha] \\ \vdots & \ddots & \vdots \\ V^{(L1)}[\boldsymbol{\xi}_\alpha] & \cdots & V^{(LL)}[\boldsymbol{\xi}_\alpha] \end{pmatrix}^- \begin{pmatrix} \Delta\boldsymbol{\xi}_\alpha^{(1)}\Delta\boldsymbol{\xi}_\alpha^{(1)\top} & \cdots & \Delta\boldsymbol{\xi}_\alpha^{(1)}\Delta\boldsymbol{\xi}_\alpha^{(L)\top} \\ \vdots & \ddots & \vdots \\ \Delta\boldsymbol{\xi}_\alpha^{(L)}\Delta\boldsymbol{\xi}_\alpha^{(1)\top} & \cdots & \Delta\boldsymbol{\xi}_\alpha^{(L)}\Delta\boldsymbol{\xi}_\alpha^{(L)\top} \end{pmatrix}
$$

$$
\begin{pmatrix} V^{(11)}[\boldsymbol{\xi}_\alpha] & \cdots & V^{(1L)}[\boldsymbol{\xi}_\alpha] \\ \vdots & \ddots & \vdots \\ V^{(L1)}[\boldsymbol{\xi}_\alpha] & \cdots & V^{(LL)}[\boldsymbol{\xi}_\alpha] \end{pmatrix}^-]
$$

$$
= \begin{pmatrix} V^{(11)}[\boldsymbol{\xi}_\alpha] & \cdots & V^{(1L)}[\boldsymbol{\xi}_\alpha] \\ \vdots & \ddots & \vdots \\ V^{(L1)}[\boldsymbol{\xi}_\alpha] & \cdots & V^{(LL)}[\boldsymbol{\xi}_\alpha] \end{pmatrix}^- \begin{pmatrix} V^{(11)}[\boldsymbol{\xi}_\alpha] & \cdots & V^{(1L)}[\boldsymbol{\xi}_\alpha] \\ \vdots & \ddots & \vdots \\ V^{(L1)}[\boldsymbol{\xi}_\alpha] & \cdots & V^{(LL)}[\boldsymbol{\xi}_\alpha] \end{pmatrix}
$$

$$
\begin{pmatrix} V^{(11)}[\boldsymbol{\xi}_\alpha] & \cdots & V^{(1L)}[\boldsymbol{\xi}_\alpha] \\ \vdots & \ddots & \vdots \\ V^{(L1)}[\boldsymbol{\xi}_\alpha] & \cdots & V^{(LL)}[\boldsymbol{\xi}_\alpha] \end{pmatrix}^- = \begin{pmatrix} V^{(11)}[\boldsymbol{\xi}_\alpha] & \cdots & V^{(1L)}[\boldsymbol{\xi}_\alpha] \\ \vdots & \ddots & \vdots \\ V^{(L1)}[\boldsymbol{\xi}_\alpha] & \cdots & V^{(LL)}[\boldsymbol{\xi}_\alpha] \end{pmatrix}^-
$$

$$
= \frac{1}{\sigma^2}\begin{pmatrix} V_0^{(11)}[\boldsymbol{\xi}_\alpha] & \cdots & V_0^{(1L)}[\boldsymbol{\xi}_\alpha] \\ \vdots & \ddots & \vdots \\ V_0^{(L1)}[\boldsymbol{\xi}_\alpha] & \cdots & V_0^{(LL)}[\boldsymbol{\xi}_\alpha] \end{pmatrix}^-, \tag{16.33}
$$

where Eq. (14.68) is used. It follows that

$$
\begin{pmatrix} V_0^{(11)}[\boldsymbol{\xi}_\alpha] & \cdots & V_0^{(1L)}[\boldsymbol{\xi}_\alpha] \\ \vdots & \ddots & \vdots \\ V_0^{(L1)}[\boldsymbol{\xi}_\alpha] & \cdots & V_0^{(LL)}[\boldsymbol{\xi}_\alpha] \end{pmatrix} \begin{pmatrix} E[\mathbf{l}_\alpha^{(1)}\mathbf{l}_\alpha^{(1)\top}] & \cdots & E[\mathbf{l}_\alpha^{(1)}\mathbf{l}_\alpha^{(L)\top}] \\ \vdots & \ddots & \vdots \\ E[\mathbf{l}_\alpha^{(L)}\mathbf{l}_\alpha^{(1)\top}] & \cdots & E[\mathbf{l}_\alpha^{(L)}\mathbf{l}_\alpha^{(L)\top}] \end{pmatrix}
$$

$$
\begin{pmatrix} V_0^{(11)}[\boldsymbol{\xi}_\alpha] & \cdots & V_0^{(1L)}[\boldsymbol{\xi}_\alpha] \\ \vdots & \ddots & \vdots \\ V_0^{(L1)}[\boldsymbol{\xi}_\alpha] & \cdots & V_0^{(LL)}[\boldsymbol{\xi}_\alpha] \end{pmatrix}
$$

$$
= \frac{1}{\sigma^2}\begin{pmatrix} V_0^{(11)}[\boldsymbol{\xi}_\alpha] & \cdots & V_0^{(1L)}[\boldsymbol{\xi}_\alpha] \\ \vdots & \ddots & \vdots \\ V_0^{(L1)}[\boldsymbol{\xi}_\alpha] & \cdots & V_0^{(LL)}[\boldsymbol{\xi}_\alpha] \end{pmatrix} \begin{pmatrix} V_0^{(11)}[\boldsymbol{\xi}_\alpha] & \cdots & V_0^{(1L)}[\boldsymbol{\xi}_\alpha] \\ \vdots & \ddots & \vdots \\ V_0^{(L1)}[\boldsymbol{\xi}_\alpha] & \cdots & V_0^{(LL)}[\boldsymbol{\xi}_\alpha] \end{pmatrix}^-
$$

$$
\begin{pmatrix} V_0^{(11)}[\boldsymbol{\xi}_\alpha] & \cdots & V_0^{(1L)}[\boldsymbol{\xi}_\alpha] \\ \vdots & \ddots & \vdots \\ V_0^{(L1)}[\boldsymbol{\xi}_\alpha] & \cdots & V_0^{(LL)}[\boldsymbol{\xi}_\alpha] \end{pmatrix} = \frac{1}{\sigma^2}\begin{pmatrix} V_0^{(11)}[\boldsymbol{\xi}_\alpha] & \cdots & V_0^{(1L)}[\boldsymbol{\xi}_\alpha] \\ \vdots & \ddots & \vdots \\ V_0^{(L1)}[\boldsymbol{\xi}_\alpha] & \cdots & V_0^{(LL)}[\boldsymbol{\xi}_\alpha] \end{pmatrix}, \tag{16.34}
$$

where again Eq. (14.68) is used. Taking out the (k, l) block, we obtain Eq. (16.31).□

Now, we are ready to show the following result.

Proposition 16.2 (Bound expression) *The matrix* M *in Eq. (16.23) has the following pseudoinverse.*

$$\mathsf{M}^- = \frac{\sigma^2}{N}\Big(\frac{1}{N}\sum_{\alpha=1}^{N}\sum_{k,l=1}^{L}\bar{W}_\alpha^{(kl)}\bar{\boldsymbol{\xi}}_\alpha^{(k)}\bar{\boldsymbol{\xi}}_\alpha^{(l)\top}\Big)^-. \tag{16.35}$$

Proof Because $E[\mathbf{l}_\alpha^{(k)}]=\mathbf{0}$, we see from Eq. (16.21) that $E[\mathbf{m}_\alpha]=\mathbf{0}$. Noise in different observations is independent, thus $E[\mathbf{m}_\alpha\mathbf{m}_\beta^\top] = E[\mathbf{m}_\alpha]E[\mathbf{m}_\beta]^\top = \mathsf{O}$ for $\alpha \neq \beta$. Therefore

$$\begin{aligned}
\mathsf{M} &= \sum_{\alpha,\beta=1}^{N} E[\mathbf{m}_\alpha\mathbf{m}_\beta^\top] = \sum_{\alpha=1}^{N} E[\mathbf{m}_\alpha\mathbf{m}_\alpha^\top]\\
&= \sum_{\alpha=1}^{N} E[\sum_{k,l,m=1}^{L}\bar{W}_\alpha^{(lm)}(\mathbf{l}_\alpha^{(k)}, V_0^{(mk)}[\boldsymbol{\xi}_\alpha]\bar{\boldsymbol{\theta}})\bar{\boldsymbol{\xi}}_\alpha^{(l)}\sum_{n,p,q=1}^{L}\bar{W}_\alpha^{(pq)}(\mathbf{l}_\alpha^{(n)}, V_0^{(qn)}[\boldsymbol{\xi}_\alpha]\bar{\boldsymbol{\theta}})\bar{\boldsymbol{\xi}}_\alpha^{(p)\top}]\\
&= \sum_{\alpha=1}^{N}\sum_{l,m,p,q=1}^{L}\bar{W}_\alpha^{(lm)}\bar{W}_\alpha^{(pq)}(\bar{\boldsymbol{\theta}}, \Big(\sum_{k,n=1}^{L}V_0^{(mk)}[\boldsymbol{\xi}_\alpha]E[\mathbf{l}_\alpha^{(k)}\mathbf{l}_\alpha^{(n)\top}]V_0^{(nq)}[\boldsymbol{\xi}_\alpha]\Big)\bar{\boldsymbol{\theta}})\bar{\boldsymbol{\xi}}_\alpha^{(l)}\bar{\boldsymbol{\xi}}_\alpha^{(p)\top}\\
&= \frac{1}{\sigma^2}\sum_{\alpha=1}^{N}\sum_{l,m,p,q=1}^{L}\bar{W}_\alpha^{(lm)}\bar{W}_\alpha^{(pq)}(\bar{\boldsymbol{\theta}}, V_0^{(mq)}[\boldsymbol{\xi}_\alpha]\bar{\boldsymbol{\theta}})\bar{\boldsymbol{\xi}}_\alpha^{(l)}\bar{\boldsymbol{\xi}}_\alpha^{(p)\top}\\
&= \frac{1}{\sigma^2}\sum_{\alpha=1}^{N}\sum_{l,p=1}^{L}\Big(\sum_{m,q=1}^{L}\bar{W}_\alpha^{(lm)}\bar{V}_\alpha^{(mq)}\bar{W}_\alpha^{(qp)}\Big)\bar{\boldsymbol{\xi}}_\alpha^{(l)}\bar{\boldsymbol{\xi}}_\alpha^{(p)\top}\\
&= \frac{N}{\sigma^2}\frac{1}{N}\sum_{\alpha=1}^{N}\sum_{l,p=1}^{L}\bar{W}_\alpha^{(lp)}\bar{\boldsymbol{\xi}}_\alpha^{(l)}\bar{\boldsymbol{\xi}}_\alpha^{(p)\top},
\end{aligned} \tag{16.36}$$

where we have used Eq. (16.31) and the identities of Eq. (14.32) for $\bar{V}_\alpha^{(kl)}$ and $\bar{W}_\alpha^{(kl)}$. Thus we obtain Eq. (16.35). □

16.5 Supplemental Note

The KCR lower bound of Eq. (16.3) was first derived by Kanatani [1] in a much more general framework of nonlinear constraint fitting. This KCR lower bound is closely related to the *Cramer-Rao lower bound* in traditional statistics, therefore in [1] the bound of Eq. (16.3) was simply called the "Cramer-Rao lower bound." Pointing out its difference from the Cramer-Rao lower bound, Chernov and Lesort [2] named it the "KCR (Kanatani-Cramer-Rao) lower bound" and gave a rigorous proof in a linear estimation framework. The details of the derivation of the KCR lower bound are given by Kanatani [3] and by Kanatani et al. [4] for single-constraint problems ($L = 1$). This chapter generalizes the theory in [4] to multiconstraint problems.

The difference of the KCR lower bound from the Cramer-Rao lower bound originates from the difference of geometric estimation, such as ellipse fitting, from traditional statistical estimation. In traditional statistics, we model a random phenomenon as a parameterized probability density $p(\boldsymbol{\xi}^{(k)}|\boldsymbol{\theta})$, which is called the *statistical model*, and estimate the parameter $\boldsymbol{\theta}$ from noisy instances $\boldsymbol{\xi}_1^{(k)}, \ldots, \boldsymbol{\xi}_N^{(k)}$ of $\boldsymbol{\xi}^{(k)}$ obtained by repeated sampling. In this book, in contrast, the constraint $(\boldsymbol{\xi}^{(k)}, \boldsymbol{\theta}) = 0$ is an *implicit* equation of $\boldsymbol{\theta}$. If we are to deal with this problem in the traditional framework of statistics, the procedure is as follows. Suppose the observation $\boldsymbol{\xi}_\alpha^{(k)}$ and the parameter $\boldsymbol{\theta}$ are both nD vectors. We introduce mD auxiliary variables $\mathsf{X}_1, \ldots, \mathsf{X}_N$ such that the true value $\bar{\boldsymbol{\xi}}_\alpha^{(k)}$ of $\boldsymbol{\xi}_\alpha^{(k)}$ can be expressed as a function of X_α and $\boldsymbol{\theta}$ in the form of $\bar{\boldsymbol{\xi}}_\alpha^{(k)} = \bar{\boldsymbol{\xi}}_\alpha^{(k)}(\mathsf{X}_\alpha, \boldsymbol{\theta})$. For example, if $\boldsymbol{\xi}_\alpha^{(k)}$ are some image measurements of the αth point X_α in the scene and $\boldsymbol{\theta}$ encodes the camera parameters, then $\bar{\boldsymbol{\xi}}_\alpha^{(k)}(\mathsf{X}_\alpha, \boldsymbol{\theta})$ are the measurements of that point we would obtain in the absence of noise by using the camera specified by $\boldsymbol{\theta}$. For ellipse fitting, X_α can be chosen to be the angle ϕ_α (from a fixed direction) of the moving radius from the center of the ellipse specified by $\boldsymbol{\theta}$. Our goal is to estimate jointly $\mathsf{X}_1, \ldots, \mathsf{X}_N$, and $\boldsymbol{\theta}$ from noisy observations $\boldsymbol{\xi}_\alpha^{(k)} = \bar{\boldsymbol{\xi}}_\alpha^{(k)} + \boldsymbol{\varepsilon}_\alpha^{(k)}$, where $\boldsymbol{\varepsilon}_\alpha^{(k)}$ indicates random noise. Let $p(\{\boldsymbol{\varepsilon}_\alpha^{(k)}\})$ be the probability density of the noise (the indices run over $k = 1, \ldots, L$ and $\alpha = 1, \ldots, N$).

The Cramer-Rao lower bound is obtained as follows (↪ Problem 16.2). First, consider $\log p(\{\boldsymbol{\xi}_\alpha^{(k)} - \bar{\boldsymbol{\xi}}_\alpha^{(k)}(\mathsf{X}_\alpha, \boldsymbol{\theta})\})$. Next, evaluate its second derivatives with respect to $\mathsf{X}_1, \ldots, \mathsf{X}_N$, and $\boldsymbol{\theta}$ (or multiply its first derivatives by each other) to define an $(mN+n) \times (mN+n)$ matrix. Then, take the expectation of that matrix with respect to the density $p(\{\boldsymbol{\xi}_\alpha^{(k)} - \bar{\boldsymbol{\xi}}_\alpha^{(k)}(\mathsf{X}_\alpha, \boldsymbol{\theta})\})$. The resulting matrix is called the *Fisher information matrix*, and its inverse is called the *Cramer-Rao lower bound* on the (joint) covariance matrix of $\mathsf{X}_1, \ldots, \mathsf{X}_N$, and $\boldsymbol{\theta}$. The auxiliary variables $\mathsf{X}_1, \ldots, \mathsf{X}_N$ are also known as the *nuisance parameters*, whereas $\boldsymbol{\theta}$ is called the *structural parameter* or *parameter of interest*. Usually, we are interested only in the covariance of $\boldsymbol{\theta}$, therefore we remove from the $(mN+n) \times (mN+n)$ joint covariance matrix the $n \times n$ lower right submatrix, which can be shown to coincide with the KCR lower bound of Eq. (16.3).

However, evaluating and inverting the large Fisher information matrix, whose size can grow quickly as the number N of observations increases, is a heavy burden both analytically and computationally. The KCR lower bound in Eq. (16.3) is expressed only in terms of the parameter $\boldsymbol{\theta}$ of interest without involving auxiliary nuisance parameters, resulting in significant analytical and computational efficiency, yet giving the same value as the Cramer-Rao lower bound.

Problems

16.1 Let $\mathbf{x}$ be a random variable sampled from the probability density $p(\mathbf{x}|\boldsymbol{\theta})$ parameterized by $\boldsymbol{\theta}$.

(1) The *score* of $\mathbf{x}$ for $\boldsymbol{\theta}$ is defined by

$$\mathbf{l} = \nabla_{\boldsymbol{\theta}} \log p(\mathbf{x}|\boldsymbol{\theta}). \tag{16.37}$$

This is also a random variable. Show that it has expectation $\mathbf{0}$:

$$E[\mathbf{l}] = \mathbf{0}. \tag{16.38}$$

(2) The *Fisher information matrix* of $\mathbf{x}$ for $\boldsymbol{\theta}$ is defined by

$$\mathsf{J} = E[\mathbf{l}\mathbf{l}^\top]. \tag{16.39}$$

Show that the following identity holds.

$$\mathsf{J} = -E[\nabla_{\boldsymbol{\theta}}^2 \log p(\mathbf{x}|\boldsymbol{\theta})]. \tag{16.40}$$

(3) If $\mathbf{x}$ is a Gaussian random variable of mean $\mathbf{m}$ and covariance matrix $\boldsymbol{\Sigma}$, show that the score $\mathbf{l}$ and the Fisher information matrix for the mean $\mathbf{m}$ are given by

$$\mathbf{l} = \boldsymbol{\Sigma}^{-1}(\mathbf{x} - \mathbf{m}), \qquad \mathsf{J} = \boldsymbol{\Sigma}^{-1}. \tag{16.41}$$

16.2 Let $\mathbf{x}$ be a random variable sampled from the probability density $p(\mathbf{x}|\boldsymbol{\theta})$ parameterized by $\boldsymbol{\theta}$, and let $\hat{\boldsymbol{\theta}} = \hat{\boldsymbol{\theta}}(\mathbf{x})$ be an *estimator* of $\boldsymbol{\theta}$, that is, a function of $\mathbf{x}$ that returns an estimate of $\boldsymbol{\theta}$. It is *unbiased* if $E[\hat{\boldsymbol{\theta}}] = \boldsymbol{\theta}$ holds identically in $\boldsymbol{\theta}$.

(1) Show that for an unbiased estimator $\hat{\boldsymbol{\theta}}$ of $\boldsymbol{\theta}$

$$E[(\hat{\boldsymbol{\theta}} - \boldsymbol{\theta})\mathbf{l}^\top] = \mathsf{I} \tag{16.42}$$

identically holds, where $\mathbf{l}$ is the score of $\mathbf{x}$ with respect to $\boldsymbol{\theta}$.

(2) The covariance matrix of the estimator $\hat{\boldsymbol{\theta}}$ is defined by

$$V[\hat{\boldsymbol{\theta}}] = E[(\hat{\boldsymbol{\theta}} - \boldsymbol{\theta})(\hat{\boldsymbol{\theta}} - \boldsymbol{\theta})^\top]. \tag{16.43}$$

Show the *Cramer-Rao inequality*

$$V[\hat{\boldsymbol{\theta}}] \succ \mathsf{J}^{-1}, \tag{16.44}$$

for an unbiased estimator $\hat{\boldsymbol{\theta}}$ (the right-hand side is called the *Cramer-Rao lower bound*), where J is the Fisher information matrix of $p(\mathbf{x}|\mathbf{x})$, and $\mathsf{A} \succ \mathsf{B}$ means that $\mathsf{A} - \mathsf{B}$ is positive semi-definite.

16.3 Let $\mathbf{x}_1$, ..., $\mathbf{x}_N$ be independent samples from the probability density $p(\mathbf{x}|\boldsymbol{\theta})$, and $\hat{\boldsymbol{\theta}} = \hat{\boldsymbol{\theta}}(\mathbf{x}_1, \ldots, \mathbf{x}_N)$ an unbiased estimator of $\boldsymbol{\theta}$.

(1) Show that the Cramer-Rao inequality has the form

$$V[\hat{\boldsymbol{\theta}}] \succ \frac{1}{N}\mathsf{J}^{-1}, \tag{16.45}$$

where J is the Fisher information matrix for $\boldsymbol{\theta}$.

(2) If $\mathbf{x}_1$, ..., $\mathbf{x}_N$ are independent samples from a Gaussian distribution of mean $\mathbf{m}$ and covariance matrix $\boldsymbol{\Sigma}$, show that the sample mean

$$\hat{\mathbf{m}} = \frac{1}{N}(\mathbf{x}_1 + \cdots + \mathbf{x}_N) \tag{16.46}$$

is an unbiased estimator of $\mathbf{m}$ and that its covariance matrix $V[\hat{\mathbf{m}}]$ satisfies the Cramer-Rao inequality with equality (such an estimator is said to be *efficient*).

References

1. K. Kanatani, *Statistical Optimization for Geometric Computation: Theory and Practice, Elsevier, Amsterdam, The Netherlands (1996)* (Reprinted by Dover, New York, U.S., 2005)
2. N. Chernov, C. Lesort, Statistical efficiency of curve fitting algorithms. Comput. Stat. Data Anal. **47**(4), 713–728 (2004)
3. K. Kanatani, Statistical optimization for geometric fitting: theoretical accuracy bound and high order error analysis. Int. J. Comput. Vis. **80**(2), 167–188 (2008)
4. K. Kanatani, Y. Sugaya, K. Kanazawa, *Ellipse Fitting for Computer Vision: Implementation and Applications* (Morgan & Claypool, San Rafael, CA, U.S., 2016)

Solutions

Problems of Chapter 2

2.1 (1) We minimize $(1/N)\sum_{\alpha=1}^{N}(\mathbf{n}, \boldsymbol{\xi}_\alpha)^2 = (1/N)(\mathbf{n}, \sum_{\alpha=1}^{N}\boldsymbol{\xi}_\alpha\boldsymbol{\xi}_\alpha^\top\mathbf{n})$. The procedure is written as follows.

1. Compute the 3×6 matrix

$$\mathsf{M} = \frac{1}{N}\sum_{\alpha=1}^{N}\boldsymbol{\xi}_\alpha\boldsymbol{\xi}_\alpha^\top.$$

2. Solve the eigenvalue problem $\mathsf{M}\mathbf{n} = \lambda\mathbf{n}$, and return the unit eigenvector $\mathbf{n}$ for the smallest eigenvalue λ.

(2) We obtain the expression:

$$V[\boldsymbol{\xi}_\alpha] = E[\Delta\boldsymbol{\xi}_\alpha\Delta\boldsymbol{\xi}_\alpha^\top] = E[\begin{pmatrix}\Delta x_\alpha \\ \Delta y_\alpha \\ 0\end{pmatrix}\begin{pmatrix}\Delta x_\alpha \\ \Delta y_\alpha \\ 0\end{pmatrix}^\top]$$

$$= \begin{pmatrix} E[\Delta x_\alpha^2] & E[\Delta x_\alpha\Delta y_\alpha] & 0 \\ E[\Delta y_\alpha\Delta y_\alpha] & E[\Delta y_\alpha^2] & 0 \\ 0 & 0 & 0 \end{pmatrix} = \sigma^2\begin{pmatrix} 1 & 0 & 0 \\ 0 & 1 & 0 \\ 0 & 0 & 0\end{pmatrix}.$$

2.2 The procedure is as follows.

1. Compute the 6×6 matrices

$$\mathsf{M} = \frac{1}{N}\sum_{\alpha=1}^{N}\boldsymbol{\xi}_\alpha\boldsymbol{\xi}_\alpha^\top, \qquad \mathsf{N} = \frac{1}{N}\sum_{\alpha=1}^{N}V_0[\boldsymbol{\xi}_\alpha].$$

2. Solve the generalized eigenvalue problem $\mathsf{M}\boldsymbol{\theta} = \lambda\mathsf{N}\boldsymbol{\theta}$, and return the unit generalized eigenvector $\boldsymbol{\theta}$ for the smallest generalized eigenvalue λ.

K. Kanatani et al., *Guide to 3D Vision Computation*, Advances in Computer Vision and Pattern Recognition, DOI 10.1007/978-3-319-48493-8

2.3 The procedure is as follows.

1. Compute the 6×6 matrices

$$\mathsf{M} = \frac{1}{N}\sum_{\alpha=1}^{N} \boldsymbol{\xi}_\alpha \boldsymbol{\xi}_\alpha^\top,$$

$$\mathsf{N} = \frac{1}{N}\sum_{\alpha=1}^{N}\Big(V_0[\boldsymbol{\xi}_\alpha] + 2\mathscr{S}[\boldsymbol{\xi}_\alpha \mathbf{e}^\top]\Big)$$
$$- \frac{1}{N^2}\sum_{\alpha=1}^{N}\Big((\boldsymbol{\xi}_\alpha, \mathsf{M}_5^- \boldsymbol{\xi}_\alpha) V_0[\boldsymbol{\xi}_\alpha] + 2\mathscr{S}[V_0[\boldsymbol{\xi}_\alpha]\mathsf{M}_5^- \boldsymbol{\xi}_\alpha \boldsymbol{\xi}_\alpha^\top]\Big).$$

2. Solve the generalized eigenvalue problem $\mathsf{M}\boldsymbol{\theta} = \lambda \mathsf{N}\boldsymbol{\theta}$, and return the unit generalized eigenvector $\boldsymbol{\theta}$ for the eigenvalue λ of the smallest absolute value.

2.4 (1) Obvious from the definition of $\mathbf{x}$ and Q.
(2) Let $(\bar{x}_\alpha, \bar{y}_\alpha)$ be the closest point on the ellipse from point (x_α, y_α). We write them as vectors $\bar{\mathbf{x}}_\alpha$ and $\mathbf{x}_\alpha$ in the form of Eq. (2.62). If we write

$$\Delta \mathbf{x}_\alpha = \mathbf{x}_\alpha - \bar{\mathbf{x}}_\alpha,$$

the distance d_α of $\mathbf{x}_\alpha$ from the ellipse is $f_0\|\Delta \mathbf{x}_\alpha\|$. We want to find $\Delta \mathbf{x}_\alpha$ that minimizes this. Because $\bar{\mathbf{x}}_\alpha$ is on the ellipse, we have the constraint

$$(\mathbf{x}_\alpha - \Delta \mathbf{x}_\alpha, \mathsf{Q}(\mathbf{x}_\alpha - \Delta \mathbf{x}_\alpha)) = 0.$$

Expanding this and ignoring quadratic terms in $\Delta \mathbf{x}_\alpha$, we obtain

$$(\mathsf{Q}\mathbf{x}_\alpha, \Delta \mathbf{x}_\alpha) = \frac{1}{2}(\mathbf{x}_\alpha, \mathsf{Q}\mathbf{x}_\alpha). \qquad (*)$$

The third components of $\mathbf{x}_\alpha$ and $\bar{\mathbf{x}}_\alpha$ are both 1, therefore the third component of $\Delta \mathbf{x}_\alpha$ is 0. This is written as $(\mathbf{k}, \Delta \mathbf{x}_\alpha) = 0$, where $\mathbf{k} = (0, 0, 1)^\top$. Introducing Lagrange multipliers, we differentiate

$$\|\Delta \mathbf{x}_\alpha\|^2 - \lambda_\alpha\Big((\mathsf{Q}\mathbf{x}_\alpha, \Delta \mathbf{x}_\alpha) - \frac{1}{2}(\mathbf{x}_\alpha, \mathsf{Q}\mathbf{x}_\alpha)\Big) - \mu(\mathbf{k}, \Delta \mathbf{x}_\alpha)$$

with respect to $\Delta \mathbf{x}_\alpha$. Letting the result be $\mathbf{0}$, we obtain

$$2\Delta \mathbf{x}_\alpha - \lambda_\alpha \mathsf{Q}\mathbf{x}_\alpha - \mu \mathbf{k} = \mathbf{0}.$$

Multiplying this by the projection matrix $\mathsf{P}_\mathbf{k}$ in Eq. (2.65) from the left, and noting that $\mathsf{P}_\mathbf{k}\Delta \mathbf{x}_\alpha = \Delta \mathbf{x}_\alpha$, $\mathsf{P}_\mathbf{k}\mathbf{k} = \mathbf{0}$, we obtain

$$\Delta \mathbf{x}_\alpha = \frac{\lambda_\alpha}{2}\mathsf{P}_\mathbf{k}\mathsf{Q}\mathbf{x}_\alpha.$$

Substituting this into Eq. $(*)$, we obtain

$$(\mathsf{Q}\mathbf{x}_\alpha, \frac{\lambda_\alpha}{2}\mathsf{P}_\mathbf{k}\mathsf{Q}\mathbf{x}_\alpha) = \frac{1}{2}(\mathbf{x}_\alpha, \mathsf{Q}\mathbf{x}_\alpha).$$

Thus λ_α is given by

$$\lambda_\alpha = \frac{(\mathbf{x}_\alpha, \mathsf{Q}\mathbf{x}_\alpha)}{(\mathsf{Q}\mathbf{x}_\alpha, \mathsf{P}_{\mathbf{k}}\mathsf{Q}\mathbf{x}_\alpha)},$$

such that $\Delta\mathbf{x}_\alpha$ is written as

$$\Delta\mathbf{x}_\alpha = \frac{(\mathbf{x}_\alpha, \mathsf{Q}\mathbf{x}_\alpha)\mathsf{P}_{\mathbf{k}}\mathsf{Q}\mathbf{x}_\alpha}{2(\mathsf{Q}\mathbf{x}_\alpha, \mathsf{P}_{\mathbf{k}}\mathsf{Q}\mathbf{x}_\alpha)}.$$

From this, we obtain

$$\|\Delta\mathbf{x}_\alpha\|^2 = \frac{(\mathbf{x}_\alpha, \mathsf{Q}\mathbf{x}_\alpha)^2\|\mathsf{P}_{\mathbf{k}}\mathsf{Q}\mathbf{x}_\alpha\|^2}{4(\mathsf{Q}\mathbf{x}_\alpha, \mathsf{P}_{\mathbf{k}}\mathsf{Q}\mathbf{x}_\alpha)^2} = \frac{(\mathbf{x}_\alpha, \mathsf{Q}\mathbf{x}_\alpha)^2}{4(\mathsf{Q}\mathbf{x}_\alpha, \mathsf{P}_{\mathbf{k}}\mathsf{Q}\mathbf{x}_\alpha)},$$

where we have noted that $\mathsf{P}_{\mathbf{k}}^2 = \mathsf{P}_{\mathbf{k}}$ from Eq. (2.65) and that $\|\mathsf{P}_{\mathbf{k}}\mathsf{Q}\mathbf{x}_\alpha\|^2 = (\mathsf{P}_{\mathbf{k}}\mathsf{Q}\mathbf{x}_\alpha, \mathsf{P}_{\mathbf{k}}\mathsf{Q}\mathbf{x}_\alpha) = (\mathsf{Q}\mathbf{x}_\alpha, \mathsf{P}_{\mathbf{k}}^2\mathsf{Q}\mathbf{x}_\alpha) = (\mathsf{Q}\mathbf{x}_\alpha, \mathsf{P}_{\mathbf{k}}\mathsf{Q}\mathbf{x}_\alpha)$ holds. Thus, $d_\alpha^2 = f_0^2\|\Delta\mathbf{x}_\alpha\|^2$ is expressed in the form of Eq. (2.64).
(3) From the definition of the vectors $\boldsymbol{\xi}$ and $\boldsymbol{\theta}$ and the matrices Q and $V_0[\boldsymbol{\xi}_\alpha]$, where $(\bar{x}_\alpha, \bar{y}_\alpha)$ in Eq. (2.15) is replaced by $(\bar{x}_\alpha, \bar{y}_\alpha)$, we can easily confirm the following identities.

$$(\mathbf{x}_\alpha, \mathsf{Q}\mathbf{x}_\alpha) = \frac{1}{f_0^2}(\boldsymbol{\xi}_\alpha, \boldsymbol{\theta}), \quad (\mathsf{Q}\mathbf{x}_\alpha, \mathsf{P}_{\mathbf{k}}\mathsf{Q}\mathbf{x}_\alpha) = \frac{1}{4f_0^2}(\boldsymbol{\theta}, V_0[\boldsymbol{\xi}_\alpha]\boldsymbol{\theta}).$$

Using this, we can rewrite Eq. (2.64) in the form of Eq. (2.66).

2.5 (1) The gradient of Eq. (2.36) with respect to $\boldsymbol{\theta}$ is written as

$$\nabla_{\boldsymbol{\theta}} J = \frac{1}{N}\sum_{\alpha=1}^{N}\frac{2(\boldsymbol{\xi}_\alpha, \boldsymbol{\theta})\boldsymbol{\xi}_\alpha}{(\boldsymbol{\theta}, V_0[\boldsymbol{\xi}_\alpha]\boldsymbol{\theta})} - \frac{1}{N}\sum_{\alpha=1}^{N}\frac{2(\boldsymbol{\xi}_\alpha, \boldsymbol{\theta})^2 V_0[\boldsymbol{\xi}_\alpha]\boldsymbol{\theta}}{(\boldsymbol{\theta}, V_0[\boldsymbol{\xi}_\alpha]\boldsymbol{\theta})^2} = 2(\mathsf{M} - \mathsf{L})\boldsymbol{\theta} = 2\mathsf{X}\boldsymbol{\theta},$$

where M, L, and X are the matrices defined in Eqs. (2.37) and (2.38).
(2) The inner product of Eq. (2.39) and $\boldsymbol{\theta}$ on both sides is $(\boldsymbol{\theta}, \mathsf{X}\boldsymbol{\theta}) = \lambda\|\boldsymbol{\theta}\|^2 = \lambda$. After the convergence, we have $W_\alpha = 1/(\boldsymbol{\theta}, V_0[\boldsymbol{\xi}_\alpha]\boldsymbol{\theta})$ from Eq. (2.40), such that

$$\begin{aligned}(\boldsymbol{\theta}, \mathsf{X}\boldsymbol{\theta}) &= (\boldsymbol{\theta}, \mathsf{M}\boldsymbol{\theta}) - (\boldsymbol{\theta}, \mathsf{L}\boldsymbol{\theta}) \\ &= \frac{1}{N}\sum_{\alpha=1}^{N}\frac{(\boldsymbol{\theta}, \boldsymbol{\xi}_\alpha\boldsymbol{\xi}_\alpha^\top\boldsymbol{\theta})}{(\boldsymbol{\theta}, V_0[\boldsymbol{\xi}_\alpha]\boldsymbol{\theta})} - \frac{1}{N}\sum_{\alpha=1}^{N}\frac{(\boldsymbol{\xi}_\alpha, \boldsymbol{\theta})^2(\boldsymbol{\theta}, V_0[\boldsymbol{\xi}_\alpha]\boldsymbol{\theta})}{(\boldsymbol{\theta}, V_0[\boldsymbol{\xi}_\alpha]\boldsymbol{\theta})^2} \\ &= \frac{1}{N}\sum_{\alpha=1}^{N}\frac{(\boldsymbol{\xi}_\alpha, \boldsymbol{\theta})^2}{(\boldsymbol{\theta}, V_0[\boldsymbol{\xi}_\alpha]\boldsymbol{\theta})} - \frac{1}{N}\sum_{\alpha=1}^{N}\frac{(\boldsymbol{\xi}_\alpha, \boldsymbol{\theta})^2}{(\boldsymbol{\theta}, V_0[\boldsymbol{\xi}_\alpha]\boldsymbol{\theta})} = 0.\end{aligned}$$

Thus $\lambda = 0$.

2.6 (1) If we define the vector

$$\Delta\hat{\mathbf{x}}_\alpha = \hat{\mathbf{x}}_\alpha - \bar{\mathbf{x}}_\alpha,$$

the sum of squares $\sum_{\alpha=1}^{N}\|\tilde{\mathbf{x}}_\alpha + \Delta\hat{\mathbf{x}}_\alpha\|^2$ is equal to the S of Eq. (2.47) divided by f_0^2. Because $\bar{\mathbf{x}}_\alpha$ satisfies the ellipse Eq., we have

$$(\hat{\mathbf{x}}_\alpha - \Delta\hat{\mathbf{x}}_\alpha, \mathsf{Q}(\hat{\mathbf{x}}_\alpha - \Delta\hat{\mathbf{x}}_\alpha)) = 0.$$

Expanding this and ignoring quadratic terms in $\Delta\hat{\mathbf{x}}_\alpha$, we obtain

$$(\mathsf{Q}\hat{\mathbf{x}}_\alpha, \Delta\hat{\mathbf{x}}_\alpha) = \frac{1}{2}(\hat{\mathbf{x}}_\alpha, \mathsf{Q}\hat{\mathbf{x}}'_\alpha). \qquad (*)$$

The third components of $\hat{\mathbf{x}}_\alpha$ and $\bar{\mathbf{x}}_\alpha$ are both 1, therefore we have the constraint $(\mathbf{k}, \Delta\hat{\mathbf{x}}_\alpha) = 0$ on $\Delta\hat{\mathbf{x}}_\alpha$, where $\mathbf{k} = (0, 0, 1)^\top$. Introducing Lagrange multipliers, we differentiate

$$\sum_{\alpha=1}^{N} \|\tilde{\mathbf{x}}_\alpha + \Delta\hat{\mathbf{x}}_\alpha\|^2 - \sum_{\alpha=1}^{N} \lambda_\alpha\Big((\mathsf{Q}\hat{\mathbf{x}}_\alpha, \Delta\hat{\mathbf{x}}_\alpha) - \frac{1}{2}(\hat{\mathbf{x}}_\alpha, \mathsf{Q}\hat{\mathbf{x}}_\alpha)\Big) - \sum_{\alpha=1}^{N} \mu_\alpha(\mathbf{k}, \Delta\hat{\mathbf{x}}_\alpha)$$

with respect to $\Delta\hat{\mathbf{x}}_\alpha$. Letting the result be $\mathbf{0}$, we obtain

$$2(\tilde{\mathbf{x}}_\alpha + \Delta\hat{\mathbf{x}}_\alpha) - \lambda_\alpha \mathsf{Q}\hat{\mathbf{x}}_\alpha - \mu_\alpha \mathbf{k} = \mathbf{0}.$$

Multiplying this by the projection matrix $\mathsf{P}_\mathbf{k}$ in Eq. (2.65) from the left, and noting that $\mathsf{P}_\mathbf{k}\tilde{\mathbf{x}}_\alpha = \tilde{\mathbf{x}}_\alpha$ from the definition of $\tilde{\mathbf{x}}_\alpha$, we obtain

$$2\tilde{\mathbf{x}}_\alpha + 2\Delta\hat{\mathbf{x}}_\alpha - \lambda_\alpha \mathsf{P}_\mathbf{k}\mathsf{Q}\hat{\mathbf{x}}_\alpha = \mathbf{0},$$

from which follows

$$\Delta\hat{\mathbf{x}}_\alpha = \frac{\lambda_\alpha}{2}\mathsf{P}_\mathbf{k}\mathsf{Q}\hat{\mathbf{x}}_\alpha - \tilde{\mathbf{x}}_\alpha.$$

Substitution of this into Eq. $(*)$ yields

$$(\mathsf{Q}\hat{\mathbf{x}}_\alpha, \frac{\lambda_\alpha}{2}\mathsf{P}_\mathbf{k}\mathsf{Q}\hat{\mathbf{x}}_\alpha - \tilde{\mathbf{x}}_\alpha) = \frac{1}{2}(\hat{\mathbf{x}}_\alpha, \mathsf{Q}\hat{\mathbf{x}}_\alpha),$$

such that λ_α is obtained in the form

$$\lambda_\alpha = \frac{(\hat{\mathbf{x}}_\alpha, \mathsf{Q}\hat{\mathbf{x}}_\alpha) + 2(\mathsf{Q}\hat{\mathbf{x}}_\alpha, \tilde{\mathbf{x}}_\alpha)}{(\mathsf{Q}\hat{\mathbf{x}}_\alpha, \mathsf{P}_\mathbf{k}\hat{\mathbf{x}}_\alpha)}.$$

Therefore we can write

$$\Delta\hat{\mathbf{x}}_\alpha = \frac{\Big((\hat{\mathbf{x}}_\alpha, \mathsf{Q}\hat{\mathbf{x}}_\alpha) + 2(\mathsf{Q}\hat{\mathbf{x}}_\alpha, \tilde{\mathbf{x}}_\alpha)\Big)\mathsf{P}_\mathbf{k}\mathsf{Q}\hat{\mathbf{x}}_\alpha}{2(\mathsf{Q}\hat{\mathbf{x}}_\alpha, \mathsf{P}_\mathbf{k}\hat{\mathbf{x}}_\alpha)} - \tilde{\mathbf{x}}_\alpha.$$

Thus, $\bar{\mathbf{x}}_\alpha$ is estimated in the following form.

$$\hat{\hat{\mathbf{x}}}_\alpha = \mathbf{x}_\alpha - \frac{\Big((\hat{\mathbf{x}}_\alpha, \mathsf{Q}\hat{\mathbf{x}}_\alpha) + 2(\mathsf{Q}\hat{\mathbf{x}}_\alpha, \tilde{\mathbf{x}}_\alpha)\Big)\mathsf{P}_\mathbf{k}\mathsf{Q}\hat{\mathbf{x}}_\alpha}{2(\mathsf{Q}\hat{\mathbf{x}}_\alpha, \mathsf{P}_\mathbf{k}\mathsf{F}\hat{\mathbf{x}}_\alpha)}.$$

Rewriting this, we obtain Eq. (2.67).

(2) From the definition of the vectors $\boldsymbol{\theta}$ and $\boldsymbol{\xi}^*_\alpha$ and the matrix $V_0[\hat{\boldsymbol{\xi}}_\alpha]$, we can confirm the following relationship.

$$(\mathsf{Q}\hat{\mathbf{x}}_\alpha, \mathsf{P}_\mathbf{k}\mathsf{Q}\hat{\mathbf{x}}_\alpha) = \frac{(\boldsymbol{\theta}, V_0[\hat{\boldsymbol{\xi}}^*_\alpha]\boldsymbol{\theta})}{4f_0^2}.$$

Therefore Eq. (2.67) can be written in the form of Eq. (2.69).

(3) Note the following relationship.

$$\begin{pmatrix} \theta_1 & \theta_2 & \theta_4 \\ \theta_2 & \theta_3 & \theta_5 \end{pmatrix} \begin{pmatrix} \hat{x}'_\alpha \\ \hat{y}'_\alpha \\ f_0 \end{pmatrix} = f_0 \begin{pmatrix} \theta_1 & \theta_2 & \theta_4 \\ \theta_2 & \theta_3 & \theta_5 \\ 0 & 0 & 0 \end{pmatrix} \begin{pmatrix} \hat{x}'_\alpha/f_0 \\ \hat{y}'_\alpha/f_0 \\ 1 \end{pmatrix} = f_0 \mathsf{P}_\mathbf{k}\mathsf{Q}\hat{\mathbf{x}}_\alpha.$$

From the relationship in the above (2), we see that the square norm of this is

$$f_0^2\|\mathsf{P_k}\mathsf{Q}\hat{\mathbf{x}}_\alpha\|^2 = f_0^2(\mathsf{P_k}\mathsf{Q}\hat{\mathbf{x}}_\alpha, \mathsf{P_k}\mathsf{Q}\hat{\mathbf{x}}_\alpha) = f_0^2(\mathsf{Q}\hat{\mathbf{x}}_\alpha, \mathsf{P_k}^2\mathsf{Q}\hat{\mathbf{x}}_\alpha) = f_0^2(\mathsf{Q}\hat{\mathbf{x}}_\alpha, \mathsf{P_k}\mathsf{Q}\hat{\mathbf{x}}_\alpha)$$
$$= \frac{1}{4}(\boldsymbol{\theta}, V_0[\hat{\boldsymbol{\xi}}_\alpha]\boldsymbol{\theta}),$$

where we have noted that $\mathsf{P}_\mathbf{k}^2 = \mathsf{P}_\mathbf{k}$ from the definition of the matrix $\mathsf{P}_\mathbf{k}$. Thus from Eq. (2.69), we can approximate the S in Eq. (2.34) in the following form.

$$S \approx \frac{1}{N}\sum_{\alpha=1}^{N}\Big((x_\alpha - \hat{\hat{x}}_\alpha)^2 + (y_\alpha - \hat{\hat{y}}_\alpha)^2\Big) = \frac{1}{N}\sum_{\alpha=1}^{N}\left\|\begin{pmatrix}\hat{\hat{x}}_\alpha \\ \hat{\hat{y}}_\alpha\end{pmatrix} - \begin{pmatrix}x_\alpha \\ y_\alpha\end{pmatrix}\right\|^2$$
$$= \frac{1}{N}\sum_{\alpha=1}^{N}\frac{4(\boldsymbol{\xi}_\alpha^*, \boldsymbol{\theta})^2}{(\boldsymbol{\theta}, V_0[\hat{\boldsymbol{\xi}}_\alpha]\boldsymbol{\theta})^2}\frac{(\boldsymbol{\theta}, V_0[\hat{\boldsymbol{\xi}}_\alpha]\boldsymbol{\theta})^2}{4} = \frac{1}{N}\sum_{\alpha=1}^{N}\frac{(\boldsymbol{\xi}_\alpha^*, \boldsymbol{\theta})^2}{(\boldsymbol{\theta}, V_0[\hat{\boldsymbol{\xi}}_\alpha]\boldsymbol{\theta})^2}.$$

Using the relationship of Problem 2.7(2), we can write this in the form of Eq. (2.43).

Problems of Chapter 3

3.1 Define $\mathbf{x}_\alpha$ and $\mathbf{x}'_\alpha$ as in Eq. (3.19), and let $\bar{\mathbf{x}}_\alpha$ and $\bar{\mathbf{x}}'_\alpha$ be the vector representations of points $(\bar{x}, \bar{y})$ and $(\bar{x}', \bar{y}')$, respectively. If we let $\Delta\mathbf{x}_\alpha = \mathbf{x}_\alpha - \bar{\mathbf{x}}_\alpha$ and $\Delta\mathbf{x}' = \mathbf{x}'_\alpha - \bar{\mathbf{x}}'_\alpha$, Eq. (3.17) is written as

$$S_\alpha = f_0^2\Big(\|\Delta\mathbf{x}_\alpha\|^2 + \|\Delta\mathbf{x}'_\alpha\|^2\Big).$$

The epipolar equation of Eq. (3.16) is written as

$$(\mathbf{x}_\alpha - \Delta\mathbf{x}_\alpha, \mathsf{F}(\mathbf{x}'_\alpha - \Delta\mathbf{x}'_\alpha)) = 0.$$

Expanding the left-hand side and ignoring second-order terms in the noise terms $\Delta\mathbf{x}_\alpha$ and $\Delta\mathbf{x}'_\alpha$, we obtain

$$(\mathsf{F}\mathbf{x}'_\alpha, \Delta\mathbf{x}_\alpha) + (\mathsf{F}^\top\mathbf{x}_\alpha, \Delta\mathbf{x}'_\alpha) = (\mathbf{x}_\alpha, \mathsf{F}\mathbf{x}'_\alpha). \qquad (*)$$

Because the third components of $\mathbf{x}_\alpha, \bar{\mathbf{x}}_\alpha, \mathbf{y}'_\alpha$, and $\bar{\mathbf{y}}'_\alpha$ are all 1, the third components of $\Delta\mathbf{x}_\alpha$ and $\Delta\mathbf{x}'_\alpha$ are both 0, which is written as $(\mathbf{k}, \Delta\mathbf{x}_\alpha) = 0$ and $(\mathbf{k}, \Delta\mathbf{x}'_\alpha) = 0$, where $\mathbf{k} = (0, 0, 1)^\top$. In order to minimize $\|\Delta\mathbf{x}_\alpha\|^2 + \|\Delta\mathbf{x}'_\alpha\|^2$, we introduce Lagrange multipliers and differentiate

$$\|\Delta\mathbf{x}_\alpha\|^2 + \|\Delta\mathbf{x}'_\alpha\|^2 - \lambda_\alpha\Big((\mathsf{F}\mathbf{x}'_\alpha, \Delta\mathbf{x}_\alpha) + (\mathsf{F}^\top\mathbf{x}_\alpha, \Delta\mathbf{x}'_\alpha) - (\mathbf{x}_\alpha, \mathsf{F}\mathbf{x}'_\alpha)\Big)$$
$$- \mu(\mathbf{k}, \Delta\mathbf{x}_\alpha) - \mu'(\mathbf{k}, \Delta\mathbf{x}'_\alpha)$$

with respect to $\Delta\mathbf{x}_\alpha$ and $\Delta\mathbf{x}'_\alpha$. Letting the result be $\mathbf{0}$, we obtain

$$2\Delta\mathbf{x}_\alpha - \lambda_\alpha\mathsf{F}\mathbf{x}'_\alpha - \mu\mathbf{k} = \mathbf{0},\ 2\Delta\mathbf{x}'_\alpha - \lambda_\alpha\mathsf{F}^\top\mathbf{x}_\alpha - \mu'\mathbf{k} = \mathbf{0}.$$

Multiplying this by the projection matrix $\mathsf{P}_\mathbf{k}$ in Eq. (3.20) from the left, and noting that $\mathsf{P}_\mathbf{k}\Delta\mathbf{x}_\alpha = \Delta\mathbf{x}_\alpha$, $\mathsf{P}_\mathbf{k}\Delta\mathbf{x}'_\alpha = \Delta\mathbf{x}'_\alpha$, and $\mathsf{P}_\mathbf{k}\mathbf{k} = \mathbf{0}$, we obtain

$$2\Delta\mathbf{x}_\alpha - \lambda_\alpha\mathsf{P}_\mathbf{k}\mathsf{F}\mathbf{x}'_\alpha = \mathbf{0},\ 2\Delta\mathbf{x}'_\alpha - \lambda_\alpha\mathsf{P}_\mathbf{k}\mathsf{F}^\top\mathbf{x}_\alpha = \mathbf{0}.$$

Thus

$$\Delta\mathbf{x}_\alpha = \frac{\lambda_\alpha}{2}\mathsf{P}_\mathbf{k}\mathsf{F}\mathbf{x}'_\alpha, \quad \Delta\mathbf{x}'_\alpha = \frac{\lambda_\alpha}{2}\mathsf{P}_\mathbf{k}\mathsf{F}^\top\mathbf{x}_\alpha$$

Substitution of these into Eq. (∗), we obtain

$$(\mathsf{F}\mathbf{x}'_\alpha, \frac{\lambda_\alpha}{2}\mathsf{P}_\mathbf{k}\mathsf{F}\mathbf{x}'_\alpha) + (\mathsf{F}^\top\mathbf{x}_\alpha, \frac{\lambda_\alpha}{2}\mathsf{P}_\mathbf{k}\mathsf{F}^\top\mathbf{x}_\alpha) = (\mathbf{x}_\alpha, \mathsf{F}\mathbf{x}'_\alpha),$$

which gives λ_α in the form

$$\frac{\lambda_\alpha}{2} = \frac{(\mathbf{x}_\alpha, \mathsf{F}\mathbf{x}'_\alpha)}{(\mathsf{F}\mathbf{x}'_\alpha, \mathsf{P}_\mathbf{k}\mathsf{F}\mathbf{x}'_\alpha) + (\mathsf{F}^\top\mathbf{x}_\alpha, \mathsf{P}_\mathbf{k}\mathsf{F}^\top\mathbf{x}_\alpha)}.$$

Hence, $\Delta\mathbf{x}_\alpha$ and $\Delta\mathbf{x}'_\alpha$ are given by

$$\Delta\mathbf{x}_\alpha = \frac{(\mathbf{x}_\alpha, \mathsf{F}\mathbf{x}'_\alpha)\mathsf{P}_\mathbf{k}\mathsf{F}\mathbf{x}'_\alpha}{(\mathsf{F}\mathbf{x}'_\alpha, \mathsf{P}_\mathbf{k}\mathsf{F}\mathbf{x}'_\alpha) + (\mathsf{F}^\top\mathbf{x}_\alpha, \mathsf{P}_\mathbf{k}\mathsf{F}^\top\mathbf{x}_\alpha)},$$

$$\Delta\mathbf{x}'_\alpha = \frac{(\mathbf{x}_\alpha, \mathsf{F}\mathbf{x}'_\alpha)\mathsf{P}_\mathbf{k}\mathsf{F}^\top\mathbf{x}_\alpha}{(\mathsf{F}\mathbf{x}'_\alpha, \mathsf{P}_\mathbf{k}\mathsf{F}\mathbf{x}'_\alpha) + (\mathsf{F}^\top\mathbf{x}_\alpha, \mathsf{P}_\mathbf{k}\mathsf{F}^\top\mathbf{x}_\alpha)}.$$

Thus $\|\Delta\mathbf{x}_\alpha\|^2 + \|\Delta\mathbf{x}'_\alpha\|^2$ is rewritten in the form

$$\|\Delta\mathbf{x}_\alpha\|^2 + \|\Delta\mathbf{x}'_\alpha\|^2 = \frac{(\mathbf{x}_\alpha, \mathsf{F}\mathbf{x}'_\alpha)^2(\|\mathsf{P}_\mathbf{k}\mathsf{F}\mathbf{x}'_\alpha\|^2 + \|\mathsf{P}_\mathbf{k}\mathsf{F}^\top\mathbf{x}_\alpha\|^2)}{((\mathsf{F}\mathbf{x}'_\alpha, \mathsf{P}_\mathbf{k}\mathsf{F}\mathbf{x}'_\alpha) + (\mathsf{F}^\top\mathbf{x}_\alpha, \mathsf{P}_\mathbf{k}\mathsf{F}^\top\mathbf{x}_\alpha))^2}$$
$$= \frac{(\mathbf{x}_\alpha, \mathsf{F}\mathbf{x}'_\alpha)^2}{(\mathsf{F}\mathbf{x}'_\alpha, \mathsf{P}_\mathbf{k}\mathsf{F}\mathbf{x}'_\alpha) + (\mathsf{F}^\top\mathbf{x}_\alpha, \mathsf{P}_\mathbf{k}\mathsf{F}^\top\mathbf{x}_\alpha)},$$

where we have noted that $\mathsf{P}_\mathbf{k}^2 = \mathsf{P}_\mathbf{k}$ from Eq. (3.20) and that $\|\mathsf{P}_\mathbf{k}\mathsf{F}\mathbf{x}'_\alpha\|^2 = (\mathsf{P}_\mathbf{k}\mathsf{F}\mathbf{x}'_\alpha, \mathsf{P}_\mathbf{k}\mathsf{F}\mathbf{x}'_\alpha) = (\mathsf{F}\mathbf{x}'_\alpha, \mathsf{P}_\mathbf{k}^2\mathsf{F}\mathbf{x}'_\alpha) = (\mathsf{F}\mathbf{x}'_\alpha, \mathsf{P}_\mathbf{k}\mathsf{F}\mathbf{x}'_\alpha)$ and $\|\mathsf{P}_\mathbf{k}\mathsf{F}^\top\mathbf{x}_\alpha\|^2 = (\mathsf{P}_\mathbf{k}\mathsf{F}^\top\mathbf{x}_\alpha, \mathsf{P}_\mathbf{k}\mathsf{F}^\top\mathbf{x}_\alpha) = (\mathsf{F}^\top\mathbf{x}_\alpha, \mathsf{P}_\mathbf{k}^2\mathsf{F}^\top\mathbf{x}_\alpha) = (\mathsf{F}^\top\mathbf{x}_\alpha, \mathsf{P}_\mathbf{k}\mathsf{F}^\top\mathbf{x}_\alpha)$. Hence, $S_\alpha = f_0^2(\|\Delta\mathbf{x}_\alpha\|^2 + \|\Delta\mathbf{x}'_\alpha\|^2)$ is approximated by Eq. (3.18).

3.2 Using the identities $\mathrm{tr}[\mathsf{A}^\top\mathsf{A}] = \|\mathsf{A}\|^2$ and $\mathrm{tr}[\mathsf{AB}] = \mathrm{tr}[\mathsf{BA}]$, we obtain

$$\|\mathsf{F}\|^2 = \mathrm{tr}[\mathsf{F}^\top\mathsf{F}] = \mathrm{tr}[\mathsf{V}\begin{pmatrix}\sigma_1 & 0 & 0\\ 0 & \sigma_2 & 0\\ 0 & 0 & \sigma_3\end{pmatrix}\mathsf{U}^\top\mathsf{U}\begin{pmatrix}\sigma_1 & 0 & 0\\ 0 & \sigma_2 & 0\\ 0 & 0 & \sigma_3\end{pmatrix}\mathsf{V}^\top]$$
$$= \mathrm{tr}[\mathsf{V}\begin{pmatrix}\sigma_1^2 & 0 & 0\\ 0 & \sigma_2^2 & 0\\ 0 & 0 & \sigma_3^2\end{pmatrix}\mathsf{V}^\top] = \mathrm{tr}[\begin{pmatrix}\sigma_1^2 & 0 & 0\\ 0 & \sigma_2^2 & 0\\ 0 & 0 & \sigma_3^2\end{pmatrix}\mathsf{V}^\top\mathsf{V}]$$
$$= \mathrm{tr}\begin{pmatrix}\sigma_1^2 & 0 & 0\\ 0 & \sigma_2^2 & 0\\ 0 & 0 & \sigma_3^2\end{pmatrix} = \sigma_1^2 + \sigma_2^2 + \sigma_3^2.$$

3.3 (1) The projected length of the vector $\mathbf{v}$ onto the direction along the unit surface normal $\mathbf{u}$ is

$$\|\mathbf{v}\| \cos\theta = \|\mathbf{u}\|\|\mathbf{v}\| \cos\theta = (\mathbf{u}, \mathbf{v}),$$

where θ is the angle made by $\mathbf{v}$ and $\mathbf{u}$ (Fig. 3.4). Thus the components of $\mathbf{v}$ in the $\mathbf{u}$-direction are $(\mathbf{u}, \mathbf{v})\mathbf{u}$. The component of $\mathbf{v}$ projected onto the plane is

$$\mathbf{u} - (\mathbf{u}, \mathbf{v})\mathbf{u} = \mathbf{u} - \mathbf{u}\mathbf{u}^\top \mathbf{v} = (\mathsf{I} - \mathbf{u}\mathbf{u}^\top)\mathbf{v} = \mathsf{P}_\mathbf{u}\mathbf{v}.$$

(2) We can easily see that

$$\begin{aligned}\mathsf{P}_\mathbf{u}^2 &= (\mathsf{I} - \mathbf{u}\mathbf{u}^\top)(\mathsf{I} - \mathbf{u}\mathbf{u}^\top) = \mathsf{I} - \mathbf{u}\mathbf{u}^\top - \mathbf{u}\mathbf{u}^\top + \mathbf{u}\mathbf{u}^\top\mathbf{u}\mathbf{u}^\top = \mathsf{I} - 2\mathbf{u}\mathbf{u}^\top + \mathbf{u}(\mathbf{u}, \mathbf{u})\mathbf{u}^\top \\ &= \mathsf{I} - 2\mathbf{u}\mathbf{u}^\top + \mathbf{u}\mathbf{u}^\top = \mathsf{I} - \mathbf{u}\mathbf{u}^\top = \mathsf{P}_\mathbf{u}.\end{aligned}$$

3.4 (1) To a first approximation, the deviation of a fraction in the presence of noise is written as the sum of the term in which only the numerator is perturbed by the noise and the term in which only the denominator is perturbed. Here, however, the numerator is 0 in the absence of noise, thus we need not consider the perturbation of the denominator. Because $(\bar{\boldsymbol{\xi}}_\alpha, \bar{\boldsymbol{\theta}}) = 0$ in the absence of noise, we see that

$$(\boldsymbol{\xi}_\alpha, \boldsymbol{\theta}) = (\bar{\boldsymbol{\xi}}_\alpha, \Delta_1\boldsymbol{\theta}) + (\Delta_1\boldsymbol{\xi}_\alpha, \bar{\boldsymbol{\theta}}) + O(\sigma^2).$$

Substituting this into the numerator of Eq. (3.23), we obtain Eq. (3.71) after expansion.
(2) Note that $(\bar{\boldsymbol{\xi}}_\alpha, \Delta_1\boldsymbol{\theta})^2 = \Delta_1\boldsymbol{\theta}^\top \bar{\boldsymbol{\xi}}_\alpha \bar{\boldsymbol{\xi}}_\alpha^\top \Delta_1\boldsymbol{\theta} = (\Delta_1\boldsymbol{\theta}, (\bar{\boldsymbol{\xi}}_\alpha \bar{\boldsymbol{\xi}}_\alpha^\top)\Delta_1\boldsymbol{\theta}$ holds in the numerator expression of Eq. (3.71). Differentiating Eq. (3.71) with respect to $\Delta_1\boldsymbol{\theta}$ and letting the result be $\mathbf{0}$, we obtain

$$\begin{aligned}&\frac{1}{N}\sum_{\alpha=1}^{N} \frac{2\bar{\boldsymbol{\xi}}_\alpha \bar{\boldsymbol{\xi}}_\alpha^\top \Delta_1\boldsymbol{\theta} + 2(\bar{\boldsymbol{\theta}}, \Delta_1\boldsymbol{\xi}_\alpha)\bar{\boldsymbol{\xi}}_\alpha}{(\bar{\boldsymbol{\theta}}, V_0[\boldsymbol{\xi}_\alpha]\bar{\boldsymbol{\theta}})} \\ &= 2\bar{\mathsf{M}}\Delta_1\boldsymbol{\theta} + 2\Big(\frac{1}{N}\sum_{\alpha=1}^{N} \frac{\bar{\boldsymbol{\xi}}_\alpha \bar{\boldsymbol{\theta}}^\top}{(\bar{\boldsymbol{\theta}}, V_0[\boldsymbol{\xi}_\alpha]\bar{\boldsymbol{\theta}})}\Delta_1\boldsymbol{\xi}_\alpha\Big) = \mathbf{0}.\end{aligned}$$

Multiplying this by $\bar{\mathsf{M}}^-$ from the left, and noting that $\bar{\mathsf{M}}^-\bar{\mathsf{M}} = \mathsf{P}_{\bar{\boldsymbol{\theta}}}$ and $\mathsf{P}_{\bar{\boldsymbol{\theta}}}\Delta_1\boldsymbol{\theta} = \Delta_1\boldsymbol{\theta}$, we obtain Eq. (3.72).
(3) Because $\bar{\mathsf{M}}$ is a symmetric matrix, $\bar{\mathsf{M}}^-$ is also. Thus $E[\Delta_1\boldsymbol{\theta}\Delta_1\boldsymbol{\theta}^\top]$ is written as

$$\begin{aligned}&E[\Delta_1\boldsymbol{\theta}\Delta_1\boldsymbol{\theta}^\top] \\ &= \bar{\mathsf{M}}^- E[\Big(\frac{1}{N}\sum_{\alpha=1}^{N} \frac{\bar{\boldsymbol{\xi}}_\alpha \bar{\boldsymbol{\theta}}^\top}{(\bar{\boldsymbol{\theta}}, V_0[\boldsymbol{\xi}_\alpha]\bar{\boldsymbol{\theta}})}\Delta_1\boldsymbol{\xi}_\alpha\Big)\Big(\frac{1}{N}\sum_{\beta=1}^{N} \Delta_1\boldsymbol{\xi}_\beta^\top \frac{\bar{\boldsymbol{\theta}} \bar{\boldsymbol{\xi}}_\beta^\top}{(\bar{\boldsymbol{\theta}}, V_0[\boldsymbol{\xi}_\alpha]\bar{\boldsymbol{\theta}})}\Big)]\bar{\mathsf{M}}^- \\ &= \bar{\mathsf{M}}^-\Big(\frac{1}{N^2}\sum_{\alpha,\beta=1}^{N} \frac{\bar{\boldsymbol{\xi}}_\alpha \bar{\boldsymbol{\theta}}^\top E[\Delta_1\boldsymbol{\xi}_\alpha \Delta_1\boldsymbol{\xi}_\beta^\top]\bar{\boldsymbol{\theta}} \bar{\boldsymbol{\xi}}_\beta^\top}{(\bar{\boldsymbol{\theta}}, V_0[\boldsymbol{\xi}_\alpha]\bar{\boldsymbol{\theta}})^2}\Big)\bar{\mathsf{M}}^-\end{aligned}$$

$$
\begin{aligned}
&= \bar{\mathsf{M}}^-\Big(\frac{1}{N^2}\sum_{\alpha,\beta=1}^{N}\frac{\bar{\xi}_\alpha\bar{\theta}^\top\sigma^2\delta_{\alpha\beta}V_0[\xi_\alpha]\bar{\theta}\bar{\xi}_\beta^\top}{(\bar{\theta}, V_0[\xi_\alpha]\bar{\theta})^2}\Big)\bar{\mathsf{M}}^- \\
&= \frac{\sigma^2}{N}\bar{\mathsf{M}}^-\Big(\frac{1}{N}\sum_{\alpha=1}^{N}\frac{\bar{\xi}_\alpha\bar{\theta}^\top V_0[\xi_\alpha]\bar{\theta}\bar{\xi}_\alpha^\top}{(\bar{\theta}, V_0[\xi_\alpha]\bar{\theta})^2}\Big)\bar{\mathsf{M}}^- \\
&= \frac{\sigma^2}{N}\bar{\mathsf{M}}^-\Big(\frac{1}{N}\sum_{\alpha=1}^{N}\frac{\bar{\xi}_\alpha(\bar{\theta}^\top, V_0[\xi_\alpha]\bar{\theta})\bar{\xi}_\alpha^\top}{(\bar{\theta}, V_0[\xi_\alpha]\bar{\theta})^2}\Big)\bar{\mathsf{M}}^- \\
&= \frac{\sigma^2}{N}\bar{\mathsf{M}}^-\Big(\frac{1}{N}\sum_{\alpha=1}^{N}\frac{\bar{\xi}_\alpha\bar{\xi}_\alpha^\top}{(\bar{\theta}, V_0[\xi_\alpha]\bar{\theta})}\Big)\bar{\mathsf{M}}^- = \frac{\sigma^2}{N}\bar{\mathsf{M}}^-\bar{\mathsf{M}}\bar{\mathsf{M}}^- = \frac{\sigma^2}{N}\bar{\mathsf{M}}^-,
\end{aligned}
$$

where $\delta_{\alpha\beta}$ is the Kronecker delta, taking 1 for $\alpha = \beta$ and 0 otherwise. We have also noted that $E[\Delta_1\xi_\alpha\Delta_1\xi_\beta] = \sigma^2\delta_{\alpha\beta}V_0[\xi_\alpha]$ from our noise assumption and used the identity $\bar{\mathsf{M}}^-\bar{\mathsf{M}}\bar{\mathsf{M}}^- = \bar{\mathsf{M}}^-$ for pseudoinverses (↪ Problem 14.3).

3.5 (1) Expanding the determinant

$$
\det[\mathsf{I} - \varepsilon\mathsf{A}] = \begin{vmatrix} 1-\varepsilon A_{11} & -\varepsilon A_{12} & -\varepsilon A_{13} \\ -\varepsilon A_{21} & 1-\varepsilon A_{22} & -\varepsilon A_{23} \\ -\varepsilon A_{31} & -\varepsilon A_{32} & 1-\varepsilon A_{13} \end{vmatrix},
$$

and taking out linear terms in ε, we obtain

$$
(1-\varepsilon A_{11})(1-\varepsilon A_{22})(1-\varepsilon A_{33}) + O(\varepsilon^2) = 1 - \varepsilon(A_{11}+A_{22}+A_{33}) + O(\varepsilon^2).
$$

Thus we obtain the following expression.

$$
\begin{aligned}
\det[\mathsf{A} - \varepsilon\mathsf{B}] &= \det[\mathsf{A}(\mathsf{I} - \varepsilon\mathsf{A}^{-1}\mathsf{B})] = \det\mathsf{A}\det[\mathsf{I} - \varepsilon\mathsf{A}^{-1}\mathsf{B}] \\
&= \det\mathsf{A}(1 - \varepsilon\mathrm{tr}[\mathsf{A}^{-1}\mathsf{B}]) + O(\varepsilon^2) = \det\mathsf{A}(1 - \varepsilon\mathrm{tr}[\frac{\mathsf{A}^\dagger\mathsf{B}}{\det\mathsf{A}}]) + O(\varepsilon^2) \\
&= \det\mathsf{A} - \varepsilon\mathrm{tr}[\mathsf{A}^\dagger\mathsf{B}] + O(\varepsilon^2).
\end{aligned}
$$

Here, we have used the relationship $\mathsf{A}^{-1} = \mathsf{A}^\dagger/\det\mathsf{A}$, assuming that A is nonsingular such that its inverse A^{-1} exists. However, the final expression does not contain A^{-1}. Inasmuch as the resulting equation is a polynomial equality in the elements of A and B, it holds whether A is nonsingular or not.
(2) Using the above relationship and ignoring high-order terms in $\Delta\theta$, we obtain

$$
\det[\mathsf{F} - \Delta\mathsf{F}] = \det\mathsf{F} - \mathrm{tr}[\mathsf{F}^\dagger\Delta\mathsf{F}] = 0.
$$

In terms of the vectors θ and $\Delta\theta$, the trace terms are written as $\mathrm{tr}[\mathsf{F}^\dagger\Delta\mathsf{F}] = \sum_{i,j=1}^{3}F_{ji}^\dagger\Delta F_{ij} = (\theta^\dagger, \Delta\theta)$. Combining this with Eq. (3.32), we obtain Eq. (3.76).
(3) Introducing a Lagrange multiplier λ for Eq. (3.76), differentiating

$$
(\Delta\theta, V_0[\theta]^-\Delta\theta) - \lambda\Big((\theta^\dagger, \Delta\theta) - \frac{1}{3}(\theta^\dagger, \theta)\Big)
$$

with respect to $\Delta\boldsymbol{\theta}$, and letting the result be $\mathbf{0}$, we obtain

$$2V_0[\boldsymbol{\theta}]^- \Delta\boldsymbol{\theta} - \lambda\boldsymbol{\theta}^\dagger = 0.$$

Multiplying this by $V_0[\boldsymbol{\theta}]$ from the left, and noting that $V_0[\boldsymbol{\theta}]V_0[\boldsymbol{\theta}]^- = \mathsf{P}_{\boldsymbol{\theta}}$ (the projection matrix defined by Eq. (3.28)) and $\mathsf{P}_{\boldsymbol{\theta}}\Delta\boldsymbol{\theta} = \Delta\boldsymbol{\theta}$, we obtain

$$\Delta\boldsymbol{\theta} = \frac{\lambda}{2}V_0[\boldsymbol{\theta}]\boldsymbol{\theta}^\dagger, \qquad (*)$$

where we have used the properties of the pseudoinverse and noted that $\boldsymbol{\theta}$ is normalized to the unit norm such that its variation $\Delta\boldsymbol{\theta}$ is orthogonal to $\boldsymbol{\theta}$ to a first approximation (we omit the details). Computing the inner product of the above expression and $\boldsymbol{\theta}^\dagger$, we obtain

$$(\boldsymbol{\theta}^\dagger, \Delta\boldsymbol{\theta}) = \frac{\lambda}{2}(\boldsymbol{\theta}^\dagger, V_0[\boldsymbol{\theta}]\boldsymbol{\theta}^\dagger).$$

From Eq. (3.76), this equals $(\boldsymbol{\theta}^\dagger, \boldsymbol{\theta})/3$. Therefore

$$\frac{\lambda}{2} = \frac{(\boldsymbol{\theta}^\dagger, \boldsymbol{\theta})}{3(\boldsymbol{\theta}^\dagger, V_0[\boldsymbol{\theta}]\boldsymbol{\theta}^\dagger)},$$

and we obtain Eq. (3.77) from Eq. $(*)$.

3.6 (1) Because $\mathsf{U} + \Delta\mathsf{U}$ is an orthogonal matrix, we have $(\mathsf{U} + \Delta\mathsf{U})(\mathsf{U} + \Delta\mathsf{U})^\top = \mathsf{I}$, and thus

$$\begin{aligned}(\mathsf{U} + \Delta\mathsf{U})(\mathsf{U} + \Delta\mathsf{U})^\top &= \mathsf{U}\mathsf{U}^\top + \mathsf{U}\Delta\mathsf{U}^\top + \Delta\mathsf{U}\mathsf{U}^\top + \Delta\mathsf{U}\Delta\mathsf{U}^\top \\ &= \mathsf{I} + (\Delta\mathsf{U}\mathsf{U}^\top)^\top + \Delta\mathsf{U}\mathsf{U}^\top + \Delta\mathsf{U}\Delta\mathsf{U}^\top.\end{aligned}$$

This is identically equal to I, therefore we obtain to a first approximation

$$(\Delta\mathsf{U}\mathsf{U}^\top)^\top + \Delta\mathsf{U}\mathsf{U}^\top = \mathsf{O}.$$

This means that $\Delta\mathsf{U}\mathsf{U}^\top$ is an antisymmetric matrix. Therefore there exist small constants $\Delta\omega_1$, $\Delta\omega_2$, and $\Delta\omega_3$ such that we can write

$$\Delta\mathsf{U}\mathsf{U}^\top = \begin{pmatrix} 0 & -\Delta\omega_3 & -\Delta\omega_2 \\ \Delta\omega_3 & 0 & -\Delta\omega_1 \\ -\Delta\omega_2 & \Delta\omega_1 & 0 \end{pmatrix}.$$

Let $\Delta\boldsymbol{\omega}$ be the vector with components $\Delta\omega_1$, $\Delta\omega_2$, and $\Delta\omega_3$, and write the matrix on the right-hand side symbolically as $(\Delta\boldsymbol{\omega}\times)$. Multiplying the above equation by U on both sides from the right, we obtain

$$\Delta\mathsf{U} = (\Delta\boldsymbol{\omega}\times)\mathsf{U}.$$

It is easy to confirm that this is the same as Eq. (3.78).
(2) Using the antisymmetric matrix $(\Delta\boldsymbol{\omega}\times)$ defined above, we see that the right-hand side of Eq. (3.78) is $\Delta\mathsf{V}^\top = ((\boldsymbol{\omega}'\times)\mathsf{V})^\top = \mathsf{V}^\top(\boldsymbol{\omega}'\times)^\top = -\mathsf{V}^\top(\boldsymbol{\omega}'\times)$. Using this and ignoring high-order small terms, we can express the small deviation $\Delta\mathsf{F}$ of F in the following form.

$$\Delta\mathsf{F} =(\Delta\boldsymbol{\omega}\times)\mathsf{U}\begin{pmatrix}\cos\phi & & \\ & \sin\phi & \\ & & 0\end{pmatrix}\mathsf{V}^\top + \mathsf{U}\begin{pmatrix}-\sin\phi\Delta\phi & & \\ & \cos\phi\Delta\phi & \\ & & 0\end{pmatrix}\mathsf{V}^\top$$
$$-\mathsf{U}\begin{pmatrix}\cos\phi & & \\ & \sin\phi & \\ & & 0\end{pmatrix}\mathsf{V}^\top(\Delta\boldsymbol{\omega}'\times)+\cdots$$
$$=(\Delta\boldsymbol{\omega}\times)\mathsf{F}+\mathsf{U}\begin{pmatrix}-\sin\phi & & \\ & \cos\phi & \\ & & 0\end{pmatrix}\mathsf{V}^\top\Delta\phi-\mathsf{F}(\Delta\boldsymbol{\omega}'\times)+\cdots$$

Taking out each element of $\Delta\mathsf{F}$, we obtain

$$\begin{aligned}\Delta F_{11} =&\Delta\omega_2 F_{31}-\Delta\omega_3 F_{21}+(U_{12}V_{12}\cos\phi-U_{11}V_{11}\sin\phi)\Delta\phi\\&+\Delta\omega_2' F_{13}-\Delta\omega_3' F_{12}+\cdots,\\ \Delta F_{12} =&\Delta\omega_2 F_{32}-\Delta\omega_3 F_{22}+(U_{12}V_{22}\cos\phi-U_{11}V_{21}\sin\phi)\Delta\phi\\&+\Delta\omega_3' F_{11}-\Delta\omega_1' F_{13}+\cdots,\\ &\vdots\\ \Delta F_{33} =&\Delta\omega_1 F_{23}-\Delta\omega_2 F_{13}+(U_{32}V_{32}\cos\phi-U_{31}V_{31}\sin\phi)\Delta\phi\\&+\Delta\omega_1' F_{32}-\Delta\omega_2' F_{31}+\cdots.\end{aligned}$$

If we define the matrices F_U and F_V by Eq. (3.36) and the vector $\boldsymbol{\theta}_\phi$ by Eq. (3.37), the vector $\Delta\boldsymbol{\theta}$ consisting of the above expressions as components can be written in the form of Eq. (3.79).

(3) If the variable $\boldsymbol{\theta}$ in Eq. (3.23) is perturbed in the form of Eq. (3.79), the deviation ΔJ of J is written to a first approximation as follows.

$$\begin{aligned}\Delta J &= (\nabla_{\boldsymbol{\theta}}J,\Delta\boldsymbol{\theta}) = (\nabla_{\boldsymbol{\theta}}J,\mathsf{F}_U\Delta\boldsymbol{\omega})+(\nabla_{\boldsymbol{\theta}}J,\boldsymbol{\theta}_\phi\Delta\phi)+(\nabla_{\boldsymbol{\theta}}J,\mathsf{F}_V\Delta\boldsymbol{\omega}')\\ &= (\mathsf{F}_U^\top\nabla_{\boldsymbol{\theta}}J,\Delta\boldsymbol{\omega})+(\nabla_{\boldsymbol{\theta}}J,\boldsymbol{\theta}_\phi)\Delta\phi+(\mathsf{F}_V^\top\nabla_{\boldsymbol{\theta}}J,\Delta\boldsymbol{\omega}').\end{aligned}$$

This means that the derivatives of J with respect to $\boldsymbol{\omega}$, $\boldsymbol{\omega}'$, and ϕ are written in the form

$$\nabla_{\boldsymbol{\omega}}J=\mathsf{F}_U^\top\nabla_{\boldsymbol{\theta}}J,\qquad \frac{\partial J}{\partial\phi}=\nabla_{\boldsymbol{\omega}'}J=\mathsf{F}_V^\top\nabla_{\boldsymbol{\theta}}J.$$

Combining this with Eq. (3.46), we can write $\nabla_{\boldsymbol{\omega}}J$, $\partial J/\partial\phi$, and $\nabla_{\boldsymbol{\omega}'}J$ in the form of Eq. (3.40).

(4) In order to derive second derivatives of J, we again differentiate $\nabla_{\boldsymbol{\theta}}J$ in the above item (2). Namely, we replace $\boldsymbol{\theta}$ by $\boldsymbol{\theta}+\Delta\boldsymbol{\theta}$ and pick out from the resulting deviation $\Delta\nabla_{\boldsymbol{\theta}}J$ of $\nabla_{\boldsymbol{\theta}}J$ those terms that contain $\Delta\boldsymbol{\theta}$. In this process, we use the Gauss-Newton approximation; that is, we ignore terms that contain $(\boldsymbol{\xi}_\alpha,\boldsymbol{\theta})$, noting that $(\boldsymbol{\xi}_\alpha,\boldsymbol{\theta})=0$ if there are no deviations of variables and assuming that deviations are small. Note that the numerator contains $(\boldsymbol{\xi}_\alpha,\boldsymbol{\theta})$ thus we need not consider the deviation of the denominator. Also, note that the numerator of L contains $(\boldsymbol{\xi}_\alpha,\boldsymbol{\theta})^2$ and therefore we need not consider the deviation of L. As a result, we obtain

$$\Delta\nabla_{\boldsymbol{\theta}}J \approx \frac{1}{N}\sum_{\alpha=1}^{N}\frac{2(\Delta\boldsymbol{\theta}, \boldsymbol{\xi}_\alpha)\boldsymbol{\xi}_\alpha}{(\boldsymbol{\theta}, V_0[\boldsymbol{\xi}_\alpha]\boldsymbol{\theta})} = \frac{2}{N}\sum_{\alpha=1}^{N}\frac{\boldsymbol{\xi}_\alpha\boldsymbol{\xi}_\alpha^\top}{(\boldsymbol{\theta}, V_0[\boldsymbol{\xi}_\alpha]\boldsymbol{\theta})}\Delta\boldsymbol{\theta} = 2\mathsf{M}\Delta\boldsymbol{\theta}.$$

This means that $2\mathsf{M}$ is the Hessian of J, that is, the second derivative with respect to $\boldsymbol{\theta}$. It follows that the second-order term $\Delta_2 J$ in $\Delta\boldsymbol{\theta}$ in the deviation ΔJ of J is written:

$$\begin{aligned}\Delta_2 J =& \frac{1}{2}(\Delta\boldsymbol{\theta}, 2\mathsf{M}\Delta\boldsymbol{\theta})\\ =&(\mathsf{F}_U\Delta\boldsymbol{\omega} + \boldsymbol{\theta}_\phi\Delta\phi + \mathsf{F}_V\Delta\boldsymbol{\omega}', \mathsf{M}(\mathsf{F}_U\Delta\boldsymbol{\omega} + \boldsymbol{\theta}_\phi\Delta\phi + \mathsf{F}_V\Delta\boldsymbol{\omega}'))\\ =&(\Delta\boldsymbol{\omega}, \mathsf{F}_U^\top\mathsf{M}\mathsf{F}_U\Delta\boldsymbol{\omega}) + (\boldsymbol{\theta}_\phi, \mathsf{M}\boldsymbol{\theta}_\phi)\Delta\phi^2 + (\Delta\boldsymbol{\omega}', \mathsf{F}_V^\top\mathsf{M}\mathsf{F}_V\Delta\boldsymbol{\omega}')\\ &+ (\Delta\boldsymbol{\omega}, \mathsf{F}_U^\top\mathsf{M}\mathsf{F}_V\Delta\boldsymbol{\omega}') + (\Delta\boldsymbol{\omega}', \mathsf{F}_V^\top\mathsf{M}\mathsf{F}_U\Delta\boldsymbol{\omega}) + (\Delta\boldsymbol{\omega}, \mathsf{F}_U^\top\mathsf{M}\boldsymbol{\theta}_\phi\Delta\phi)\\ &+ (\boldsymbol{\theta}_\phi\Delta\phi, \mathsf{M}\mathsf{F}_U\Delta\boldsymbol{\omega}) + (\Delta\boldsymbol{\omega}, \mathsf{F}_V^\top\mathsf{M}\boldsymbol{\theta}_\phi\Delta\phi) + (\boldsymbol{\theta}_\phi\Delta\phi, \mathsf{M}\mathsf{F}_V\Delta\boldsymbol{\omega}).\end{aligned}$$

The second derivatives of J are defined via the expression:

$$\begin{aligned}\Delta_2 J =& \frac{1}{2}(\Delta\boldsymbol{\omega}, \nabla_{\boldsymbol{\omega}\boldsymbol{\omega}}J\Delta\boldsymbol{\omega}) + \frac{1}{2}\nabla_\phi J\Delta\phi^2 + \frac{1}{2}(\Delta\boldsymbol{\omega}', \nabla_{\boldsymbol{\omega}'\boldsymbol{\omega}}J\Delta\boldsymbol{\omega}')\\ &+ \frac{1}{2}(\Delta\boldsymbol{\omega}, \nabla_{\boldsymbol{\omega}\boldsymbol{\omega}'}J\Delta\boldsymbol{\omega}') + \frac{1}{2}(\Delta\boldsymbol{\omega}', \nabla_{\boldsymbol{\omega}'\boldsymbol{\omega}}J\Delta\boldsymbol{\omega})\\ &+ (\frac{\partial\nabla_{\boldsymbol{\omega}}J}{\partial\phi}, \Delta\boldsymbol{\omega})\Delta\phi + (\frac{\partial\nabla_{\boldsymbol{\omega}'}J}{\partial\phi}, \Delta\boldsymbol{\omega}')\Delta\phi.\end{aligned}$$

From this, we obtain Eq. (3.41) and $\nabla_{\boldsymbol{\omega}'\boldsymbol{\omega}}J = (\nabla_{\boldsymbol{\omega}\boldsymbol{\omega}'}J)^\top$.

3.7 (1) From Eq. (3.32), the rank constraint $\det\mathsf{F} = 0$ is written as $(\boldsymbol{\theta}^\dagger, \boldsymbol{\theta}) = 0$. This means that $\boldsymbol{\theta}$ is orthogonal to $\boldsymbol{\theta}^\dagger$; that is, the projection of $\boldsymbol{\theta}$ along $\boldsymbol{\theta}^\dagger$ is $\boldsymbol{\theta}$ itself. Thus $\mathsf{P}_{\boldsymbol{\theta}^\dagger}\boldsymbol{\theta} = \boldsymbol{\theta}$. The rank constraint $\det\mathsf{F} = 0$ defines a (hyper)surface in the space of $\boldsymbol{\theta}$. Differentiating

$$\det\mathsf{F} = \theta_1\theta_5\theta_9 + \theta_2\theta_6\theta_7 + \theta_3\theta_4\theta_8 - \theta_3\theta_5\theta_7 - \theta_2\theta_4\theta_9 - \theta_1\theta_6\theta_8,$$

and using the vector $\boldsymbol{\theta}^\dagger$ in Eq. (3.32), we obtain $\nabla_{\boldsymbol{\theta}}\det\mathsf{F} = 3\boldsymbol{\theta}^\dagger$. Thus $\boldsymbol{\theta}^\dagger$ is the surface normal to that surface. At the point on this surface where J takes its minimum, the gradient $\nabla_{\boldsymbol{\theta}}J$ should be in the direction of $\boldsymbol{\theta}^\dagger$; if the gradient $\nabla_{\boldsymbol{\theta}}J$ is not orthogonal to this surface, we can proceed along this surface in some direction to decrease J. Thus $\mathsf{P}_{\boldsymbol{\theta}^\dagger}\nabla_{\boldsymbol{\theta}}J = 0$ holds. From Eq. (3.46), we can write this as

$$\mathsf{P}_{\boldsymbol{\theta}^\dagger}\mathsf{X}\boldsymbol{\theta} = \mathbf{0}.$$

Because $\mathsf{P}_{\boldsymbol{\theta}^\dagger}\boldsymbol{\theta} = \boldsymbol{\theta}$, we can write this as

$$\mathsf{P}_{\boldsymbol{\theta}^\dagger}\mathsf{X}\mathsf{P}_{\boldsymbol{\theta}^\dagger}\boldsymbol{\theta} = \mathbf{0}.$$

In terms of the matrix Y in Eq. (3.51), we can write this in the form of Eq. (3.80); we use this form because symmetric matrices are computationally convenient for eigenvalue analysis.

(2) From the rank constraint $\mathsf{P}_{\boldsymbol{\theta}^\dagger}\boldsymbol{\theta} = \boldsymbol{\theta}$, we see that

$$\begin{aligned}(\boldsymbol{\theta}, \mathsf{Y}\boldsymbol{\theta}) &= (\boldsymbol{\theta}, \mathsf{P}_{\boldsymbol{\theta}^\dagger}\mathsf{X}\mathsf{P}_{\boldsymbol{\theta}^\dagger}\boldsymbol{\theta}) = (\mathsf{P}_{\boldsymbol{\theta}^\dagger}\boldsymbol{\theta}, \mathsf{X}\mathsf{P}_{\boldsymbol{\theta}^\dagger}\boldsymbol{\theta}) = (\boldsymbol{\theta}, \mathsf{X}\boldsymbol{\theta}) \\ &= \frac{1}{N}\sum_{\alpha=1}^{N}\frac{(\boldsymbol{\theta}, \boldsymbol{\xi}_\alpha\boldsymbol{\xi}_\alpha^\top\boldsymbol{\theta})}{(\boldsymbol{\theta}, V_0[\boldsymbol{\xi}_\alpha]\boldsymbol{\theta})} - \frac{1}{N}\sum_{\alpha=1}^{N}\frac{(\boldsymbol{\xi}_\alpha, \boldsymbol{\theta})^2}{(\boldsymbol{\theta}, V_0[\boldsymbol{\xi}_\alpha]\boldsymbol{\theta})^2}(\boldsymbol{\theta}, V_0[\boldsymbol{\xi}_\alpha]\boldsymbol{\theta}) \\ &= \frac{1}{N}\sum_{\alpha=1}^{N}\frac{(\boldsymbol{\xi}_\alpha, \boldsymbol{\theta})^2}{(\boldsymbol{\theta}, V_0[\boldsymbol{\xi}_\alpha]\boldsymbol{\theta})} - \frac{1}{N}\sum_{\alpha=1}^{N}\frac{(\boldsymbol{\xi}_\alpha, \boldsymbol{\theta})^2}{(\boldsymbol{\theta}, V_0[\boldsymbol{\xi}_\alpha]\boldsymbol{\theta})} = 0.\end{aligned}$$

(3) If the smallest two eigenvalues of Y are both 0 at the time of the convergence of the extended FNS iterations, $\hat{\boldsymbol{\theta}}$ is an eigenvector of Y for eigenvalue 0 because it is a linear combination of eigenvectors of Y for eigenvalue 0. On the other hand, $\mathsf{P}_{\boldsymbol{\theta}^\dagger}\boldsymbol{\theta}^\dagger = \mathbf{0}$ holds from the definition of the projection matrix. Thus from Eq. (3.51), $\boldsymbol{\theta}^\dagger$ is also an eigenvector of Y for eigenvalue 0. The projection of Eq. (3.53) is a linear combination of $\hat{\boldsymbol{\theta}}$ and $\boldsymbol{\theta}^\dagger$. Therefore $\boldsymbol{\theta}'$ is also an eigenvector of Y for eigenvalue 0. A projection of a projection is itself, thus $\mathsf{P}_{\boldsymbol{\theta}^\dagger}\boldsymbol{\theta}' = \boldsymbol{\theta}'$ holds. At the time of convergence, $\boldsymbol{\theta} = \boldsymbol{\theta}'$ holds therefore $\boldsymbol{\theta}$ is an eigenvector of Y for eigenvalue 0. At the same time, it satisfies the rank constraint $\mathsf{P}_{\boldsymbol{\theta}^\dagger}\boldsymbol{\theta} = \boldsymbol{\theta}$.

3.8 (1) If we define vectors

$$\Delta\hat{\mathbf{x}}_\alpha = \hat{\mathbf{x}}_\alpha - \bar{\mathbf{x}}_\alpha, \qquad \Delta\hat{\mathbf{x}}'_\alpha = \hat{\mathbf{x}}'_\alpha - \bar{\mathbf{x}}'_\alpha,$$

the sum of squares $\sum_{\alpha=1}^{N}\left(\|\tilde{\mathbf{x}}_\alpha + \Delta\hat{\mathbf{x}}_\alpha\|^2 + \|\tilde{\mathbf{x}}'_\alpha + \Delta\hat{\mathbf{x}}'_\alpha\|^2\right)$ is equal to the S in Eq. (3.60) divided by f_0^2. The epipolar equation of Eq. (3.16) is written in the form

$$(\hat{\mathbf{x}}_\alpha - \Delta\hat{\mathbf{x}}_\alpha, \mathsf{F}(\hat{\mathbf{x}}'_\alpha - \Delta\hat{\mathbf{x}}'_\alpha)) = 0.$$

Expanding the right-hand side and ignoring quadratic terms in $\Delta\hat{\mathbf{x}}_\alpha$ and $\Delta\hat{\mathbf{x}}'_\alpha$, we obtain

$$(\mathsf{F}\hat{\mathbf{x}}'_\alpha, \Delta\hat{\mathbf{x}}_\alpha) + (\mathsf{F}^\top\hat{\mathbf{x}}_\alpha, \Delta\hat{\mathbf{x}}'_\alpha) = (\hat{\mathbf{x}}_\alpha, \mathsf{F}\hat{\mathbf{x}}'_\alpha). \qquad (*)$$

Because the third components of $\hat{\mathbf{x}}_\alpha$, $\hat{\mathbf{x}}'_\alpha$, $\bar{\mathbf{x}}_\alpha$, and $\bar{\mathbf{x}}'_\alpha$ are all 0, we can write the constraint on $\Delta\hat{\mathbf{x}}_\alpha$ and $\Delta\hat{\mathbf{x}}'_\alpha$ as $(\mathbf{k}, \Delta\hat{\mathbf{x}}_\alpha) = 0$ and $(\mathbf{k}, \Delta\hat{\mathbf{x}}'_\alpha) = 0$, where $\mathbf{k} = (0, 0, 1)^\top$. Introducing Lagrange multipliers, differentiating

$$\begin{aligned}&\sum_{\alpha=1}^{N}\left(\|\tilde{\mathbf{x}}_\alpha + \Delta\hat{\mathbf{x}}_\alpha\|^2 + \|\tilde{\mathbf{x}}'_\alpha + \Delta\hat{\mathbf{x}}'_\alpha\|^2\right) \\ &\quad - \sum_{\alpha=1}^{N}\lambda_\alpha\left((\mathsf{F}\hat{\mathbf{x}}'_\alpha, \Delta\hat{\mathbf{x}}_\alpha) + (\mathsf{F}^\top\hat{\mathbf{x}}_\alpha, \Delta\hat{\mathbf{x}}'_\alpha) - (\hat{\mathbf{x}}_\alpha, \mathsf{F}\hat{\mathbf{x}}'_\alpha)\right) \\ &\quad - \sum_{\alpha=1}^{N}\mu_\alpha(\mathbf{k}, \Delta\hat{\mathbf{x}}_\alpha) - \sum_{\alpha=1}^{N}\mu'_\alpha(\mathbf{k}, \Delta\hat{\mathbf{x}}'_\alpha),\end{aligned}$$

with respect to $\Delta\hat{\mathbf{x}}_\alpha$, and $\Delta\hat{\mathbf{x}}'_\alpha$, and letting the result be $\mathbf{0}$, we obtain

$$2(\tilde{\mathbf{x}}_\alpha + \Delta\hat{\mathbf{x}}_\alpha) - \lambda_\alpha\mathsf{F}\hat{\mathbf{x}}'_\alpha - \mu_\alpha\mathbf{k} = \mathbf{0}, \quad 2(\tilde{\mathbf{x}}'_\alpha + \Delta\hat{\mathbf{x}}'_\alpha) - \lambda_\alpha\mathsf{F}^\top\hat{\mathbf{x}}_\alpha - \mu'_\alpha\mathbf{k} = \mathbf{0}.$$

Multiplying this by the projection matrix $\mathsf{P}_{\mathbf{k}}$ of Eq. (3.20) from the left, and noting that $\mathsf{P}_{\mathbf{k}}\tilde{\mathbf{x}}_\alpha = \tilde{\mathbf{x}}_\alpha$ and $\mathsf{P}_{\mathbf{k}}\tilde{\mathbf{x}}' = \tilde{\mathbf{x}}'_\alpha$ hold from the definition of $\tilde{\mathbf{x}}_\alpha$ and $\tilde{\mathbf{x}}'_\alpha$, we obtain

$$2\tilde{\mathbf{x}}_\alpha + 2\Delta\hat{\mathbf{x}}_\alpha - \lambda_\alpha \mathsf{P}_{\mathbf{k}}\mathsf{F}\hat{\mathbf{x}}'_\alpha = \mathbf{0},\ 2\tilde{\mathbf{x}}_\alpha + 2\Delta\hat{\mathbf{x}}'_\alpha - \lambda_\alpha \mathsf{P}_{\mathbf{k}}\mathsf{F}^\top\hat{\mathbf{x}}_\alpha = \mathbf{0}.$$

Thus

$$\Delta\hat{\mathbf{x}}_\alpha = \frac{\lambda_\alpha}{2}\mathsf{P}_{\mathbf{k}}\mathsf{F}\hat{\mathbf{x}}'_\alpha - \tilde{\mathbf{x}}_\alpha, \qquad \Delta\hat{\mathbf{x}}'_\alpha = \frac{\lambda_\alpha}{2}\mathsf{P}_{\mathbf{k}}\mathsf{F}^\top\hat{\mathbf{x}}_\alpha - \tilde{\mathbf{x}}'_\alpha.$$

Substituting these into Eq. (∗), we obtain

$$(\mathsf{F}\hat{\mathbf{x}}'_\alpha, \frac{\lambda_\alpha}{2}\mathsf{P}_{\mathbf{k}}\mathsf{F}\hat{\mathbf{x}}'_\alpha - \tilde{\mathbf{x}}_\alpha) + (\mathsf{F}^\top\hat{\mathbf{x}}_\alpha, \frac{\lambda_\alpha}{2}\mathsf{P}_{\mathbf{k}}\mathsf{F}^\top\hat{\mathbf{x}}_\alpha - \tilde{\mathbf{x}}'_\alpha) = (\hat{\mathbf{x}}_\alpha, \mathsf{F}\hat{\mathbf{x}}'_\alpha)$$

which gives λ_α in the form

$$\frac{\lambda_\alpha}{2} = \frac{(\hat{\mathbf{x}}_\alpha, \mathsf{F}\hat{\mathbf{x}}'_\alpha) + (\mathsf{F}\hat{\mathbf{x}}'_\alpha, \tilde{\mathbf{x}}_\alpha) + (\mathsf{F}^\top\hat{\mathbf{x}}_\alpha, \tilde{\mathbf{x}}'_\alpha)}{(\mathsf{F}\hat{\mathbf{x}}'_\alpha, \mathsf{P}_{\mathbf{k}}\mathsf{F}\hat{\mathbf{x}}'_\alpha) + (\mathsf{F}^\top\hat{\mathbf{x}}_\alpha, \mathsf{P}_{\mathbf{k}}\mathsf{F}^\top\hat{\mathbf{x}}_\alpha)}.$$

Thus we obtain

$$\Delta\hat{\mathbf{x}}_\alpha = \frac{\Big((\hat{\mathbf{x}}_\alpha, \mathsf{F}\hat{\mathbf{x}}'_\alpha) + (\mathsf{F}\hat{\mathbf{x}}'_\alpha, \tilde{\mathbf{x}}_\alpha) + (\mathsf{F}^\top\hat{\mathbf{x}}_\alpha, \tilde{\mathbf{x}}'_\alpha)\Big)\mathsf{P}_{\mathbf{k}}\mathsf{F}\hat{\mathbf{x}}'_\alpha}{(\mathsf{F}\hat{\mathbf{x}}'_\alpha, \mathsf{P}_{\mathbf{k}}\mathsf{F}\hat{\mathbf{x}}'_\alpha) + (\mathsf{F}^\top\hat{\mathbf{x}}_\alpha, \mathsf{P}_{\mathbf{k}}\mathsf{F}^\top\hat{\mathbf{x}}_\alpha)} - \tilde{\mathbf{x}}_\alpha,$$

$$\Delta\hat{\mathbf{x}}'_\alpha = \frac{\Big((\hat{\mathbf{x}}_\alpha, \mathsf{F}\hat{\mathbf{x}}'_\alpha) + (\mathsf{F}\hat{\mathbf{x}}'_\alpha, \tilde{\mathbf{x}}_\alpha) + (\mathsf{F}^\top\hat{\mathbf{x}}_\alpha, \tilde{\mathbf{x}}'_\alpha)\Big)\mathsf{P}_{\mathbf{k}}\mathsf{F}^\top\hat{\mathbf{x}}_\alpha}{(\mathsf{F}\hat{\mathbf{x}}'_\alpha, \mathsf{P}_{\mathbf{k}}\mathsf{F}\hat{\mathbf{x}}'_\alpha) + (\mathsf{F}^\top\hat{\mathbf{x}}_\alpha, \mathsf{P}_{\mathbf{k}}\mathsf{F}^\top\hat{\mathbf{x}}_\alpha)} - \tilde{\mathbf{x}}'_\alpha.$$

It follows that $\bar{\mathbf{x}}_\alpha$ and $\bar{\mathbf{x}}'_\alpha$ are estimated in the form

$$\hat{\hat{\mathbf{x}}}_\alpha = \mathbf{x}_\alpha - \frac{\Big((\hat{\mathbf{x}}_\alpha, \mathsf{F}\hat{\mathbf{x}}'_\alpha) + (\mathsf{F}\hat{\mathbf{x}}'_\alpha, \tilde{\mathbf{x}}_\alpha) + (\mathsf{F}^\top\hat{\mathbf{x}}_\alpha, \tilde{\mathbf{x}}'_\alpha)\Big)\mathsf{P}_{\mathbf{k}}\mathsf{F}\hat{\mathbf{x}}'_\alpha}{(\mathsf{F}\hat{\mathbf{x}}'_\alpha, \mathsf{P}_{\mathbf{k}}\mathsf{F}\hat{\mathbf{x}}'_\alpha) + (\mathsf{F}^\top\hat{\mathbf{x}}_\alpha, \mathsf{P}_{\mathbf{k}}\mathsf{F}^\top\hat{\mathbf{x}}_\alpha)},$$

$$\hat{\hat{\mathbf{x}}}'_\alpha = \mathbf{x}'_\alpha - \frac{\Big((\hat{\mathbf{x}}_\alpha, \mathsf{F}\hat{\mathbf{x}}'_\alpha) + (\mathsf{F}\hat{\mathbf{x}}'_\alpha, \tilde{\mathbf{x}}_\alpha) + (\mathsf{F}^\top\hat{\mathbf{x}}_\alpha, \tilde{\mathbf{x}}'_\alpha)\Big)\mathsf{P}_{\mathbf{k}}\mathsf{F}^\top\hat{\mathbf{x}}_\alpha}{(\mathsf{F}\hat{\mathbf{x}}'_\alpha, \mathsf{P}_{\mathbf{k}}\mathsf{F}\hat{\mathbf{x}}'_\alpha) + (\mathsf{F}^\top\hat{\mathbf{x}}_\alpha, \mathsf{P}_{\mathbf{k}}\mathsf{F}^\top\hat{\mathbf{x}}_\alpha)}.$$

Rewriting these, we obtain Eq. (3.82).

(2) From the definition of the vectors $\boldsymbol{\theta}$ and $\boldsymbol{\xi}^*_\alpha$ and the matrix $V_0[\hat{\boldsymbol{\xi}}_\alpha]$, we can easily confirm the following identities.

$$(\hat{\mathbf{x}}_\alpha, \mathsf{F}\hat{\mathbf{x}}'_\alpha) + (\mathsf{F}\hat{\mathbf{x}}'_\alpha, \tilde{\mathbf{x}}_\alpha) + (\mathsf{F}^\top\hat{\mathbf{x}}_\alpha, \tilde{\mathbf{x}}'_\alpha) = \frac{(\boldsymbol{\xi}^*_\alpha, \boldsymbol{\theta})}{f_0^2},$$

$$(\mathsf{F}\hat{\mathbf{x}}'_\alpha, \mathsf{P}_{\mathbf{k}}\mathsf{F}\hat{\mathbf{x}}'_\alpha) + (\mathsf{F}^\top\hat{\mathbf{x}}_\alpha, \mathsf{P}_{\mathbf{k}}\mathsf{F}^\top\hat{\mathbf{x}}_\alpha) = \frac{(\boldsymbol{\theta}, V_0[\hat{\boldsymbol{\xi}}_\alpha]\boldsymbol{\theta})}{f_0^2}.$$

Thus Eq. (3.82) is written in the form of Eq. (3.84).

(3) Note the following relationships.

$$\mathsf{P}_{\mathbf{k}}\mathsf{F}\hat{\mathbf{x}}'_\alpha = \begin{pmatrix} F_{11} & F_{12} & F_{13} \\ F_{21} & F_{22} & F_{23} \\ 0 & 0 & 0 \end{pmatrix} \begin{pmatrix} \hat{x}'_\alpha/f_0 \\ \hat{y}'_\alpha/f_0 \\ 1 \end{pmatrix},$$

$$\mathsf{P}_{\mathbf{k}}\mathsf{F}^\top\hat{\mathbf{x}}_\alpha = \begin{pmatrix} F_{11} & F_{21} & F_{33} \\ F_{12} & F_{22} & F_{32} \\ 0 & 0 & 0 \end{pmatrix} \begin{pmatrix} \hat{x}_\alpha/f_0 \\ \hat{y}_\alpha/f_0 \\ 1 \end{pmatrix}.$$

These are vectors whose third component is 0, therefore we have

$$\|\mathsf{P}_{\mathbf{k}}\mathsf{F}\hat{\mathbf{x}}'_\alpha\|^2 = \frac{1}{f_0^2}\left\|\begin{pmatrix} F_{11} & F_{12} & F_{13} \\ F_{21} & F_{22} & F_{23} \end{pmatrix}\begin{pmatrix} \hat{x}'_\alpha \\ \hat{y}'_\alpha \\ f_0 \end{pmatrix}\right\|^2,$$

$$\|\mathsf{P}_{\mathbf{k}}\mathsf{F}^\top\hat{\mathbf{x}}_\alpha\|^2 = \frac{1}{f_0^2}\left\|\begin{pmatrix} F_{11} & F_{21} & F_{33} \\ F_{12} & F_{22} & F_{32} \end{pmatrix}\begin{pmatrix} \hat{x}_\alpha \\ \hat{y}_\alpha \\ f_0 \end{pmatrix}\right\|^2.$$

Because $\mathsf{P}_{\mathbf{k}}^2 = \mathsf{P}_{\mathbf{k}}^2$ from the definition of the projection matrix $\mathsf{P}_{\mathbf{k}}$, we have

$$\|\mathsf{P}_{\mathbf{k}}\mathsf{F}\hat{\mathbf{x}}'_\alpha\|^2 = (\mathsf{P}_{\mathbf{k}}\mathsf{F}\hat{\mathbf{x}}'_\alpha, \mathsf{P}_{\mathbf{k}}\mathsf{F}\hat{\mathbf{x}}'_\alpha) = (\mathsf{F}\hat{\mathbf{x}}'_\alpha, \mathsf{P}_{\mathbf{k}}^2\mathsf{F}\hat{\mathbf{x}}'_\alpha) = (\mathsf{F}\hat{\mathbf{x}}'_\alpha, \mathsf{P}_{\mathbf{k}}\mathsf{F}\hat{\mathbf{x}}'_\alpha),$$

$$\|\mathsf{P}_{\mathbf{k}}\mathsf{F}^\top\hat{\mathbf{x}}_\alpha\|^2 = (\mathsf{P}_{\mathbf{k}}\mathsf{F}^\top\hat{\mathbf{x}}_\alpha, \mathsf{P}_{\mathbf{k}}\mathsf{F}^\top\hat{\mathbf{x}}_\alpha) = (\mathsf{F}^\top\hat{\mathbf{x}}_\alpha, \mathsf{P}_{\mathbf{k}}^2\mathsf{F}^\top\hat{\mathbf{x}}_\alpha) = (\mathsf{F}^\top\hat{\mathbf{x}}_\alpha, \mathsf{P}_{\mathbf{k}}\mathsf{F}^\top\hat{\mathbf{x}}_\alpha).$$

Thus from Eq. (3.82), we can approximate the S in Eq. (3.15) in the form:

$$\begin{aligned}
S \approx& \frac{1}{N}\sum_{\alpha=1}^{N}\Big((x_\alpha - \hat{x}_\alpha)^2 + (y_\alpha - \hat{y}_\alpha)^2 + (x'_\alpha - \hat{x}'_\alpha)^2 + (y'_\alpha - \hat{y}'_\alpha)^2\Big) \\
=& \frac{1}{N}\sum_{\alpha=1}^{N}\left(\left\|\begin{pmatrix} \hat{x}_\alpha \\ \hat{y}_\alpha \end{pmatrix} - \begin{pmatrix} x_\alpha \\ y_\alpha \end{pmatrix}\right\|^2 + \left\|\begin{pmatrix} \hat{x}'_\alpha \\ \hat{y}'_\alpha \end{pmatrix} - \begin{pmatrix} x'_\alpha \\ y'_\alpha \end{pmatrix}\right\|^2\right) \\
=& \frac{1}{N}\sum_{\alpha=1}^{N}\left(\frac{(\hat{\mathbf{x}}_\alpha, \mathsf{F}\hat{\mathbf{x}}'_\alpha) + (\mathsf{F}\hat{\mathbf{x}}'_\alpha, \tilde{\mathbf{x}}_\alpha) + (\mathsf{F}^\top\hat{\mathbf{x}}_\alpha, \tilde{\mathbf{x}}'_\alpha)}{(\mathsf{F}\hat{\mathbf{x}}'_\alpha, \mathsf{P}_{\mathbf{k}}\hat{\mathbf{x}}'_\alpha) + (\mathsf{F}^\top\hat{\mathbf{x}}_\alpha, \mathsf{P}_{\mathbf{k}}\mathsf{F}^\top\hat{\mathbf{x}}_\alpha)}\right)^2 (f_0^2\|\mathsf{P}_{\mathbf{k}}\mathsf{F}\hat{\mathbf{x}}'_\alpha\|^2 \\
&+ f_0^2\|\mathsf{P}_{\mathbf{k}}\mathsf{F}^\top\hat{\mathbf{x}}_\alpha\|^2) \\
=& \frac{f_0^2}{N}\sum_{\alpha=1}^{N}\frac{\Big((\hat{\mathbf{x}}_\alpha, \mathsf{F}\hat{\mathbf{x}}'_\alpha) + (\mathsf{F}\hat{\mathbf{x}}'_\alpha, \tilde{\mathbf{x}}_\alpha) + (\mathsf{F}^\top\hat{\mathbf{x}}_\alpha, \tilde{\mathbf{x}}'_\alpha)\Big)^2}{(\mathsf{F}\hat{\mathbf{x}}'_\alpha, \mathsf{P}_{\mathbf{k}}\hat{\mathbf{x}}'_\alpha) + (\mathsf{F}^\top\hat{\mathbf{x}}_\alpha, \mathsf{P}_{\mathbf{k}}\mathsf{F}^\top\hat{\mathbf{x}}_\alpha)}
\end{aligned}$$

Using the result of item (2) above, we can rewrite this in the form of Eq. (3.56).

Problems of Chapter 4

4.1 Letting C ($\neq 0$) be the proportionality constant implied by Eq. (4.2), we can rewrite it as

$$\begin{aligned}
\frac{x}{f_0} &= C(P_{11}X + P_{12}Y + P_{13}Z + P_{14}), \\
\frac{y}{f_0} &= C(P_{21}X + P_{22}Y + P_{23}Z + P_{24}), \\
1 &= C(P_{31}X + P_{32}Y + P_{33}Z + P_{34}).
\end{aligned}$$

The third equation gives $C = 1/(P_{31}X + P_{32}Y + P_{33}Z + P_{34})$. Substituting this into the first and second equations, we obtain Eq. (4.1).

4.2 (1) The square norm $\|\mathsf{A}\mathbf{x}-\mathbf{b}\|^2$ is expanded into

$$\begin{aligned}\|\mathsf{A}\mathbf{x}-\mathbf{b}\|^2 &= (\mathsf{A}\mathbf{x}-\mathbf{b}, \mathsf{A}\mathbf{x}-\mathbf{b}) = (\mathsf{A}\mathbf{x}, \mathsf{A}\mathbf{x}) - 2(\mathsf{A}\mathbf{x}, \mathbf{b}) + (\mathbf{b}, \mathbf{b}) \\ &= (\mathbf{x}, \mathsf{A}^\top\mathsf{A}\mathbf{x}) - 2(\mathbf{x}, \mathsf{A}^\top\mathbf{b}) + \|\mathbf{b}\|^2.\end{aligned}$$

Differentiating this with respect to each component of $\mathbf{x}$, and letting the result be $\mathbf{0}$, we obtain Eq. (4.15).
(2) If the left-hand side matrix of Eq. (4.15) is nonsingular, we multiply the right-hand side by its inverse from the left to obtain

$$\mathbf{x} = (\mathsf{A}^\top\mathsf{A})^{-1}\mathsf{A}^\top\mathbf{b} = \mathsf{A}^-\mathbf{b}.$$

4.3 (1) The four equations of Eq. (4.5) are linearly dependent if and only if the determinant of the 4×4 coefficient matrix vanishes; that is,

$$\begin{vmatrix} f_0P_{11}-xP_{31} & f_0P_{12}-xP_{32} & f_0P_{13}-xP_{33} & f_0P_{14}-xP_{34} \\ f_0P_{21}-yP_{31} & f_0P_{22}-yP_{32} & f_0P_{23}-yP_{33} & f_0P_{24}-yP_{34} \\ f_0P'_{11}-x'P'_{31} & f_0P'_{12}-x'P'_{32} & f_0P'_{13}-x'P'_{33} & f_0P'_{14}-x'P'_{34} \\ f_0P'_{21}-y'P'_{31} & f_0P'_{22}-y'P'_{32} & f_0P'_{23}-y'P'_{33} & f_0P'_{24}-y'P'_{34} \end{vmatrix} = 0.$$

We add to this a diagonal block consisting of the identity matrix. The determinant is unchanged if the added nondiagonal block consists of 0. Hence, the left side of the above equation is rewritten in the form

$$\begin{vmatrix} P_{11}-xP_{31}/f_0 & P_{12}-xP_{32}/f_0 & P_{13}-xP_{33}/f_0 & P_{14}-xP_{34}/f_0 & 0 & 0 \\ P_{21}-yP_{31}/f_0 & P_{22}-yP_{32}/f_0 & P_{23}-yP_{31}/f_0 & P_{24}-yP_{33}/f_0 & 0 & 0 \\ P'_{11}-x'P'_{31}/f_0 & P'_{12}-x'P'_{32}/f_0 & P'_{13}-x'P'_{33}/f_0 & P'_{14}-x'P'_{34}/f_0 & 0 & 0 \\ P'_{21}-y'P'_{31}/f_0 & P'_{22}-y'P'_{32}/f_0 & P'_{23}-y'P'_{33}/f_0 & P'_{24}-y'P'_{34}/f_0 & 0 & 0 \\ P_{31} & P_{32} & P_{33} & P_{34} & 1 & 0 \\ P'_{31} & P'_{32} & P'_{33} & P'_{34} & 0 & 1 \end{vmatrix}$$

$$= \begin{vmatrix} P_{11} & P_{12} & P_{13} & P_{14} & x/f_0 & 0 \\ P_{21} & P_{22} & P_{23} & P_{24} & y/f_0 & 0 \\ P_{31} & P_{32} & P_{33} & P_{34} & 1 & 0 \\ P'_{11} & P'_{12} & P'_{13} & P'_{14} & 0 & x'/f_0 \\ P'_{21} & P'_{22} & P'_{23} & P'_{24} & 0 & y'/f_0 \\ P'_{31} & P'_{32} & P'_{33} & P'_{34} & 0 & 1 \end{vmatrix},$$

where we have multiplied the fifth row by x/f_0 and y/f_0 and added the resulting rows to the first and the second rows, respectively. Also, we have multiplied the sixth row by x'/f_0 and y'/f_0 and added the resulting rows to the third and the fourth rows, respectively. These operations do not change the determinant. We have finally interchanged rows; the sign alters each time two rows are interchanged.

(2) Doing cofactor expansion on the left-hand side of Eq. (4.17) with respect to the fifth column and then doing cofactor expansion of the result with respect to the sixth column, we obtain

$$
\begin{vmatrix} P_{21} & P_{22} & P_{23} & P_{24} & 0 \\ P_{31} & P_{32} & P_{33} & P_{34} & 0 \\ P'_{11} & P'_{12} & P'_{13} & P'_{14} & x'/f_0 \\ P'_{21} & P'_{22} & P'_{23} & P'_{24} & y'/f_0 \\ P'_{31} & P'_{32} & P'_{33} & P'_{34} & 1 \end{vmatrix} \frac{x}{f_0} - \begin{vmatrix} P_{11} & P_{12} & P_{13} & P_{14} & 0 \\ P_{31} & P_{32} & P_{33} & P_{34} & 0 \\ P'_{11} & P'_{12} & P'_{13} & P'_{14} & x'/f_0 \\ P'_{21} & P'_{22} & P'_{23} & P'_{24} & y'/f_0 \\ P'_{31} & P'_{32} & P'_{33} & P'_{34} & 1 \end{vmatrix} \frac{y}{f_0} + \begin{vmatrix} P_{11} & P_{12} & P_{13} & P_{14} & 0 \\ P_{21} & P_{22} & P_{23} & P_{24} & 0 \\ P'_{11} & P'_{12} & P'_{13} & P'_{14} & x'/f_0 \\ P'_{21} & P'_{22} & P'_{23} & P'_{24} & y'/f_0 \\ P'_{31} & P'_{32} & P'_{33} & P'_{34} & 1 \end{vmatrix}
$$

$$
= \left(\begin{vmatrix} P_{21} & P_{22} & P_{23} & P_{24} \\ P_{31} & P_{32} & P_{33} & P_{34} \\ P'_{21} & P'_{22} & P'_{23} & P'_{24} \\ P'_{31} & P'_{32} & P'_{33} & P'_{34} \end{vmatrix} \frac{x'}{f_0} - \begin{vmatrix} P_{21} & P_{22} & P_{23} & P_{24} \\ P_{31} & P_{32} & P_{33} & P_{34} \\ P'_{11} & P'_{12} & P'_{13} & P'_{14} \\ P'_{31} & P'_{32} & P'_{33} & P'_{34} \end{vmatrix} \frac{y'}{f_0} + \begin{vmatrix} P_{21} & P_{22} & P_{23} & P_{24} \\ P_{31} & P_{32} & P_{33} & P_{34} \\ P'_{11} & P'_{12} & P'_{13} & P'_{14} \\ P'_{21} & P'_{22} & P'_{23} & P'_{24} \end{vmatrix} \right) \frac{x}{f_0}
$$

$$
- \left(\begin{vmatrix} P_{11} & P_{12} & P_{13} & P_{14} \\ P_{31} & P_{32} & P_{33} & P_{34} \\ P'_{21} & P'_{22} & P'_{23} & P'_{24} \\ P'_{31} & P'_{32} & P'_{33} & P'_{34} \end{vmatrix} \frac{x'}{f_0} - \begin{vmatrix} P_{11} & P_{12} & P_{13} & P_{14} \\ P_{31} & P_{32} & P_{33} & P_{34} \\ P'_{11} & P'_{12} & P'_{13} & P'_{14} \\ P'_{31} & P'_{32} & P'_{33} & P'_{34} \end{vmatrix} \frac{y'}{f_0} + \begin{vmatrix} P_{11} & P_{12} & P_{13} & P_{14} \\ P_{31} & P_{32} & P_{33} & P_{34} \\ P'_{11} & P'_{12} & P'_{13} & P'_{14} \\ P'_{21} & P'_{22} & P'_{23} & P'_{24} \end{vmatrix} \right) \frac{y}{f_0}
$$

$$
+ \left(\begin{vmatrix} P_{11} & P_{12} & P_{13} & P_{14} \\ P_{21} & P_{22} & P_{23} & P_{24} \\ P'_{21} & P'_{22} & P'_{23} & P'_{24} \\ P'_{31} & P'_{32} & P'_{33} & P'_{34} \end{vmatrix} \frac{x'}{f_0} - \begin{vmatrix} P_{11} & P_{12} & P_{13} & P_{14} \\ P_{21} & P_{22} & P_{23} & P_{24} \\ P'_{11} & P'_{12} & P'_{13} & P'_{14} \\ P'_{31} & P'_{32} & P'_{33} & P'_{34} \end{vmatrix} \frac{y'}{f_0} + \begin{vmatrix} P_{11} & P_{12} & P_{13} & P_{14} \\ P_{21} & P_{22} & P_{23} & P_{24} \\ P'_{11} & P'_{12} & P'_{13} & P'_{14} \\ P'_{21} & P'_{22} & P'_{23} & P'_{24} \end{vmatrix} \right)
$$

$$
= F_{11}\Big(\frac{x}{f_0}\Big)\Big(\frac{x'}{f_0}\Big) + F_{12}\Big(\frac{x}{f_0}\Big)\Big(\frac{y'}{f_0}\Big) + F_{13}\Big(\frac{x}{f_0}\Big) + F_{21}\Big(\frac{y}{f_0}\Big)\Big(\frac{x'}{f_0}\Big) + F_{22}\Big(\frac{y}{f_0}\Big)\Big(\frac{y'}{f_0}\Big)
$$

$$
+ F_{23}\Big(\frac{y}{f_0}\Big) + F_{31}\Big(\frac{x}{f_0}\Big) + F_{32}\Big(\frac{x}{f_0}\Big) + F_{33},
$$

where we define (with appropriate interchanges of rows)

$$
F_{11} = \begin{vmatrix} P_{21} & P_{22} & P_{23} & P_{24} \\ P_{31} & P_{32} & P_{33} & P_{34} \\ P'_{21} & P'_{22} & P'_{23} & P'_{24} \\ P'_{31} & P'_{32} & P'_{33} & P'_{34} \end{vmatrix}, \quad F_{12} = \begin{vmatrix} P_{21} & P_{22} & P_{23} & P_{24} \\ P_{31} & P_{32} & P_{33} & P_{34} \\ P'_{31} & P'_{32} & P'_{33} & P'_{34} \\ P'_{11} & P'_{12} & P'_{13} & P'_{14} \end{vmatrix}, \quad F_{33} = \begin{vmatrix} P_{21} & P_{22} & P_{23} & P_{24} \\ P_{31} & P_{32} & P_{33} & P_{34} \\ P'_{11} & P'_{12} & P'_{13} & P'_{14} \\ P'_{21} & P'_{22} & P'_{23} & P'_{24} \end{vmatrix},
$$

$$
F_{21} = \begin{vmatrix} P_{31} & P_{32} & P_{33} & P_{34} \\ P_{11} & P_{12} & P_{13} & P_{14} \\ P'_{21} & P'_{22} & P'_{23} & P'_{24} \\ P'_{31} & P'_{32} & P'_{33} & P'_{34} \end{vmatrix}, \quad F_{22} = \begin{vmatrix} P_{11} & P_{12} & P_{13} & P_{14} \\ P_{31} & P_{32} & P_{33} & P_{34} \\ P'_{11} & P'_{12} & P'_{13} & P'_{14} \\ P'_{31} & P'_{32} & P'_{33} & P'_{34} \end{vmatrix}, \quad F_{23} = \begin{vmatrix} P_{31} & P_{32} & P_{33} & P_{34} \\ P_{11} & P_{12} & P_{13} & P_{14} \\ P'_{11} & P'_{12} & P'_{13} & P'_{14} \\ P'_{21} & P'_{22} & P'_{23} & P'_{24} \end{vmatrix},
$$

$$
F_{31} = \begin{vmatrix} P_{11} & P_{12} & P_{13} & P_{14} \\ P_{21} & P_{22} & P_{23} & P_{24} \\ P'_{21} & P'_{22} & P'_{23} & P'_{24} \\ P'_{31} & P'_{32} & P'_{33} & P'_{34} \end{vmatrix}, \quad F_{32} = \begin{vmatrix} P_{11} & P_{12} & P_{13} & P_{14} \\ P_{21} & P_{22} & P_{23} & P_{24} \\ P'_{31} & P'_{32} & P'_{33} & P'_{34} \\ P'_{11} & P'_{12} & P'_{13} & P'_{14} \end{vmatrix}, \quad F_{33} = \begin{vmatrix} P_{11} & P_{12} & P_{13} & P_{14} \\ P_{21} & P_{22} & P_{23} & P_{24} \\ P'_{11} & P'_{12} & P'_{13} & P'_{14} \\ P'_{21} & P'_{22} & P'_{23} & P'_{24} \end{vmatrix}.
$$

Using the thus-defined matrix $\mathsf{F} = (F_{ij})$, we can write Eq. (4.17) in the form of Eq. (4.7).

4.4 (1) Define $\mathbf{x}$ and $\mathbf{x}'$ as in Eq. (4.19), and let $\bar{\mathbf{x}}$ and $\bar{\mathbf{x}}'$ be similarly defined vectors obtained by replacing (x, y) and (x', y') by $(\bar{x}, \bar{y})$ and $(\bar{x}', \bar{y}')$, respectively. If we let

$$\Delta\mathbf{x} = \mathbf{x} - \bar{\mathbf{x}}, \qquad \Delta\mathbf{x}' = \mathbf{x}' - \bar{\mathbf{x}}',$$

we see that $\|\Delta\mathbf{x}\|^2 + \|\Delta\mathbf{x}'\|^2$ equals the value S of Eq. (4.8) divided by f_0^2. The epipolar equation of Eq. (4.7) is written as

$$(\mathbf{x} - \Delta\mathbf{x}, \mathsf{F}(\mathbf{x}' - \Delta\mathbf{x}')) = 0.$$

Doing Taylor expansion and ignoring second-order terms in the noise components $\Delta\mathbf{x}$ and $\Delta\mathbf{x}'$, we obtain

$$(\mathsf{F}\mathbf{x}', \Delta\mathbf{x}) + (\mathsf{F}^\top\mathbf{x}, \Delta\mathbf{x}') = (\mathbf{x}, \mathsf{F}\mathbf{x}'). \qquad (*)$$

The third components of $\Delta\mathbf{x}$ and $\Delta\mathbf{x}'$ are 0, because noise occurs in the image plane. This constraint is written as $(\mathbf{k}, \Delta\mathbf{x}) = 0$ and $(\mathbf{k}, \Delta\mathbf{x}') = 0$, where $\mathbf{k} = (0, 0, 1)^\top$. Introducing Lagrange multipliers, we differentiate

$$\|\Delta\mathbf{x}\|^2 + \|\Delta\mathbf{x}'\|^2 - \lambda\Big((\mathsf{F}\mathbf{x}', \Delta\mathbf{x}) + (\mathsf{F}^\top\mathbf{x}, \Delta\mathbf{x}') - (\mathbf{x}, \mathsf{F}\mathbf{x}')\Big) - \mu(\mathbf{k}, \Delta\mathbf{x}) - \mu'(\mathbf{k}, \Delta\mathbf{x}')$$

with respect to $\Delta\mathbf{x}$ and $\Delta\mathbf{x}'$. Letting the result be $\mathbf{0}$, we obtain

$$2\Delta\mathbf{x} - \lambda\mathsf{F}\mathbf{x}' - \mu\mathbf{k} = \mathbf{0}, \qquad 2\Delta\mathbf{x}' - \lambda\mathsf{F}^\top\mathbf{x} - \mu'\mathbf{k} = \mathbf{0}.$$

Multiplying Eq. (4.19) by $\mathsf{P}_\mathbf{k}$ from the left on both sides, and noting that $\mathsf{P}_\mathbf{k}\Delta\mathbf{x} = \Delta\mathbf{x}$, $\mathsf{P}_\mathbf{k}\Delta\mathbf{x}' = \Delta\mathbf{x}'$, and $\mathsf{P}_\mathbf{k}\mathbf{k} = \mathbf{0}$, we obtain

$$2\Delta\mathbf{x} - \lambda\mathsf{P}_\mathbf{k}\mathsf{F}\mathbf{x}' = \mathbf{0}, \qquad 2\Delta\mathbf{x}' - \lambda\mathsf{P}_\mathbf{k}\mathsf{F}^\top\mathbf{x} = \mathbf{0}.$$

Thus

$$\Delta\mathbf{x} = \frac{\lambda}{2}\mathsf{P}_\mathbf{k}\mathsf{F}\mathbf{x}', \qquad \Delta\mathbf{x}' = \frac{\lambda}{2}\mathsf{P}_\mathbf{k}\mathsf{F}^\top\mathbf{x}.$$

Substitution of these into Eq. $(*)$ yields

$$(\mathsf{F}\mathbf{x}', \frac{\lambda}{2}\mathsf{P}_\mathbf{k}\mathsf{F}\mathbf{x}') + (\mathsf{F}^\top\mathbf{x}, \frac{\lambda}{2}\mathsf{P}_\mathbf{k}\mathsf{F}^\top\mathbf{x}) = (\mathbf{x}, \mathsf{F}\mathbf{x}'),$$

which gives λ in the form

$$\frac{\lambda}{2} = \frac{(\mathbf{x}, \mathsf{F}\mathbf{x}')}{(\mathsf{F}\mathbf{x}', \mathsf{P}_\mathbf{k}\mathsf{F}\mathbf{x}') + (\mathsf{F}^\top\mathbf{x}, \mathsf{P}_\mathbf{k}\mathsf{F}^\top\mathbf{x})}.$$

Thus we obtain $\Delta\mathbf{x}$ and $\Delta\mathbf{x}'$ in the form

$$\Delta\mathbf{x} = \frac{(\mathbf{x}, \mathsf{F}\mathbf{x}')\mathsf{P}_\mathbf{k}\mathsf{F}\mathbf{x}'}{(\mathsf{F}\mathbf{x}', \mathsf{P}_\mathbf{k}\mathsf{F}\mathbf{x}') + (\mathsf{F}^\top\mathbf{x}, \mathsf{P}_\mathbf{k}\mathsf{F}^\top\mathbf{x})}, \qquad \Delta\mathbf{x}' = \frac{(\mathbf{x}, \mathsf{F}\mathbf{x}')\mathsf{P}_\mathbf{k}\mathsf{F}^\top\mathbf{x}}{(\mathsf{F}\mathbf{x}', \mathsf{P}_\mathbf{k}\mathsf{F}\mathbf{x}') + (\mathsf{F}^\top\mathbf{x}, \mathsf{P}_\mathbf{k}\mathsf{F}^\top\mathbf{x})}.$$

This means that $\bar{\mathbf{x}}$ and $\bar{\mathbf{x}}'$ are estimated to be

$$\hat{\mathbf{x}} = \mathbf{x} - \frac{(\mathbf{x}, \mathsf{F}\mathbf{x}')\mathsf{P}_\mathbf{k}\mathsf{F}\mathbf{x}'}{(\mathsf{F}\mathbf{x}', \mathsf{P}_\mathbf{k}\mathsf{F}\mathbf{x}') + (\mathsf{F}^\top\mathbf{x}, \mathsf{P}_\mathbf{k}\mathsf{F}^\top\mathbf{x})}, \qquad \hat{\mathbf{x}}' = \mathbf{x}' - \frac{(\mathbf{x}, \mathsf{F}\mathbf{x}')\mathsf{P}_\mathbf{k}\mathsf{F}^\top\mathbf{x}}{(\mathsf{F}\mathbf{x}', \mathsf{P}_\mathbf{k}\mathsf{F}\mathbf{x}') + (\mathsf{F}^\top\mathbf{x}, \mathsf{P}_\mathbf{k}\mathsf{F}^\top\mathbf{x})}.$$

Rewriting these, we obtain Eq. (4.18).

(2) From the definitions of the vectors $\boldsymbol{\theta}$ and $\boldsymbol{\xi}$ and the matrix $V_0[\boldsymbol{\xi}]$, we can confirm the following identities.

$$(\mathbf{x}, \mathsf{F}\mathbf{x}') = \frac{(\boldsymbol{\xi}, \boldsymbol{\theta})}{f_0^2}, \qquad (\mathsf{F}\mathbf{x}', \mathsf{P}_\mathbf{k}\mathsf{F}\mathbf{x}') + (\mathsf{F}^\top\mathbf{x}, \mathsf{P}_\mathbf{k}\mathsf{F}^\top\mathbf{x}) = \frac{(\boldsymbol{\theta}, V_0[\boldsymbol{\xi}]\boldsymbol{\theta})}{f_0^2}.$$

Therefore Eq. (4.18) can be written in the form of Eq. (4.20).
(3) Let $\bar{\mathbf{x}}$ and $\bar{\mathbf{x}}'$ be the vectors obtained from the vectors $\mathbf{x}$ and $\mathbf{x}'$ of Eq. (4.19) by replacing (x, y) and (x', y'), respectively, by $(\tilde{x}, \tilde{y})$ and $(\tilde{x}', \tilde{y}')$ defined by Eq. (4.14). If we let

$$\Delta\hat{\mathbf{x}} = \hat{\mathbf{x}} - \bar{\mathbf{x}}, \; 2\hat{\mathbf{x}}' = \hat{\mathbf{x}}' - \bar{\mathbf{x}}',$$

we see that $\|\tilde{\mathbf{x}} + \Delta\hat{\mathbf{x}}\|^2 + \|\tilde{\mathbf{x}}' + \Delta\hat{\mathbf{x}}'\|^2$ equals the value S of Eq. (4.13) divided by f_0^2. The epipolar equation of Eq. (4.7) can be written as

$$(\hat{\mathbf{x}} - \Delta\hat{\mathbf{x}}, \mathsf{F}(\hat{\mathbf{x}}' - \Delta\hat{\mathbf{x}}')) = 0.$$

Doing Taylor expansion and ignoring second-order terms in $\Delta\hat{\mathbf{x}}$ and $\Delta\hat{\mathbf{x}}'$, we obtain

$$(\mathsf{F}\hat{\mathbf{x}}', \Delta\hat{\mathbf{x}}) + (\mathsf{F}^\top\hat{\mathbf{x}}, \Delta\hat{\mathbf{x}}') = (\hat{\mathbf{x}}, \mathsf{F}\hat{\mathbf{x}}'). \qquad (**)$$

Because noise occurs in the image, we have constraints $(\mathbf{k}, \Delta\hat{\mathbf{x}}) = 0$ and $(\mathbf{k}, \Delta\hat{\mathbf{x}}') = 0$. Introducing Lagrange multipliers, we differentiate

$$\|\tilde{\mathbf{x}} + \Delta\hat{\mathbf{x}}\|^2 + \|\tilde{\mathbf{x}}' + \Delta\hat{\mathbf{x}}'\|^2 - \lambda\Big((\mathsf{F}\hat{\mathbf{x}}', \Delta\hat{\mathbf{x}}) + (\mathsf{F}^\top\hat{\mathbf{x}}, \Delta\hat{\mathbf{x}}')\Big) - \mu(\mathbf{k}, \Delta\hat{\mathbf{x}}) - \mu'(\mathbf{k}, \Delta\hat{\mathbf{x}}')$$

with respect to $\Delta\hat{\mathbf{x}}$ and $\Delta\hat{\mathbf{x}}'$. Letting the result be $\mathbf{0}$, we obtain

$$2(\tilde{\mathbf{x}} + \Delta\hat{\mathbf{x}}) - \lambda\mathsf{F}\hat{\mathbf{x}}' - \mu\mathbf{k} = \mathbf{0}, \qquad 2(\tilde{\mathbf{x}}' + \Delta\hat{\mathbf{x}}') - \lambda\mathsf{F}^\top\hat{\mathbf{x}} - \mu'\mathbf{k} = \mathbf{0}.$$

Multiplying this by $\mathsf{P}_\mathbf{k}$ from the left on both sides, and noting that the definition of $\tilde{\mathbf{x}}$ and $\tilde{\mathbf{x}}'$ implies $\mathsf{P}_\mathbf{k}\tilde{\mathbf{x}} = \tilde{\mathbf{x}}$ and $\mathsf{P}_\mathbf{k}\tilde{\mathbf{x}}' = \tilde{\mathbf{x}}'$, we obtain

$$2\tilde{\mathbf{x}} + 2\Delta\hat{\mathbf{x}} - \lambda\mathsf{P}_\mathbf{k}\mathsf{F}\hat{\mathbf{x}}' = \mathbf{0}, \qquad 2\tilde{\mathbf{x}} + 2\Delta\hat{\mathbf{x}}' - \lambda\mathsf{P}_\mathbf{k}\mathsf{F}^\top\hat{\mathbf{x}} = \mathbf{0}.$$

Thus

$$\Delta\hat{\mathbf{x}} = \frac{\lambda}{2}\mathsf{P}_\mathbf{k}\mathsf{F}\hat{\mathbf{x}}' - \tilde{\mathbf{x}}, \qquad \Delta\hat{\mathbf{x}}' = \frac{\lambda}{2}\mathsf{P}_\mathbf{k}\mathsf{F}^\top\hat{\mathbf{x}} - \tilde{\mathbf{x}}'.$$

Substitution of this into Eq. $(**)$ yields

$$(\mathsf{F}\hat{\mathbf{x}}', \frac{\lambda}{2}\mathsf{P}_\mathbf{k}\mathsf{F}\hat{\mathbf{x}}' - \tilde{\mathbf{x}}) + (\mathsf{F}^\top\hat{\mathbf{x}}, \frac{\lambda}{2}\mathsf{P}_\mathbf{k}\mathsf{F}^\top\hat{\mathbf{x}} - \tilde{\mathbf{x}}') = (\hat{\mathbf{x}}, \mathsf{F}\hat{\mathbf{x}}'),$$

which gives λ in the form

$$\frac{\lambda}{2} = \frac{(\hat{\mathbf{x}}, \mathsf{F}\hat{\mathbf{x}}') + (\mathsf{F}\hat{\mathbf{x}}', \tilde{\mathbf{x}}) + (\mathsf{F}^\top\hat{\mathbf{x}}, \tilde{\mathbf{x}}')}{(\mathsf{F}\hat{\mathbf{x}}', \mathsf{P}_\mathbf{k}\mathsf{F}\hat{\mathbf{x}}') + (\mathsf{F}^\top\hat{\mathbf{x}}, \mathsf{P}_\mathbf{k}\mathsf{F}^\top\hat{\mathbf{x}})}.$$

Therefore we obtain

$$\Delta\hat{\mathbf{x}} = \frac{\Big((\hat{\mathbf{x}}, \mathsf{F}\hat{\mathbf{x}}') + (\mathsf{F}\hat{\mathbf{x}}', \tilde{\mathbf{x}}) + (\mathsf{F}^\top\hat{\mathbf{x}}, \tilde{\mathbf{x}}')\Big)\mathsf{P}_\mathbf{k}\mathsf{F}\hat{\mathbf{x}}'}{(\mathsf{F}\hat{\mathbf{x}}', \mathsf{P}_\mathbf{k}\mathsf{F}\hat{\mathbf{x}}') + (\mathsf{F}^\top\hat{\mathbf{x}}, \mathsf{P}_\mathbf{k}\mathsf{F}^\top\hat{\mathbf{x}})} - \tilde{\mathbf{x}},$$

$$\Delta\hat{\mathbf{x}}' = \frac{\Big((\hat{\mathbf{x}}, \mathsf{F}\hat{\mathbf{x}}') + (\mathsf{F}\hat{\mathbf{x}}', \tilde{\mathbf{x}}) + (\mathsf{F}^\top\hat{\mathbf{x}}, \tilde{\mathbf{x}}')\Big)\mathsf{P}_\mathbf{k}\mathsf{F}^\top\hat{\mathbf{x}}}{(\mathsf{F}\hat{\mathbf{x}}', \mathsf{P}_\mathbf{k}\mathsf{F}\hat{\mathbf{x}}') + (\mathsf{F}^\top\hat{\mathbf{x}}, \mathsf{P}_\mathbf{k}\mathsf{F}^\top\hat{\mathbf{x}})} - \tilde{\mathbf{x}}'.$$

This means that $\bar{\mathbf{x}}$ and $\bar{\mathbf{x}}'$ are estimated to be

$$\hat{\hat{\mathbf{x}}} = \mathbf{x} - \frac{\Big((\hat{\mathbf{x}}, \mathsf{F}\hat{\mathbf{x}}') + (\mathsf{F}\hat{\mathbf{x}}', \tilde{\mathbf{x}}) + (\mathsf{F}^\top\hat{\mathbf{x}}, \tilde{\mathbf{x}}')\Big)\mathsf{P}_\mathbf{k}\mathsf{F}\hat{\mathbf{x}}'}{(\mathsf{F}\hat{\mathbf{x}}', \mathsf{P}_\mathbf{k}\mathsf{F}\hat{\mathbf{x}}') + (\mathsf{F}^\top\hat{\mathbf{x}}, \mathsf{P}_\mathbf{k}\mathsf{F}^\top\hat{\mathbf{x}})},$$

$$\hat{\hat{\mathbf{x}}}' = \mathbf{x}' - \frac{\Big((\hat{\mathbf{x}}, \mathsf{F}\hat{\mathbf{x}}') + (\mathsf{F}\hat{\mathbf{x}}', \tilde{\mathbf{x}}) + (\mathsf{F}^\top\hat{\mathbf{x}}, \tilde{\mathbf{x}}')\Big)\mathsf{P}_\mathbf{k}\mathsf{F}^\top\hat{\mathbf{x}}}{(\mathsf{F}\hat{\mathbf{x}}', \mathsf{P}_\mathbf{k}\mathsf{F}\hat{\mathbf{x}}') + (\mathsf{F}^\top\hat{\mathbf{x}}, \mathsf{P}_\mathbf{k}\mathsf{F}^\top\hat{\mathbf{x}})}.$$

Rewriting these, we obtain Eq. (4.21).

(4) From the definitions of the vectors $\boldsymbol{\theta}$ and $\boldsymbol{\xi}^*$ and the matrix $V_0[\hat{\boldsymbol{\xi}}]$, we can confirm the following identities.

$$(\hat{\mathbf{x}}, \mathsf{F}\hat{\mathbf{x}}') + (\mathsf{F}\hat{\mathbf{x}}', \tilde{\mathbf{x}}) + (\mathsf{F}^\top\hat{\mathbf{x}}, \tilde{\mathbf{x}}') = \frac{(\boldsymbol{\xi}^*, \boldsymbol{\theta})}{f_0^2},$$

$$(\mathsf{F}\hat{\mathbf{x}}', \mathsf{P}_\mathbf{k}\mathsf{F}\hat{\mathbf{x}}') + (\mathsf{F}^\top\hat{\mathbf{x}}, \mathsf{P}_\mathbf{k}\mathsf{F}^\top\hat{\mathbf{x}}) = \frac{(\boldsymbol{\theta}, V_0[\hat{\boldsymbol{\xi}}]\boldsymbol{\theta})}{f_0^2}.$$

Therefore Eq. (4.21) can be written in the form of Eq. (4.23).

Problems of Chapter 5

5.1 (1) The (signed) volume of the parallelepiped defined by vectors $\mathbf{a}$, $\mathbf{b}$, and $\mathbf{c}$ is $|\mathbf{a}, \mathbf{b}, \mathbf{c}|$. If we let $\mathbf{a}' = \mathsf{A}\mathbf{a}$, $\mathbf{b}' = \mathsf{A}\mathbf{b}$, and $\mathbf{c}' = \mathsf{A}\mathbf{c}$, then

$$|\mathbf{a}', \mathbf{b}', \mathbf{c}'| = |\mathsf{A}||\mathbf{a}, \mathbf{b}, \mathbf{c}|,$$

which is rewritten as

$$(\mathbf{a}' \times \mathbf{b}', \mathbf{c}') = |\mathsf{A}|(\mathbf{a} \times \mathbf{b}, \mathbf{c}) = (|\mathsf{A}|(\mathbf{a} \times \mathbf{b}), \mathsf{A}^{-1}\mathbf{c}') = (|\mathsf{A}|(\mathsf{A}^{-1})^\top(\mathbf{a} \times \mathbf{b}), \mathbf{c}').$$

This should hold for an arbitrary $\mathbf{c}$. Thus

$$\mathbf{a}' \times \mathbf{b}' = |\mathsf{A}|(\mathsf{A}^{-1})^\top(\mathbf{a} \times \mathbf{b}).$$

Rewriting this, we obtain

$$(\mathsf{A}\mathbf{a}) \times \mathbf{b}' = |\mathsf{A}|(\mathsf{A}^{-1})^\top(\mathbf{a}\times)\mathsf{A}^{-1}\mathbf{b}'.$$

This should hold for an arbitrary $\mathbf{b}'$. Thus

$$(\mathsf{A}\mathbf{a})\times = |\mathsf{A}|(\mathsf{A}^{-1})^\top(\mathbf{a}\times)\mathsf{A}^{-1}.$$

If we let $(\mathsf{A}^{-1})^\top = \mathsf{T}$, we have $\mathsf{A} = (\mathsf{T}^{-1})^\top$ and $|\mathsf{A}| = 1/|\mathsf{T}|$. Thus the above equation is written as

$$((\mathsf{T}^{-1})^\top\mathbf{a})\times = \frac{1}{|\mathsf{T}|}\mathsf{T}(\mathbf{a}\times)\mathsf{T}^\top.$$

Multiplying this by $|\mathsf{T}|$ on both sides, and multiplying this by $(\mathsf{T}^{-1})^\top$ $(= (\mathsf{T}^\top)^{-1})$ from the right, we obtain Eq. (5.26).

(2) If we note that $(\mathbf{t} \times \mathsf{R})^\top = \mathsf{R}^\top(\mathbf{t}\times)^\top = -\mathsf{R}^\top(\mathbf{t}\times)$, we obtain

$$(\mathbf{t} \times \mathsf{R})^\top\mathbf{t} = -\mathsf{R}^\top(\mathbf{t} \times \mathbf{t}) = \mathbf{0}.$$

We also have

$$(\mathbf{t}\times \mathsf{R})\mathsf{R}^\top \mathbf{t} = (\mathbf{t}\times)\mathsf{R}\mathsf{R}^\top \mathbf{t} = \mathbf{t}\times\mathbf{t} = \mathbf{0}.$$

Thus if the vectors $\mathbf{e}$ and $\mathbf{e}'$ are defined as in Eq. (5.27), the relations $\mathsf{F}^\top\mathbf{e} = \mathbf{0}$ and $\mathsf{F}\mathbf{e}' = \mathbf{0}$ hold.

(3) Using Eq. (2.26), we can rewrite (5.12) in terms of the vector $\mathbf{e}$ of Eq. (5.12) in the form:

$$\begin{aligned}
\mathsf{F} &= \begin{pmatrix} f_0 & 0 & 0 \\ 0 & f_0 & 0 \\ 0 & 0 & f \end{pmatrix} (\mathbf{t}\times)\mathsf{R} \begin{pmatrix} f_0 & 0 & 0 \\ 0 & f_0 & 0 \\ 0 & 0 & f' \end{pmatrix} \\
&\simeq (\begin{pmatrix} 1/f_0 & 0 & 0 \\ 0 & 1/f_0 & 0 \\ 0 & 0 & 1/f \end{pmatrix} \mathbf{t}) \times \begin{pmatrix} 1/f_0 & 0 & 0 \\ 0 & 1/f_0 & 0 \\ 0 & 0 & 1/f \end{pmatrix} \mathsf{R} \begin{pmatrix} f_0 & 0 & 0 \\ 0 & f_0 & 0 \\ 0 & 0 & f' \end{pmatrix} \\
&\simeq \mathbf{e} \times \begin{pmatrix} 1/f_0 & 0 & 0 \\ 0 & 1/f_0 & 0 \\ 0 & 0 & 1/f \end{pmatrix} \mathsf{R} \begin{pmatrix} f_0 & 0 & 0 \\ 0 & f_0 & 0 \\ 0 & 0 & f' \end{pmatrix}.
\end{aligned}$$

Noting that $(\mathbf{t}\times\mathsf{R})^\top = \mathsf{R}^\top(\mathbf{t}\times)^\top = -\mathsf{R}^\top(\mathbf{t}\times)$, we can express $\mathsf{F}^\top$ in terms of the vector $\mathbf{e}'$ of (5.27) in the form:

$$\begin{aligned}
\mathsf{F}^\top &= -\begin{pmatrix} f_0 & 0 & 0 \\ 0 & f_0 & 0 \\ 0 & 0 & f' \end{pmatrix} \mathsf{R}^\top(\mathbf{t}\times) \begin{pmatrix} f_0 & 0 & 0 \\ 0 & f_0 & 0 \\ 0 & 0 & f \end{pmatrix} \\
&\simeq (\begin{pmatrix} 1/f_0 & 0 & 0 \\ 0 & 1/f_0 & 0 \\ 0 & 0 & 1/f' \end{pmatrix} \mathsf{R}^\top\mathbf{t}) \times \begin{pmatrix} 1/f_0 & 0 & 0 \\ 0 & 1/f_0 & 0 \\ 0 & 0 & 1/f' \end{pmatrix} \mathsf{R}^\top \begin{pmatrix} f_0 & 0 & 0 \\ 0 & f_0 & 0 \\ 0 & 0 & f \end{pmatrix} \\
&\simeq \mathbf{e}' \times \begin{pmatrix} 1/f_0 & 0 & 0 \\ 0 & 1/f_0 & 0 \\ 0 & 0 & 1/f' \end{pmatrix} \mathsf{R}^\top \begin{pmatrix} f_0 & 0 & 0 \\ 0 & f_0 & 0 \\ 0 & 0 & f \end{pmatrix}.
\end{aligned}$$

(4) From Eq. (5.28), we obtain

$$\mathsf{F}\begin{pmatrix} 1/f_0 & 0 & 0 \\ 0 & 1/f_0 & 0 \\ 0 & 0 & 1/f' \end{pmatrix} \simeq \mathbf{e} \times \begin{pmatrix} 1/f_0 & 0 & 0 \\ 0 & 1/f_0 & 0 \\ 0 & 0 & 1/f \end{pmatrix} \mathsf{R}.$$

Multiplying each side by its transpose from the right, we obtain

$$\begin{aligned}
&\mathsf{F}\begin{pmatrix} 1/f_0 & 0 & 0 \\ 0 & 1/f_0 & 0 \\ 0 & 0 & 1/f' \end{pmatrix}\begin{pmatrix} 1/f_0 & 0 & 0 \\ 0 & 1/f_0 & 0 \\ 0 & 0 & 1/f' \end{pmatrix}\mathsf{F}^\top \\
\simeq\, &\mathbf{e} \times \begin{pmatrix} 1/f_0 & 0 & 0 \\ 0 & 1/f_0 & 0 \\ 0 & 0 & 1/f \end{pmatrix} \mathsf{R}\mathsf{R}^\top \begin{pmatrix} 1/f_0 & 0 & 0 \\ 0 & 1/f_0 & 0 \\ 0 & 0 & 1/f \end{pmatrix} (\mathbf{e}\times)^\top,
\end{aligned}$$

which is rewritten in the form of Eq. (5.30). Similarly, we obtain from Eq. (5.29)

$$\mathsf{F}\begin{pmatrix} 1/f_0 & 0 & 0 \\ 0 & 1/f_0 & 0 \\ 0 & 0 & 1/f \end{pmatrix} \simeq \mathbf{e}' \times \begin{pmatrix} 1/f_0 & 0 & 0 \\ 0 & 1/f_0 & 0 \\ 0 & 0 & 1/f' \end{pmatrix} \mathsf{R}^\top.$$

Multiplying each side by its transpose from the right, we obtain

$$\mathsf{F}^\top \begin{pmatrix} 1/f_0 & 0 & 0 \\ 0 & 1/f_0 & 0 \\ 0 & 0 & 1/f \end{pmatrix} \begin{pmatrix} 1/f_0 & 0 & 0 \\ 0 & 1/f_0 & 0 \\ 0 & 0 & 1/f \end{pmatrix} \mathsf{F}$$
$$\simeq \mathbf{e}' \times \begin{pmatrix} 1/f_0 & 0 & 0 \\ 0 & 1/f_0 & 0 \\ 0 & 0 & 1/f' \end{pmatrix} \mathsf{R}^\top \mathsf{R} \begin{pmatrix} 1/f_0 & 0 & 0 \\ 0 & 1/f_0 & 0 \\ 0 & 0 & 1/f' \end{pmatrix} (\mathbf{e}'\times)^\top,$$

which is rewritten in the form of Eq. (5.31).

(5) If we define ξ and η by Eq. (5.32) and use the vector $\mathbf{k} = (0, 0, 1)^\top$, we can write the Kruppa equations of Eqs. (5.30) and (5.31) in the form:

$$\mathsf{F}(\mathsf{I} + \eta \mathbf{k}\mathbf{k}^\top)\mathsf{F}^\top \simeq \mathbf{e} \times (\mathsf{I} + \xi \mathbf{k}\mathbf{k}^\top) \times \mathbf{e},$$
$$\mathsf{F}^\top(\mathsf{I} + \xi \mathbf{k}\mathbf{k}^\top)\mathsf{F} \simeq \mathbf{e}' \times (\mathsf{I} + \eta \mathbf{k}\mathbf{k}^\top) \times \mathbf{e}'.$$

By elementwise comparison, we can confirm the identity

$$\mathbf{e} \times \mathsf{I} \times \mathbf{e} = -(\mathbf{e}\times)^2 (\mathbf{e}, \mathbf{e})\mathsf{I} - \mathbf{e}\mathbf{e}^\top = \mathsf{P}_{\mathbf{e}}.$$

Noting that $\mathbf{e} \times \mathbf{k}\mathbf{k}^\top \times \mathbf{e} = (\mathbf{e} \times \mathbf{k})(\mathbf{e} \times \mathbf{k})^\top$, we can rewrite the above first Kruppa equation in the form

$$\mathsf{F}\mathsf{F}^\top + \eta(\mathsf{F}\mathbf{k})(\mathsf{F}\mathbf{k})^\top \simeq \mathsf{P}_{\mathbf{e}} + \xi(\mathbf{e} \times \mathbf{k})(\mathbf{e} \times \mathbf{k})^\top.$$

Similarly, the second Kruppa equation is written in the form

$$\mathsf{F}^\top\mathsf{F} + \xi(\mathsf{F}^\top\mathbf{k})(\mathsf{F}^\top\mathbf{k})^\top \simeq \mathsf{P}_{\mathbf{e}'} + \eta(\mathbf{e}' \times \mathbf{k})(\mathbf{e}' \times \mathbf{k})^\top.$$

Multiplying these by $\mathbf{k}$ from the right on both sides, and noting $(\mathbf{e}\times\mathbf{k})^\top\mathbf{k} = (\mathbf{e}\times\mathbf{k}, \mathbf{k}) = 0$, we obtain Eqs. (5.33) and (5.34).

(6) Computing the inner product of Eq. (5.33) and $\mathbf{k}$ on both sides, we obtain

$$(\mathbf{k}, \mathsf{F}\mathsf{F}^\top\mathbf{k}) + \eta(\mathbf{k}, \mathsf{F}\mathbf{k})(\mathbf{k}, \mathsf{F}\mathbf{k}) = c(\mathbf{k}, \mathsf{P}_{\mathbf{e}}\mathbf{k}).$$

Computing the inner product of Eq. (5.33) and $\mathsf{F}\mathbf{k}$ on both sides, we obtain

$$(\mathsf{F}\mathbf{k}, \mathsf{F}\mathsf{F}^\top\mathbf{k}) + \eta(\mathbf{k}, \mathsf{F}\mathbf{k})(\mathsf{F}\mathbf{k}, \mathsf{F}\mathbf{k}) = c(\mathsf{F}\mathbf{k}, \mathsf{P}_{\mathbf{e}}\mathbf{k}).$$

Note the identities $(\mathbf{k}, \mathsf{F}\mathsf{F}^\top\mathbf{k}) = (\mathsf{F}^\top\mathbf{k}, \mathsf{F}^\top\mathbf{k}) = \|\mathsf{F}^\top\mathbf{k}\|^2$ and $(\mathsf{F}\mathbf{k}, \mathsf{F}\mathsf{F}^\top\mathbf{k}) = (\mathbf{k}, \mathsf{F}^\top\mathsf{F}\,\mathsf{F}^\top\mathbf{k})$. Also note the identities

$$(\mathbf{k}, \mathsf{P}_{\mathbf{e}}\mathbf{k}) = (\mathbf{k}, \mathbf{k} - (\mathbf{k}, \mathbf{e})\mathbf{e}) = (\mathbf{k}, \mathbf{k}) - (\mathbf{k}, \mathbf{e})^2 = 1 - \cos^2\theta = \sin^2\theta = \|\mathbf{e} \times \mathbf{k}\|^2,$$
$$(\mathsf{F}\mathbf{k}, \mathsf{P}_{\mathbf{e}}\mathbf{k}) = (\mathsf{F}\mathbf{k}, \mathbf{k} - (\mathbf{k}, \mathbf{e})\mathbf{e}) = (\mathsf{F}\mathbf{k}, \mathbf{k}) - (\mathbf{k}, \mathbf{e})(\mathsf{F}\mathbf{k}, \mathbf{e}) = (\mathbf{k}, \mathsf{F}\mathbf{k}) - (\mathbf{k}, \mathbf{e})(\mathbf{k}, \mathsf{F}^\top\mathbf{e})$$
$$= (\mathbf{k}, \mathsf{F}\mathbf{k}),$$

where θ is the angle made by the unit vectors $\mathbf{k}$ and $\mathbf{e}$. By definition, $\mathbf{e}$ is a unit eigenvector of $\mathsf{F}^\top$ for eigenvalue 0. Using these, we can rewrite the first two equalities above in the form:

$$\|\mathsf{F}^\top\mathbf{k}\|^2 + \eta(\mathbf{k}, \mathsf{F}\mathbf{k})^2 = c\|\mathbf{e} \times \mathbf{k}\|^2,$$
$$(\mathbf{k}, \mathsf{F}^\top\mathsf{F}\mathsf{F}^\top\mathbf{k}) + \eta(\mathbf{k}, \mathsf{F}\mathbf{k})\|\mathsf{F}\mathbf{k}\|^2 = c(\mathbf{k}, \mathsf{F}^\top\mathbf{k}).$$

Regarding these as simultaneous linear equations in η and c and solving them, we can write η in the form of the second equation of Eq. (5.13). Similarly, the inner products of Eq. (5.34) with $\mathbf{k}$ and $\mathsf{F}^\top\mathbf{k}$ on both sides give the following two equations.

$$\|\mathsf{F}\mathbf{k}\|^2 + (\mathbf{k}, \mathsf{F}\mathbf{k})^2\xi = c'\|\mathbf{e}' \times \mathbf{k}\|^2,$$
$$(\mathbf{k}, \mathsf{F}\mathsf{F}^\top\mathsf{F}\mathbf{k}) + (\mathbf{k}, \mathsf{F}\mathbf{k})\|\mathsf{F}^\top\mathbf{k}\|^2\xi = c'(\mathbf{k}, \mathsf{F}\mathbf{k}).$$

Regarding these as simultaneous linear equations in ξ and c' and solving them, we can write ξ in the form of the first equation of Eq. (5.13). Once ξ and η are determined, we can express f and f' in the form of Eq. (5.14), which is obtained by rewriting Eq. (5.32).

5.2 We can see that for any vector $\mathbf{x}$ the quadratic forms $(\mathbf{x}, \mathsf{A}^\top\mathsf{A}\mathbf{x})$ and $(\mathbf{x}, \mathsf{A}\mathsf{A}^\top\mathbf{x})$ are nonnegative as follows.

$$(\mathbf{x}, \mathsf{A}^\top\mathsf{A}\mathbf{x}) = (\mathsf{A}\mathbf{x}, \mathsf{A}\mathbf{x}) = \|\mathsf{A}\mathbf{x}\|^2 \geq 0, \quad (\mathbf{x}, \mathsf{A}\mathsf{A}^\top\mathbf{x}) = (\mathsf{A}^\top\mathbf{x}, \mathsf{A}^\top\mathbf{x}) = \|\mathsf{A}^\top\mathbf{x}\|^2 \geq 0.$$

Namely, $\mathsf{A}^\top\mathsf{A}$ and $\mathsf{A}\mathsf{A}^\top$ are both positive semi-definite. Thus their eigenvalues are all nonnegative.

5.3 By definition, $\mathbf{x}_\alpha$ indicates the direction of point P_α viewed from the first camera and is normalized such that its Z component is 1. Thus if Z_α is the depth (i.e., distance from the XY plane) of the point P_α, the 3D position of P_α is $Z_\alpha\mathbf{x}_\alpha$ (Fig. 5.2). Similarly, $\mathbf{x}'_\alpha$ indicates the direction of point P_α viewed from the second camera and is normalized such that its Z component is 1. Thus if Z'_α is the depth with respect to the second camera (i.e., distance from the $X'_cY'_c$ plane) of the point P_α, the 3D position of P_α with respect to the $X'_cY'_cZ'_c$ coordinate system is $Z'_\alpha\mathbf{x}'_\alpha$. Inasmuch as the $X'_cY'_cZ'_c$ coordinate system is translated by $\boldsymbol{\theta}$ and rotated by R relative to the XYZ coordinate system, the 3D position of P_α with respect to the XYZ coordinate system is $\mathbf{t} + Z'_\alpha\mathsf{R}\mathbf{x}'_\alpha$. Therefore the following relationship holds (Fig. 5.2).

$$Z_\alpha\mathbf{x}_\alpha = \mathbf{t} + Z'_\alpha\mathsf{R}\mathbf{x}'_\alpha.$$

The vector product of this and $\mathbf{t}$ on both sides gives

$$Z_\alpha\mathbf{t} \times \mathbf{x}_\alpha = Z'_\alpha\mathbf{t} \times \mathsf{R}\mathbf{x}'_\alpha = Z'_\alpha\mathsf{E}\mathbf{x}'_\alpha.$$

The inner product of this and $\mathbf{t} \times \mathbf{x}_\alpha$ on both sides gives

$$Z_\alpha\|\mathbf{t} \times \mathbf{x}_\alpha\|^2 = Z'_\alpha(\mathbf{t} \times \mathbf{x}_\alpha, \mathsf{E}\mathbf{x}'_\alpha) = Z'_\alpha|\mathbf{t}, \mathbf{x}_\alpha, \mathsf{E}\mathbf{x}'_\alpha|.$$

Thus we obtain

$$|\mathbf{t}, \mathbf{x}_\alpha, \mathsf{E}\mathbf{x}'_\alpha| = \frac{Z_\alpha}{Z'_\alpha}\|\mathbf{t} \times \mathbf{x}_\alpha\|^2,$$

indicating that Z_α and Z'_α have the same sign if and only if $|\mathbf{t}, \mathbf{x}_\alpha, \mathsf{E}\mathbf{x}'_\alpha| > 0$.

5.4 (1) Using the identity $\mathrm{tr}[\mathsf{AB}] = \mathrm{tr}[\mathsf{BA}]$ $(= \sum_{i,j=1}^{3} A_{ij}B_{ji})$ for any matrices A and B, we can write Eq. (5.23) as

$$\begin{aligned}
&\|c\mathsf{E} - \mathbf{t} \times \mathsf{R}\|^2 = \mathrm{tr}[(c\mathsf{E} - \mathbf{t} \times \mathsf{R})^\top (c\mathsf{E} - \mathbf{t} \times \mathsf{R})] \\
&= c^2 \mathrm{tr}[\mathsf{E}^\top \mathsf{E}] - 2c\mathrm{tr}[\mathsf{E}^\top (\mathbf{t} \times \mathsf{R})] + \mathrm{tr}[(\mathbf{t} \times \mathsf{R})^\top (\mathbf{t} \times \mathsf{R})] \\
&= c^2 \|\mathsf{E}\|^2 - 2c\mathrm{tr}[((\mathbf{t}\times)^\top \mathsf{E})^\top \mathsf{R}] + \mathrm{tr}[(\mathbf{t} \times \mathsf{R})(\mathbf{t} \times \mathsf{R})^\top] \\
&= c^2 \|\mathsf{E}\|^2 - 2c\mathrm{tr}[(-\mathbf{t} \times \mathsf{E})^\top \mathsf{R}] + \mathrm{tr}[(\mathbf{t}\times)\mathsf{R}\mathsf{R}^\top (\mathbf{t}\times)^\top] \\
&= c^2 \|\mathsf{E}\|^2 - 2c\mathrm{tr}[\mathsf{K}^\top \mathsf{R}] + \|\mathbf{t} \times\|^2.
\end{aligned}$$

Because the first and the third terms do not depend on R (actually, the third term is $2\|\mathbf{t}\|^2 = 2$), the above expression is minimized by the rotation R that maximizes $\mathrm{tr}[\mathsf{K}^\top \mathsf{R}]$ (recall that we are assuming $c > 0$).
(2) Because $\mathrm{tr}[\mathsf{T}\boldsymbol{\Lambda}] = T_{11}\sigma_1 + T_{22}\sigma_2$ and T is an orthogonal matrix consisting of mutually orthogonal rows and columns of the unit norm, we see that $|T_{11}| \leq 1$ and $|T_{22}| \leq 1$. Therefore $\mathrm{tr}[\mathsf{T}\boldsymbol{\Lambda}]$ takes its maximum when $T_{11} = T_{22} = 1$. If two diagonal elements of T are 1, the other elements of the rows and columns that contain them are 0. It follows that the third row and the third column also consist of 0 other than T_{33}, therefore $T_{33} = \pm 1$. If $T_{33} = 1$, then $\mathsf{T} = \mathsf{I}$, and if $T_{33} = -1$, then $\mathsf{T} = \mathrm{diag}(1, 1, -1)$.
(3) From the identity $\mathrm{tr}[\mathsf{AB}] = \mathrm{tr}[\mathsf{BA}]$, we can rewrite $\mathrm{tr}[\mathsf{K}^\top \mathsf{R}]$ as

$$\mathrm{tr}[\mathsf{K}^\top \mathsf{R}] = \mathrm{tr}[\mathsf{V}\boldsymbol{\Lambda}\mathsf{U}^\top \mathsf{R}] = \mathrm{tr}[\mathsf{U}^\top \mathsf{R}\mathsf{V}\boldsymbol{\Lambda}].$$

$\mathsf{U}^\top \mathsf{R}\mathsf{V}$ is an orthogonal matrix, therefore this expression takes its maximum when $\mathsf{U}^\top \mathsf{R}\mathsf{V} = \mathsf{I}$ or $\mathsf{U}^\top \mathsf{R}\mathsf{V} = \mathrm{diag}(1, 1, -1)$, which means $\mathsf{R} = \mathsf{U}\mathsf{V}^\top$ or $\mathsf{R} = \mathsf{U}\mathrm{diag}(1, 1, -1)$ $\mathsf{V}^\top$. Although U and V in Eq. (5.19) are orthogonal matrices, they may not necessarily be a rotation matrix with determinant 1. If $\det(\mathsf{U}\mathsf{V}^\top)$ $(= \det \mathsf{U} \det \mathsf{V})$ is 1, then $\mathsf{U}\mathsf{V}^\top$ is a rotation matrix, and if $\det(\mathsf{U}\mathsf{V}^\top)$ is -1, then $\mathsf{U}\mathrm{diag}(1, 1, -1)\mathsf{V}^\top$ is a rotation matrix. These two cases are combined in the form of Eq. (5.20).

Problems of Chapter 6

6.1 The procedure is as follows.

1. Compute the 9×9 matrix

$$\mathsf{M} = \frac{1}{N} \sum_{\alpha=1}^{N} \sum_{k=1}^{3} \boldsymbol{\xi}_\alpha^{(k)} \boldsymbol{\xi}_\alpha^{(k)\top}.$$

2. Solve the eigenvalue problem $\mathsf{M}\boldsymbol{\theta} = \lambda \boldsymbol{\theta}$, and return the unit eigenvector $\boldsymbol{\theta}$ for the smallest eigenvalue λ.

6.2 The procedure is as follows.

1. Compute the following 9×9 matrices

$$\mathsf{M} = \frac{1}{N} \sum_{\alpha=1}^{N} \sum_{k=1}^{3} \boldsymbol{\xi}_\alpha^{(k)} \boldsymbol{\xi}_\alpha^{(k)\top}, \qquad \mathsf{N} = \frac{1}{N} \sum_{\alpha=1}^{N} \sum_{k=1}^{3} V_0^{(kk)}[\boldsymbol{\xi}_\alpha].$$

2. Solve the generalized eigenvalue problem $\mathsf{M}\boldsymbol{\theta} = \lambda\mathsf{N}\boldsymbol{\theta}$, and return the unit generalized eigenvector $\boldsymbol{\theta}$ for the smallest generalized eigenvalue λ.

6.3 The procedure is as follows.

1. Compute the 9×9 matrices

$$\mathsf{M} = \frac{1}{N}\sum_{\alpha=1}^{N}\sum_{k=1}^{3}\boldsymbol{\xi}_\alpha^{(k)}\boldsymbol{\xi}_\alpha^{(k)\top},$$

$$\mathsf{N} = \frac{1}{N}\sum_{\alpha=1}^{N}\sum_{k=1}^{3}V_0^{(kk)}[\boldsymbol{\xi}_\alpha]$$
$$-\frac{1}{N^2}\sum_{\alpha=1}^{N}\sum_{k,l=1}^{3}\Big((\boldsymbol{\xi}_\alpha^{(k)}, \mathsf{M}_8^-\boldsymbol{\xi}_\alpha^{(l)})V_0^{(kl)}[\boldsymbol{\xi}_\alpha] + 2\mathscr{S}[V_0^{(kl)}[\boldsymbol{\xi}_\alpha]\mathsf{M}_8^-\boldsymbol{\xi}_\alpha^{(k)}\boldsymbol{\xi}_\alpha^{(l)\top}]\Big).$$

2. Solve the generalized eigenvalue problem $\mathsf{M}\boldsymbol{\theta} = \lambda\mathsf{N}\boldsymbol{\theta}$, and return the unit generalized unit eigenvector $\boldsymbol{\theta}$ for the generalized eigenvalue λ of the smallest absolute value.

6.4 If we define vectors $\mathbf{p}_\alpha$ and $\bar{\mathbf{p}}_\alpha$ by

$$\mathbf{p} = \begin{pmatrix} x_\alpha \\ y_\alpha \\ x'_\alpha \\ y'_\alpha \end{pmatrix}, \qquad \bar{\mathbf{p}} = \begin{pmatrix} \bar{x}_\alpha \\ \bar{y}_\alpha \\ \bar{x}'_\alpha \\ \bar{y}'_\alpha \end{pmatrix},$$

and let $\Delta\mathbf{p}_\alpha = \mathbf{p}_\alpha - \bar{\mathbf{p}}_\alpha$, Eq. (6.29) is written as $\|\Delta\mathbf{p}\|^2$. Write the value of $\boldsymbol{\xi}_\alpha^{(k)}$ for $\mathbf{p}_\alpha$ as a function $\boldsymbol{\xi}^{(k)}(\mathbf{p}_\alpha)$. Then, $\bar{\boldsymbol{\xi}}_\alpha^{(k)} = \boldsymbol{\xi}^{(k)}(\bar{\mathbf{p}}_\alpha) = \boldsymbol{\xi}^{(k)}(\mathbf{p}_\alpha - \Delta\mathbf{p}_\alpha)$. The definition of $\mathsf{T}_\alpha^{(k)}$ of Eq. (6.10) implies that if Δx_α, Δy_α, $\Delta x'_\alpha$, and $\Delta y'_\alpha$ are small (i.e., if $\Delta\mathbf{p}_\alpha$ is small), we can write

$$\bar{\boldsymbol{\xi}}_\alpha^{(k)} = \boldsymbol{\xi}^{(k)}(\mathbf{p}_\alpha - \Delta\mathbf{p}_\alpha) = \boldsymbol{\xi}^{(k)}(\mathbf{p}_\alpha) - \mathsf{T}_\alpha^{(k)}\Delta\mathbf{p}_\alpha + \cdots,$$

where $\cdots$ means higher-order terms in $\Delta\mathbf{p}$. Thus if we ignore higher-order terms in $\Delta\mathbf{p}$, we can write the constraint $(\bar{\boldsymbol{\xi}}_\alpha^{(k)}, \boldsymbol{\theta}) = 0$ as

$$(\mathsf{T}_\alpha^{(k)}\Delta\mathbf{p}_\alpha, \boldsymbol{\theta}) = (\boldsymbol{\xi}_\alpha^{(k)}, \boldsymbol{\theta}). \qquad (*)$$

To minimize $\|\Delta\mathbf{p}_\alpha\|^2$, we introduce Lagrange multipliers and differentiate

$$\|\Delta\mathbf{p}_\alpha\|^2 - \sum_{k=1}^{3}\lambda_\alpha^{(k)}\Big((\mathsf{T}_\alpha^{(k)}\Delta\mathbf{p}_\alpha, \boldsymbol{\theta}) - (\boldsymbol{\xi}_\alpha^{(k)}, \boldsymbol{\theta})\Big)$$

with respect to $\Delta\mathbf{p}_\alpha$. Letting the result be $\mathbf{0}$, we obtain

$$2\Delta\mathbf{p}_\alpha - \sum_{k=1}^{3}\lambda_\alpha^{(k)}\mathsf{T}_\alpha^{(k)\top}\boldsymbol{\theta} = \mathbf{0}.$$

Therefore

$$\Delta \mathbf{p}_\alpha = \frac{1}{2}\sum_{k=1}^{3} \lambda_\alpha^{(k)} \mathsf{T}_\alpha^{(k)\top}\boldsymbol{\theta}.$$

Substitution of this into Eq. (∗) yields

$$\frac{1}{2}(\mathsf{T}_\alpha^{(k)}\sum_{l=1}^{3}\lambda_\alpha^{(l)}\mathsf{T}_\alpha^{(l)\top}\boldsymbol{\theta}, \boldsymbol{\theta}) = (\boldsymbol{\xi}_\alpha^{(k)}, \boldsymbol{\theta}).$$

From Eq. (6.15), this is rewritten in the form

$$\frac{1}{2}\sum_{l=1}^{3}(\boldsymbol{\theta}, V_0^{(kl)}[\boldsymbol{\xi}_\alpha]\boldsymbol{\theta})\lambda_\alpha^{(l)} = (\boldsymbol{\xi}_\alpha^{(k)}, \boldsymbol{\theta}).$$

This defines simultaneous linear equations in $\lambda_\alpha^{(1)}$, $\lambda_\alpha^{(2)}$, and $\lambda_\alpha^{(3)}$. Considering that the coefficient matrix $\Big((\boldsymbol{\theta}, V_0^{(kl)}[\hat{\boldsymbol{\xi}}_\alpha]\boldsymbol{\theta})\Big)$ has determinant 0, we solve them by least squares. The solution is written using the pseudoinverse $\hat{W}_\alpha^{(kl)} = \Big((\boldsymbol{\theta}, V_0^{(kl)}[\hat{\boldsymbol{\xi}}_\alpha]\boldsymbol{\theta})\Big)_2^-$ of rank 2 in the form

$$\frac{\lambda_\alpha^{(k)}}{2} = \sum_{l=1}^{3} W_\alpha^{(kl)}(\boldsymbol{\xi}_\alpha^{(l)}, \boldsymbol{\theta}).$$

Thus $\Delta \mathbf{p}_\alpha$ is given by

$$\Delta \mathbf{p}_\alpha = \sum_{k,l=1}^{3} W_\alpha^{(kl)}(\boldsymbol{\xi}_\alpha^{(l)}, \boldsymbol{\theta})\mathsf{T}_\alpha^{(k)\top}\boldsymbol{\theta},$$

and $\|\Delta \mathbf{p}_\alpha\|^2$ is written as

$$\begin{aligned}
\|\Delta \mathbf{p}_\alpha\|^2 &= (\sum_{k,l=1}^{3} W_\alpha^{(kl)}(\boldsymbol{\xi}_\alpha^{(l)}, \boldsymbol{\theta})\mathsf{T}_\alpha^{(k)\top}\boldsymbol{\theta}, \sum_{m,n=1}^{3} W_\alpha^{(mn)}(\boldsymbol{\xi}_\alpha^{(n)}, \boldsymbol{\theta})\mathsf{T}_\alpha^{(m)\top}\boldsymbol{\theta}) \\
&= \sum_{k,l,m,n=1}^{3} W_\alpha^{(kl)}W_\alpha^{(mn)}(\boldsymbol{\xi}_\alpha^{(l)}, \boldsymbol{\theta})(\boldsymbol{\xi}_\alpha^{(n)}, \boldsymbol{\theta})(\mathsf{T}_\alpha^{(k)\top}\boldsymbol{\theta}, \mathsf{T}_\alpha^{(m)\top}\boldsymbol{\theta}) \\
&= \sum_{k,l,m,n=1}^{3} W_\alpha^{(kl)}W_\alpha^{(mn)}(\boldsymbol{\xi}_\alpha^{(l)}, \boldsymbol{\theta})(\boldsymbol{\xi}_\alpha^{(n)}, \boldsymbol{\theta})(\boldsymbol{\theta}, \mathsf{T}_\alpha^{(k)}\mathsf{T}_\alpha^{(m)\top}\boldsymbol{\theta}) \\
&= \sum_{k,l,m,n=1}^{3} W_\alpha^{(kl)}W_\alpha^{(mn)}(\boldsymbol{\xi}_\alpha^{(l)}, \boldsymbol{\theta})(\boldsymbol{\xi}_\alpha^{(n)}, \boldsymbol{\theta})(\boldsymbol{\theta}, V_0^{(km)}[\boldsymbol{\xi}_\alpha]\boldsymbol{\theta}) \\
&= \sum_{k,l=1}^{3} W_\alpha^{(kl)}(\boldsymbol{\xi}_\alpha^{(k)}, \boldsymbol{\theta})(\boldsymbol{\xi}_\alpha^{(l)}, \boldsymbol{\theta}),
\end{aligned}$$

where we have noted that if we let $V_\alpha^{(kl)} = (\boldsymbol{\theta}, V_0^{(kl)}[\boldsymbol{\xi}_\alpha]\boldsymbol{\theta})$, the following pseudoinverse identity holds ($\hookrightarrow$ Problem 14.3).

$$\sum_{l,m=1}^{3} W_\alpha^{(kl)} V_\alpha^{(lm)} W_\alpha^{(mn)} = W_\alpha^{(km)}.$$

6.5 (1) Let W_α be the matrix whose (k,l) element is $W_\alpha^{(kl)}$, and V_α the matrix whose (k,l) element is $V_\alpha^{(kl)} = (\boldsymbol{\theta}, V_0^{(kl)}[\boldsymbol{\xi}_\alpha]\boldsymbol{\theta})$. Because $\mathsf{W}_\alpha = (\mathsf{V}_\alpha)_2^-$, the identity

$$\mathsf{V}_\alpha \mathsf{W}_\alpha = \mathsf{W}_\alpha \mathsf{V}_\alpha = \mathsf{P}_{\mathcal{N}}$$

holds, where $\mathsf{P}_{\mathcal{N}}$ is the projection matrix onto the domain of W_α (the space orthogonal to the eigenvector for eigenvalue 0). The fact that V_α is singular is due to the linear dependence of $\boldsymbol{\xi}_\alpha^{(1)}$, $\boldsymbol{\xi}_\alpha^{(2)}$, and $\boldsymbol{\xi}_\alpha^{(3)}$ irrespective of $\boldsymbol{\theta}$. Therefore $\mathsf{P}_{\mathcal{N}}$ does not depend on $\boldsymbol{\theta}$. Differentiating the above relation with respect to θ_i, we obtain

$$\frac{\partial \mathsf{V}_\alpha}{\partial \theta_i}\mathsf{W}_\alpha + \mathsf{V}_\alpha \frac{\partial \mathsf{W}_\alpha}{\partial \theta_i} = \mathsf{O}.$$

Multiplying this by W_α from the left on both sides, and noting that $\partial \mathsf{W}_\alpha/\partial\theta_i$ and W have the same domain and that $\mathsf{W}_\alpha \mathsf{V}_\alpha = \mathsf{P}_{\mathcal{N}}$ and $\mathsf{P}_{\mathcal{N}} \partial \mathsf{V}_\alpha/\partial\theta_i = \partial \mathsf{W}_\alpha/\partial\theta_i$ hold, we obtain

$$\frac{\partial \mathsf{W}_\alpha}{\partial \theta_i} = -\mathsf{W}_\alpha \frac{\partial \mathsf{V}_\alpha}{\partial \theta_i} \mathsf{W}_\alpha.$$

Equating the (k,l) element on both sides, we can write

$$\frac{\partial W_\alpha^{(kl)}}{\partial \theta_i} = -\sum_{m,n=1}^{3} W_\alpha^{(km)} \frac{\partial V_\alpha^{(mn)}}{\partial \theta_i} W_\alpha^{(nl)}.$$

Note that the derivative of $V_\alpha^{(mn)} = (\boldsymbol{\theta}, V_0^{(mn)}[\boldsymbol{\xi}_\alpha]\boldsymbol{\theta})$ with respect to $\boldsymbol{\theta}$ is $2V_0^{(mn)}[\boldsymbol{\xi}_\alpha]\boldsymbol{\theta}$. Using the nabla operator $\nabla_{\boldsymbol{\theta}}$ for $\partial/\partial\theta_i$, we can write the result in the form of Eq. (6.55).
(2) From Eq. (6.31), the gradient of J is written as

$$\begin{aligned}
\nabla_{\boldsymbol{\theta}} J =& \frac{1}{N}\sum_{\alpha=1}^{N}\sum_{k,l=1}^{3}\Big(2W_\alpha^{(kl)}(\boldsymbol{\xi}_\alpha^{(l)}, \boldsymbol{\theta})\boldsymbol{\xi}_\alpha^{(k)} + \nabla_{\boldsymbol{\theta}} W_\alpha^{(kl)}(\boldsymbol{\xi}_\alpha^{(k)}, \boldsymbol{\theta})(\boldsymbol{\xi}_\alpha^{(l)}, \boldsymbol{\theta})\Big) \\
=& \frac{1}{N}\sum_{\alpha=1}^{N}\sum_{k,l=1}^{3}\Big(2W_\alpha^{(kl)}\boldsymbol{\xi}_\alpha^{(k)}\boldsymbol{\xi}_\alpha^{(l)\top}\boldsymbol{\theta} \\
&- 2\sum_{m,n=1}^{3} W_\alpha^{(km)} W_\alpha^{(ln)} V_0^{(mn)}[\boldsymbol{\xi}_\alpha]\boldsymbol{\theta}(\boldsymbol{\xi}_\alpha^{(k)}, \boldsymbol{\theta})(\boldsymbol{\xi}_\alpha^{(l)}, \boldsymbol{\theta})\Big) \\
=& \frac{2}{N}\sum_{\alpha=1}^{N}\sum_{k,l=1}^{3} W_\alpha^{(kl)}\boldsymbol{\xi}_\alpha^{(k)}\boldsymbol{\xi}_\alpha^{(l)\top}\boldsymbol{\theta}
\end{aligned}$$

$$-\frac{2}{N}\sum_{\alpha=1}^{N}\sum_{k,l,m,n=1}^{3} W_\alpha^{(km)}W_\alpha^{(ln)}V_0^{(mn)}[\boldsymbol{\xi}_\alpha](\boldsymbol{\xi}_\alpha^{(k)},\boldsymbol{\theta})(\boldsymbol{\xi}_\alpha^{(l)},\boldsymbol{\theta})\boldsymbol{\theta}$$

$$=\frac{2}{N}\sum_{\alpha=1}^{N}\sum_{k,l=1}^{3} W_\alpha^{(kl)}\boldsymbol{\xi}_\alpha^{(k)}\boldsymbol{\xi}_\alpha^{(l)\top}\boldsymbol{\theta}$$

$$-\frac{2}{N}\sum_{\alpha=1}^{N}\sum_{m,n=1}^{3} v_\alpha^{(m)}v_\alpha^{(n)}V_0^{(mn)}[\boldsymbol{\xi}_\alpha]\boldsymbol{\theta} = 2(\mathsf{M}-\mathsf{L})\boldsymbol{\theta} = 2\mathsf{X}\boldsymbol{\theta}.$$

(3) Computing the inner product of (6.35) and $\boldsymbol{\theta}$, we obtain $(\boldsymbol{\theta},\mathsf{X}\boldsymbol{\theta}) = \lambda\|\boldsymbol{\theta}\|^2 = \lambda$. When the iterations have converged, Eq. (6.36) implies $W_\alpha^{(kl)} = \left((\boldsymbol{\theta}, V_0^{(kl)}[\boldsymbol{\theta}_\alpha]\boldsymbol{\theta})\right)_2^-$. Therefore the following holds from Eqs. (6.32) and (6.34).

$$(\boldsymbol{\theta},\mathsf{X}\boldsymbol{\theta}) = \frac{1}{N}\sum_{\alpha=1}^{N}\sum_{k,l=1}^{3} W_\alpha^{(kl)}(\boldsymbol{\xi}_\alpha^{(k)},\boldsymbol{\theta})(\boldsymbol{\xi}_\alpha^{(l)},\boldsymbol{\theta}) - \frac{1}{N}\sum_{\alpha=1}^{N}\sum_{k,l=1}^{3} v_\alpha^{(k)}v_\alpha^{(l)}(\boldsymbol{\theta}, V_0^{(kl)}[\boldsymbol{\xi}_\alpha]\boldsymbol{\theta}).$$

If we let $V_\alpha^{(kl)} = (\boldsymbol{\theta}, V_0^{(kl)}[\boldsymbol{\xi}_\alpha]\boldsymbol{\theta})$, Eq. (6.33) implies that

$$\sum_{k,l=1}^{3} v_\alpha^{(k)}v_\alpha^{(l)}V_\alpha^{(kl)} = \sum_{k,l=1}^{3}\Big(\sum_{m=1}^{3} W_\alpha^{(km)}(\boldsymbol{\xi}_\alpha^{(m)},\boldsymbol{\theta})\Big)\Big(\sum_{n=1}^{3} W_\alpha^{(ln)}(\boldsymbol{\xi}_\alpha^{(n)},\boldsymbol{\theta})\Big)V_\alpha^{(kl)}$$

$$=\sum_{m,n=1}^{3}\Big(\sum_{k,l=1}^{3} W_\alpha^{(mk)}V_\alpha^{(kl)}W_\alpha^{(ln)}\Big)(\boldsymbol{\xi}_\alpha^{(m)},\boldsymbol{\theta})(\boldsymbol{\xi}_\alpha^{(n)},\boldsymbol{\theta}) = \sum_{m,n=1}^{3} W_\alpha^{(mn)}(\boldsymbol{\xi}_\alpha^{(m)},\boldsymbol{\theta})(\boldsymbol{\xi}_\alpha^{(n)},\boldsymbol{\theta}),$$

where we have noted that if we let V_α and W_α be the matrices whose (k, l) element is $V_\alpha^{(kl)}$ and $W_\alpha^{(kl)}$, respectively, then $\mathsf{W}_\alpha = (\mathsf{V}_\alpha)_2^-$ by definition, and the pseudoinverse identity $\mathsf{W}_\alpha\mathsf{V}_\alpha\mathsf{W}_\alpha = \mathsf{W}_\alpha$ holds ($\hookrightarrow$ Problem 14.3). Thus we see that $(\boldsymbol{\theta},\mathsf{X}\boldsymbol{\theta}) = 0$ and therefore $\lambda = 0$.

6.6 If we let $\Delta\hat{\mathbf{p}}_\alpha = \hat{\mathbf{p}}_\alpha - \bar{\mathbf{p}}_\alpha$, we can write Eq. (6.45) as $S = (1/N)\sum_{\alpha=1}^{N}\|\tilde{\mathbf{p}}_\alpha + \Delta\hat{\mathbf{p}}_\alpha\|^2$. As in the solution of Problem 6.4, we write the value $\boldsymbol{\xi}_\alpha^{(k)}$ for $\mathbf{p}_\alpha$ as a function $\boldsymbol{\xi}^{(k)}(\mathbf{p}_\alpha)$. Then, $\bar{\boldsymbol{\xi}}_\alpha^{(k)} = \boldsymbol{\xi}^{(k)}(\hat{\mathbf{p}}_\alpha - \Delta\hat{\mathbf{p}}_\alpha)$. If $\Delta\hat{\mathbf{p}}_\alpha$ is small, we can write

$$\bar{\boldsymbol{\xi}}_\alpha^{(k)} = \boldsymbol{\xi}^{(k)}(\hat{\mathbf{p}}_\alpha - \Delta\hat{\mathbf{p}}_\alpha) = \hat{\boldsymbol{\xi}}_\alpha^{(k)} - \hat{\mathsf{T}}_\alpha^{(k)}\Delta\hat{\mathbf{p}}_\alpha + \cdots,$$

where $\cdots$ indicates higher-order terms in $\Delta\hat{\mathbf{p}}$. Ignoring the higher-order terms, we can write the constraint $(\bar{\boldsymbol{\xi}}_\alpha^{(k)},\boldsymbol{\theta}) = 0$ as

$$(\hat{\mathsf{T}}_\alpha^{(k)}\Delta\hat{\mathbf{p}}_\alpha,\boldsymbol{\theta}) = (\hat{\boldsymbol{\xi}}_\alpha^{(k)},\boldsymbol{\theta}). \qquad (*)$$

Multiplying S by N, and introducing Lagrange multipliers, we differentiate

$$\sum_{\alpha=1}^{N}\|\tilde{\mathbf{p}}_\alpha + \Delta\hat{\mathbf{p}}_\alpha\|^2 - \sum_{k=1}^{3}\lambda_\alpha^{(k)}\Big((\hat{\mathsf{T}}_\alpha^{(k)}\Delta\hat{\mathbf{p}}_\alpha,\boldsymbol{\theta}) - (\hat{\boldsymbol{\xi}}_\alpha^{(k)},\boldsymbol{\theta})\Big)$$

with respect to $\Delta\hat{\mathbf{p}}_\alpha$. Letting the result be $\mathbf{0}$, we obtain

$$2(\tilde{\mathbf{p}}_\alpha + \Delta\hat{\mathbf{p}}_\alpha) - \sum_{k=1}^{3} \lambda_\alpha^{(k)} \hat{\mathsf{T}}_\alpha^{(k)\top}\boldsymbol{\theta} = \mathbf{0}.$$

Thus

$$\Delta\hat{\mathbf{p}}_\alpha = \sum_{k=1}^{3} \frac{\lambda_\alpha^{(k)}}{2} \hat{\mathsf{T}}_\alpha^{(k)\top}\boldsymbol{\theta} - \tilde{\mathbf{p}}_\alpha.$$

Substitution of this into Eq. (∗) yields

$$(\hat{\mathsf{T}}_\alpha^{(k)}\Big(\sum_{l=1}^{3} \frac{\lambda_\alpha^{(l)}}{2} \hat{\mathsf{T}}_\alpha^{(l)\top}\boldsymbol{\theta} - \tilde{\mathbf{p}}_\alpha\Big), \boldsymbol{\theta}) = (\hat{\boldsymbol{\xi}}_\alpha^{(k)}, \boldsymbol{\theta}),$$

which is rewritten in the form

$$\sum_{l=1}^{3} \frac{\lambda_\alpha^{(l)}}{2} (\boldsymbol{\theta}, \hat{\mathsf{T}}_\alpha^{(k)} \hat{\mathsf{T}}_\alpha^{(l)\top}\boldsymbol{\theta}) = (\hat{\boldsymbol{\xi}}_\alpha^{(k)} + \hat{\mathsf{T}}_\alpha^{(k)}\tilde{\mathbf{p}}, \boldsymbol{\theta}).$$

From the definitions of $V_0^{(kl)}[\hat{\boldsymbol{\xi}}_\alpha]$ and $\boldsymbol{\xi}_\alpha^{*(k)}$, this is written as

$$\frac{1}{2}\sum_{l=1}^{3} (\boldsymbol{\theta}, V_0^{(kl)}[\boldsymbol{\xi}_\alpha]\boldsymbol{\theta})\lambda_\alpha^{(l)} = (\boldsymbol{\xi}_\alpha^{*(k)}, \boldsymbol{\theta}).$$

This defines simultaneous linear equations in $\lambda_\alpha^{(1)}$, $\lambda_\alpha^{(2)}$, and $\lambda_\alpha^{(3)}$. Considering that the coefficient matrix $\Big((\boldsymbol{\theta}, V_0^{(kl)}[\hat{\boldsymbol{\xi}}_\alpha]\boldsymbol{\theta})\Big)$ has determinant 0, we solve them by least squares. The solution is written using the pseudoinverse $\hat{W}_\alpha^{(kl)} = \Big((\boldsymbol{\theta}, V_0^{(kl)}[\hat{\boldsymbol{\xi}}_\alpha]\boldsymbol{\theta})\Big)_2^-$ of rank 2 in the form

$$\frac{\lambda_\alpha^{(k)}}{2} = \sum_{l=1}^{3} \hat{W}_\alpha^{(kl)} (\boldsymbol{\xi}_\alpha^{*(k)}, \boldsymbol{\theta}).$$

Thus S is written as

$$\begin{aligned}
S &= \frac{1}{N}\sum_{\alpha=1}^{N} \|\tilde{\mathbf{p}}_\alpha + \Delta\hat{\mathbf{p}}_\alpha\|^2 = \frac{1}{N}\sum_{\alpha=1}^{N} \|\sum_{k=1}^{3} \frac{\lambda_\alpha^{(k)}}{2} \mathsf{T}_\alpha^{(k)\top}\boldsymbol{\theta}\|^2 \\
&= \frac{1}{N}\sum_{\alpha=1}^{N} (\sum_{k=1}^{3} \frac{\lambda_\alpha^{(k)}}{2} \mathsf{T}_\alpha^{(k)\top}\boldsymbol{\theta}, \sum_{l=1}^{3} \frac{\lambda_\alpha^{(l)}}{2} \mathsf{T}_\alpha^{(l)\top}\boldsymbol{\theta}) \\
&= \frac{1}{N}\sum_{\alpha=1}^{N} \sum_{k,l=1}^{3} \frac{\lambda_\alpha^{(k)}\lambda_\alpha^{(l)}}{4} (\boldsymbol{\theta}, \mathsf{T}_\alpha^{(k)}\mathsf{T}_\alpha^{(l)\top}\boldsymbol{\theta}) = \frac{1}{N}\sum_{\alpha=1}^{N} \sum_{k,l=1}^{3} \frac{\lambda_\alpha^{(k)}\lambda_\alpha^{(l)}}{4} (\boldsymbol{\theta}, V_0^{(kl)}[\hat{\boldsymbol{\xi}}_\alpha]\boldsymbol{\theta}) \\
&= \frac{1}{N}\sum_{\alpha=1}^{N} \sum_{k,l=1}^{3} \Big(\sum_{m=1}^{3} \hat{W}_\alpha^{(km)} (\boldsymbol{\xi}_\alpha^{*(m)}, \boldsymbol{\theta})\Big)\Big(\sum_{n=1}^{3} \hat{W}_\alpha^{(ln)} (\boldsymbol{\xi}_\alpha^{*(n)}, \boldsymbol{\theta})\Big)(\boldsymbol{\theta}, V_0^{(kl)}[\hat{\boldsymbol{\xi}}_\alpha]\boldsymbol{\theta})
\end{aligned}$$

$$= \frac{1}{N}\sum_{\alpha=1}^{N}\sum_{k,l,m,n=1}^{3} \hat{W}_\alpha^{(km)}(\boldsymbol{\theta}, V_0^{(kl)}[\hat{\boldsymbol{\xi}}_\alpha]\boldsymbol{\theta})\hat{W}_\alpha^{(ln)}(\boldsymbol{\xi}_\alpha^{*(m)}, \boldsymbol{\theta})(\boldsymbol{\xi}_\alpha^{*(n)}, \boldsymbol{\theta})$$

$$= \frac{1}{N}\sum_{\alpha=1}^{N}\sum_{k,l=1}^{3} \hat{W}_\alpha^{(kl)}(\boldsymbol{\xi}_\alpha^{(k)*}, \boldsymbol{\theta})(\boldsymbol{\xi}_\alpha^{(l)*}, \boldsymbol{\theta}),$$

where we have noted that if we let $\hat{V}_\alpha^{(kl)} = (\boldsymbol{\theta}, V_0^{(kl)}[\hat{\boldsymbol{\xi}}_\alpha]\boldsymbol{\theta})$, the following pseudoinverse identity holds ($\hookrightarrow$ Problem 14.3).

$$\sum_{l,m=1}^{3} \hat{W}_\alpha^{(kl)}\hat{V}_\alpha^{(lm)}\hat{W}_\alpha^{(mn)} = \hat{W}_\alpha^{(kn)}.$$

Problems of Chapter 7

7.1 Suppose point (x, y) is mapped to (x', y') by a homography defined by matrix H, and if the point (x', y') is mapped to (x'', y'') by another homography defined by matrix H':

$$\begin{pmatrix} x'/f_0 \\ y'/f_0 \\ 1 \end{pmatrix} \simeq \mathsf{H}\begin{pmatrix} x/f_0 \\ y/f_0 \\ 1 \end{pmatrix}, \qquad \begin{pmatrix} x''/f_0 \\ y''/f_0 \\ 1 \end{pmatrix} \simeq \mathsf{H}'\begin{pmatrix} x'/f_0 \\ y'/f_0 \\ 1 \end{pmatrix}$$

their composition is

$$\begin{pmatrix} x''/f_0 \\ y''/f_0 \\ 1 \end{pmatrix} \simeq \mathsf{H}'\mathsf{H}\begin{pmatrix} x/f_0 \\ y/f_0 \\ 1 \end{pmatrix},$$

which is a homography defined by the product $\mathsf{H}'\mathsf{H}$. If the result is further mapped by another homography defined by H'', the composition is a homography defined by the product $\mathsf{H}''\mathsf{H}'\mathsf{H}$. Because $\mathsf{H}''(\mathsf{H}'\mathsf{H}) = (\mathsf{H}''\mathsf{H}')\mathsf{H}$, the associativity holds; that is, the result does not depend on which product is computed first. Evidently, the identity matrix I defines the identity mapping, and the inverse H^{-1} defines the inverse mapping of the homography defined by H.

7.2 If point (x, y) is mapped to point (x', y') by a homography defined by a nonsingular matrix H, the following holds.

$$\begin{pmatrix} x/f_0 \\ y/f_0 \\ 1 \end{pmatrix} \simeq \mathsf{H}^{-1}\begin{pmatrix} x'/f_0 \\ y'/f_0 \\ 1 \end{pmatrix}.$$

If the point (x, y) is on line $ax + by + cf_0 = 0$, we see that

$$ax + by + cf_0 = f_0(\begin{pmatrix} a \\ b \\ c \end{pmatrix}, \begin{pmatrix} x/f_0 \\ y/f_0 \\ 1 \end{pmatrix}) \simeq (\begin{pmatrix} a \\ b \\ c \end{pmatrix}, \mathsf{H}^{-1}\begin{pmatrix} x'/f_0 \\ y'/f_0 \\ 1 \end{pmatrix})$$

$$= ((\mathsf{H}^{-1})^\top\begin{pmatrix} a \\ b \\ c \end{pmatrix}, \begin{pmatrix} x'/f_0 \\ y'/f_0 \\ 1 \end{pmatrix}).$$

This means that if a', b', and c' are defined by

$$\begin{pmatrix} a' \\ b' \\ c' \end{pmatrix} = \mathsf{H}^{-1} \begin{pmatrix} a \\ b \\ c \end{pmatrix},$$

the point (x', y') satisfies $a'x' + b'y' + c'f_0 = 0$. Thus line $ax + by + cf_0 = 0$ is mapped to line $a'x' + b'y' + c'f_0 = 0$.

7.3 If we let $\Delta\hat{\mathbf{p}} = \hat{\mathbf{p}} - \bar{\mathbf{p}}$, we can write Eq. (7.11) as $S = \|\tilde{\mathbf{p}} + \Delta\hat{\mathbf{p}}\|^2$. As in the solution of Problem 6.4, we write the value $\boldsymbol{\xi}^{(k)}$ for $\mathbf{p}$ as a function $\boldsymbol{\xi}^{(k)}(\mathbf{p})$. Then, $\bar{\boldsymbol{\xi}}^{(k)} = \boldsymbol{\xi}^{(k)}(\hat{\mathbf{p}} - \Delta\hat{\mathbf{p}})$. If $\Delta\hat{\mathbf{p}}$ is small, we can write

$$\bar{\boldsymbol{\xi}}^{(k)} = \boldsymbol{\xi}^{(k)}(\hat{\mathbf{p}} - \Delta\hat{\mathbf{p}}) = \hat{\boldsymbol{\xi}}^{(k)} - \hat{\mathsf{T}}^{(k)}\Delta\hat{\mathbf{p}}, + \cdots$$

where $\cdots$ indicates higher-order terms in $\Delta\hat{\mathbf{p}}$. Ignoring the higher-order terms, we can write the constraint $(\bar{\boldsymbol{\xi}}^{(k)}, \boldsymbol{\theta}) = 0$ as

$$(\hat{\mathsf{T}}^{(k)}\Delta\hat{\mathbf{p}}, \boldsymbol{\theta}) = (\hat{\boldsymbol{\xi}}^{(k)}, \boldsymbol{\theta}). \qquad (*)$$

Introducing Lagrange multipliers, we differentiate

$$\|\tilde{\mathbf{p}} + \Delta\hat{\mathbf{p}}\|^2 - \sum_{k=1}^{3} \lambda^{(k)}\Big((\hat{\mathsf{T}}^{(k)}\Delta\hat{\mathbf{p}}, \boldsymbol{\theta}) - (\hat{\boldsymbol{\xi}}^{(k)}, \boldsymbol{\theta})\Big)$$

with respect to $\Delta\hat{\mathbf{p}}$. Letting the result be $\mathbf{0}$, we obtain

$$2(\tilde{\mathbf{p}} + \Delta\hat{\mathbf{p}}) - \sum_{k=1}^{3} \lambda^{(k)}\hat{\mathsf{T}}^{(k)\top}\boldsymbol{\theta} = \mathbf{0}.$$

Therefore

$$\Delta\hat{\mathbf{p}} = \sum_{k=1}^{3} \frac{\lambda^{(k)}}{2}\hat{\mathsf{T}}^{(k)\top}\boldsymbol{\theta} - \tilde{\mathbf{p}}.$$

Substitution of this into Eq. $(*)$ yields

$$(\hat{\mathsf{T}}^{(k)}\Big(\sum_{l=1}^{3} \frac{\lambda^{(l)}}{2}\hat{\mathsf{T}}^{(l)\top}\boldsymbol{\theta} - \tilde{\mathbf{p}}\Big), \boldsymbol{\theta}) = (\hat{\boldsymbol{\xi}}^{(k)}, \boldsymbol{\theta}),$$

which is rewritten in the form

$$\sum_{l=1}^{3} \frac{\lambda^{(l)}}{2}(\boldsymbol{\theta}, \hat{\mathsf{T}}^{(k)}\hat{\mathsf{T}}^{(l)\top}\boldsymbol{\theta}) = (\hat{\boldsymbol{\xi}}^{(k)} + \hat{\mathsf{T}}^{(k)}\tilde{\mathbf{p}}, \boldsymbol{\theta}).$$

From the definitions of $V_0^{(kl)}[\hat{\boldsymbol{\xi}}]$ and $\boldsymbol{\xi}^{*(k)}$, this is written as

$$\frac{1}{2}\sum_{l=1}^{3}(\boldsymbol{\theta}, V_0^{(kl)}[\boldsymbol{\xi}]\boldsymbol{\theta})\lambda^{(l)} = (\boldsymbol{\xi}^{*(k)}, \boldsymbol{\theta}).$$

This defines simultaneous linear equations in $\lambda^{(1)}$, $\lambda^{(2)}$, and $\lambda^{(3)}$. Considering that the coefficient matrix $\Big((\boldsymbol{\theta}, V_0^{(kl)}[\hat{\boldsymbol{\xi}}]\boldsymbol{\theta})\Big)$ has determinant 0, we solve them by least squares. The solution is written using the pseudoinverse $\hat{W}^{(kl)} = \Big((\boldsymbol{\theta}, V_0^{(kl)}[\hat{\boldsymbol{\xi}}]\boldsymbol{\theta})\Big)_2^-$ of rank 2 in the form

$$\frac{\lambda^{(k)}}{2} = \sum_{l=1}^{3} \hat{W}^{(kl)}(\boldsymbol{\xi}^{*(l)}, \boldsymbol{\theta}).$$

Thus $\bar{\mathbf{p}} = \hat{\mathbf{p}} - \Delta\hat{\mathbf{p}}$ is estimated in the form

$$\hat{\hat{\mathbf{p}}} = \hat{\mathbf{p}} - \Big(\sum_{k=1}^{3} \frac{\lambda^{(k)}}{2}\hat{\mathsf{T}}^{(k)\top}\boldsymbol{\theta} - \tilde{\mathbf{p}}\Big) = \mathbf{p} - \sum_{k=1}^{3}\sum_{l=1}^{3} \hat{W}^{(kl)}(\boldsymbol{\xi}^{*(l)}, \boldsymbol{\theta})\hat{\mathsf{T}}^{(k)\top}\boldsymbol{\theta}.$$

7.4 If point (x, y) corresponds to (x', y'), the epipolar equation and the homography relation are written, respectively, in the form

$$(\begin{pmatrix} x/f_0 \\ y/f_0 \\ 1 \end{pmatrix}, \mathsf{F}\begin{pmatrix} x'/f_0 \\ y'/f_0 \\ 1 \end{pmatrix}) = 0, \qquad \begin{pmatrix} x'/f_0 \\ y'/f_0 \\ 1 \end{pmatrix} \simeq \mathsf{H}\begin{pmatrix} x/f_0 \\ y/f_0 \\ 1 \end{pmatrix}.$$

Therefore the following relation holds.

$$(\begin{pmatrix} x/f_0 \\ y/f_0 \\ 1 \end{pmatrix}, \mathsf{FH}\begin{pmatrix} x/f_0 \\ y/f_0 \\ 1 \end{pmatrix}) = 0.$$

This holds for an arbitrary point (x, y). Thus the quadratic form defined by FH is identically 0. A quadratic form is determined by the symmetric part of the coefficient matrix. This means that the symmetric part $(\mathsf{FH} + (\mathsf{FH})^\top)/2$ of FH equals O. Therefore Eq. (7.13) holds.

Problems of Chapter 8

8.1 (1) The (i, j) element of the matrix $\mathsf{A} = \mathsf{I} + \mathbf{a}\mathbf{b}^\top$ is $A_{ij} = \delta_{ij} + a_i b_j$, where δ_{ij} is the Kronecker delta, taking 1 for $i = j$ and 0 otherwise. Its determinant is by definition $|\mathsf{A}| = \sum_{i,j,k=1}^{3} \varepsilon_{ijk} A_{1i} A_{2j} A_{3k}$, where ε_{ijk} is the permutation signature, taking 1 if (i, j, k) is an even permutation of $(1, 2, 3)$ (i.e., obtained by an even number of interchanges), -1 if it is an odd permutation (i.e., obtained by an odd number of interchanges), and 0 otherwise. Thus $|\mathsf{A}|$ is

$$\begin{aligned}
|\mathsf{A}| &= \sum_{i,j,k=1}^{3} \varepsilon_{ijk}(\delta_{1i} + a_1 b_i)(\delta_{2j} + a_2 b_j)(\delta_{3k} + a_3 b_k) \\
&= \sum_{i,j,k=1}^{3} \varepsilon_{ijk}(\delta_{1i}\delta_{2j}\delta_{3k} + \delta_{2j}\delta_{3k}a_1 b_i + \delta_{1i}\delta_{3k}a_2 b_j + \delta_{1i}\delta_{2j}a_3 b_k + \delta_{1i}a_2 a_3 b_j b_k \\
&\quad + \delta_{2j}a_1 a_3 b_i b_k + \delta_{3k}a_1 a_2 b_i b_j + a_1 a_2 a_3 b_i b_j b_k)
\end{aligned}$$

$$
\begin{aligned}
=&\varepsilon_{123} + \sum_{i=1}^{3} a_1\varepsilon_{i23}b_i + \sum_{j=1}^{3} a_2\varepsilon_{1j3}b_j + \sum_{k=1}^{3} a_3\varepsilon_{12k}b_k + a_2a_3 \sum_{j,k=1}^{3} \varepsilon_{1jk}b_jb_k \\
&+ a_1a_3 \sum_{i,k=1}^{3} \varepsilon_{i2k}b_ib_k + a_1a_2 \sum_{i,j=1}^{3} \varepsilon_{ij3}b_ib_j + a_1a_2a_3 \sum_{i,j,k=1}^{3} \varepsilon_{ijk}b_ib_jb_k \\
=&1 + a_1b_1 + a_2b_2 + a_3b_3,
\end{aligned}
$$

where we have noted that the quadratic and cubic terms are canceled due to the permutation symbol ε_{ijk}. In the sums $\sum_{j,k=1}^{3} \varepsilon_{1jk}b_jb_k$ and $\sum_{i,j,k=1}^{3} \varepsilon_{ijk}b_ib_jb_k$, for example, the terms of b_jb_k and b_ib_j have opposite signs and are canceled.
(2) This can be shown as follows.

$$
\begin{aligned}
\Big(\mathsf{I} - \frac{\mathbf{ab}^\top}{1+(\mathbf{a},\mathbf{b})}\Big)(\mathsf{I} + \mathbf{ab}^\top)^{-1} &= \mathsf{I} + \mathbf{ab}^\top - \frac{\mathbf{ab}^\top}{1+(\mathbf{a},\mathbf{b})} - \frac{\mathbf{ab}^\top\mathbf{ab}^\top}{1+(\mathbf{a},\mathbf{b})} \\
&= \mathsf{I} + \frac{(\mathbf{a},\mathbf{b})\mathbf{ab}^\top}{1+(\mathbf{a},\mathbf{b})} - \frac{(\mathbf{a},\mathbf{b})\mathbf{ab}^\top}{1+(\mathbf{a},\mathbf{b})} = \mathsf{I}.
\end{aligned}
$$

(3) If $\mathsf{I} + \mathbf{ab}^\top$ is an orthogonal matrix, $|\mathsf{I} + \mathbf{ab}^\top| = 1 + (\mathbf{a},\mathbf{b}) = 1$ holds. Because $(\mathbf{a},\mathbf{b}) = 0$, vectors $\mathbf{a}$ and $\mathbf{b}$ are orthogonal to each other. On the other hand, if $\mathsf{I}+\mathbf{ab}^\top$ is an orthogonal matrix, its inverse $(\mathsf{I} + \mathbf{ab}^\top)^{-1} = \mathsf{I} - \mathbf{ab}^\top$ equals $(\mathsf{I} + \mathbf{ab}^\top)^\top = \mathsf{I} + \mathbf{ba}^\top$. Thus $\mathbf{ab}^\top - \mathbf{ba}^\top = \mathsf{O}$. Comparing the nondiagonal elements on both sides, we see that $\mathbf{a} \times \mathbf{b} = \mathbf{0}$, meaning that $\mathbf{a}$ and $\mathbf{b}$ are parallel to each other. Two vectors are mutually orthogonal and parallel if and only if either or both are $\mathbf{0}$. This contradicts our assumption that $\mathbf{a} \neq \mathbf{0}$ and $\mathbf{b} \neq \mathbf{0}$. Therefore $\mathsf{I} + \mathbf{ab}^\top$ is an orthogonal matrix.

8.2 (1) From the definition of Eq. (8.9), we can write

$$
\mathbf{x} = \begin{pmatrix} f_0 & 0 & 0 \\ 0 & f_0 & 0 \\ 0 & 0 & f \end{pmatrix}\begin{pmatrix} x/f_0 \\ y/f_0 \\ 1 \end{pmatrix}, \qquad \mathbf{x}' = \begin{pmatrix} f_0 & 0 & 0 \\ 0 & f_0 & 0 \\ 0 & 0 & f' \end{pmatrix}\begin{pmatrix} x'/f_0 \\ y'/f_0 \\ 1 \end{pmatrix}.
$$

Hence, the following holds.

$$
\mathbf{x}' = \begin{pmatrix} f_0 & 0 & 0 \\ 0 & f_0 & 0 \\ 0 & 0 & f' \end{pmatrix}\mathsf{H}\begin{pmatrix} x/f_0 \\ y/f_0 \\ 1 \end{pmatrix} \simeq \begin{pmatrix} f_0 & 0 & 0 \\ 0 & f_0 & 0 \\ 0 & 0 & f' \end{pmatrix}\mathsf{H}\begin{pmatrix} 1/f_0 & 0 & 0 \\ 0 & 1/f_0 & 0 \\ 0 & 0 & 1/f \end{pmatrix}\mathbf{x} = \tilde{\mathsf{H}}\mathbf{x}.
$$

(2) As shown in Fig. 5.2, the position vector of the 3D point (X, Y, Z) is a scalar multiple of the vector $\mathbf{x}$, which points to the point (x, y) on the image plane from the origin O. Hence, the 3D position is $c\mathbf{x}$ for some constant c. Similarly, the 3D position of that point with respect to the second camera is $c'\mathbf{x}'$ for some constant c'. Inasmuch as the second camera is translated by $\boldsymbol{\theta}$ and rotated by R relative to the first camera (= the world coordinate system), it is equal to $\mathbf{t} + c'\mathsf{R}\mathbf{x}'$ with respect to the XYZ coordinate system. Therefore

$$
c\mathbf{x} = \mathbf{t} + c'\mathsf{R}\mathbf{x}'.
$$

Computing the inner product of this and $\mathbf{n}$ on both sides and using the equation of the plane of Eq. (8.1), we see that $(\mathbf{n}, c\mathbf{x}) = h$. Thus $c = h/(\mathbf{n}, \mathbf{x})$, therefore we can write

$$\mathbf{x}' = \frac{1}{c'}\mathsf{R}^\top(c\mathbf{x} - \mathbf{t}) = \frac{1}{c'}\mathsf{R}^\top(\frac{h\mathbf{x}}{(\mathbf{n}, \mathbf{x})} - \mathbf{t}) = \frac{1}{c'h(\mathbf{n}, \mathbf{x})}\mathsf{R}^\top(\mathbf{x} - \frac{\mathbf{t}\mathbf{n}^\top}{h}\mathbf{x})$$
$$= \frac{1}{c'(\mathbf{n}, \mathbf{x})}\mathsf{R}^\top(\mathsf{I} - \frac{\mathbf{t}\mathbf{n}^\top}{h})\mathbf{x}.$$

Comparing this with Eq. (8.10), we can write $\tilde{\mathsf{H}}$ as

$$\tilde{\mathsf{H}} \simeq \mathsf{R}^\top(\mathsf{I} - \frac{\mathbf{t}\mathbf{n}^\top}{h}).$$

Because $|\mathsf{R}| = 1$, the determinant of the right-hand side is $1 - (\mathbf{n}, \mathbf{t})/h$ from Eq. (8.7). Therefore $\tilde{\mathsf{H}}$ is normalized to the form of Eq. (8.11).
(3) Because the plane $n_1X + n_2Y + n_3Z = h$ divides the entire 3D space into two: $n_1X+n_2Y+n_3Z > h$ on one side, and $n_1X+n_2Y+n_3Z < h$ on the other side. We are assuming that $h > 0$, therefore the origin O (= the viewpoint of the first camera) is on the side of $n_1X+n_2Y+n_3Z < h$. We are also assuming that the camera translation $\mathbf{t}$ = (t_i) takes place on one side of the plane, therefore $n_1t_1 + n_2t_2 + n_3t_3 < h$, or $(\mathbf{n}, \mathbf{t})$ $< h$. Thus, $k > 0$.

8.3 Because U and $\mathsf{V}^\top$ are both orthogonal matrices, they have determinant 1, and thus $|\tilde{\mathsf{H}}| = |\mathsf{U}|\sigma_1\sigma_2\sigma_3|\mathsf{V}^\top| = \sigma_1\sigma_2\sigma_3$. This is normalized to 1 by Eq. (8.3). If $\sigma_1 = \sigma_2 = \sigma_3 = 1$, then $\tilde{\mathsf{H}} = \mathsf{U}\mathsf{V}^\top$ is an orthogonal matrix. If $\tilde{\mathsf{H}}$ is orthogonal, Eq. (8.11) implies that $\mathsf{I} - \mathbf{t}\mathbf{n}^\top/h$ is also orthogonal. However, $\mathbf{n}$ is a unit vector and $\mathbf{t}$ is not $\mathbf{0}$ by our assumption. Therefore $\mathsf{I} - \mathbf{t}\mathbf{n}^\top/h$ is not orthogonal. ($\hookrightarrow$ Problem 8.1(3)), thus $\sigma_1 = \sigma_2 = \sigma_3 = 1$ does not hold.

8.4 From Eq. (8.11), we obtain

$$k^2\tilde{\mathsf{H}}^\top\tilde{\mathsf{H}} = (\mathsf{I} - \frac{\mathbf{n}\mathbf{t}^\top}{h})(\mathsf{I} - \frac{\mathbf{t}\mathbf{n}^\top}{h}).$$

Multiplying this by $\mathsf{V}^\top$ from the left and V from the right, and noting that $\|\boldsymbol{\tau}\|^2 = \|\mathbf{t}\|^2 = 1$, we obtain

$$k^2\mathsf{V}^\top\tilde{\mathsf{H}}^\top\tilde{\mathsf{H}}\mathsf{V} = \mathsf{V}^\top(\mathsf{I} - \frac{\mathbf{n}\mathbf{t}^\top}{h})\mathsf{V}\mathsf{V}^\top(\mathsf{I} - \frac{\mathbf{t}\mathbf{n}^\top}{h})\mathsf{V} = (\mathsf{I} - \boldsymbol{\nu}\boldsymbol{\tau}^\top)(\mathsf{I} - \boldsymbol{\tau}\boldsymbol{\nu}^\top)$$
$$= \mathsf{I} - \boldsymbol{\tau}\boldsymbol{\nu}^\top - \boldsymbol{\nu}\boldsymbol{\tau}^\top + \boldsymbol{\nu}\boldsymbol{\nu}^\top.$$

From Eq. (8.2), we have

$$k^2\mathsf{V}^\top\tilde{\mathsf{H}}^\top\tilde{\mathsf{H}}\mathsf{V} = k^2\begin{pmatrix} \sigma_1^2 & 0 & 0 \\ 0 & \sigma_2^2 & 0 \\ 0 & 0 & \sigma_3^2 \end{pmatrix}.$$

Therefore we obtain Eq. (8.13).

8.5 (1) Because $\nu_2 = 0$, Eq. (8.13) implies

$$1 - 2\nu_1\tau_1 + \nu_1^2 = k^2\sigma_1^2, \quad 1 = k^2\sigma_2^2, \quad 1 - 2\nu_3\tau_3 + \nu_3^2 = k^2\sigma_3^2$$
$$-\nu_1\tau_2 = 0, \quad -\tau_2\nu_3 = 0, \quad -\tau_3\nu_1 - \nu_3\tau_1 + \nu_3\nu_1 = 0. \qquad (*1)$$

Substituting $k^2 = 1/\sigma_2^2$ obtained from the second equation into the first and third equations, we can write τ_1 and τ_3 in the form

$$\tau_1 = \frac{1 - (\sigma_1/\sigma_2)^2 + \nu_1^2}{2\nu_1}, \; \tau_3 = \frac{1 - (\sigma_3/\sigma_2)^2 + \nu_3^2}{2\nu_3}. \qquad (*2)$$

Substituting these into the sixth equation of Eq. (∗1), we obtain

$$\frac{\nu_1}{2\nu_3}(1 - (\sigma_3/\sigma_2)^2 + \nu_3^2) + \frac{\nu_3}{2\nu_1}(1 - (\sigma_1/\sigma_2)^2 + \nu_1^2) = \nu_3\nu_1,$$

which is rewritten as

$$\Big(\Big(\frac{\sigma_1}{\sigma_2}\Big)^2 - 1\Big)\nu_3^2 = \Big(1 - \Big(\frac{\sigma_3}{\sigma_2}\Big)^2\Big)\nu_1^2. \qquad (*3)$$

Because $\sigma_1 \geq \sigma_2 \geq \sigma_3$, the two sides are either both positive or both 0. Because $\nu_1 \neq 0$ and $\nu_3 \neq 0$, both sides are 0 only when $\sigma_1 = \sigma_2 = \sigma_3$, which is a contradiction (↪ Problem 8.3). Therefore both sides equal some constant $\mu > 0$, and $\sigma_1 > \sigma_2 > \sigma_3$ holds. Hence we can write ν_1^2 and ν_3^2 as

$$\nu_1^2 = \mu\Big(\Big(\frac{\sigma_1}{\sigma_2}\Big)^2 - 1\Big), \; \nu_3^2 = \mu\Big(1 - \Big(\frac{\sigma_3}{\sigma_2}\Big)^2\Big). \qquad (*4)$$

Substituting these into the numerators in Eq. (∗2), we obtain

$$\tau_1 = \frac{(\sigma_1/\sigma_2)^2 - 1}{2\nu_1}(\mu - 1), \; \tau_3 = \frac{1 - (\sigma_3/\sigma_2)^2}{2\nu_3}(\mu + 1). \qquad (*5)$$

Squaring both sides and substituting Eq. (∗4), we obtain

$$\tau_1^2 = \frac{1}{4\mu}\Big(\Big(\frac{\sigma_1}{\sigma_2}\Big)^2 - 1\Big)(\mu - 1)^2, \; \tau_3^2 = \frac{1}{4\mu}\Big(1 - \Big(\frac{\sigma_3}{\sigma_2}\Big)^2\Big)(\mu + 1)^2.$$

Because $\boldsymbol{\tau} = (\tau_i)$ is obtained by multiplying the unit vector $\mathbf{t}$ by $\mathsf{V}^\top$, which is an orthogonal matrix, $\boldsymbol{\tau}$ is also a unit vector with $\tau_1^2 + \tau_3^2 = 1$. Thus the following holds.

$$(\sigma_1^2 - \sigma_3)\mu^2 - 2(\sigma_1^2 + \sigma_3^2)\mu + (\sigma_1^2 - \sigma_3) = 0.$$

This is a quadratic equation in μ, whose solution is given by

$$\mu = \frac{\sigma_1 - \sigma_3}{\sigma_1 + \sigma_3}, \; \frac{\sigma_1 + \sigma_3}{\sigma_1 - \sigma_3}.$$

Inasmuch as $\mathsf{V}^\top$ is an orthogonal matrix, we see from Eq. (8.12) that $(\boldsymbol{\nu}, \boldsymbol{\tau}) = (\mathbf{n}, \mathbf{t})/h$, which is less than 1 by assumption. From Eq. (∗5), we obtain

$$(\boldsymbol{\nu}, \boldsymbol{\tau}) = \nu_1\tau_1 + \nu_3\tau_3 = 1 - \frac{1}{2\sigma_2^2}(\sigma_1^2 + \sigma_1^3 - (\sigma_1^2 - \sigma_3^2)\mu). \qquad (*6)$$

This equals $1 - \sigma_1\sigma_3/\sigma_2^2$ (< 1) for $\mu = (\sigma_1 - \sigma_3)/(\sigma_1 + \sigma_3)$ and $1 + \sigma_1\sigma_3/\sigma_2^2$ (> 1) for $\mu = (\sigma_1 + \sigma_3)/(\sigma_1 + \sigma_3)$. Thus the former solution is chosen. It follows from Eq. ($*4$) that ν_1 and ν_3 are given by Eq. (8.14) and its simultaneous sign reversal, and $\nu_2 = 0$ by assumption. From Eq. ($*5$), τ_1 and τ_3 are given by Eq. (8.14) and its simultaneous sign reversal. From the fourth and fifth equations of Eq. ($*1$), we see that $\tau_2 = 0$.
(2) We consider the following four cases.

Case 1: $\nu_1 \neq 0$, $\nu_2 \neq 0$, $\nu_3 \neq 0$.
From Eq. (8.13), we obtain

$$-\nu_2\tau_3 - \tau_2\nu_3 + \nu_2\nu_3 = 0, \quad -\nu_3\tau_1 - \tau_3\nu_1 + \nu_3\nu_1 = 0, \quad -\nu_1\tau_2 - \tau_1\nu_2 + \nu_1\nu_2 = 0.$$

Thus

$$\begin{pmatrix} 0 & \nu_3 & \nu_2 \\ \nu_3 & 0 & \nu_1 \\ \nu_2 & \nu_1 & 0 \end{pmatrix}\begin{pmatrix} \tau_1 \\ \tau_2 \\ \tau_3 \end{pmatrix} = \begin{pmatrix} \nu_2\nu_3 \\ \nu_3\nu_1 \\ \nu_1\nu_2 \end{pmatrix}.$$

The determinant of the matrix on the left is $2\nu_1\nu_2\nu_3$, which is not 0. Therefore the inverse exists, and the solution is given by

$$\begin{pmatrix} \tau_1 \\ \tau_2 \\ \tau_3 \end{pmatrix} = \frac{1}{2\nu_1\nu_2\nu_3}\begin{pmatrix} -\nu_1^2 & \nu_1\nu_2 & \nu_1\nu_3 \\ \nu_2\nu_1 & -\nu_2^2 & \nu_2\nu_3 \\ \nu_3\nu_1 & \nu_3\nu_2 & -\nu_3^2 \end{pmatrix}\begin{pmatrix} \nu_2\nu_3 \\ \nu_3\nu_1 \\ \nu_1\nu_2 \end{pmatrix} = \frac{1}{2}\begin{pmatrix} \nu_1 \\ \nu_2 \\ \nu_3 \end{pmatrix}.$$

Thus $2\tau_i = \nu_i$, which means that the diagonal elements on the right-hand side of Eq. (8.13) are all 1. This implies that $\sigma_1 = \sigma_2 = \sigma_3$, a contradiction ($\hookrightarrow$ Problem 8.3).

Case 2: $\nu_1 = 0$, $\nu_2 \neq 0$, $\nu_3 \neq 0$.
The equations for $\nu_1 \neq 0$, $\nu_2 = 0$, and $\nu_3 \neq 0$ hold if ν_1 and ν_3 are replaced by ν_2 and ν_3, respectively. Thus Eq. ($*3$) becomes

$$\left(\left(\frac{\sigma_2}{\sigma_1}\right)^2 - 1\right)\nu_3^2 = \left(1 - \left(\frac{\sigma_3}{\sigma_1}\right)^2\right)\nu_2^2.$$

Because $\sigma_1 \geq \sigma_2 \geq \sigma_3$, the left-hand side is 0 or negative, and the right-hand side is 0 or positive. Therefore both sides are 0, which means $\sigma_1 = \sigma_2 = \sigma_3$, a contradiction ($\hookrightarrow$ Problem 8.3).

Case 3: $\nu_1 \neq 0$, $\nu_2 \neq 0$, $\nu_3 = 0$.
In this case, Eq. ($*3$) becomes

$$\left(\left(\frac{\sigma_1}{\sigma_3}\right)^2 - 1\right)\nu_2^2 = \left(1 - \left(\frac{\sigma_2}{\sigma_3}\right)^2\right)\nu_1^2.$$

The left-hand side is positive or 0, and the right-hand side is negative or 0. Therefore both sides are 0, meaning $\sigma_1 = \sigma_2 = \sigma_3$, a contradiction ($\hookrightarrow$ Problem 8.3).

Case 4: $\nu_1 = 0,\ \nu_2 \neq 0,\ \nu_3 = 0$.

Comparing the diagonal elements of Eq. (8.13) on both sides, we see that $k^2\sigma_1^2 = 1$ and $k^2\sigma_3^2 = 1$. Thus $\sigma_1 = \sigma_3$. Because $\sigma_1 \geq \sigma_2 \geq \sigma_3$, this means $\sigma_1 = \sigma_2 = \sigma_3$, a contradiction ($\hookrightarrow$ Problem 8.3).

(3) Because $\boldsymbol{\nu} = \mathsf{V}^\top \mathbf{n}/h \neq \mathbf{0}$ from our assumption, we consider the following two cases.

Case 1: $\nu_1 \neq 0,\ \nu_2 = \nu_3 = 0$.

Comparing the nondiagonal elements of Eq. (8.13), we see that $\tau_2\nu_1 = 0$ and $\tau_3\nu_1 = 0$. Thus $\tau_2 = \tau_3 = 0$. Because $\boldsymbol{\tau}$ is a unit vector, we have $\tau_1 = \pm 1$. In the end, however, the sign reversal of a solution is also regarded as a solution, therefore we may let $\tau_1 = -1$ here. Comparing the diagonal elements of Eq. (8.13), we see that

$$k^2\sigma_1^2 = 1 + 2\nu_1 + \nu_1^2, \qquad k^2\sigma_2 = 1, \qquad k^2\sigma_3 = 1.$$

The right-hand side of the first equation is $(1+\nu_1)^2$, but because $(\boldsymbol{\nu}, \boldsymbol{\tau}) = (\mathbf{n}, \mathbf{t})/h < 1$ ($\hookrightarrow$ Problem 8.2(3)), we have $-nu_1 < 1$; that is, $1 + \nu_1 > 0$. Therefore we obtain $k\sigma_1 = 1 + \nu_1$. The second and third equations imply $\sigma_2 = \sigma_3$ and $k = 1/\sigma_2$ $(= 1/\sigma_3)$. Thus,

$$\nu_1 = k\sigma_1 - 1 = \frac{\sigma_1 - \sigma_2}{\sigma_2},$$

and $\nu_2 = \nu_3 = 0$. Also, $\tau_1 = -1$ and $\tau_2 = \tau_3 = 0$. This solution is a special case of Eqs. (8.14) and (8.16) for $\sigma_2 = \sigma_3$.

Case 2: $\nu_1 = \nu_2 = 0,\ \nu_3 \neq 0$.

Comparing the nondiagonal elements of Eq. (8.13), we see that $\tau_1\nu_3 = 0$ and $\tau_2\nu_3 = 0$. Hence, $\tau_1 = \tau_2 = 0$. Because $\boldsymbol{\tau}$ is a unit vector and because we reverse the sign in the end, we can let $\tau_1 = 1$ here. Comparing the diagonal elements of Eq. (8.13), we see that

$$k^2\sigma_1^2 = 1, \qquad k^2\sigma_2 = 1, \qquad k^2\sigma_3 = 1 - 2\nu_3 + \nu_3^2.$$

The first and second equations imply $\sigma_1 = \sigma_2$ and $k = 1/\sigma_1$ $(= 1/\sigma_2)$. The right-hand side of the third equation is $(1-\nu_3)^2$, but because $(\boldsymbol{\nu}, \boldsymbol{\tau}) = (\mathbf{n}, \mathbf{t})/h < 1$ ($\hookrightarrow$ Problem 8.2(3)), we have $\nu_3 < 1$; that is, $1 - \nu_3 > 0$. Hence, $k\sigma_3 = 1 - \nu_3$, which means

$$\nu_3 = 1 - k\sigma_3 = \frac{\sigma_1 - \sigma_3}{\sigma_1},$$

and $\nu_1 = \nu_2 = 0$. Also, $\tau_1 = \tau_2 = 0$ and $\tau_3 = 1$. This solution and its sign reversal are a special case of Equations (8.14) and (8.15) for $\sigma_1 = \sigma_2$.

(4) Because V is an orthogonal matrix, we can write from Eq. (8.12) the vectors $\mathbf{t}$ and $\mathbf{n}$ in the form

$$\mathbf{t} = \mathsf{V}\boldsymbol{\tau} = \tau_1\mathbf{v}_1 + \tau_2\mathbf{v}_2 + \tau_3\mathbf{v}_3, \mathbf{n} = h\mathsf{V}\boldsymbol{\nu} = h(\nu_1\mathbf{v}_1 + \nu_2\mathbf{v}_2 + \nu_3\mathbf{v}_3),$$

where $\mathbf{v}_1$, $\mathbf{v}_2$, and $\mathbf{v}_3$ are the first, second, and third columns of V, respectively. Because $\mathbf{t}$ and $\mathbf{n}$ are both unit vectors, we obtain Eqs. (8.5) and (8.6) from Eqs. (8.14) and (8.15), using the normalization operation $\mathscr{N}[\,\cdot\,]$. The sign reversal of all the equations also yields a solution.
(5) Because $\mathbf{n} = h(\nu_1\mathbf{v}_1 + \nu_2\mathbf{v}_2 + \nu_3\mathbf{v}_3)$ is a unit vector, we have $h = 1/(\nu_1^2 + \nu_2^2 + \nu_1^3)$. From Eq. (8.14), we see that

$$\nu_1^2 + \nu_2^2 + \nu_1^3 = \frac{1}{\sigma_2^2}\frac{\sigma_1 - \sigma_3}{\sigma_1 + \sigma_3}(\sigma_1^2 - \sigma_3^2) = \frac{1}{\sigma_2^2}(\sigma_1 - \sigma_3)^2.$$

We are assuming that $h > 0$ and $\sigma_1 \geq \sigma_3$, thus h is written in the form of Eq. (8.5).
(6) Transposing Eq. (8.11) on both sides, we obtain

$$\tilde{\mathsf{H}}^\top = \frac{1}{k}\Big(\mathsf{I} - \frac{\mathbf{n}\mathbf{t}^\top}{h}\Big)\mathsf{R}.$$

Thus from Eq. (8.8) we obtain

$$\mathsf{R} = k\Big(\mathsf{I} - \frac{\mathbf{n}\mathbf{t}^\top}{h}\Big)^{-1}\tilde{\mathsf{H}}^\top = k\Big(\mathsf{I} + \frac{\mathbf{n}\mathbf{t}^\top/h}{1 - (\mathbf{n}, \mathbf{t})/h}\Big)\tilde{\mathsf{H}}^\top.$$

From Eq. (8.12), we have $(\mathbf{n}, \mathbf{t})/h = (\boldsymbol{\nu}, \boldsymbol{\tau})$, and Eq. (∗6) implies

$$(\boldsymbol{\nu}, \boldsymbol{\tau}) = \nu_1\tau_1 + \nu_3\tau_3 = 1 - \frac{\sigma_1\sigma_3}{\sigma_2^2} = 1 - \frac{1}{\sigma_2^3},$$

where we have noted that $\tilde{\mathsf{H}}$ has determinant 1 and hence $\sigma_1\sigma_2\sigma_3 = 1$ from Eq. (8.4). It follows that $1 - (\mathbf{n}, \mathbf{t})/h = 1/\sigma_2^3$. From Eq. (8.11), we obtain $k = \sqrt[3]{1 - (\mathbf{n}, \mathbf{t})/h} = \sqrt[3]{1/\sigma_2^3} = 1/\sigma_2$. Thus R is written in the form of Eq. (8.6).

Problems of Chapter 9

9.1 If we use Eq. (9.5) to translate the quadratic curve of Eq. (9.1) by $-x_c$ and $-y_c$ in the x- and y -directions, respectively, the expression of Eq. (9.7) becomes

$$Ax^2 + 2Bxy + Cy^2 = c, c = Ax_c^2 + 2Bx_cy_c + Cy_c^2 - f_0F.$$

There exists an angle θ such that if the xy-coordinate system is rotated around the origin by θ, the curve is expressed in the new $x'y'$-coordinate system in the canonical form

$$\lambda_1 x'^2 + \lambda_2 y'^2 = c.$$

As is well known in linear algebra, λ_1 and λ_2 are the eigenvalues of the matrix $\begin{pmatrix} A & B \\ B & C \end{pmatrix}$. This quadratic curve represents an ellipse if and only if λ_1 and λ_2 have the same sign; that is, $\lambda_1\lambda_2 > 0$. The eigenvalues λ_1 and λ_2 are the roots of the secular equation

$$\begin{vmatrix} \lambda - A & -B \\ -B & \lambda - C \end{vmatrix} = \lambda^2 - (A + C)\lambda + (AC - B^2) = 0.$$

From the well-known relation between the roots and the coefficients, the identity $\lambda_1\lambda_2 = AC - B^2$ holds. Hence, the quadratic curve is an ellipse when $AC - B^2 > 0$, and it is a real ellipse if λ_1 and λ_2 have the same sign as c, and an imaginary conic for different signs. We can also see that the quadratic curve is a parabola when one of λ_1 and λ_2 is 0, that is, $\lambda_1\lambda_2 = 0$ and that it is a hyperbola when λ_1 and λ_2 have different signs, that is, $\lambda_1\lambda_2 < 0$. These conditions correspond to $AC - B^2 = 0$ and $AC - B^2 < 0$, respectively.

9.2 (1) Expanding the determinant, we can easily see that the third-order term is λ^3 and that the second-order term is $(A_{11} + A_{22} + A_{33})\lambda^2$, that is, $\lambda^2\mathrm{tr}[\mathsf{A}]$. We can also see that the coefficient of the first-order term equals the sum of the minors defined by removing the row and the column that contain the diagonal element of A, which equals $A^\dagger_{11} + A^\dagger_{22} + A^\dagger_{33} = \mathrm{tr}[\mathsf{A}^\dagger]$. Evidently, the constant term is $|\mathsf{A}|$, as we can easily see by letting $\lambda = 0$.
(2) From the identity $\mathsf{A}^{-1} = \mathsf{A}^\dagger/|\mathsf{A}|$, we obtain

$$|\lambda\mathsf{A} + \mathsf{B}| = |\mathsf{A}| \cdot |\lambda\mathsf{I} + \mathsf{A}^{-1}\mathsf{B}| = |\mathsf{A}| \cdot \left|\lambda\mathsf{I} + \frac{\mathsf{A}^\dagger}{|\mathsf{A}|}\mathsf{B}\right|.$$

Evidently, the third-order term in λ is $|\mathsf{A}|\lambda^3$, and the coefficient of the second-order term is $|\mathsf{A}|\mathrm{tr}[(\mathsf{A}^\dagger/|\mathsf{A}|)\mathsf{B}] = \mathrm{tr}[\mathsf{A}^\dagger\mathsf{B}]$ from the result of the above item (1). The constant term is $|\mathsf{B}|$, as we see by letting $\lambda = 0$. Now, we can write

$$|\lambda\mathsf{A} + \mathsf{B}| = \lambda^3|\frac{\mathsf{B}}{\lambda} + \mathsf{A}|.$$

Hence, the coefficient of λ on the left-hand side equals the coefficient of $1/\lambda^2$ of $1/\lambda^2$ of $|\mathsf{B}/\lambda + \mathsf{A}|$ の $1/\lambda^2$, which is $\mathrm{tr}[\mathsf{B}^\dagger\mathsf{A}]$ (=$\mathrm{tr}[\mathsf{A}\mathsf{B}^\dagger]$) from the above result. From these results, we obtain Eq. (9.35). In this derivation, we are assuming that A has its inverse A^{-1} and that $\lambda \neq 0$. However, Eq. (9.35) is a polynomial in A and λ. Therefore this is an identity in A and λ, which holds including the cases of $|\mathsf{A}| = 0$ and $\lambda = 0$.

9.3 Arranging Eq. (2.1) in x, we obtain

$$Ax^2 + 2(By + Df_0)x + (Cy^2 + 2f_0Ey + f_0^2F) = 0,$$

which can be factorized in x in the form

$$A(x - \alpha)(y - \beta) = 0,$$

$$\alpha, \beta = \frac{-(By + Df_0) \pm \sqrt{(By + Df_0)^2 - A(Cy^2 + 2f_0Ey + f_0^2F)}}{A}.$$

Because $|\mathsf{Q}| = 0$, this quadratic equation is degenerated into the product of two linear terms. Therefore the inside of the square root

$$(B^2 - AC)y^2 + 2f_0(BD - AE)y + f_0^2(D^2 - AF)$$

is the square of a linear expression in y. Because $B^2 - AC > 0$, it has the form

$$(\sqrt{B^2 - AC}y + f_0\frac{BD - AE}{\sqrt{B^2 - AC}})^2.$$

Thus the entire quadratic equation is factorized into the form

$$A\left(x - \frac{-(By + Df_0) + \sqrt{B^2 - AC}y + f_0(BD - AE)/\sqrt{B^2 - AC}}{A}\right)$$
$$\times \left(x - \frac{-(By + Df_0) - \sqrt{B^2 - AC}y - f_0(BD - AE)/\sqrt{B^2 - AC}}{A}\right) = 0,$$

from which we obtain Eq. (9.36).

9.4 If $n_2 \neq 0$, we substitute $y = -(n_1x + n_3f_0)/n_2$ into Eq. (9.1) to obtain

$$Ax^2 - 2Bx(n_1x + n_3f_0)/n_2 + C(n_1x + n_3f_0)^2/n_2^2$$
$$+2f_0(Dx - E(n_1x + n_3f_0)/n_2) + f_0^2F = 0.$$

Rearranging this, we obtain the following quadratic Equation.

$$(An_2^2 - 2Bn_1n_2 + Cn_1^2)x^2 + 2f_0(Dn_2^2 + Cn_1n_3 - Bn_2n_3 - En_1n_2)x$$
$$+(Cn_3^2 - 2En_2n_3 + Fn_2^2)f_0^2 = 0.$$

Letting x_i, $i = 1, 2$, be the two roots, we compute $y_i = -(n_1x_i + n_3f_0)/n_2$. If $n_2 \approx 0$, we substitute $x = -(n_2y + n_3f_0)/n_1$ into Eq. (9.1) to obtain

$$A(n_2x + n_3f_0)^2/n_1^2 - 2Bx(n_2y + n_3f_0)/n_1 + Cy^2$$
$$+2f_0(-D(n_2y + n_3f_0)/n_1 + Ey) + f_0^2F = 0.$$

Rearranging this, we obtain the following quadratic equation.

$$(An_2^2 - 2Bn_1n_2 + Cn_1^2)y^2 + 2f_0(En_1^2 + An_2n_3 - Bn_1n_3 - Dn_1n_2)y$$
$$+(An_3^2 - 2Dn_1n_3 + Fn_1^2)f_0^2 = 0.$$

Letting y_i, $i = 1, 2$, be the two roots, we compute $x_i = -(n_2x_i + n_3f_0)/n_1$. In actual computation, instead of considering the cases of $n_2 \neq 0$ and $n_2 \approx 0$, it is more practical to do the former computation for $|n_2| \geq |n_1|$ and the latter computation for $|n_2| < |n_1|$. If the quadratic equation has imaginary roots, the line does not intersect the ellipse. For a double root, we obtain the tangent point.

9.5 The line $n_1x + n_2y + n_3f_0 = 0$ has its normal $(n_1, n_2)^\top$. The normal to curve $F(x, y) = 0$ at (x_0, y_0) is given by $\nabla F = (\partial F/\partial x, \partial F/\partial y)$. Therefore the normal to the curve of Eq. (9.1) at (x_0, y_0) is given by $(n_1, n_2)^\top$ with

$$n_1 = Ax_0 + By_0 + f_0D, \qquad n_2 = Bx_0 + Cy_0 + f_0E.$$

It follows that the line that passes through (x_0, y_0) and has normal $(n_1, n_2)^\top$ is

$$n_1(x - x_0) + n_2(x - x_0) + n_3f_0 = 0.$$

Expanding this and noting that $Ax_0^2 + 2Bx_0y_0 + Cy_0^2 + 2f_0(Dx_0 + Ey_0) + f_0^2F = 0$, we obtain the line $n_1x + n_2y + n_3f_0 = 0$ given by Eq. (9.8).

9.6 In Procedure 2.6, we replace (x_α, y_α) by (a, b) and remove Step 4 for ellipse update to obtain the following procedure.

1. Express the ellipse using the 6D vector $\boldsymbol{\theta}$ as in Eq. (2.3). Let $J_0^* = \infty$ (a sufficiently large number), $\hat{a} = a$, $\hat{b} = b$, and $\tilde{a} = \tilde{b} = 0$.
2. Let $V_0[\hat{\boldsymbol{\xi}}]$ be the normalized covariance matrix obtained from $V_0[\boldsymbol{\xi}_\alpha]$ in Eq. (2.15) by replacing $\bar{x}_\alpha$ and $\bar{y}_\alpha$ by $\hat{a}$ and $\hat{b}$, respectively.
3. Compute the following $\boldsymbol{\xi}^*$.

$$\boldsymbol{\xi}^* = \begin{pmatrix} \hat{a}^2 + 2\hat{a}\tilde{a} \\ 2(\hat{a}\hat{b} + \hat{b}\tilde{a} + \hat{a}\tilde{b}) \\ \hat{b}^2 + 2\hat{b}\tilde{b} \\ 2f_0(\hat{a} + \tilde{a}) \\ 2f_0(\hat{b} + \tilde{b}) \\ f_0 \end{pmatrix}.$$

4. Update $\tilde{a}$, $\tilde{b}$, $\hat{a}$, and $\hat{b}$ to

$$\begin{pmatrix} \tilde{a} \\ \tilde{b} \end{pmatrix} \leftarrow \frac{2(\boldsymbol{\xi}^*, \boldsymbol{\theta})^2}{(\boldsymbol{\theta}, V_0[\hat{\boldsymbol{\xi}}]\boldsymbol{\theta})} \begin{pmatrix} \theta_1 & \theta_2 & \theta_4 \\ \theta_2 & \theta_3 & \theta_5 \end{pmatrix} \begin{pmatrix} \hat{a} \\ \hat{b} \\ f_0 \end{pmatrix}, \quad \hat{a} \leftarrow a - \tilde{a}, \quad \hat{b} \leftarrow b - \tilde{b}.$$

5. Compute the following J^*.

$$J^* = \tilde{a}^2 + \tilde{b}^2.$$

   If $J^* \approx J_0$, return $(\hat{a}, \hat{b})$ and stop. Else, let $J_0 \leftarrow J^*$, and go back to Step 2.

9.7 If we transpose $\mathsf{A}\mathsf{A}^{-1} = \mathsf{I}$ on both sides, we have $(\mathsf{A}^{-1})^\top\mathsf{A}^\top = \mathsf{I}$. This means that $(\mathsf{A}^{-1})^\top$ is the inverse of $\mathsf{A}^\top$. Therefore $(\mathsf{A}^{-1})^\top = (\mathsf{A}^\top)^{-1}$.

9.8 (1) The matrix Q and its normalization $\bar{\mathsf{Q}}$ have the form:

$$\mathsf{Q} = \begin{pmatrix} 1 & 0 & 0 \\ 0 & \alpha & 0 \\ 0 & 0 & -\gamma/f_0^2 \end{pmatrix}, \quad \bar{\mathsf{Q}} = \kappa \begin{pmatrix} 1 & 0 & 0 \\ 0 & \alpha & 0 \\ 0 & 0 & -\gamma/f^2 \end{pmatrix}, \quad \kappa = \Big(\frac{f}{\sqrt{\alpha\gamma}}\Big)^{2/3}.$$

Because the major axis is along the x-axis, the supporting plane is inclined along the y-axis. If the inclination angle is θ, the unit surface normal to the supporting plane is $\mathbf{n} = (0, \sin\theta, \cos\theta)^\top$ with the sign of θ indeterminate. If the camera is rotated around the X-axis by $-\theta$, the supporting plane becomes parallel to the image plane. Such a camera rotation is given by

$$\mathsf{R} = \begin{pmatrix} 1 & 0 & 0 \\ 0 & \cos\theta & \sin\theta \\ 0 & -\sin\theta & \cos\theta \end{pmatrix}.$$

After the camera rotation, we observe a circle in the form

$$x^2 + (y + c)^2 = \rho^2, c \geq 0, \rho > 0.$$

The corresponding normalized matrix is

$$\bar{\mathsf{Q}}' = \kappa' \begin{pmatrix} 1 & 0 & 0 \\ 0 & 1 & c/f \\ 0 & c/f & (c^2 - \rho^2)/f^2 \end{pmatrix}, \kappa' = \left(\frac{f}{\rho}\right)^{2/3}.$$

Thus we obtain from Eq. (9.24) the equality

$$\kappa' \begin{pmatrix} 1 & 0 & 0 \\ 0 & 1 & c/f \\ 0 & c/f & (c^2 - \rho^2)/f^2 \end{pmatrix}$$
$$= \kappa \begin{pmatrix} 1 & 0 & 0 \\ 0 & \cos\theta & -\sin\theta \\ 0 & \sin\theta & \cos\theta \end{pmatrix} \begin{pmatrix} 1 & 0 & 0 \\ 0 & \alpha & 0 \\ 0 & 0 & -\gamma/f^2 \end{pmatrix} \begin{pmatrix} 1 & 0 & 0 \\ 0 & \cos\theta & \sin\theta \\ 0 & -\sin\theta & \cos\theta \end{pmatrix}.$$

Comparing the (1, 1) elements on both sides, we see that $\kappa = \kappa'$ and hence $\rho = \sqrt{\alpha\gamma}$. Comparing the remaining elements, we obtain

$$\begin{pmatrix} 1 & c/f \\ c/f & (c^2 - \rho^2)/f^2 \end{pmatrix} = \begin{pmatrix} \cos\theta & -\sin\theta \\ \sin\theta & \cos\theta \end{pmatrix} \begin{pmatrix} \alpha & 0 \\ 0 & -\gamma/f^2 \end{pmatrix} \begin{pmatrix} \cos\theta & \sin\theta \\ -\sin\theta & \cos\theta \end{pmatrix}.$$

This implies that the matrix on the left-hand side has eigenvalues α and $-\gamma/f^2$ with corresponding unit eigenvectors $(\cos\theta, \sin\theta)^\top$ and $(-\sin\theta, \cos\theta)^\top$. Because the trace and the determinant are invariant under the above congruent transformation, which is actually SVD, we obtain

$$1 + \frac{c^2 - \rho^2}{f^2} = \alpha - \frac{\gamma}{f^2}, \frac{c^2 - \rho^2}{f^2} - \frac{c^2}{f^2} = -\frac{\alpha\gamma}{f^2}.$$

Hence, $\rho = \sqrt{\alpha\gamma}$ and $c = \sqrt{(\alpha - 1)(\gamma + f^2)}$. Because $(\cos\theta, \sin\theta)^\top$ is the eigenvector for eigenvalue α, the following holds.

$$\begin{pmatrix} 1 & c/f \\ c/f & (c^2 - \rho^2)/f^2 \end{pmatrix} \begin{pmatrix} \cos\theta \\ \sin\theta \end{pmatrix} = \alpha \begin{pmatrix} \cos\theta \\ \sin\theta \end{pmatrix}.$$

Thus, we obtain

$$\tan\theta = \frac{\alpha - 1}{c/f} = \sqrt{\frac{\alpha - 1}{1 + \gamma/f^2}},$$

from which we obtain Eq. (9.38). Comparing the radius ρ of the circle on the image plane, which is at distance f from the viewpoint O (Fig. 5.1), and the true radius r on the supporting plane, we see that the distance h to the supporting plane is $h = fr/\rho$. Therefore we obtain Eq. (9.39).

(2) If the observed ellipse is not in the form of Eq. (9.37), we rotate the camera by some rotation R such that it has the form of Eq. (9.37). This process corresponds to the following diagonalization of the normalized matrix $\bar{\mathsf{Q}}$ in the form of Eq. (9.24).

$$\mathsf{R}^\top \bar{\mathsf{Q}} \mathsf{R} = \begin{pmatrix} \lambda_1 & 0 & 0 \\ 0 & \lambda_2 & 0 \\ 0 & 0 & \lambda_3 \end{pmatrix}.$$

The determinant of $\bar{\mathsf{Q}}$ is normalized to -1, therefore we have $\lambda_1\lambda_2\lambda_3 = -1$. We are assuming that the three eigenvalues are arranged in the order $\lambda_2 \geq \lambda_1 > 0 > \lambda_3$, therefore Eq. (9.21) implies that this ellipse has the form

$$x^2 + \frac{\lambda_2}{\lambda_1} y^2 = -f^2 \frac{\lambda_3}{\lambda_1}.$$

Comparing this with Eq. (9.37), we see that $\alpha = \lambda_2/\lambda_1$ (≥ 1) and $\gamma = f^2\lambda_3/\lambda_1$ (> 0). Thus, the inclination angle θ is given by Eq. (9.38). The y- and the z-axes after the camera rotation are, respectively, in the directions $\mathbf{u}_2$ and $\mathbf{u}_3$, which are the second and the third columns of the matrix R in Eq. (9.24). Thus the unit surface normal $\mathbf{n}$ to the supporting plane is given by $\mathbf{n} = \mathbf{u}_2 \sin\theta + \mathbf{u}_3 \cos\theta$. Rewriting the α and γ in Eqs. (9.38) and (9.39) in terms of λ_1, λ_2, and λ_3, we obtain Eq. (9.19).

9.9 Because $(x, y, f)^\top$ is the direction of the line of sight, we can write $X = cx$, $Y = cy$, and $Z = cf$ for some c. This point is on the supporting plane $n_1X + n_2Y + n_3Z = h$, thus $c(n_1x + n_2y + n_3f) = h$ holds. Hence, $c = h/(n_1x + n_2y + n_3f)$, and X, Y, and Z are given by Eq. (9.40).

9.10 (1) If the supporting plane is parallel to the image plane, the circle is imaged in the form of $(x - x_c)^2 + (y - y_c)^2 = r^2$ for some (x_c, y_c) and ρ. This circle corresponds to the matrix $\bar{\mathsf{Q}}$ in the form

$$\bar{\mathsf{Q}} \simeq \begin{pmatrix} 1 & 0 & -x_c/f \\ 0 & 1 & -y_c/f \\ -x_c/f & -y_c/f & (x_c^2 + y_c^2 - r^2)/f^2 \end{pmatrix}.$$

Because $(x_c, y_c) = (x_C, y_C)$, that is, the center of the circle in the scene is projected to the center of the imaged circle, the right-hand side of Eq. (9.27) is $(0, 0, -r^2/f)^\top$, which is a vector orthogonal to the image plane.

(2) Suppose we observe a circle on the supporting plane with unit surface normal $\mathbf{n}$. If the camera is rotated by R around the viewpoint, the unit surface normal is $\mathbf{n}' = \mathsf{R}^\top$ relative to the rotated camera. Suppose the projected center (x_C, y_C) moves to (x'_C, y'_C) after the camera rotation. Their ray directions $(x'_C, y'_C, f)^\top$ and (x_C, y_C), (x'_C, y'_C) are related by

$$\begin{pmatrix} x'_C \\ y'_C \\ f \end{pmatrix} \simeq \mathsf{R}^\top \begin{pmatrix} x_C \\ y_C \\ f \end{pmatrix}$$

(see Fig. 9.3). Multiplying Eq. (9.27) by $\mathsf{R}^\top$ on both sides, and noting Eq. (9.24), we obtain

$$\mathbf{n}' = \mathsf{R}^\top \mathbf{n} \simeq \mathsf{R}^\top \bar{\mathsf{Q}} \begin{pmatrix} x_C \\ y_C \\ f \end{pmatrix} \simeq (\mathsf{R}^\top \bar{\mathsf{Q}} \mathsf{R}) \mathsf{R}^\top \begin{pmatrix} x_C \\ y_C \\ f \end{pmatrix} \simeq \bar{\mathsf{Q}}' \begin{pmatrix} x'_C \\ y'_C \\ f \end{pmatrix},$$

which means that Eq. (9.27) also holds after the camera rotation.

9.11 The unit vector along the Z-axis is $\mathbf{k} = (0, 0, 1)^\top$. Let Ω be the angle made by $\mathbf{n}$ and $\mathbf{k}$ (positive for rotating $\mathbf{n}$ toward $\mathbf{k}$ screw-wise). It is computed by

$$\Omega = \sin^{-1} \|\mathbf{n} \times \mathbf{k}\|.$$

The unit vector orthogonal to both $\mathbf{n}$ and $\mathbf{k}$ is given by

$$\mathbf{l} = \mathcal{N}[\mathbf{n} \times \mathbf{k}].$$

Thus we rotate the camera around $\mathbf{l}$ by angle Ω screw-wise. The corresponding rotation matrix R is given by Eq. (3.48).

9.12 Let $I(i, j)$ be the value of the pixel with integer coordinates (i, j). For noninteger coordinates (x, y), let (i, j) be their integer parts, and (ξ, η) $(= (x - i, y - j))$ their fraction parts. We compute $I(i, j)$ by the bilinear interpolation:

$$\begin{aligned} I(x, y) =& (1 - \xi)(1 - \eta)I(i, j) + \xi(1 - \eta)I(i + 1, j) + (1 - \xi)\eta I(i, j + 1) \\ &+ \xi\eta I(i + 1, j + 1). \end{aligned}$$

This means that we first do linear interpolation in the j-direction, combining $I(i, j)$ and $I(i, j + 1)$ in the ratio $\eta : 1 - \eta$, and then combining these in the i-direction in the ratio $\xi : 1 - \xi$. We obtain the same result if we change the order, linearly interpolating first in the i-direction and then in the j-direction.

Solutions of Chapter 10

10.1 If we let

$$T_{pq} = \sum_{i,j,k,l,m=1}^{3} \varepsilon_{ljp}\varepsilon_{mkq} T_i^{lm} x_{0(i)} x_{1(j)} x_{2(k)},$$

the following summation vanishes.

$$\sum_{p=1}^{3} T_{pq} x_{1(p)} = \sum_{i,j,k,l,m,p=1}^{3} \varepsilon_{ljp}\varepsilon_{mkq} T_i^{lm} x_{0(i)} x_{1(j)} x_{2(k)} x_{1(p)} = 0.$$

This is because the sign of ε_{ljp} is reversed by interchanging j and p but the value of $x_{1(j)} x_{1(p)}$ remains the same. Thus only two among T_{1q}, T_{2q}, and T_{3q} are linearly independent (e.g., T_{3q} can be expressed as a linear combination of T_{1q} and T_{2q}). Similarly, the following holds.

$$\sum_{q=1}^{3} T_{pq} x_{2(q)} = \sum_{i,j,k,l,m,p=1}^{3} \varepsilon_{ljp}\varepsilon_{mkq} T_i^{lm} x_{0(i)} x_{1(j)} x_{2(k)} x_{2(q)} = 0.$$

Therefore only two among T_{p1}, T_{p2}, and T_{p3} are linearly independent. Thus, only four among nine T_{pq} are linearly independent (e.g., each of T_{13}, T_{23}, T_{31}, T_{32}, and T_{33} can be expressed as a linear combination of T_{11}, T_{12}, T_{21}, and T_{22}).

10.2 Suppose each observed position $\mathbf{x}_\kappa$ is already corrected to $\hat{\mathbf{x}}_\kappa$ (initially, we let $\hat{\mathbf{x}}_\kappa = \mathbf{x}_\kappa$). We consider how to correct it to an optimal position $\bar{\mathbf{x}}_\kappa$. Instead of directly estimating $\bar{\mathbf{x}}_\kappa$, we let

$$\bar{\mathbf{x}}_\kappa = \hat{\mathbf{x}}_\kappa - \Delta\hat{\mathbf{x}}_\kappa, \qquad \kappa = 0, 1, 2,$$

and estimate the additional correction $\Delta\hat{\mathbf{x}}_\kappa$. The reprojection error E of Eq. (10.4) is written as

$$E = \sum_{\kappa=0}^{2} \|\tilde{\mathbf{x}}_\kappa + \Delta\hat{\mathbf{x}}_\kappa\|^2, \qquad (*1)$$

where

$$\tilde{\mathbf{x}}_\kappa = \mathbf{x}_\kappa - \hat{\mathbf{x}}_\kappa, \qquad \kappa = 0, 1, 2.$$

The trilinear constraint of Eq. (10.2) is written as

$$\sum_{i,j,k,l,m=1}^{3} \varepsilon_{ljp}\varepsilon_{mkq}T_i^{lm}(\hat{x}_{0(i)} - \Delta\hat{x}_{0(i)})(\hat{x}_{1(j)} - \Delta\hat{x}_{1(j)})(\hat{x}_{2(k)} - \Delta\hat{x}_{2(k)}) = 0.$$

Expanding this and ignoring second-order terms in $\Delta\hat{\mathbf{x}}_\kappa$, we obtain

$$\sum_{i,j,k,l,m=1}^{3} \varepsilon_{ljp}\varepsilon_{mkq}T_i^{lm}\Big(\Delta\hat{x}_{0(i)}\hat{x}_{1(j)}\hat{x}_{2(k)} + \hat{x}_{0(i)}\Delta\hat{x}_{1(j)}\hat{x}_{2(k)} + \hat{x}_{0(i)}\hat{x}_{1(j)}\Delta\hat{x}_{2(k)}\Big)$$

$$= \sum_{i,j,k,l,m=1}^{3} \varepsilon_{ljp}\varepsilon_{mkq}T_i^{lm}\hat{x}_{0(i)}\hat{x}_{1(j)}\hat{x}_{2(k)}. \qquad (*2)$$

From the definition of $\mathbf{x}_\kappa$ of Eq. (10.1), the third component of $\Delta\mathbf{x}_\kappa$ is identically 0. This constraint is written in the form

$$\sum_{k=1}^{3} k_i \Delta\hat{x}_{\kappa(i)} = 0, \qquad \kappa = 0, 1, 2, \qquad (*3)$$

where k_i is the ith component of $\mathbf{k} \equiv (0, 0, 1)^\top$. Dividing Eq. (∗1) by 2 and introducing Lagrange multipliers for Eqs. (∗2) and (∗3), we differentiate

$$\frac{1}{2}\sum_{\kappa=0}^{2} \|\tilde{\mathbf{x}}_\kappa + \Delta\hat{\mathbf{x}}_\kappa\|^2 - \sum_{i,j,k,l,m,p,q=1}^{3} \lambda_{pq}\varepsilon_{ljp}\varepsilon_{mkq}T_i^{lm}\Big(\Delta\hat{x}_{0(i)}\hat{x}_{1(j)}\hat{x}_{2(k)} + \hat{x}_{0(i)}\Delta\hat{x}_{1(j)}\hat{x}_{2(k)}$$

$$+ \hat{x}_{0(i)}\hat{x}_{1(j)}\Delta\hat{x}_{2(k)}\Big) - \sum_{\kappa=0}^{2}\sum_{i=1}^{3} \mu_\kappa k_i \Delta\hat{x}_{\kappa(i)}$$

with respect to $\Delta\hat{x}_{0(i)}$, $\Delta\hat{x}_{1(i)}$, and $\Delta\hat{x}_{2(i)}$. Letting the result be 0, we obtain

$$\Delta\hat{x}_{0(i)} = \sum_{j,k,l,m,p,q=1}^{3} \lambda_{pq}\varepsilon_{ljp}\varepsilon_{mkq}T_i^{lm}\hat{x}_{1(j)}\hat{x}_{2(k)} + \mu_0 k_i - \tilde{x}_{0(i)},$$

$$\Delta\hat{x}_{1(i)} = \sum_{j,k,l,m,p,q=1}^{3} \lambda_{pq}\varepsilon_{ljp}\varepsilon_{mkq}T_i^{lm}\hat{x}_{0(i)}\hat{x}_{2(k)} + \mu_1 k_i - \tilde{x}_{1(i)},$$

$$\Delta\hat{x}_{2(i)} = \sum_{j,k,l,m,p,q=1}^{3} \lambda_{pq}\varepsilon_{ljp}\varepsilon_{mkq}T_i^{lm}\hat{x}_{0(i)}\hat{x}_{1(j)} + \mu_2 k_i - \tilde{x}_{2(i)}.$$

Multiplying this by the matrix $\mathsf{P}_\mathbf{k}$ of Eq. (10.6) from the left on both sides, and noting that $\mathsf{P}_\mathbf{k}\Delta\hat{\mathbf{x}}_\kappa = \Delta\hat{\mathbf{x}}_\kappa$, $\mathsf{P}_\mathbf{k}\tilde{\mathbf{x}}_\kappa = \tilde{\mathbf{x}}_\kappa$, and $\mathsf{P}_\mathbf{k}\mathbf{k} = \mathbf{0}$, we obtain

$$\Delta\hat{x}_{0(s)} = \sum_{i,j,k,l,m,p,q=1}^{3} \lambda_{pq}\varepsilon_{ljp}\varepsilon_{mkq}T_i^{lm}P_\mathbf{k}^{si}\hat{x}_{1(j)}\hat{x}_{2(k)} - \tilde{x}_{0(s)} = \sum_{p,q=1}^{3} P_{pqs}\lambda_{pq} - \tilde{x}_{0(s)},$$

$$\Delta\hat{x}_{1(s)} = \sum_{i,j,k,l,m,p,q=1}^{3} \lambda_{pq}\varepsilon_{ljp}\varepsilon_{mkq}T_i^{lm}\hat{x}_{0(i)}P_\mathbf{k}^{sj}\hat{x}_{2(k)} - \tilde{x}_{1(s)} = \sum_{p,q=1}^{3} Q_{pqs}\lambda_{pq} - \tilde{x}_{1(s)},$$

$$\Delta\hat{x}_{2(s)} = \sum_{i,j,k,l,m,p,q=1}^{3} \lambda_{pq}\varepsilon_{ljp}\varepsilon_{mkq}T_i^{lm}\hat{x}_{0(i)}\hat{x}_{1(j)}P_\mathbf{k}^{sk} - \tilde{x}_{2(s)} = \sum_{p,q=1}^{3} R_{pqs}\lambda_{pq} - \tilde{x}_{2(s)},$$

$$(*4)$$

where P_{pqs}, Q_{pqs}, and R_{pqs} are defined by Eq. (10.5). Substituting these into Eq. (∗2), we obtain Eq. (10.8), using the definition of Eq. (10.7). Solving this for λ_{pq}, and substituting them into Eq. (∗4), we can determine $\Delta\hat{\mathbf{x}}_\kappa$. Thus $\bar{\mathbf{x}}_\kappa$ is given by $\hat{\mathbf{x}}_\kappa - \Delta\hat{\mathbf{x}}_\kappa$. However, because second-order terms were ignored in Eq. (∗2), the trilinear constraint may not exactly hold. Hence we newly regard this solution as $\hat{\mathbf{x}}_\kappa$ in the form of (10.9) and repeat the same procedure. The ignored terms decrease each time. In the end, $\Delta\hat{\mathbf{x}}_\kappa$ becomes $\mathbf{0}$, and the trilinear constraint exactly holds. From Eq. (∗1), the reprojection error E is given by Eq. (10.10). If its decrease is sufficiently small, we stop the iterations.

10.3 We can write Eq. (10.23) as

$$\begin{pmatrix} f_0P_{0(11)} - x_0P_{0(31)} & f_0P_{0(12)} - x_0P_{0(32)} & f_0P_{0(13)} - x_0P_{0(33)} \\ f_0P_{0(21)} - y_0P_{0(31)} & f_0P_{0(22)} - y_0P_{0(32)} & f_0P_{0(23)} - y_0P_{0(33)} \\ f_0P_{1(11)} - x_1P_{1(31)} & f_0P_{1(12)} - x_1P_{1(32)} & f_0P_{1(13)} - x_1P_{1(33)} \\ f_0P_{1(21)} - y_1P_{1(31)} & f_0P_{1(22)} - y_1P_{1(32)} & f_0P_{1(23)} - y_1P_{1(33)} \\ f_0P_{2(11)} - x_2P_{2(31)} & f_0P_{2(12)} - x_2P_{2(32)} & f_0P_{2(13)} - x_2P_{2(33)} \\ ff_0P_{2(21)} - y_2P_{2(31)} & f_0P_{2(22)} - y_2P_{2(32)} & f_0P_{2(23)} - y_2P_{2(33)} \end{pmatrix}$$

$$\left.\begin{matrix} f_0P_{0(14)} - x_0P_{0(34)} \\ f_0P_{0(24)} - y_0P_{0(34)} \\ f_0P_{1(14)} - x_1P_{1(34)} \\ f_0P_{1(24)} - y_1P_{1(34)} \\ f_0P_{2(14)} - x_2P_{2(34)} \\ f_0P_{2(24)} - y_2P_{2(34)} \end{matrix}\right) \begin{pmatrix} X \\ Y \\ Z \\ 1 \end{pmatrix} = \begin{pmatrix} 0 \\ 0 \\ 0 \\ 0 \\ 0 \\ 0 \end{pmatrix}.$$

This has a unique solution (X, Y, Z) if and only if three among the six equations are linearly independent and others are written as their linear combinations, or equivalently the matrix on the left-hand side has rank 3. This means that its arbitrary 4×4 minors are 0. Therefore we can obtain the constraint among the three images by extracting, from among the six rows, four rows that contain quantities of the three images and let the resulting minor be 0. For example, if we extract the first, second, third, and fifth rows, we obtain

$$\begin{vmatrix} f_0P_{0(11)} - x_0P_{0(31)} & f_0P_{0(12)} - x_0P_{0(32)} & f_0P_{0(13)} - x_0P_{0(33)} & f_0P_{0(14)} - x_0P_{0(34)} \\ f_0P_{0(21)} - y_0P_{0(31)} & f_0P_{0(22)} - y_0P_{0(32)} & f_0P_{0(23)} - y_0P_{0(33)} & f_0P_{0(24)} - y_0P_{0(34)} \\ f_0P_{1(11)} - x_1P_{1(31)} & f_0P_{1(12)} - x_1P_{1(32)} & f_0P_{1(13)} - x_1P_{1(33)} & f_0P_{1(14)} - x_1P_{1(34)} \\ f_0P_{2(11)} - x_2P_{2(31)} & f_0P_{2(12)} - x_2P_{2(32)} & f_0P_{2(13)} - x_2P_{2(33)} & f_0P_{2(14)} - x_2P_{2(34)} \end{vmatrix} = 0.$$

Dividing each element by f_0, writing x_0/f_0, y_0/f_0, x_1/f_0, and x_2/f_0 as $x_{0(1)}$, $x_{0(2)}$, $x_{1(1)}$, and $x_{2(1)}$, respectively, we can rewrite the above equation as

$$\begin{vmatrix} P_{0(11)}-x_{0(1)}P_{0(31)} & P_{0(12)}-x_{0(1)}P_{0(32)} & P_{0(13)}-x_{0(1)}P_{0(33)} & P_{0(14)}-x_{0(1)}P_{0(34)} \\ P_{0(21)}-x_{0(2)}P_{0(31)} & P_{0(22)}-x_{0(2)}P_{0(32)} & P_{0(23)}-x_{0(2)}P_{0(33)} & P_{0(24)}-x_{0(2)}P_{0(34)} \\ P_{1(11)}-x_{1(1)}P_{1(31)} & P_{1(12)}-x_{1(1)}P_{1(32)} & P_{1(13)}-x_{1(1)}P_{1(33)} & P_{1(14)}-x_{1(1)}P_{1(34)} \\ P_{2(11)}-x_{2(1)}P_{2(31)} & P_{2(12)}-x_{2(1)}P_{2(32)} & P_{2(13)}-x_{2(1)}P_{2(33)} & P_{2(14)}-x_{2(1)}P_{2(34)} \end{vmatrix}$$

$$= \begin{vmatrix} P_{0(11)}-x_{0(1)}P_{0(31)} & P_{0(12)}-x_{0(1)}P_{0(32)} & P_{0(13)}-x_{0(1)}P_{0(33)} & P_{0(14)}-x_{0(1)}P_{0(34)} & 0 & 0 & 0 \\ P_{0(21)}-x_{0(2)}P_{0(31)} & P_{0(22)}-x_{0(2)}P_{0(32)} & P_{0(23)}-x_{0(2)}P_{0(33)} & P_{0(24)}-x_{0(2)}P_{0(34)} & 0 & 0 & 0 \\ P_{1(11)}-x_{1(1)}P_{1(31)} & P_{1(12)}-x_{1(1)}P_{1(32)} & P_{1(13)}-x_{1(1)}P_{1(33)} & P_{1(14)}-x_{1(1)}P_{1(34)} & 0 & 0 & 0 \\ P_{2(11)}-x_{2(1)}P_{2(31)} & P_{2(12)}-x_{2(1)}P_{2(32)} & P_{2(13)}-x_{2(1)}P_{2(33)} & P_{2(14)}-x_{2(1)}P_{2(34)} & 0 & 0 & 0 \\ P_{0(31)} & P_{0(32)} & P_{0(33)} & P_{0(34)} & 1 & 0 & 0 \\ P_{1(31)} & P_{1(32)} & P_{1(33)} & P_{1(34)} & 0 & 1 & 0 \\ P_{2(31)} & P_{2(32)} & P_{2(33)} & P_{2(34)} & 0 & 0 & 1 \end{vmatrix}$$

$$= \begin{vmatrix} P_{0(11)} & P_{0(12)} & P_{0(13)} & P_{0(14)} & x_{0(1)} & 0 & 0 \\ P_{0(21)} & P_{0(22)} & P_{0(23)} & P_{0(24)} & x_{0(2)} & 0 & 0 \\ P_{1(11)} & P_{1(12)} & P_{1(13)} & P_{1(14)} & 0 & x_{1(1)} & 0 \\ P_{2(11)} & P_{2(12)} & P_{2(13)} & P_{2(14)} & 0 & 0 & x_{2(1)} \\ P_{0(31)} & P_{0(32)} & P_{0(33)} & P_{0(34)} & 1 & 0 & 0 \\ P_{1(31)} & P_{1(32)} & P_{1(33)} & P_{1(34)} & 0 & 1 & 0 \\ P_{2(31)} & P_{2(32)} & P_{2(33)} & P_{2(34)} & 0 & 0 & 1 \end{vmatrix}$$

$$= \begin{vmatrix} P_{0(21)} & P_{0(22)} & P_{0(23)} & P_{0(24)} & 0 & 0 \\ P_{1(11)} & P_{1(12)} & P_{1(13)} & P_{1(14)} & x_{1(1)} & 0 \\ P_{2(11)} & P_{2(12)} & P_{2(13)} & P_{2(14)} & 0 & x_{2(1)} \\ P_{0(31)} & P_{0(32)} & P_{0(33)} & P_{0(34)} & 0 & 0 \\ P_{1(31)} & P_{1(32)} & P_{1(33)} & P_{1(34)} & 1 & 0 \\ P_{2(31)} & P_{2(32)} & P_{2(33)} & P_{2(34)} & 0 & 1 \end{vmatrix} x_{0(1)}$$

$$- \begin{vmatrix} P_{0(11)} & P_{0(12)} & P_{0(13)} & P_{0(14)} & 0 & 0 \\ P_{1(11)} & P_{1(12)} & P_{1(13)} & P_{1(14)} & x_{1(1)} & 0 \\ P_{2(11)} & P_{2(12)} & P_{2(13)} & P_{2(14)} & 0 & x_{2(1)} \\ P_{0(31)} & P_{0(32)} & P_{0(33)} & P_{0(34)} & 0 & 0 \\ P_{1(31)} & P_{1(32)} & P_{1(33)} & P_{1(34)} & 1 & 0 \\ P_{2(31)} & P_{2(32)} & P_{2(33)} & P_{2(34)} & 0 & 1 \end{vmatrix} x_{0(2)}$$

$$
+\begin{vmatrix}
P_{0(11)} & P_{0(12)} & P_{0(13)} & P_{0(14)} & 0 & 0 \\
P_{0(21)} & P_{0(22)} & P_{0(23)} & P_{0(24)} & 0 & 0 \\
P_{1(11)} & P_{1(12)} & P_{1(13)} & P_{1(14)} & x_{1(1)} & 0 \\
P_{2(11)} & P_{2(12)} & P_{2(13)} & P_{2(14)} & 0 & x_{2(1)} \\
P_{1(31)} & P_{1(32)} & P_{1(33)} & P_{1(34)} & 1 & 0 \\
P_{2(31)} & P_{2(32)} & P_{2(33)} & P_{2(34)} & 0 & 1
\end{vmatrix} = 0, \qquad (*)
$$

where in going from the first expression to the second we have used the fact that the determinant does not change by adding a diagonal block consisting of the identity matrix if either of the nondiagonal blocks consists of only 0. In going from the second to the third expression, the fifth row is multiplied by $x_{0(1)}$ and added to the first row, the sixth row is multiplied by $x_{0(2)}$ and added to the second row, and the seventh row is multiplied by $x_{2(1)}$ and added to the fourth row ($\hookrightarrow$ Problem 4.3(2)). The fourth expression is obtained by cofactor expansion of the third expression with respect to the fifth column. The first term of the fourth expression is rewritten by cofactor expansion with respect to the fifth and sixth columns in the following form ($\hookrightarrow$ Problem 4.3(2)).

$$
\left(-\begin{vmatrix}
P_{0(21)} & P_{0(22)} & P_{0(23)} & P_{0(24)} & 0 \\
P_{2(11)} & P_{2(12)} & P_{2(13)} & P_{2(14)} & x_{2(1)} \\
P_{0(31)} & P_{0(32)} & P_{0(33)} & P_{0(34)} & 0 \\
P_{1(31)} & P_{1(32)} & P_{1(33)} & P_{1(34)} & 0 \\
P_{2(31)} & P_{2(32)} & P_{2(33)} & P_{2(34)} & 1
\end{vmatrix} x_{1(1)} + \begin{vmatrix}
P_{0(21)} & P_{0(22)} & P_{0(23)} & P_{0(24)} & 0 \\
P_{1(11)} & P_{1(12)} & P_{1(13)} & P_{1(14)} & 0 \\
P_{2(11)} & P_{2(12)} & P_{2(13)} & P_{2(14)} & x_{2(1)} \\
P_{0(31)} & P_{0(32)} & P_{0(33)} & P_{0(34)} & 0 \\
P_{2(31)} & P_{2(32)} & P_{2(33)} & P_{2(34)} & 1
\end{vmatrix}\right) x_{0(1)}
$$

$$
= \left(\begin{vmatrix}
P_{0(21)} & P_{0(22)} & P_{0(23)} & P_{0(24)} \\
P_{0(31)} & P_{0(32)} & P_{0(33)} & P_{0(34)} \\
P_{1(31)} & P_{1(32)} & P_{1(33)} & P_{1(34)} \\
P_{2(31)} & P_{2(32)} & P_{2(33)} & P_{2(34)}
\end{vmatrix} x_{2(1)} - \begin{vmatrix}
P_{0(21)} & P_{0(22)} & P_{0(23)} & P_{0(24)} \\
P_{2(11)} & P_{2(12)} & P_{2(13)} & P_{2(14)} \\
P_{0(31)} & P_{0(32)} & P_{0(33)} & P_{0(34)} \\
P_{1(31)} & P_{1(32)} & P_{1(33)} & P_{1(34)}
\end{vmatrix}\right) x_{0(1)} x_{1(1)}
$$

$$
+ \left(\begin{vmatrix}
P_{0(21)} & P_{0(22)} & P_{0(23)} & P_{0(24)} \\
P_{1(11)} & P_{1(12)} & P_{1(13)} & P_{1(14)} \\
P_{0(31)} & P_{0(32)} & P_{0(33)} & P_{0(34)} \\
P_{2(31)} & P_{2(32)} & P_{2(33)} & P_{2(34)}
\end{vmatrix} x_{2(1)} + \begin{vmatrix}
P_{0(21)} & P_{0(22)} & P_{0(23)} & P_{0(24)} \\
P_{1(11)} & P_{1(12)} & P_{1(13)} & P_{1(14)} \\
P_{2(11)} & P_{2(12)} & P_{2(13)} & P_{2(14)} \\
P_{0(31)} & P_{0(32)} & P_{0(33)} & P_{0(34)}
\end{vmatrix}\right) x_{0(1)}
$$

$$
= T_1^{33} x_{0(1)} x_{1(1)} x_{2(1)} - T_1^{31} x_{0(1)} x_{1(1)} - T_1^{13} x_{0(1)} x_{2(1)} + T_1^{11} x_{0(1)}
$$

$$
= T_1^{33} x_{0(1)} x_{1(1)} x_{2(1)} - T_1^{31} x_{0(1)} x_{1(1)} x_{2(3)} - T_1^{13} x_{0(1)} x_{1(3)} x_{2(1)} + T_1^{11} x_{0(1)} x_{1(3)} x_{2(3)}
$$

$$
= \sum_{j,k,l,m=1}^{3} \varepsilon_{lj2} \varepsilon_{mk2} T_1^{lm} x_{0(1)} x_{1(j)} x_{2(k)}.
$$

Here, we have used the definition of T_i^{jk} of Eq. (10.3), the fact that $x_{\kappa(3)} = 1$, and the properties of the permutation signature ε_{ijk}. Similarly expanding the second and third terms of Eq. $(*)$, we obtain, respectively,

$$
\sum_{j,k,l,m=1}^{3} \varepsilon_{lj2} \varepsilon_{mk2} T_2^{lm} x_{0(1)} x_{1(j)} x_{2(k)}, \qquad \sum_{j,k,l,m=1}^{3} \varepsilon_{lj2} \varepsilon_{mk2} T_3^{lm} x_{0(1)} x_{1(j)} x_{2(k)}.
$$

Thus, Eq. (∗) represents the trilinear constraint of Eq. (10.2) for $p = 2$ and $q = 2$; that is,

$$\sum_{i,j,k,l,m=1}^{3} \varepsilon_{lj2}\varepsilon_{mk2}T_i^{lm}x_{0(1)}x_{1(j)}x_{2(k)} = 0.$$

This is the result of extracting the first, second, third, and fifth rows of Eq. (10.23). Extracting other combinations of the rows, we obtain the trilinear constraint for other p and q.

10.4 Suppose each observed position $\mathbf{x}_\kappa$ is already corrected to $\hat{\mathbf{x}}_\kappa$ (initially, we let $\hat{\mathbf{x}}_\kappa = \mathbf{x}_\kappa$). We consider how to correct it to an optimal position $\bar{\mathbf{x}}_\kappa$. Instead of directly estimating $\bar{\mathbf{x}}_\kappa$, we let

$$\bar{\mathbf{x}}_\kappa = \hat{\mathbf{x}}_\kappa - \Delta\hat{\mathbf{x}}_\kappa, \qquad \kappa = 0, \ldots, M-1,$$

and estimate the additional correction $\Delta\hat{\mathbf{x}}_\kappa$. The reprojection error E of Eq. (10.24) is written as

$$E = \sum_{\kappa=0}^{M-1} \|\tilde{\mathbf{x}}_\kappa + \Delta\hat{\mathbf{x}}_\kappa\|^2, \qquad (*1)$$

where

$$\tilde{\mathbf{x}}_\kappa = \mathbf{x}_\kappa - \hat{\mathbf{x}}_\kappa, \qquad \kappa = 0, \ldots, M-1.$$

The constraint of Eq. (10.25) is written as

$$\sum_{i,j,k,l,m=1}^{3} \varepsilon_{ljp}\varepsilon_{mkq}T_{(\kappa)i}^{lm}(\hat{x}_{\kappa(i)} - \Delta\hat{x}_{\kappa(i)})(\hat{x}_{\kappa+1(j)} - \Delta\hat{x}_{\kappa+1(j)})(\hat{x}_{\kappa+2(k)} - \Delta\hat{x}_{\kappa+2(k)}) = 0.$$

Expanding this and ignoring second-order terms in $\Delta\hat{\mathbf{x}}_\kappa$, we obtain

$$\sum_{i,j,k,l,m=1}^{3} \varepsilon_{ljp}\varepsilon_{mkq}T_{(\kappa)i}^{lm}\Big(\Delta\hat{x}_{\kappa(i)}\hat{x}_{\kappa+1(j)}\hat{x}_{\kappa+2(k)} + \hat{x}_{\kappa(i)}\Delta\hat{x}_{\kappa+1(j)}\hat{x}_{\kappa+2(k)}$$
$$+\hat{x}_{\kappa(i)}\hat{x}_{\kappa+1(j)}\Delta\hat{x}_{\kappa+2(k)}\Big)$$
$$= \sum_{i,j,k,l,m=1}^{3} \varepsilon_{ljp}\varepsilon_{mkq}T_{(\kappa)i}^{lm}\hat{x}_{\kappa(i)}\hat{x}_{\kappa+1(j)}\hat{x}_{\kappa+2(k)}. \qquad (*2)$$

From the definition of $\mathbf{x}_\kappa$, the third component of $\Delta\mathbf{x}_\kappa$ is identically 0. This constraint is written in the form

$$\sum_{k=1}^{3} k_i\Delta\hat{x}_{\kappa(i)} = 0, \qquad \kappa = 0, \ldots, M-1, \qquad (*3)$$

where k_i is the ith component of $\mathbf{k} \equiv (0, 0, 1)^\top$. Dividing Eq. (∗1) by 2 and introducing Lagrange multipliers for Eqs. (∗2) and (∗3), we differentiate

$$\frac{1}{2}\sum_{\kappa=0}^{M-1} \|\tilde{\mathbf{x}}_\kappa + \Delta\hat{\mathbf{x}}_\kappa\|^2 - \sum_{\kappa=0}^{M-3}\sum_{i,j,k,l,m,p,q=1}^{3} \lambda_{\kappa(pq)}\varepsilon_{ljp}\varepsilon_{mkq}T_{(\kappa)i}^{lm}\Big(\Delta\hat{x}_{\kappa(i)}\hat{x}_{\kappa+1(j)}\hat{x}_{\kappa+2(k)}$$
$$+\hat{x}_{\kappa(i)}\Delta\hat{x}_{\kappa+1(j)}\hat{x}_{\kappa+2(k)} + \hat{x}_{\kappa(i)}\hat{x}_{\kappa+1(j)}\Delta\hat{x}_{\kappa+2(k)}\Big) - \sum_{\kappa=0}^{M-1}\sum_{i=1}^{3}\mu_\kappa k_i \Delta\hat{x}_{\kappa(i)}$$

with respect to $\Delta\hat{x}_{\kappa(n)}$. Letting the result be 0, we obtain

$$\Delta\hat{x}_{\kappa(n)} = \sum_{j,k,l,m,p,q=1}^{3} \lambda_{\kappa(pq)}\varepsilon_{ljp}\varepsilon_{mkq}T_{(\kappa)n}^{lm}\hat{x}_{\kappa+1(j)}\hat{x}_{\kappa+2(k)} + \mu_\kappa k_n - \tilde{x}_{\kappa(n)}.$$

Multiplying this by the matrix $\mathsf{P}_\mathbf{k}$ of Eq. (10.6) on both sides from the left, and noting that $\mathsf{P}_\mathbf{k}\Delta\hat{\mathbf{x}}_\kappa = \Delta\hat{\mathbf{x}}_\kappa$, $\mathsf{P}_\mathbf{k}\tilde{\mathbf{x}}_\kappa = \tilde{\mathbf{x}}_\kappa$, and $\mathsf{P}_\mathbf{k}\mathbf{k} = \mathbf{0}$, we obtain

$$\Delta\hat{x}_{\kappa(s)} = \sum_{p,q=1}^{3} P_{\kappa(pqs)}\lambda_{\kappa(pq)} + \sum_{p,q=1}^{3} Q_{\kappa(pqs)}\lambda_{\kappa-1(pq)} + \sum_{p,q=1}^{3} R_{\kappa(pqs)}\lambda_{\kappa-2(pq)}, \quad (*4)$$

where $P_{\kappa(pqs)}$, $Q_{\kappa(pqs)}$, and $R_{\kappa(pqs)}$ are defined by Eq. (10.26). Substituting this into Eq. (∗2), we obtain Eq. (10.28), using Eq. (10.27). Solving this for $\lambda_{\kappa(pq)}$, and substituting them into Eq. (∗4), we can determine $\Delta\hat{\mathbf{x}}_\kappa$. Hence, $\bar{\mathbf{x}}_\kappa$ is given by $\hat{\mathbf{x}}_\kappa - \Delta\hat{\mathbf{x}}_\kappa$. However, because second-order terms were ignored in Eq. (∗2), the constraint may not exactly hold. Thus we newly regard this solution as $\hat{\mathbf{x}}_\kappa$ in the form of Eq. (10.29) and repeat the same procedure. The ignored terms decrease each time. In the end, $\Delta\hat{\mathbf{x}}_\kappa$ becomes $\mathbf{0}$, and the constraint exactly holds. From Eq. (∗1), the reprojection error E is given by Eq. (10.31). If its decrease is sufficient small, we stop the iterations.

Problems of Chapter 11

11.1 If the coordinate system of the κth camera coincides with the world coordinate system, the relationship of Eq. (5.1) holds. Considering the displacement $(x_{\alpha\kappa} - u_{0\kappa}, y_{\alpha\kappa} - v_{0\kappa})$ from the principal point, we obtain

$$x_{\alpha\kappa} - u_{0\kappa} = f_\kappa\frac{X_\alpha}{Z_\alpha}, \qquad y_{\alpha\kappa} - v_{0\kappa} = f_\kappa\frac{Y_\alpha}{Z_\alpha},$$

which is rewritten as

$$\begin{pmatrix} x_{\alpha\kappa}/f_0 \\ y_{\alpha\kappa}/f_0 \\ 1 \end{pmatrix} \simeq \begin{pmatrix} x_{\alpha\kappa} \\ y_{\alpha\kappa} \\ f_0 \end{pmatrix} = \begin{pmatrix} f_\kappa X_\alpha/Z_\alpha + u_{0\kappa} \\ f_\kappa Y_\alpha/Z_\alpha + v_{0\kappa} \\ f_0 \end{pmatrix} \simeq \begin{pmatrix} f_\kappa X_\alpha + u_{0\kappa}Z_\alpha \\ f_\kappa Y_\alpha + v_{0\kappa}Z_\alpha \\ f_0 Z_\alpha \end{pmatrix}$$
$$= \begin{pmatrix} f_\kappa & 0 & u_{0\kappa} \\ 0 & f_\kappa & v_{0\kappa} \\ 0 & 0 & f_0 \end{pmatrix}\begin{pmatrix} X_\alpha \\ Y_\alpha \\ Z_\alpha \end{pmatrix}.$$

If the κth camera translates by $\mathbf{t}_\kappa$ and rotates by R_κ relative to the world coordinate system, the above equation is modified, using the relationships of Eqs. (5.4) and (5.5), to the form

$$\begin{pmatrix} x_{\alpha\kappa}/f_0 \\ y_{\alpha\kappa}/f_0 \\ 1 \end{pmatrix} \simeq \begin{pmatrix} f_\kappa & 0 & u_{0\kappa} \\ 0 & f_\kappa & v_{0\kappa} \\ 0 & 0 & f_0 \end{pmatrix} \begin{pmatrix} \mathsf{R}_\kappa^\top & -\mathsf{R}_\kappa^\top \mathbf{t}_\kappa \end{pmatrix} \begin{pmatrix} X_\alpha \\ Y_\alpha \\ Z_\alpha \\ 1 \end{pmatrix}.$$

11.2 (1) The derivatives of Eq. (11.12) with respect to X_α, Y_α, and Z_α are

$$\frac{\partial p_{\alpha\kappa}}{\partial X_\beta} = \delta_{\alpha\beta} P_{\kappa(11)}, \quad \frac{\partial p_{\alpha\kappa}}{\partial Y_\beta} = \delta_{\alpha\beta} P_{\kappa(12)}, \quad \frac{\partial p_{\alpha\kappa}}{\partial Z_\beta} = \delta_{\alpha\beta} P_{\kappa(13)},$$
$$\frac{\partial q_{\alpha\kappa}}{\partial X_\beta} = \delta_{\alpha\beta} P_{\kappa(21)}, \quad \frac{\partial q_{\alpha\kappa}}{\partial Y_\beta} = \delta_{\alpha\beta} P_{\kappa(22)}, \quad \frac{\partial q_{\alpha\kappa}}{\partial Z_\beta} = \delta_{\alpha\beta} P_{\kappa(23)},$$
$$\frac{\partial r_{\alpha\kappa}}{\partial X_\beta} = \delta_{\alpha\beta} P_{\kappa(31)}, \quad \frac{\partial r_{\alpha\kappa}}{\partial Y_\beta} = \delta_{\alpha\beta} P_{\kappa(32)}, \quad \frac{\partial r_{\alpha\kappa}}{\partial Z_\beta} = \delta_{\alpha\beta} P_{\kappa(33)},$$

which are written in the form of Eq. (11.16).
(2) The derivative of the matrix P_κ in Eq. (11.3) with respect to f_λ is

$$\frac{\partial \mathsf{P}_\kappa}{\partial f_\lambda} = \delta_{\kappa\lambda} \begin{pmatrix} 1 & 0 & 0 \\ 0 & 1 & 0 \\ 0 & 0 & 0 \end{pmatrix} \mathsf{R}_\kappa^\top \begin{pmatrix} \mathsf{I} & -\mathbf{t}_\kappa \end{pmatrix} = \delta_{\kappa\lambda} \begin{pmatrix} 1 & 0 & 0 \\ 0 & 1 & 0 \\ 0 & 0 & 0 \end{pmatrix} \mathsf{K}_\kappa^{-1} \Big(\mathsf{K}_\kappa \mathsf{R}_\kappa^\top \begin{pmatrix} \mathsf{I} & -\mathbf{t}_\kappa \end{pmatrix} \Big)$$
$$= \delta_{\kappa\lambda} \begin{pmatrix} 1 & 0 & 0 \\ 0 & 1 & 0 \\ 0 & 0 & 0 \end{pmatrix} \frac{1}{f_\kappa} \begin{pmatrix} 1 & 0 & -u_{0\kappa}/f_0 \\ 0 & 1 & -v_{0\kappa}/f_0 \\ 0 & 0 & f/f_0 \end{pmatrix} \mathsf{P}_\kappa = \frac{\delta_{\kappa\lambda}}{f_\kappa} \begin{pmatrix} 1 & 0 & -u_{0\kappa}/f_0 \\ 0 & 1 & -v_{0\kappa}/f_0 \\ 0 & 0 & 0 \end{pmatrix} \mathsf{P}_\kappa$$
$$= \frac{\delta_{\kappa\lambda}}{f} \begin{pmatrix} P_{\kappa(11)} - u_{0\kappa} P_{\kappa(31)}/f_0 & P_{\kappa(12)} - u_{0\kappa} P_{\kappa(32)}/f_0 & P_{\kappa(13)} - u_{0\kappa} P_{\kappa(33)}/f_0 & P_{\kappa(14)} - u_{0\kappa} P_{\kappa(34)}/f_0 \\ P_{\kappa(21)} - v_{0\kappa} P_{\kappa(31)}/f_0 & P_{\kappa(22)} - v_{0\kappa} P_{\kappa(32)}/f_0 & P_{\kappa(23)} - v_{0\kappa} P_{\kappa(33)}/f_0 & P_{\kappa(24)} - v_{0\kappa} P_{\kappa(34)}/f_0 \\ 0 & 0 & 0 & 0 \end{pmatrix},$$

where we have noted that the inverse of the matrix K_κ in Eq. (11.3) is given by

$$\mathsf{K}_\kappa^{-1} = \frac{1}{f_\kappa} \begin{pmatrix} 1 & 0 & -u_{0\kappa}/f_0 \\ 0 & 1 & -v_{0\kappa}/f_0 \\ 0 & 0 & f/f_0 \end{pmatrix}.$$

Substituting the above result into Eq. (11.12), we can express the derivatives of $p_{\alpha\kappa}$, $q_{\alpha\kappa}$, and $r_{\alpha\kappa}$ with respect to f_λ in the form of Eq. (11.17).

(3) The derivative of the matrix P_κ in Eq. (11.3) is

$$\frac{\partial \mathsf{P}_\kappa}{\partial u_{0\lambda}} = \delta_{\kappa\lambda}\begin{pmatrix} 0 & 0 & 1 \\ 0 & 0 & 0 \\ 0 & 0 & 0 \end{pmatrix}\mathsf{R}_\kappa^\top\left(\mathsf{I} - \mathbf{t}_\kappa\right) = \delta_{\kappa\lambda}\begin{pmatrix} 0 & 0 & 1 \\ 0 & 0 & 0 \\ 0 & 0 & 0 \end{pmatrix}\mathsf{K}_\kappa^{-1}\Big(\mathsf{K}_\kappa\mathsf{R}_\kappa^\top\left(\mathsf{I} - \mathbf{t}_\kappa\right)\Big)$$

$$= \delta_{\kappa\lambda}\begin{pmatrix} 0 & 0 & 1 \\ 0 & 0 & 0 \\ 0 & 0 & 0 \end{pmatrix}\frac{1}{f_\kappa}\begin{pmatrix} 1 & 0 & -u_{0\kappa}/f_0 \\ 0 & 1 & -v_{0\kappa}/f_0 \\ 0 & 0 & f_\kappa/f_0 \end{pmatrix}\mathsf{P}_\kappa$$

$$= \frac{\delta_{\kappa\lambda}}{f_0}\begin{pmatrix} P_{\kappa(31)} & P_{\kappa(32)} & P_{\kappa(33)} & P_{\kappa(34)} \\ 0 & 0 & 0 & 0 \\ 0 & 0 & 0 & 0 \end{pmatrix}.$$

Similarly, the derivative with respect to $v_{0\lambda}$ is

$$\frac{\partial \mathsf{P}_\kappa}{\partial v_{0\lambda}} = \delta_{\kappa\lambda}\begin{pmatrix} 0 & 0 & 0 \\ 0 & 0 & 1 \\ 0 & 0 & 0 \end{pmatrix}\mathsf{R}_\kappa^\top\left(\mathsf{I} - \mathbf{t}_\kappa\right) = \frac{\delta_{\kappa\lambda}}{f_0}\begin{pmatrix} 0 & 0 & 0 & 0 \\ P_{\kappa(31)} & P_{\kappa(32)} & P_{\kappa(33)} & P_{\kappa(34)} \\ 0 & 0 & 0 & 0 \end{pmatrix}.$$

Substituting these into Eq. (11.12), we can express the derivatives of $p_{\alpha\kappa}$, $q_{\alpha\kappa}$, and $r_{\alpha\kappa}$ with respect to $u_{0\lambda}$ and $v_{0\lambda}$ in the form of Eq. (11.18).
(4) The translation $\mathbf{t}$ is involved in the fourth column of the matrix P_κ in Eq. (11.3) in the form

$$\begin{pmatrix} P_{\kappa(14)} \\ P_{\kappa(24)} \\ P_{\kappa(34)} \end{pmatrix} = -\mathsf{K}_\kappa\mathsf{R}_\kappa^\top\mathbf{t}_\kappa =$$

$$-\begin{pmatrix} (f_\kappa R_{\kappa(11)}+u_{0\kappa}R_{\kappa(13)})t_{\kappa 1}+(f_\kappa R_{\kappa(21)}+u_{0\kappa}R_{\kappa(23)})t_{\kappa 2}+(f_\kappa R_{\kappa(31)}+u_{0\kappa}R_{\kappa(33)})t_{\kappa 3} \\ (f_\kappa R_{\kappa(12)}+v_{0\kappa}R_{\kappa(13)})t_{\kappa 1}+(f_\kappa R_{\kappa(22)}+v_{0\kappa}R_{\kappa(23)})t_{\kappa 2}+(f_\kappa R_{\kappa(32)}+v_{0\kappa}R_{\kappa(33)})t_{\kappa 3} \\ f_0(R_{\kappa(13)}t_{\kappa 1}+R_{\kappa(23)}t_{\kappa 2}+R_{\kappa(33)}t_{\kappa 3}) \end{pmatrix},$$

from which we obtain

$$\frac{\partial}{\partial t_{\lambda 1}}\begin{pmatrix} P_{\kappa(14)} \\ P_{\kappa(24)} \\ P_{\kappa(34)} \end{pmatrix} = -\delta_{\kappa\lambda}\begin{pmatrix} f_\kappa R_{\kappa(11)} + u_{0\kappa}R_{\kappa(13)} \\ f_\kappa R_{\kappa(12)} + v_{0\kappa}R_{\kappa(13)} \\ f_0 R_{\kappa(13)} \end{pmatrix},$$

$$\frac{\partial}{\partial t_{\lambda 2}}\begin{pmatrix} P_{\kappa(14)} \\ P_{\kappa(24)} \\ P_{\kappa(34)} \end{pmatrix} = -\delta_{\kappa\lambda}\begin{pmatrix} f_\kappa R_{\kappa(21)} + u_{0\kappa}R_{\kappa(23)} \\ f_\kappa R_{\kappa(22)} + v_{0\kappa}R_{\kappa(23)} \\ f_0 R_{\kappa(23)} \end{pmatrix},$$

$$\frac{\partial}{\partial t_{\lambda 3}}\begin{pmatrix} P_{\kappa(14)} \\ P_{\kappa(24)} \\ P_{\kappa(34)} \end{pmatrix} = -\delta_{\kappa\lambda}\begin{pmatrix} f_\kappa R_{\kappa(31)} + u_{0\kappa}R_{\kappa(33)} \\ f_\kappa R_{\kappa(32)} + v_{0\kappa}R_{\kappa(33)} \\ f_0 R_{\kappa(33)} \end{pmatrix}.$$

Substituting these into Eq. (11.12) and using Eq. (11.20), we can express the derivatives of $p_{\alpha\kappa}$, $q_{\alpha\kappa}$, and $r_{\alpha\kappa}$ with respect to $t_{\lambda 1}$, $t_{\lambda 2}$, and $t_{\lambda 3}$ in the form of Eq. (11.19).
(5) If we add a small rotation specified by $\Delta\boldsymbol{\omega}_\kappa = (\Delta\omega_{\kappa 1}, \Delta\omega_{\kappa 2}, \Delta\omega_{\kappa 3})^\top$ to the rotation R_κ, its change is to a first approximation $\Delta\mathsf{R}_\kappa = \Delta\boldsymbol{\omega}_\kappa \times \mathsf{R}_\kappa$ ($\hookrightarrow$ Problem

3.5(1)), where "$\Delta\boldsymbol{\omega}_\kappa \times$" is the antisymmetric matrix defined by Eq. (5.10). Therefore the change of the matrix P_κ in Eq. (11.3) is given to a first approximation by

$$\Delta\mathsf{P}_\kappa = \mathsf{K}_\kappa(\Delta\boldsymbol{\omega}_\kappa \times \mathsf{R}_\kappa)^\top \begin{pmatrix} \mathsf{I} & -\mathbf{t}_\kappa \end{pmatrix}$$
$$= \mathsf{K}_\kappa\mathsf{R}_\kappa^\top \begin{pmatrix} 0 & \Delta\omega_{\kappa 3} & -\Delta\omega_{\kappa 2} & \Delta\omega_{\kappa 2}t_{\kappa 3} - \Delta\omega_{\kappa 3}t_{\kappa 2} \\ -\Delta\omega_{\kappa 3} & 0 & \Delta\omega_{\kappa 1} & \Delta\omega_{\kappa 3}t_{\kappa 1} - \Delta\omega_{\kappa 1}t_{\kappa 3} \\ \Delta\omega_{\kappa 2} & -\Delta\omega_{\kappa 1} & 0 & \Delta\omega_{\kappa 1}t_{\kappa 2} - \Delta\omega_{\kappa 2}t_{\kappa 1} \end{pmatrix},$$

where we have noted the identities $(\boldsymbol{\omega} \times \mathsf{R})^\top = \mathsf{R}^\top(\boldsymbol{\omega}\times)^\top = -\mathsf{R}^\top\boldsymbol{\omega}\times$ and $(\boldsymbol{\omega}\times)\mathbf{t} = \boldsymbol{\omega} \times \mathbf{t}$. From the above expression, we obtain the derivatives $\partial\mathsf{P}_\kappa/\partial\omega_{\lambda 1}$, $\partial\mathsf{P}_\kappa/\partial\omega_{\lambda 2}$, and $\partial\mathsf{P}_\kappa/\partial\omega_{\lambda 3}$ in the form

$$\frac{\partial\mathsf{P}_\kappa}{\partial\omega_{\lambda 1}} = \delta_{\kappa\lambda}\begin{pmatrix} 0 & -f_\kappa R_{\kappa(31)} - u_{0\kappa}R_{\kappa(33)} & f_\kappa R_{\kappa(21)} + u_{0\kappa}R_{\kappa(23)} \\ 0 & -f_\kappa R_{\kappa(32)} - v_{0\kappa}R_{\kappa(33)} & f_\kappa R_{\kappa(22)} + v_{0\kappa}R_{\kappa(23)} \\ 0 & -f_0 R_{\kappa(33)} & f_0 R_{\kappa(23)} \end{pmatrix.$$
$$\begin{matrix} f_\kappa(t_{\kappa 2}R_{\kappa(31)} - t_{\kappa 3}R_{\kappa(21)}) + u_{0\kappa}(t_{\kappa 2}R_{\kappa(33)} - t_{\kappa 3}R_{\kappa(23)}) \\ f_\kappa(t_{\kappa 2}R_{\kappa(32)} - t_{\kappa 3}R_{\kappa(22)}) + v_{0\kappa}(t_{\kappa 2}R_{\kappa(33)} - t_{\kappa 3}R_{\kappa(23)}) \\ f_0(t_{\kappa 2}R_{\kappa(33)} - t_{\kappa 3}R_{\kappa(23)}) \end{matrix}\Bigg),$$

$$\frac{\partial\mathsf{P}_\kappa}{\partial\omega_{\lambda 2}} = \delta_{\kappa\lambda}\Bigg(\begin{matrix} f_\kappa R_{\kappa(31)} + u_{0\kappa}R_{\kappa(33)} & 0 & -f_\kappa R_{\kappa(11)} - u_{0\kappa}R_{\kappa(13)} \\ f_\kappa R_{\kappa(32)} + v_{0\kappa}R_{\kappa(33)} & 0 & -f_\kappa R_{\kappa(12)} - v_{0\kappa}R_{\kappa(13)} \\ f_0 R_{\kappa(33)} & 0 & -f_0 R_{\kappa(13)} \end{matrix}$$
$$\begin{matrix} f_\kappa(t_{\kappa 3}R_{\kappa(11)} - t_{\kappa 1}R_{\kappa(31)}) + u_{0\kappa}(t_{\kappa 3}R_{\kappa(13)} - t_{\kappa 1}R_{\kappa(33)}) \\ f_\kappa(t_{\kappa 3}R_{\kappa(12)} - t_{\kappa 1}R_{\kappa(32)}) + v_{0\kappa}(t_{\kappa 3}R_{\kappa(13)} - t_{\kappa 1}R_{\kappa(33)}) \\ f_0(t_{\kappa 3}R_{\kappa(13)} - t_{\kappa 1}R_{\kappa(33)}) \end{matrix}\Bigg),$$

$$\frac{\partial\mathsf{P}_\kappa}{\partial\omega_{\lambda 3}} = \delta_{\kappa\lambda}\Bigg(\begin{matrix} -f_\kappa R_{\kappa(21)} - u_{0\kappa}R_{\kappa(23)} & f_\kappa R_{\kappa(11)} + u_{0\kappa}R_{\kappa(13)} & 0 \\ -f_\kappa R_{\kappa(22)} - v_{0\kappa}R_{\kappa(23)} & f_\kappa R_{\kappa(12)} + v_{0\kappa}R_{\kappa(13)} & 0 \\ -f_0 R_{\kappa(23)} & f_0 R_{\kappa(13)} & 0 \end{matrix}$$
$$\begin{matrix} f_\kappa(t_{\kappa 1}R_{\kappa(21)} - t_{\kappa 2}R_{\kappa(11)}) + u_{0\kappa}(t_{\kappa 1}R_{\kappa(23)} - t_{\kappa 2}R_{\kappa(13)}) \\ f_\kappa(t_{\kappa 1}R_{\kappa(22)} - t_{\kappa 2}R_{\kappa(12)}) + v_{0\kappa}(t_{\kappa 1}R_{\kappa(23)} - t_{\kappa 2}R_{\kappa(13)}) \\ f_0(t_{\kappa 1}R_{\kappa(23)} - t_{\kappa 2}R_{\kappa(13)}) \end{matrix}\Bigg).$$

Substituting these into Eq. (11.12), we can express the derivatives of $p_{\alpha\kappa}$, $q_{\alpha\kappa}$, and $r_{\alpha\kappa}$ with respect to $\omega_{\lambda 1}$, $\omega_{\lambda 2}$, and $\omega_{\lambda 3}$ in the form of Eq. (11.21).

11.3 (1) Equation (11.31) is decomposed to the following two equations.

$$\mathsf{A}\mathbf{x} + \mathsf{B}\mathbf{y} = \mathbf{a}, \qquad \mathsf{C}\mathbf{x} + \mathsf{D}\mathbf{y} = \mathbf{b}.$$

We obtain Eq. (11.33) by solving the second equation for $\mathbf{y}$. Substituting this into the first equation, we obtain Eq. (11.32). (2) The solution of Eq. (11.32) is given by

$$\mathbf{x} = (\mathsf{A} - \mathsf{B}\mathsf{D}^{-1}\mathsf{C})^{-1}(\mathbf{a} - \mathsf{B}\mathsf{D}^{-1}\mathbf{b}).$$

Substituting this into Eq. (11.33), we obtain

$$\mathbf{y} = -\mathsf{D}^{-1}\mathsf{C}(\mathsf{A} - \mathsf{B}\mathsf{D}^{-1}\mathsf{C})^{-1}(\mathbf{a} - \mathsf{B}\mathsf{D}^{-1}\mathbf{b}) + \mathsf{D}^{-1}\mathbf{b}.$$

Thus, $\mathbf{x}$ and $\mathbf{y}$ are given by

$$\begin{pmatrix} \mathbf{x} \\ \mathbf{y} \end{pmatrix} = \begin{pmatrix} (\mathsf{A} - \mathsf{B}\mathsf{D}^{-1}\mathsf{C})^{-1} & -(\mathsf{A} - \mathsf{B}\mathsf{D}^{-1}\mathsf{C})^{-1}\mathsf{B}\mathsf{D}^{-1} \\ -\mathsf{D}^{-1}\mathsf{C}(\mathsf{A} - \mathsf{B}\mathsf{D}^{-1}\mathsf{C})^{-1} & \mathsf{D}^{-1} + \mathsf{D}^{-1}\mathsf{C}(\mathsf{A} - \mathsf{B}\mathsf{D}^{-1}\mathsf{C})^{-1}\mathsf{B}\mathsf{D}^{-1} \end{pmatrix} \begin{pmatrix} \mathbf{a} \\ \mathbf{b} \end{pmatrix}.$$

Comparing this with Eq. (11.31), we obtain the inversion formula of Eq. (11.34).

Problems of Chapter 12

12.1 (1) If we let $\mathbf{m}_1$, $\mathbf{m}_2$, and $\mathbf{m}_3$ be the columns of the $2M \times 3$ motion matrix M, the αth columns of Eq. (12.8) on both sides are written as

$$\begin{pmatrix} x_{\alpha 1} \\ y_{\alpha 1} \\ \vdots \\ y_{\alpha M} \end{pmatrix} = X_\alpha \mathbf{m}_1 + Y_\alpha \mathbf{m}_2 + Z_\alpha \mathbf{m}_3.$$

This implies that the trajectory of the αth point is included in the 3D subspace spanned by $\mathbf{m}_1$, $\mathbf{m}_2$, and $\mathbf{m}_3$.
(2) As is well known, the orthonormal basis of the space spanned by the N trajectories is given by the unit eigenvectors of the matrix

$$\sum_{\alpha=1}^{N} \begin{pmatrix} x_{\alpha 1} \\ y_{\alpha 1} \\ \vdots \\ y_{\alpha M} \end{pmatrix} \begin{pmatrix} x_{\alpha 1} \\ y_{\alpha 1} \\ \vdots \\ y_{\alpha M} \end{pmatrix}^\top = \mathsf{W}\mathsf{W}^\top,$$

which is known in statistics as the "covariance matrix." Let $\mathsf{W} = \mathsf{U}\boldsymbol{\Sigma}\mathsf{V}^\top$ be the SVD of W, where $\boldsymbol{\Sigma}$ is the diagonal matrix of the form of Eq. (12.10). We see that

$$\mathsf{W}\mathsf{W}^\top = \mathsf{U}\boldsymbol{\Sigma}\mathsf{V}^\top\mathsf{V}\boldsymbol{\Sigma}\mathsf{U}^\top = \mathsf{U}\boldsymbol{\Sigma}^2\mathsf{U}^\top = \sum_{i=1}^{3} \sigma_i^2 \mathbf{u}_i \mathbf{u}_i^\top.$$

This implies that $\mathbf{u}_i$ is the unit eigenvector of this matrix for eigenvalue σ_i^2.

12.2 Let $\lambda_1 \geq \cdots \geq \lambda_n$ (≥ 0) be the eigenvalues of A, and $\mathbf{u}_1, \ldots, \mathbf{u}_n$ the corresponding orthonormal basis of the unit eigenvectors. In terms of these, the matrix A is expressed in the form

$$\mathsf{A} = \sum_{i=1}^{n} \lambda_i \mathbf{u}_i \mathbf{u}_i^\top.$$

Let

$$\mathsf{B} = \sum_{i=1}^{n} \sqrt{\lambda_i} \mathbf{u}_i \mathbf{u}_i^\top.$$

Because $(\mathbf{u}_i, \mathbf{u}_j) = \delta_{ij}$ (the Kronecker delta), we can see that $\mathsf{A} = \mathsf{B}\mathsf{B}^\top$ as follows.

$$\begin{aligned}\mathsf{B}\mathsf{B}^\top &= \sum_{i=1}^{n} \sqrt{\lambda_i}\mathbf{u}_i\mathbf{u}_i^\top \sum_{j=1}^{n} \sqrt{\lambda_j}\mathbf{u}_j\mathbf{u}_j^\top = \sum_{i,j=1}^{n} \sqrt{\lambda_i\lambda_j}(\mathbf{u}_i, \mathbf{u}_j)\mathbf{u}_i\mathbf{u}_j^\top \\ &= \sum_{i,j=1}^{n} \delta_{ij}\sqrt{\lambda_i\lambda_j}\mathbf{u}_i\mathbf{u}_j^\top = \sum_{i=1}^{n} \lambda_i\mathbf{u}_i\mathbf{u}_i^\top = \mathsf{A}.\end{aligned}$$

12.3 (1) The metric condition of Eq. (12.44) is written as

$$\begin{pmatrix} 1 & t_{x\kappa}^2 \\ 1 & t_{y\kappa}^2 \\ 0 & t_{x\kappa}t_{x\kappa} \end{pmatrix} \begin{pmatrix} 1/\zeta_\kappa^2 \\ \beta_\kappa^2 \end{pmatrix} = \begin{pmatrix} (\mathbf{u}_{\kappa(1)}, \mathsf{T}\mathbf{u}_{\kappa(1)}) \\ (\mathbf{u}_{\kappa(1)}, \mathsf{T}\mathbf{u}_{\kappa(2)}) \\ (\mathbf{u}_{\kappa(2)}, \mathsf{T}\mathbf{u}_{\kappa(3)}) \end{pmatrix}.$$

Because there are three equations for two unknowns, we solve them by least squares. Multiplying this on both sides by the transpose of the coefficient matrix on the left-hand side from the left, we obtain the following normal equation (↪ Problem 4.4).

$$\begin{pmatrix} 2 & t_{x\kappa}^2 + t_{y\kappa}^2 \\ t_{x\kappa}^2 + t_{y\kappa}^2 & t_{x\kappa}^4 + t_{y\kappa}^4 + t_{x\kappa}^2 t_{y\kappa}^2 \end{pmatrix} \begin{pmatrix} 1/\zeta_\kappa^2 \\ \beta_\kappa^2 \end{pmatrix}$$
$$= \begin{pmatrix} (\mathbf{u}_{\kappa(1)}, \mathsf{T}\mathbf{u}_{\kappa(1)}) + (\mathbf{u}_{\kappa(2)}, \mathsf{T}\mathbf{u}_{\kappa(2)}) \\ t_{x\kappa}^2(\mathbf{u}_{\kappa(1)}, \mathsf{T}\mathbf{u}_{\kappa(1)}) + t_{x\kappa}t_{y\kappa}(\mathbf{u}_{\kappa(1)}, \mathsf{T}\mathbf{u}_{\kappa(2)}) + t_{y\kappa}^2(\mathbf{u}_{\kappa(2)}, \mathsf{T}\mathbf{u}_{\kappa(2)}) \end{pmatrix}.$$

Solving this, we can determine ζ_κ and β_κ (considering the correspondence with the paraperspective case, we can let $\zeta_\kappa > 0$ and $\beta_\kappa \geq 0$). If $\beta_\kappa^2 < 0$, we let $\beta_\kappa = 0$. However, if it happens that $t_{x\kappa} \approx 0$ and $t_{y\kappa} \approx 0$, we let $\beta_\kappa = 0$ and solve only the first equation for $1/\zeta_\kappa^2$ to obtain

$$\frac{1}{\zeta_\kappa^2} = \frac{(\mathbf{u}_{\kappa(1)}, \mathsf{T}\mathbf{u}_{\kappa(1)}) + (\mathbf{u}_{\kappa(2)}, \mathsf{T}\mathbf{u}_{\kappa(2)})}{2}.$$

If $1/\zeta_\kappa^2 \leq 0$, we let ζ_κ be a sufficient large value. From Eq. (12.45), $g_{x\kappa}$ and $g_{y\kappa}$ are given by

$$g_{x\kappa} = \zeta_\kappa t_{x\kappa}, \qquad g_{y\kappa} = \zeta_\kappa t_{y\kappa}.$$

(2) The camera matrix $\mathbf{\Pi}_\kappa$ is determined by the motion matrix M computed by Eq. (12.42) considering the definition of the matrix M of Eq. (12.6). Substituting the values ζ_κ, β_κ, $g_{x\kappa}$, and $g_{y\kappa}$ obtained in the above item (1) into $\mathbf{\Pi}_\kappa$, we can write from Eq. (12.27) the matrix C_κ of the κth camera in the form

$$\mathsf{C}_\kappa = \begin{pmatrix} 1/\zeta_\kappa & 0 & -\beta_\kappa g_{x\kappa}/\zeta \\ 0 & 1/\zeta_\kappa & -\beta_\kappa g_{y\kappa}/\zeta \end{pmatrix}.$$

From Eq. (12.23), we obtain $\mathbf{\Pi}_\kappa = \mathsf{C}_\kappa\mathsf{R}_\kappa^\top$ (both sides are 2×3 matrices). Transposing this on both sides, we obtain $\mathbf{\Pi}_\kappa^\top = \mathsf{R}_\kappa\mathsf{C}_\kappa^\top$ (both sides are 3×2 matrices). Let $\boldsymbol{\pi}_{\kappa(1)}$ and $\boldsymbol{\pi}_{\kappa(2)}$ be the first and second columns of $\mathbf{\Pi}_\kappa^\top$, respectively, and $\mathbf{r}_{\kappa(1)}$, $\mathbf{r}_{\kappa(2)}$, and $\mathbf{r}_{\kappa(3)}$ the first, second, and third columns of R_κ, respectively. Then, from $\mathbf{\Pi}_\kappa^\top = \mathsf{R}_\kappa\mathsf{C}_\kappa^\top$ we obtain

$$\boldsymbol{\pi}_{\kappa(1)} = \frac{1}{\zeta_\kappa}(\mathbf{r}_{\kappa(1)} - \beta_\kappa g_{x\kappa}\mathbf{r}_{\kappa(3)}), \quad \boldsymbol{\pi}_{\kappa(2)} = \frac{1}{\zeta_\kappa}(\mathbf{r}_{\kappa(2)} - \beta_\kappa g_{y\kappa}\mathbf{r}_{\kappa(3)}).$$

Consider their vector product $\boldsymbol{\pi}_{\kappa(1)} \times \boldsymbol{\pi}_{\kappa(2)}$. Because $\{\mathbf{r}_{\kappa(1)}, \mathbf{r}_{\kappa(2)}, \mathbf{r}_{\kappa(3)}\}$ are a right-handed orthonormal system and $\mathbf{r}_{\kappa(1)} \times \mathbf{r}_{\kappa(2)} = \mathbf{r}_{\kappa(3)}$, and so on hold, we obtain

$$\boldsymbol{\pi}_{\kappa(1)} \times \boldsymbol{\pi}_{\kappa(2)} = \frac{1}{\zeta_\kappa^2}(\beta_\kappa g_{x\kappa}\mathbf{r}_{\kappa(1)} + \beta_\kappa g_{y\kappa}\mathbf{r}_{\kappa(2)} + \mathbf{r}_{\kappa(3)}).$$

Regarding the above three equations as simultaneous linear equations in $\mathbf{r}_{\kappa(1)}, \mathbf{r}_{\kappa(2)}$, and $\mathbf{r}_{\kappa(3)}$ and solving them, we obtain

$$\mathbf{r}_{\kappa(1)} = \zeta_\kappa \boldsymbol{\pi}_{\kappa(1)} + \beta_\kappa g_{x\kappa}\mathbf{r}_{\kappa(3)}$$
$$\mathbf{r}_{\kappa(2)} = \zeta_\kappa \boldsymbol{\pi}_{\kappa(2)} + \beta_\kappa g_{y\kappa}\mathbf{r}_{\kappa(3)}$$
$$\mathbf{r}_{\kappa(3)} = \frac{\zeta_\kappa \boldsymbol{\pi}_{\kappa(1)} \times \boldsymbol{\pi}_{\kappa(2)} - \beta_\kappa(g_{x\kappa}\boldsymbol{\pi}_{\kappa(1)} + g_{y\kappa}\boldsymbol{\pi}_{\kappa(2)})}{1 + \beta_\kappa^2(g_{x\kappa}^2 + g_{y\kappa}^2)}.$$

These determine the matrix R_κ, but because the computation of ζ_κ, β_κ, $g_{x\kappa}$, $g_{y\kappa}$, $\boldsymbol{\pi}_{\kappa(1)}$, and $\boldsymbol{\pi}_{\kappa(2)}$ involves approximation, the resulting R_κ may not be an exact rotation matrix. Hence, we compute the SVD of the matrix consisting of columns $\mathbf{r}_{\kappa(1)}, \mathbf{r}_{\kappa(2)}$, and $\mathbf{r}_{\kappa(3)}$ in the form

$$\begin{pmatrix} \mathbf{r}_{\kappa(1)} & \mathbf{r}_{\kappa(2)} & \mathbf{r}_{\kappa(3)} \end{pmatrix} = \mathsf{U}\boldsymbol{\Lambda}\mathsf{V}^\top,$$

where $\boldsymbol{\Lambda}$ is the diagonal matrix consisting of the singular values as diagonal elements. This should be a rotation matrix if $\boldsymbol{\Lambda} = \mathsf{I}$. If this is not so, we force R_κ to be a rotation matrix in the form

$$\mathsf{R}_\kappa = \mathsf{U}\mathsf{V}^\top.$$

Problems of Chapter 13

13.1 Using the definition in Eq. (13.80), we can write Eq. (13.3) in the form

$$\mathsf{A}^{(\alpha)} = \mathsf{C}^{(\alpha)}\mathsf{C}^{(\alpha)\top}.$$

Using the SVD of Eq. (13.81), we can rewrite this as

$$\mathsf{A}^{(\alpha)} = \mathsf{U}^{(\alpha)}\boldsymbol{\Sigma}^{(\alpha)2}\mathsf{U}^{(\alpha)\top}.$$

This implies that $\mathsf{U}^{(\alpha)}$ is the $M \times 4$ matrix consisting of the eigenvectors of $\mathsf{A}^{(\alpha)}$ as columns. Hence, the first column of $\mathsf{U}^{(\alpha)}$ is the unit eigenvector $\boldsymbol{\xi}_\alpha$ of $\mathsf{A}^{(\alpha)}$ for the largest eigenvalue.

13.2 (1) The columns of $\mathsf{V}_{N\times L}$ are mutually orthogonal unit vectors, therefore we have $\mathsf{V}_{N\times L}^\top \mathsf{V}_{N\times L} = \mathsf{I}_L$ (= the $L \times L$ identity matrix). Thus we see from Eq. (13.7) that

$$\mathsf{W}\mathsf{W}^\top = \mathsf{U}_{3M\times L}\boldsymbol{\Sigma}_L\mathsf{V}_{N\times L}^\top\mathsf{V}_{N\times L}\boldsymbol{\Sigma}_L\mathsf{U}_{3M\times L}^\top = \mathsf{U}_{3M\times L}\boldsymbol{\Sigma}_L^2\mathsf{U}_{3M\times L}^\top.$$

This implies that $\mathsf{U}_{3M\times L}$ is the matrix whose columns are the unit eigenvectors of $\mathsf{W}\mathsf{W}^\top$ for eigenvalues $\sigma_1^2, \ldots, \sigma_L^2$.

(2) If we let $\bar{\mathbf{u}}_1, \ldots, \bar{\mathbf{u}}_{3M}$ be the unit eigenvectors of $\mathsf{W}\mathsf{W}^\top$ for eigenvalues $\lambda_1 \geq \cdots \geq \lambda_{3M}$, we can write $\mathsf{W}\mathsf{W}^\top$ in the form

$$\mathsf{W}\mathsf{W}^\top = \sum_{k=1}^{3M} \lambda_k \bar{\mathbf{u}}_k \bar{\mathbf{u}}_k^\top.$$

Thus

$$\mathsf{W}\mathsf{W}^\top \mathbf{u}_i = \sum_{k=1}^{3M} \lambda_k (\mathbf{u}_i, \bar{\mathbf{u}}_k) \bar{\mathbf{u}}_k.$$

We can express $\mathbf{u}_i$ in terms of $\bar{\mathbf{u}}_1, \ldots, \bar{\mathbf{u}}_{3M}$ in the form $\sum_{k=1}^{3M} (\mathbf{u}_i, \bar{\mathbf{u}}_k)\bar{\mathbf{u}}_k$. Because $\lambda_5, \ldots, \lambda_{3M}$ approach 0 as iterations proceed, the projection of $\mathsf{W}\mathsf{W}^\top \mathbf{u}_i$ onto the orthogonal complement $\mathscr{L}_4^\perp$ (= the $(3M-4)$D space spanned by $\bar{\mathbf{u}}_5, \ldots, \bar{\mathbf{u}}_{3M}$) of $\mathscr{L}_4$ is smaller than $\mathbf{u}_i$. Therefore, after the Schmidt orthogonalization, we obtain a better orthonormal basis of $\mathscr{L}_4$ than $\mathbf{u}_1, \ldots, \mathbf{u}_4$. In doing this, we can compute $\mathsf{W}\mathsf{W}^\top \mathbf{u}_i$ more efficiently by multiplying $\mathsf{W}^\top \mathbf{u}_i$ by W than by multiplying $\mathbf{u}_i$ by $\mathsf{W}\mathsf{W}^\top$.
(3) First, $\mathbf{u}_1$ is normalized to unit norm. Suppose $\mathbf{u}_1, \ldots, \mathbf{u}_k$ are an orthonormal system. The projection of $\mathbf{u}_{k+1}$ onto the space they span is $\sum_{i=1}^{k} (\mathbf{u}_i, \mathbf{u}_k)\mathbf{u}_i$. Therefore we subtract this and normalize $\mathbf{u}_{k+1} - \sum_{i=1}^{k} (\mathbf{u}_i, \mathbf{u}_k)\mathbf{u}_i$ to unit norm; we obtain an extended orthonormal system $\mathbf{u}_1, \ldots, \mathbf{u}_{k+1}$. Doing this for $k = 2, 3, 4$, we can obtain the orthonormal basis of $\mathscr{L}_4$.

13.3 Identifying U with M means using Eq. (13.9). Then, Eq. (13.6) is written as

$$\mathsf{W} = \begin{pmatrix} \mathbf{u}_1 & \mathbf{u}_2 & \mathbf{u}_3 & \mathbf{u}_4 \end{pmatrix} \begin{pmatrix} \mathsf{X}_1 & \cdots & \mathsf{X}_N \end{pmatrix}.$$

Hence the α column $\mathbf{p}_\alpha$ of W is given by

$$\mathbf{p}_\alpha = X_{\alpha 1}\mathbf{u}_1 + X_{\alpha 2}\mathbf{u}_2 + X_{\alpha 3}\mathbf{u}_3 + X_{\alpha 4}\mathbf{u}_4.$$

In other words, $X_{\alpha 1}, \ldots, X_{\alpha 4}$ are the coefficients of the expansion of $\mathbf{p}_\alpha$ in terms of $\mathbf{u}_1, \ldots, \mathbf{u}_4$. Thus they are given by Eq. (13.84).

13.4 From Eq. (13.23), we can write

$$\|\mathbf{q}_{\kappa(1)}\|^2 = \sum_{\alpha=1}^{N} \frac{z_{\alpha\kappa}^2 x_{\alpha\kappa}^2}{f_0^2}, \quad \|\mathbf{q}_{\kappa(2)}\|^2 = \sum_{\alpha=1}^{N} \frac{z_{\alpha\kappa}^2 y_{\alpha\kappa}^2}{f_0^2}, \quad \|\mathbf{q}_{\kappa(3)}\|^2 = \sum_{\alpha=1}^{N} z_{\alpha\kappa}^2.$$

Therefore Eq. (13.25) becomes

$$\sum_{i=1}^{3} \|\mathbf{q}_{\kappa(i)}\|^2 = \sum_{\alpha=1}^{N} z_{\alpha\kappa}^2 \|\mathbf{x}_{\alpha\kappa}\|^2 = \sum_{\alpha=1}^{N} \xi_{\alpha\kappa}^2 = \|\boldsymbol{\xi}_\kappa\|^2 = 1.$$

13.5 If we let $v_{k\alpha}$ be the αth component of the vector $\mathbf{v}_k$, we obtain from Eq. (13.23)

$$\sum_{k=1}^{4}(\mathbf{q}_{\kappa(1)}, \mathbf{v}_k)^2 = \sum_{k=1}^{4}\Big(\sum_{\alpha=1}^{N}\frac{z_{\alpha\kappa}x_{\alpha\kappa}v_{k\alpha}}{f_0}\Big)\Big(\sum_{\beta=1}^{N}\frac{z_{\beta\kappa}x_{\beta\kappa}v_{k\beta}}{f_0}\Big)$$

$$= \sum_{\alpha,\beta=1}^{N}\frac{z_{\alpha\kappa}z_{\beta\kappa}x_{\alpha\kappa}x_{\beta\kappa}}{f_0^2}\sum_{k=1}^{4}v_{k\alpha}v_{k\beta} = \sum_{\alpha,\beta=1}^{N}\frac{z_{\alpha\kappa}z_{\beta\kappa}x_{\alpha\kappa}x_{\beta\kappa}}{f_0^2}(\mathbf{v}_\alpha, \mathbf{v}_\beta),$$

$$\sum_{k=1}^{4}(\mathbf{q}_{\kappa(2)}, \mathbf{v}_k)^2 = \sum_{k=1}^{4}\Big(\sum_{\alpha=1}^{N}\frac{z_{\alpha\kappa}y_{\alpha\kappa}v_{k\alpha}}{f_0}\Big)\Big(\sum_{\beta=1}^{N}\frac{z_{\beta\kappa}y_{\beta\kappa}v_{k\beta}}{f_0}\Big)$$

$$= \sum_{\alpha,\beta=1}^{N}\frac{z_{\alpha\kappa}z_{\beta\kappa}y_{\alpha\kappa}y_{\beta\kappa}}{f_0^2}\sum_{k=1}^{4}v_{k\alpha}v_{k\beta} = \sum_{\alpha,\beta=1}^{N}\frac{z_{\alpha\kappa}z_{\beta\kappa}y_{\alpha\kappa}y_{\beta\kappa}}{f_0^2}(\mathbf{v}_\alpha, \mathbf{v}_\beta),$$

$$\sum_{k=1}^{4}(\mathbf{q}_{\kappa(3)}, \mathbf{v}_k)^2 = \sum_{k=1}^{4}\Big(\sum_{\alpha=1}^{N}z_{\alpha\kappa}v_{k\alpha}\Big)\Big(\sum_{\beta=1}^{N}z_{\beta\kappa}v_{k\beta}\Big)$$

$$= \sum_{\alpha,\beta=1}^{N}z_{\alpha\kappa}z_{\beta\kappa}\sum_{k=1}^{4}v_{k\alpha}v_{k\beta} = \sum_{\alpha,\beta=1}^{N}z_{\alpha\kappa}z_{\beta\kappa}(\mathbf{v}_\alpha, \mathbf{v}_\beta).$$

Therefore if we define the variable $\xi_{\alpha\kappa}$ by Eq. (13.21) and the matrix $\mathsf{B}^{(\kappa)} = (B^{(\kappa)}_{\alpha\beta})$ by Eq. (13.26), we obtain

$$\sum_{i=1}^{3}\sum_{k=1}^{4}(\mathbf{q}_{\kappa(i)}, \mathbf{v}_k)^2 = \sum_{\alpha,\beta=1}^{N}z_{\alpha\kappa}z_{\beta\kappa}(\mathbf{x}_{\alpha\kappa}, \mathbf{x}_{\beta\kappa})(\mathbf{v}_\alpha, \mathbf{v}_\beta)$$

$$= \sum_{\alpha,\beta=1}^{N}\frac{\xi_{\alpha\kappa}}{\|\mathbf{x}_{\alpha\kappa}\|}\frac{\xi_{\beta\kappa}}{\|\mathbf{x}_{\beta\kappa}\|}(\mathbf{x}_{\alpha\kappa}, \mathbf{x}_{\beta\kappa})(\mathbf{v}_\alpha, \mathbf{v}_\beta) = \sum_{\alpha,\beta=1}^{N}B^{(\kappa)}_{\alpha\beta}\xi_{\alpha\kappa}\xi_{\beta\kappa} = (\boldsymbol{\xi}_\kappa, \mathsf{B}^{(\kappa)}\boldsymbol{\xi}_\kappa).$$

13.6 We can rewrite Eq. (13.26) in the form

$$B^{(\kappa)}_{\alpha\beta} = \Big(\frac{x_{\alpha\kappa}\mathbf{v}_\alpha}{f_0\|\mathbf{x}_{\alpha\kappa}\|}, \frac{x_{\beta\kappa}\mathbf{v}_\beta}{f_0\|\mathbf{x}_{\beta\kappa}\|}\Big) + \Big(\frac{y_{\alpha\kappa}\mathbf{v}_\alpha}{f_0\|\mathbf{x}_{\alpha\kappa}\|}, \frac{y_{\beta\kappa}\mathbf{v}_\beta}{f_0\|\mathbf{x}_{\beta\kappa}\|}\Big) + \Big(\frac{\mathbf{v}_\alpha}{\|\mathbf{x}_{\alpha\kappa}\|}, \frac{\mathbf{v}_\beta}{\|\mathbf{x}_{\beta\kappa}\|}\Big).$$

Hence, if we define $\mathsf{C}^{(\kappa 1)}$, $\mathsf{C}^{(\kappa 2)}$, $\mathsf{C}^{(\kappa 3)}$, and $\mathsf{C}^{(\kappa)}$ by Eqs. (13.85) and (13.86), we can write $\mathsf{B}^{(\kappa)}$ as

$$\mathsf{B}^{(\kappa)} = \mathsf{C}^{(\kappa 1)}\mathsf{C}^{\kappa(1)\top} + \mathsf{C}^{(\kappa 2)}\mathsf{C}^{\kappa(2)\top} + \mathsf{C}^{(\kappa 3)}\mathsf{C}^{\kappa(3)\top} = \mathsf{C}^{(\kappa)}\mathsf{C}^{(\kappa)\top}.$$

Using the SVD of Eq. (13.87), we can rewrite this in the form

$$\mathsf{B}^{(\alpha)} = \mathsf{U}^{(\kappa)}\boldsymbol{\Sigma}^{(\kappa)2}\mathsf{U}^{(\kappa)\top}.$$

This implies that $\mathsf{U}^{(\kappa)}$ is the $N \times 4$ matrix whose columns are the eigenvectors of $\mathsf{B}^{(\alpha)}$. Therefore its first column gives $\boldsymbol{\xi}_\kappa$.

13.7 (1) Because the columns of $\mathsf{W}_{3M\times L}$ are mutually orthogonal unit vectors, we have $\mathsf{U}_{3M\times L}^\top \mathsf{U}_{3M\times L} = \mathsf{I}_L$ (= the $L \times L$ identity matrix). Thus we see from Eq. (13.7) that

$$\mathsf{W}^\top \mathsf{W} = \mathsf{V}_{N\times L}\boldsymbol{\Sigma}_L \mathsf{U}_{3M\times L}^\top \mathsf{U}_{3M\times L}\boldsymbol{\Sigma}_L \mathsf{V}_{N\times L}^\top = \mathsf{V}_{N\times L}\boldsymbol{\Sigma}_L^2 \mathsf{V}_{N\times L}^\top.$$

This implies that $\mathsf{V}_{N\times L}$ is the matrix whose columns are the unit eigenvectors of $\mathsf{W}^\top\mathsf{W}$ for eigenvalues $\sigma_1^2, \ldots, \sigma_L^2$.
(2) If we let $\bar{\mathbf{v}}_1, \ldots, \bar{\mathbf{v}}_N$ be the unit eigenvectors of $\mathsf{W}^\top\mathsf{W}$ for eigenvalues $\lambda_1 \geq \cdots \geq \lambda_N$, we can write $\mathsf{W}^\top\mathsf{W}$ in the form

$$\mathsf{W}^\top\mathsf{W} = \sum_{k=1}^{N} \lambda_k \bar{\mathbf{v}}_k \bar{\mathbf{v}}_k^\top.$$

Therefore

$$\mathsf{W}^\top\mathsf{W}\mathbf{v}_i = \sum_{k=1}^{N} \lambda_k (\mathbf{v}_i, \bar{\mathbf{v}}_k)\bar{\mathbf{v}}_k.$$

We can express $\mathbf{v}_i$ in terms of $\bar{\mathbf{v}}_1, \ldots, \bar{\mathbf{v}}_N$ in the form $\sum_{k=1}^{N}(\mathbf{v}_i, \bar{\mathbf{v}}_k)\bar{\mathbf{v}}_k$. Because $\lambda_5, \ldots, \lambda_N$ approach 0 as the iterations proceed, the projection of $\mathsf{W}^\top\mathsf{W}\mathbf{v}_i$ onto the orthogonal complement $\mathscr{L}_4^{*\perp}$ (= the $(N-4)$D space spanned by $\bar{\mathbf{v}}_5, \ldots, \bar{\mathbf{v}}_N$) of $\mathscr{L}_4^*$ is smaller than $\mathbf{v}_i$. Thus, after the Schmidt orthogonalization, we obtain a better orthonormal basis of $\mathscr{L}_4^*$ than $\mathbf{v}_1, \ldots, \mathbf{v}_4$. In doing this, we can compute $\mathsf{W}^\top\mathsf{W}\mathbf{v}_i$ more efficiently by multiplying $\mathsf{W}\mathbf{v}_i$ by $\mathsf{W}^\top$ than by multiplying $\mathbf{v}_i$ by $\mathsf{W}^\top\mathsf{W}$.

13.8 Identifying $\mathsf{V}^\top$ with S means using Eq. (13.10). Then Eq. (13.6) is written as

$$\mathsf{W} = \begin{pmatrix} \mathsf{P}_1 \\ \vdots \\ \mathsf{P}_M \end{pmatrix} \begin{pmatrix} \mathbf{v}_1^\top \\ \mathbf{v}_2^\top \\ \mathbf{v}_3^\top \\ \mathbf{v}_4^\top \end{pmatrix},$$

whose transpose is

$$\mathsf{W}^\top = \begin{pmatrix} \mathbf{v}_1 & \mathbf{v}_2 & \mathbf{v}_3 & \mathbf{v}_4 \end{pmatrix} \begin{pmatrix} \mathsf{P}_1^\top & \cdots & \mathsf{P}_M^\top \end{pmatrix}.$$

Thus the $3(\kappa - 1) + i$th column $\mathbf{q}_{\kappa(i)}$ of $\mathsf{W}^\top$ is given by

$$\mathbf{q}_{\kappa(i)} = P_{\kappa(i1)}\mathbf{v}_1 + P_{\kappa(i2)}\mathbf{v}_2 + P_{\kappa(i3)}\mathbf{v}_3 + P_{\kappa(i4)}\mathbf{v}_4.$$

In other words, $P_{\kappa(ij)}$ are the coefficients of the expansion of $\mathbf{q}_{\kappa(i)}$ in terms of $\mathbf{u}_1, \ldots, \mathbf{u}_4$. They are given by Eq. (13.90).

13.9 (1) This can be shown as follows.

$$\begin{pmatrix} K_{11} & K_{12} & K_{13} \\ 0 & K_{22} & K_{23} \\ 0 & 0 & K_{33} \end{pmatrix} \begin{pmatrix} K'_{11} & K'_{12} & K'_{13} \\ 0 & K'_{22} & K'_{23} \\ 0 & 0 & K'_{33} \end{pmatrix}$$
$$= \begin{pmatrix} K_{11}K'_{11} & K_{11}K'_{12} + K_{12}K'_{22} & K_{11}K'_{13} + K_{12}K'_{23} + K_{13}K'_{33} \\ 0 & K_{22}K'_{22} & K_{22}K'_{23} + K_{23}K'_{33} \\ 0 & 0 & K_{33}K'_{33} \end{pmatrix}.$$

(2) Because the determinant of the upper triangular matrix $\mathsf{K} = (K_{ij})$ is the product of its diagonal elements, all the diagonal elements are not 0 if K is nonsingular. Therefore it has the inverse:

$$\begin{pmatrix} K_{11} & K_{12} & K_{13} \\ 0 & K_{22} & K_{23} \\ 0 & 0 & K_{33} \end{pmatrix}^{-1} = \frac{1}{K_{11}K_{22}K_{33}} \begin{pmatrix} K_{22}K_{33} & -K_{12}K_{33} & K_{12}K_{23} - K_{22}K_{13} \\ 0 & K_{11}K_{33} & -K_{11}K_{23} \\ 0 & 0 & K_{11}K_{22} \end{pmatrix}.$$

13.10 Equation (13.92) is rewritten as follows.

$$\begin{pmatrix} a_{11} & a_{12} & a_{13} \\ a_{12} & a_{22} & a_{23} \\ a_{13} & a_{23} & a_{33} \end{pmatrix} = \begin{pmatrix} x_{11}^2 & x_{11}x_{12} & x_{11}x_{13} \\ x_{11}x_{12} & x_{12}^2 + x_{22}^2 & x_{12}x_{13} + x_{22}x_{23} \\ x_{11}x_{13} & x_{12}x_{13} + x_{22}x_{23} & x_{13}^2 + x_{23}^2 + x_{33}^2 \end{pmatrix}.$$

Comparing the (1,1) elements on both sides, we obtain $a_{11} = x_{11}^2$, from which the first equation of Eq. (13.93) is obtained. The sign of x_{11} is chosen to be positive. Next, comparing the (1,2) elements, we obtain $a_{12} = x_{11}x_{12}$, from which the second equation is obtained. Similarly, comparing the (1,3) elements, we obtain $a_{13} = x_{11}x_{13}$, from which the third equation is obtained. Comparing the (2,2) elements, we obtain $a_{22} = x_{12}^2 + x_{22}^2$, from which the fourth equation is obtained, where the sign of x_{22} is chosen to be positive. Comparing the (2,3) elements, we obtain $a_{23} = x_{12}x_{13} + x_{22}x_{23}$, from which the fifth equation is obtained. Finally, comparing (3,3) elements, we obtain $a_{33} = x_{13}^2 + x_{23}^2 + x_{33}^2$, from which the sixth equation is obtained, where the sign of x_{33} is chosen to be positive. Comparing this way in turn, we obtain the expressions of Eq. (13.93), where new values are expressed in terms of already computed values. Note that Eq. (13.14) holds irrespective of the choice of the signs of the diagonal elements x_{11}, x_{22}, and x_{33}. Here we choose positive signs for convenience. This process is straightforwardly extended to the general $n \times n$ matrix case.

13.11 Let $\mathsf{P}_\kappa = \begin{pmatrix} \mathsf{Q} & \mathbf{q} \end{pmatrix}$; that is, let Q be the first 3×3 part of P_κ, and $\mathbf{q}$ its last column. To remove sign indeterminacy, we change the signs of Q and $\mathbf{q}$ if $\det \mathsf{Q} < 0$. We want to obtain an upper triangular matrix K_κ, a rotation matrix R_κ, and a translation vector $\mathbf{t}_\kappa$ such that

$$\mathsf{Q} = \mathsf{K}_\kappa \mathsf{R}_\kappa^\top, \qquad \mathbf{q} = \mathsf{K}_\kappa \mathsf{R}_\kappa^\top \mathbf{t}_\kappa.$$

First, from $\mathbf{q} = -\mathsf{Q}\mathbf{t}_\kappa$ we obtain $\mathbf{t}_\kappa$ in the form of Eq. (13.94). Next, from $\mathsf{R}_\kappa^\top \mathsf{R}_\kappa = \mathsf{I}$ we see that

$$\mathsf{Q}\mathsf{Q}^\top = \mathsf{K}_\kappa \mathsf{K}_\kappa^\top.$$

The inverse of this is

$$(\mathsf{Q}\mathsf{Q}^\top)^{-1} = (\mathsf{K}_\kappa^-)^\top \mathsf{K}_\kappa^{-1},$$

where we note that the inverse of the transpose equals the transpose of the inverse. Computing the Choleski decomposition of $(\mathsf{Q}\mathsf{Q}^\top)^{-1}$ and expressing it in the form of Eq. (13.95) in terms of an upper triangular matrix C, we obtain Eq. (13.96). Hence we can write

$$\mathsf{Q} = \mathsf{C}^{-1} \mathsf{R}_\kappa^\top.$$

Transposing this on both sides, we obtain $\mathsf{Q}^\top = \mathsf{R}_\kappa(\mathsf{C}^\top)^{-1}$, from which R_κ is given by Eq. (13.97). The computed R_κ is a rotation matrix. In fact,

$$\mathsf{R}_\kappa\mathsf{R}_\kappa^\top = \mathsf{Q}^\top\mathsf{C}^\top\mathsf{C}\mathsf{Q} = \mathsf{Q}^\top(\mathsf{Q}\mathsf{Q}^\top)^{-1}\mathsf{Q} = \mathsf{Q}^\top(\mathsf{Q}^\top)^{-1}\mathsf{Q}^{-1}\mathsf{Q} = \mathsf{I}.$$

We can also see that $\det\mathsf{R}_\kappa > 0$, because $\det\mathsf{R}_\kappa = \det\mathsf{Q}^\top\det\mathsf{C}^\top = \det\mathsf{Q}\det\mathsf{C}$. Recall that $\det\mathsf{Q} > 0$ by our choice of sign and that $\det\mathsf{C} > 0$ by the choice of the positive sign for the diagonal elements in the course of the Choleski decomposition.

Problems of Chapter 14

14.1 From the rule of matrix multiplication, we observe that

$$\mathsf{A} = \begin{pmatrix}\mathbf{u}_1 & \cdots & \mathbf{u}_r\end{pmatrix}\begin{pmatrix}\sigma_1 & & \\ & \ddots & \\ & & \sigma_r\end{pmatrix}\begin{pmatrix}\mathbf{v}_1^\top \\ \vdots \\ \mathbf{v}_r\end{pmatrix} = \sum_{i=1}^r \sigma_i\mathbf{u}_i\mathbf{v}_i^\top.$$

14.2 This is immediately obtained from the above result.

14.3 Because $\{\mathbf{u}_i\}$ and $\{\mathbf{v}_i\}$ are orthonormal sets, $(\mathbf{u}_i, \mathbf{u}_j) = \delta_{ij}$ and $(\mathbf{v}_i, \mathbf{v}_j) = \delta_{ij}$ hold, where δ_{ij} is the Kronecker delta (1 for $i = j$ and 0 otherwise). Thus we observe that

$$\begin{aligned}\mathsf{A}\mathsf{A}^-\mathsf{A} &= \sum_{i=1}^r \sigma_i\mathbf{u}_i\mathbf{v}_i^\top \sum_{j=1}^r \frac{1}{\sigma_j}\mathbf{v}_j\mathbf{u}_j^\top \sum_{k=1}^r \sigma_k\mathbf{u}_k\mathbf{v}_k^\top = \sum_{i,j,k=1}^r \frac{\sigma_i\sigma_k}{\sigma_j}\mathbf{u}_i(\mathbf{v}_i, \mathbf{v}_j)(\mathbf{u}_j, \mathbf{u}_k)\mathbf{v}_k^\top \\ &= \sum_{i,j,k=1}^r \frac{\sigma_i\sigma_k}{\sigma_j}\delta_{ij}\delta_{jk}\mathbf{u}_i\mathbf{v}_k^\top = \sum_{i=1}^r \sigma_i\mathbf{u}_i\mathbf{v}_i^\top = \mathsf{A},\end{aligned}$$

$$\begin{aligned}\mathsf{A}^-\mathsf{A}\mathsf{A}^- &= \sum_{i=1}^r \frac{1}{\sigma_i}\mathbf{v}_i\mathbf{u}_i^\top \sum_{j=1}^r \sigma_j\mathbf{u}_j\mathbf{v}_j^\top \sum_{k=1}^r \frac{1}{\sigma_k}\mathbf{v}_k\mathbf{u}_k^\top = \sum_{i,j,k=1}^r \frac{\sigma_j}{\sigma_i\sigma_k}\mathbf{v}_i(\mathbf{u}_i, \mathbf{u}_j)(\mathbf{v}_j, \mathbf{v}_k)\mathbf{u}_k^\top \\ &= \sum_{i,j,k=1}^r \frac{\sigma_j}{\sigma_i\sigma_k}\delta_{ij}\delta_{jk}\mathbf{v}_i\mathbf{u}_k^\top = \sum_{i=1}^r \frac{1}{\sigma_i}\mathbf{v}_i\mathbf{u}_i^\top = \mathsf{A}^-.\end{aligned}$$

14.4 We see that

$$\mathsf{A}\mathsf{A}^- = \sum_{i=1}^r \sigma_i\mathbf{u}_i\mathbf{v}_i^\top \sum_{j=1}^r \frac{1}{\sigma_j}\mathbf{v}_j\mathbf{u}_j^\top = \sum_{i,j=1}^r \frac{\sigma_i}{\sigma_j}\mathbf{u}_i(\mathbf{v}_i, \mathbf{v}_j)\mathbf{u}_j^\top = \sum_{i,j=1}^r \frac{\sigma_i}{\sigma_j}\delta_{ij}\mathbf{u}_i\mathbf{u}_j^\top = \sum_{i=1}^r \mathbf{u}_i\mathbf{u}_i^\top,$$

$$\mathsf{A}^-\mathsf{A} = \sum_{i=1}^r \frac{1}{\sigma_i}\mathbf{v}_i\mathbf{u}_i^\top \sum_{j=1}^r \sigma_j\mathbf{u}_j\mathbf{v}_j^\top = \sum_{i,j=1}^r \frac{\sigma_j}{\sigma_i}\mathbf{v}_i(\mathbf{u}_i, \mathbf{u}_j)\mathbf{v}_j^\top = \sum_{i,j=1}^r \frac{\sigma_j}{\sigma_i}\delta_{ij}\mathbf{v}_i\mathbf{v}_j^\top = \sum_{i=1}^r \mathbf{v}_i\mathbf{v}_i^\top.$$

Therefore

$$\mathsf{A}\mathsf{A}^-\mathbf{u}_k = \sum_{i=1}^{r} \mathbf{u}_i(\mathbf{u}_i, \mathbf{u}_k) = \sum_{i=1}^{r} \delta_{i,j}\mathbf{u}_i = \begin{cases} \mathbf{u}_k & 1 \le k \le r \\ \mathbf{0} & r < k \end{cases},$$

$$\mathsf{A}^-\mathsf{A}\mathbf{v}_k = \sum_{i=1}^{r} \mathbf{v}_i(\mathbf{v}_i, \mathbf{v}_k) = \sum_{i=1}^{r} \delta_{i,j}\mathbf{v}_i = \begin{cases} \mathbf{v}_k & 1 \le k \le r \\ \mathbf{0} & r < k \end{cases}.$$

This means that $\mathsf{A}\mathsf{A}^-$ is the projection matrix onto the space spanned by $\mathbf{u}_1$, ..., $\mathbf{u}_r$ and that $\mathsf{A}^-\mathsf{A}$ is the projection matrix onto the space spanned by $\mathbf{v}_1$, ..., $\mathbf{v}_r$.

Problems of Chapter 15

15.1 Note the following relation.

$$(\mathsf{A}^{-1} - \mathsf{A}^{-1}\delta\mathsf{A}\mathsf{A}^{-1})(\mathsf{A} + \delta\mathsf{A}) = \mathsf{I} - \mathsf{A}^{-1}\delta + \mathsf{A}^{-1}\delta\mathsf{A} - \mathsf{A}^{-1}\delta\mathsf{A}\mathsf{A}^{-1}\delta\mathsf{A} = \mathsf{I} + O(\delta\mathsf{A})^2).$$

Thus

$$(\mathsf{A} + \delta\mathsf{A})^{-1} = (\mathsf{A}^{-1} - \mathsf{A}^{-1}\delta\mathsf{A}\mathsf{A}^{-1})$$

in the first order, that is, with $O(\delta\mathsf{A})^2$ omitted, therefore the first variation of $\delta\mathsf{A}^{-1}$ is $-\mathsf{A}^{-1}\delta\mathsf{A}\mathsf{A}^{-1}$. Recall that

$$|\mathsf{A} + \delta\mathsf{A}| = |\mathsf{A}| + \mathrm{tr}[\mathsf{A}^{\dagger}\delta\mathsf{A}] + O(\delta\mathsf{A})^2,$$

where $\mathsf{A}^{\dagger} = |\mathsf{A}|\mathsf{A}^{-1}$ is the cofactor of the matrix A (↪ Problem 3.5(1)). Therefore the first variation $\delta|\mathsf{A}|$ of $|\mathsf{A}|$ is $|\mathsf{A}|\mathrm{tr}[\mathsf{A}^{-1}\delta\mathsf{A}]$.

15.2 The probability density of a Gaussian distribution of mean $\mathbf{m}$ and covariance matrix $\boldsymbol{\Sigma}$ is

$$p(\mathbf{x}) = \frac{1}{\sqrt{(2\pi)^n|\boldsymbol{\Sigma}|}} \exp\Big(-\frac{1}{2}(\mathbf{x} - \mathbf{m}, \boldsymbol{\Sigma}^{-1}(\mathbf{x} - \mathbf{m}))\Big),$$

where n is the dimension of $\mathbf{x}$. The likelihood of $\mathbf{x}_1$, ..., $\mathbf{x}_N$ is $p(\mathbf{x}_1)\cdots p(\mathbf{x}_N)$, and maximizing this is equivalent to minimizing

$$J = -\log p(\mathbf{x}_1)\cdots p(\mathbf{x}_N) = \frac{nN}{2}\log(2\pi) + \frac{N}{2}\log|\boldsymbol{\Sigma}| + \frac{1}{2}\sum_{\alpha=1}^{N}(\mathbf{x}_\alpha - \mathbf{m}, \boldsymbol{\Sigma}^{-1}(\mathbf{x}_\alpha - \mathbf{m})).$$

Differentiation of this with respect to $\mathbf{m}$ is

$$\nabla_{\mathbf{m}} J = \boldsymbol{\Sigma}^{-1}\sum_{\alpha=1}^{N}(\mathbf{x}_\alpha - \mathbf{m}).$$

Letting this be $\mathbf{0}$, we obtain $\sum_{\alpha=1}^{N}(\mathbf{x}_\alpha - \mathbf{m}) = \mathbf{0}$. Thus $\hat{\mathbf{m}}$ is obtained in the form of Eq. (15.53). The first variation of J for $\boldsymbol{\Sigma} \to \boldsymbol{\Sigma} + \delta\boldsymbol{\Sigma}$ is

$$\begin{aligned}
\delta J_{\boldsymbol{\Sigma}} &= \frac{N}{2}\frac{\delta|\boldsymbol{\Sigma}|}{|\boldsymbol{\Sigma}|} - \frac{1}{2}\sum_{\alpha=1}^{N}(\mathbf{x}_\alpha - \mathbf{m}, \boldsymbol{\Sigma}^{-1}\delta\boldsymbol{\Sigma}\boldsymbol{\Sigma}^{-1}(\mathbf{x}_\alpha - \mathbf{m})) \\
&= \frac{N}{2}\mathrm{tr}[\boldsymbol{\Sigma}^{-1}\delta\boldsymbol{\Sigma}] - \frac{1}{2}\sum_{\alpha=1}^{N}\mathrm{tr}[(\mathbf{x}_\alpha - \mathbf{m})(\mathbf{x}_\alpha - \mathbf{m}))^\top\boldsymbol{\Sigma}^{-1}\delta\boldsymbol{\Sigma}\boldsymbol{\Sigma}^{-1}] \\
&= \frac{N}{2}\mathrm{tr}[\boldsymbol{\Sigma}^{-1}\delta\boldsymbol{\Sigma}] - \frac{1}{2}\mathrm{tr}[\boldsymbol{\Sigma}^{-1}\sum_{\alpha=1}^{N}(\mathbf{x}_\alpha - \mathbf{m})(\mathbf{x}_\alpha - \mathbf{m})^\top\boldsymbol{\Sigma}^{-1}\delta\boldsymbol{\Sigma}] \\
&= \frac{1}{2}\mathrm{tr}[\Big(N\boldsymbol{\Sigma}^{-1} - \boldsymbol{\Sigma}^{-1}\sum_{\alpha=1}^{N}(\mathbf{x}_\alpha - \mathbf{m})(\mathbf{x}_\alpha - \mathbf{m})^\top\boldsymbol{\Sigma}^{-1}\Big)\delta\boldsymbol{\Sigma}],
\end{aligned}$$

where we have used Eq. (15.52). Because $\delta J_{\boldsymbol{\Sigma}}$ should vanish for an arbitrary symmetric matrix $\delta\boldsymbol{\Sigma}$, we obtain

$$\boldsymbol{\Sigma}^{-1} = \frac{1}{N}\boldsymbol{\Sigma}^{-1}\sum_{\alpha=1}^{N}(\mathbf{x}_\alpha - \mathbf{m})(\mathbf{x}_\alpha - \mathbf{m})^\top\boldsymbol{\Sigma}^{-1}.$$

Substituting $\hat{\mathbf{m}}$ for $\mathbf{m}$ and multiplying this by $\boldsymbol{\Sigma}$ from both sides, we obtain the maximum likelihood estimate $\hat{\boldsymbol{\Sigma}}$ in the form of Eq. (15.53).

Problems of Chapter 16

16.1 (1) Note that

$$\nabla_{\boldsymbol{\theta}} \log p = \frac{\nabla_{\boldsymbol{\theta}} p}{p},$$

where we write p for $p(\mathbf{x}|\boldsymbol{\theta})$ for short. The expectation of $\mathbf{l}$ is

$$E[\mathbf{l}] = \int p\mathbf{l}d\mathbf{x} = \int p\nabla_{\boldsymbol{\theta}} \log p d\mathbf{x} = \int \nabla_{\boldsymbol{\theta}} p d\mathbf{x} = \nabla_{\boldsymbol{\theta}} \int p d\mathbf{x} = 0,$$

because $\int p d\mathbf{x} = 1$ identically holds.

(2) Differentiating $\partial \log p/\partial\theta_i = (1/p)\partial p/\partial\theta_i$ with respect to θ_j, we obtain

$$\frac{\partial^2 \log p}{\partial\theta_i\partial\theta_j} = -\frac{\partial \log p}{\partial\theta_i}\frac{\partial \log p}{\partial\theta_j} + \frac{1}{p}\frac{\partial^2 p}{\partial\theta_i\partial\theta_j}.$$

Therefore

$$\begin{aligned}
E[\frac{\partial^2 \log p}{\partial\theta_i\partial\theta_j}] &= -\int p\frac{\partial \log p}{\partial\theta_i}\frac{\partial \log p}{\partial\theta_j}d\mathbf{x} + \int \frac{\partial^2 p}{\partial\theta_i\partial\theta_j}d\mathbf{x} \\
&= -\int p l_i l_j d\mathbf{x} + \frac{\partial^2}{\partial\theta_i\partial\theta_j}\int p d\mathbf{x} = -E[l_i l_j],
\end{aligned}$$

which proves Eq. (16.40).

(3) The probability density of $\mathbf{x}$ is

$$p = C\exp[-((\mathbf{x}-\mathbf{m}),\ \boldsymbol{\Sigma}^{-1}(\mathbf{x}-\mathbf{m}))/2],$$

where C is a normalization constant. Because

$$\log p = -\frac{1}{2}((\mathbf{x}-\mathbf{m}),\ \boldsymbol{\Sigma}^{-1}(\mathbf{x}-\mathbf{m})) + \text{constant},$$

we obtain Eq. (16.41).

16.2 (1) Consider an infinitesimal variation of $\boldsymbol{\theta}$ to $\boldsymbol{\theta}+\delta\boldsymbol{\theta}$ of $E[\hat{\boldsymbol{\theta}}-\boldsymbol{\theta}]=\mathbf{0}$, which is an identically in $\boldsymbol{\theta}$. We obtain

$$\begin{aligned}
\delta\int(\hat{\boldsymbol{\theta}}-\boldsymbol{\theta})p\mathrm{d}\mathbf{x} &= -\int\delta\boldsymbol{\theta}p\mathrm{d}\mathbf{x} + \int(\hat{\boldsymbol{\theta}}-\boldsymbol{\theta})(\nabla_{\boldsymbol{\theta}}p,\ \delta\boldsymbol{\theta})\mathrm{d}\mathbf{x} \\
&= -\delta\boldsymbol{\theta}\int p\mathrm{d}\mathbf{x} + \int(\hat{\boldsymbol{\theta}}-\boldsymbol{\theta})(p\nabla_{\boldsymbol{\theta}}\log p,\ \delta\boldsymbol{\theta})\mathrm{d}\mathbf{x} \\
&= -\delta\boldsymbol{\theta} + \int(\hat{\boldsymbol{\theta}}-\boldsymbol{\theta})(\mathbf{l},\ \delta\boldsymbol{\theta})p\mathrm{d}\mathbf{x} \\
&= -\delta\boldsymbol{\theta} + E[(\hat{\boldsymbol{\theta}}-\boldsymbol{\theta})\mathbf{l}^\top]\delta\boldsymbol{\theta},
\end{aligned}$$

where we have used the identity $\nabla_{\boldsymbol{\theta}}p = p\nabla_{\boldsymbol{\theta}}\log p$. This is identically 0 for arbitrary variation $\delta\boldsymbol{\theta}$, therefore we obtain Eq. (16.42).

(2) From Eqs. (16.39), (16.42), and (16.43), we obtain

$$E[\begin{pmatrix}\hat{\boldsymbol{\theta}}-\boldsymbol{\theta}\\ \mathbf{l}\end{pmatrix}\begin{pmatrix}\hat{\boldsymbol{\theta}}-\boldsymbol{\theta}\\ \mathbf{l}\end{pmatrix}^\top] = \begin{pmatrix}V[\hat{\boldsymbol{\theta}}] & \mathsf{I}\\ \mathsf{I} & \mathsf{J}\end{pmatrix}.$$

Because the left-hand side is positive semi-definite, the following matrix is also semi-definite.

$$\begin{pmatrix}\mathsf{I} & -\mathsf{J}^{-1}\\ & \mathsf{J}^{-1}\end{pmatrix}\begin{pmatrix}V[\hat{\boldsymbol{\theta}}] & \mathsf{I}\\ \mathsf{I} & \mathsf{J}\end{pmatrix}\begin{pmatrix}\mathsf{I} & \\ -\mathsf{J}^{-1} & \mathsf{J}^{-1}\end{pmatrix} = \begin{pmatrix}V[\hat{\boldsymbol{\theta}}]-\mathsf{J}^{-1} & \\ & \mathsf{J}^{-1}\end{pmatrix}.$$

Because J^{-1} is positive semi-definite from the definition of Eq. (16.39), the positive semi-definiteness of the above matrix implies Eq. (16.44).

16.3 (1) If we regard the direct sum

$$\tilde{\mathbf{x}} = \mathbf{x}_1 \oplus \cdots \oplus \mathbf{x}_N$$

as a new random variable, the independence of $\mathbf{x}_1, \ldots, \mathbf{x}_N$ implies that the probability density of $\tilde{\mathbf{x}}$ is

$$\tilde{p}(\tilde{\mathbf{x}}|\boldsymbol{\theta}) = p(\mathbf{x}_1|\boldsymbol{\theta})\cdots p(\mathbf{x}_N|\boldsymbol{\theta}).$$

An estimator $\hat{\boldsymbol{\theta}} = \hat{\boldsymbol{\theta}}(\mathbf{x}_1, \ldots, \mathbf{x}_N)$ of $\boldsymbol{\theta}$ is regarded as a function of $\tilde{\mathbf{x}}$. The score of $\tilde{\mathbf{x}}$ is

$$\tilde{\mathbf{l}} = \nabla_{\boldsymbol{\theta}}\log\tilde{p}(\tilde{\mathbf{x}}|\boldsymbol{\theta}) = \nabla_{\boldsymbol{\theta}}\sum_{\alpha=1}^{N}\log p(\mathbf{x}_\alpha|\boldsymbol{\theta}) = \sum_{\alpha=1}^{N}\mathbf{l}_\alpha,$$

where $\mathbf{l}_\alpha$ is the score of $\mathbf{x}_\alpha$. The Fisher information matrix for $\tilde{p}(\tilde{\mathbf{x}}|\boldsymbol{\theta})$ is

$$\tilde{\mathsf{J}} = E[\tilde{\mathbf{l}}\tilde{\mathbf{l}}^\top] = E[\Big(\sum_{\alpha=1}^{N}\mathbf{l}_\alpha\Big)\Big(\sum_{\beta=1}^{N}\mathbf{l}_\beta\Big)^\top] = \sum_{\alpha,\beta=1} E[\mathbf{l}_\alpha\mathbf{l}_\beta^\top].$$

Because $\mathbf{x}_1, \ldots, \mathbf{x}_N$ are independent and $E[\mathbf{l}_\alpha] = \mathbf{0}$ (see Eq. (16.9)), we have $E[\mathbf{l}_\alpha\mathbf{l}_\alpha^\top] = E[\mathbf{l}_\alpha]E[\mathbf{l}_\alpha]^\top = \mathsf{O}$ for $\alpha \neq \beta$. Therefore the above expression reduces to $\sum_{\alpha=1} E[\mathbf{l}_\alpha\mathbf{l}_\alpha^\top]$ = $N\mathsf{J}$, where J is the Fisher information matrix for $p(\mathbf{x}|\boldsymbol{\theta})$. Thus the Cramer-Rao inequality $V[\hat{\boldsymbol{\theta}}] \succ \tilde{\mathsf{J}}^{-1}$ (see Eq. (16.44)) reduces to Eq. (16.45).
(2) Because

$$E[\hat{\mathbf{m}}] = \frac{1}{N}(E[\mathbf{x}_1] + \cdots + E[\mathbf{x}_N]) = \mathbf{m},$$

the sample mean $\hat{\mathbf{m}}$ is an unbiased estimator of $\mathbf{m}$. The covariance matrix of $\hat{\mathbf{m}}$ is

$$V[\hat{\mathbf{m}}] = \frac{1}{N^2}(V[\mathbf{x}_1] + \cdots + V[\mathbf{x}_N]) = \frac{1}{N}\boldsymbol{\Sigma}.$$

Because the Fisher information matrix J of $\mathbf{x}_\alpha$ equals $\boldsymbol{\Sigma}^{-1}$ (see Eq. (16.41)), the Cramer-Rao inequality of Eq. (16.45) reduces to

$$V[\hat{\mathbf{m}}] \succ \frac{1}{N}\boldsymbol{\Sigma},$$

which is satisfied with equality, meaning that $\hat{\mathbf{m}}$ is an efficient estimator.

Index

K. Kanatani et al., *Guide to 3D Vision Computation*, Advances in Computer Vision and Pattern Recognition, DOI 10.1007/978-3-319-48493-8

FSC
www.fsc.org
MIX
Papier | Fördert gute Waldnutzung
FSC® C083411

Zeitfracht Medien GmbH
Ferdinand-Jühlke-Straße 7
99095 Erfurt, Deutschland
produktsicherheit@kolibri360.de